T0189019

Blockchain for 5G-Enabled IoT

Sudeep Tanwar
Editor

Blockchain for 5G-Enabled IoT

The new wave for Industrial Automation

Editor
Sudeep Tanwar
Department of Computer Science
and Engineering
Institute of Technology, Nirma University
Ahmedabad, Gujarat, India

ISBN 978-3-030-67492-2 ISBN 978-3-030-67490-8 (eBook)
https://doi.org/10.1007/978-3-030-67490-8

This Springer imprint is published by the registered company Springer Nature Switzerland AG
The registered company address is: Gewerbestrasse 11, 6330 Cham, Switzerland

Preface

IoT has made ubiquitous computing a reality by extending Internet connectivity in various applications deployed across the globe. IoT connects billions of objects for high-speed data transfer, especially in a 5G-enabled industrial environment for information collection and processing. Most of the issues such as access control mechanism, time to fetch the data from different devices, and protocols used may not be applicable in the future as these protocols are based upon a centralized mechanism. This centralized mechanism may have a single point of failure alongwith the computational overhead. So, there is a need for an efficient decentralized access control mechanism for D2D communication in various industrial sectors, for example, sensors in different regions may collect and process the data for making intelligent decisions. In such an environment, security and privacy are major concerns as most of the solutions are based upon the centralized control mechanism.

To mitigate the aforementioned issues, this edited book includes the following:

- An in-depth analysis of state-of-the-art proposals having 5G-enabled IoT as a backbone for blockchain-based industrial automation in applications such as Smart City, Smart Home, Healthcare 4.0, Smart Agriculture, Autonomous Vehicles, and Supply Chain Management.
- From the existing proposals, it has been observed that blockchain can revolutionize most of the current and future industrial applications in different sectors by providing fine-grained access control.
- Open issues and challenges of 5G-enabled IoT for blockchain-based industrial automation are analyzed. Finally, a comparison of existing proposals concerning various parameters is presented, which allows end-users to select a proposal based on its merits over the others.
- Case studies to demonstrate the adoption of blockchain for 5G-enabled IoT, which make the readers aware of future challenges associated with this adoption, especially for smart industrial applications.
- Layered architecture of 5G-enabled IoT for blockchain-based industrial automation.

The book is organized into five parts. The first part is focused on the background and preliminaries of blockchain and 5G-enabled IoT, which includes five chapters. The second part discusses the enabling technologies and architecture for 5G-enabled IoT, which has five chapters. The third part illustrates the AI-assisted secure 5G-enabled IoT with well-structured five chapters. The fourth part highlights the 5G-enabled IoT models, solutions, and standards, which has four chapters. Finally, the last part focuses on the next-generation 5G-enabled IoT for industrial automation with four chapters.

Part I: Background and Preliminaries
Chapter 1 presents an introduction to the blockchain and 5G-enabled IoT. The aim of this chapter is to provide a systematic view of the blockchain and 5G-enabled IoT with a perspective of industrial automation. This chapter also gives a comparative study of the different hurdles in applying blockchain-based solutions for 5G-enabled IoT applications. Furthermore, this chapter expects to expound and underscore the key parts of the utilization of blockchain for 5G and IoT. This chapter also presents an inside-out review of the best-in-class proposition under the selected domain.

Chapter 2 intends to guide researchers and stakeholders for the overall improvement in the functioning of the blockchain and 5G IoT in industrial automation. At the end of the chapter, the authors summarize findings to describe the advantages and limitations of existing mechanisms and provide insights into possible research directions.

Chapter 3 presents the foundational ideas of both 5G and blockchain technology along with their complementary strengths and weaknesses in various application domains. These points are supported and followed by apparent attractiveness of application areas so that appropriateness of 5G with blockchain can open up new research directions as well as future service-oriented applications for upcoming communicational network systems.

Chapter 4 introduces the basic architecture and main features of blockchain along with how the blockchain can be integrated with 5G-enabled IoT. Subsequently, the security requirements to manage 5G-enabled IoT devices are illustrated in this chapter. Additionally, the opportunities, applications, issues & challenges, limitations, and research directions of blockchain-based 5G-enabled IoT are explored which will be helpful to the researchers to dive into the area of IoT and blockchain.

Chapter 5 presents a brief introduction to each of these emerging technologies, their impact on industrial automation, and their applications in different industries. This chapter is divided into six major sections, namely, introduction, the rise of industrial automation, IoT, the emergence of the 5G wireless network, blockchain technology; the next best thing, and blockchain, and 5G-enabled IoT use cases in the finance sector.

Part II: Enabling Technologies and Architecture for 5G-Enabled IoT
Chapter 6 presents an overview of the key enabling technologies such as Cloud Computing, Heterogeneous Network (HetNet), Device to Device (D2D) Communication, and Software-Defined Networking (SDN), which have become essential technologies to achieve better efficiency in any industrial automation applications.

Furthermore, in this chapter, the current state of the art in the context of IoT application requirements and related cellular communication technologies are reviewed. Then, a comparative analysis of various communication technologies is presented along with emerging IoT applications. At the end of the chapter, a case study on 5G-enabled IoT with the challenges and future research trends for the deployment of various IoT services and applications are discussed.

Chapter 7 enlightens the benefits of 5G-enabled IoT system by integrating it with the cloud ecosystem. Then, a case study is presented, which reflects the benefits of using the cloud ecosystem for the 5G-based IoT infrastructure for an effective decision-making process via intelligent communication mechanism and handling security in the IoT framework.

Chapter 8 analyzes the prospects of 5G-based IoT networks used for industrial automation and also explores their technical software capabilities as required by modern applications and technologies. Moreover, this chapter examines the integration of 5G-based IoT networks into the field of industrial automation, their architecture, existing technologies used in them, and blockchain technology, which plays a key role in ensuring the security and reliability of these technologies. Moreover, the optimization methods are also discussed, which are used in the solution of resource allocation problems and also highlighted the importance of blockchain technology.

Chapter 9 discusses the problems that will arise with the standardization of the two technologies. Amid many possible alternatives, this chapter discusses the use of blockchain to solve all of these problems in the area of 5G IoT. From this chapter, the end-user will benefit from the capabilities of 5G, followed by the convergence of blockchain with IoT applications. At the end, this chapter discusses 5G-enabled IoT application architectures and their characteristics.

Chapter 10 presents a detailed design for the development of intelligent data analytics and mobile computer-assisted healthcare systems. The proposed advanced Proof-of-stake (PoS) consensus algorithm provides better performance than other existing algorithms. Moreover, in this chapter, the authors designed an eHealth program that deploys several instances of a three-layer Patient Agent software: Sensing, Near processing, and FAR processing layer. It also defines how to implement the Patient Agent on a 5G unit.

Part III: AI-Assisted Secure 5G-Enabled IoT

Chapter 11 discusses the important facts about Cloud Computing and Edge Standards, Security fundamentals for 5G network, and other security measures. Then, the architecture of 5G-enabled IoT, security threats in 5G-enabled IoT, security analysis, privacy threats in 5G-enabled IoT, security and privacy threats in specific domains, and challenges and opportunities are discussed. Moreover, this chapter discusses the 5G-enabled IoT that contains five layers: recognition layer, connectivity/edge computing layer, support layer, application layer, and business layer. The challenges in 5G-enabled IoT concerning security and privacy strategies to the edge paradigms domain are reviewed.

Chapter 12 provides a detailed method for data encryption and schemes, which make it highly sophisticated to decrypt data by hackers. Later, this chapter discusses some case studies for a better understanding of data security and privacy in a 5G-enabled IoT network. This chapter helps the readers to understand how data is embezzled and provides solutions in the corresponding field of research.

Chapter 13 gives an overview of adversarial AI techniques for IoT-enabled 5G networks for detecting and classifying threats automatically, and enabling secure transactions using blockchain. Then the chapter discusses case studies that are helpful to understand real-time attacks on traffic direction and road signal identification. Finally, this chapter discusses how to check the automation systems deployed in industries and how they are preventing the system from attacks and also safeguard the data.

Chapter 14 focuses on IoT platforms, the convergence of machine learning with IoT platforms, the convergence of data mining with IoT platforms, and the convergence of big data analytics with IoT platforms. This concept includes how machine learning enhances the efficiency of 5G networks, how data mining furnishes data for 5G networks, and how big data analytics reduce the time consumption of 5G networks. In this chapter, two case studies are presented which will bestow a closer look at the mechanism of 5G networks with the help of these revolutionary technologies. The first case study is about smart cities in which the role of 5G networks is highlighted, and the second case study is about mobile networks where the concept of Mobile Social Networks (MSNs) is elaborated. This chapter is a complete package of information that allows users to explore new things and how technology is improving every second and how humans have to adapt to the change for their survival in the world.

Chapter 15 develops a Protected Health Information (PHI) system using blockchain-based smart contracts to assist safe data analysis, data sharing, data transfer, and management to handle the secured health information of the smart telerehabilitation app. The app called Autism Telerehabilitation App (ATA) uses a private blockchain based on the Ethereum protocol to write history and records of all the Electronic Health Records (EHRs) of patients from the smart device. Furthermore, this ATA would provide medical interventions and real-time patient observation by sending alerts to the patient and medical specialists. Besides, it can secure and maintain the record of who has initiated these activities. This proposed blockchain with ATA offers high data security of all the stakeholders.

Part IV: 5G-Enabled IoT Models, Solutions, and Standards

Chapter 16 proposes a blockchain-assisted app-based system, supported by a cloud environment for the elderly healthcare system to provide a convenient, adaptable, and efficient platform to address healthcare issues of the elderly. Furthermore, an architecture is proposed which targets at facilitating necessary medical services to the user with features like prescription, diet plans, and medicine intake details from the doctor's end. This chapter shows how the patient's records are added to the database with the help of a QR code scan on the patient's Aadhar, the patient's medical history of their previous visits to different doctors, and symptoms observed on those visits. Prescriptions given would be well maintained and easily accessible for future reference by any doctor or patient by simply scanning the QR code on the Aadhaar.

Chapter 17 provides a brief background of the technologies (blockchain, IoT, edge computing) and explores the deep learning techniques for resource management in upcoming technologies: future generation cellular networks, IoT, and edge computing. Then, this chapter discusses the current deep learning techniques' potential to facilitate the efficient deployment of deep learning with blockchain onto upcoming emerging technologies. This chapter provides an encyclopedia review of deep learning techniques and concludes the analysis by pinpointing the current research challenges and directions for future research.

Chapter 18 briefly introduces the impact 5G-enabled IoT has on industries and industrial automation. Industrial growth has known no limits in the last few years, but this chapter only includes those categories which are expected to go through a major revolution with the advent of 5G and its integration with IoT. 5G-enabled IoT is expected to make healthcare much more advanced, bring usable self-driving vehicles closer to reality, and evolve many typical industrial products and systems into their smarter versions, like smart homes, smart cities, smart agriculture, smart supply chain management, etc. All these mentioned changes have been briefly discussed in the chapter.

Chapter 19 a decentralized application has been proposed, which uses a smart contract to facilitate the authentication and verification of documents by leveraging the blockchain technology. In contrast to the traditional way of storing the entire input digital document, the proposed approach creates a unique fingerprint of every input document by using a cryptographic hash function. This fingerprint is stored on the blockchain network to verify the document in future. This blockchain-based solution can be used by organizations to authenticate documents that they generate and allow other entities to verify them.

Part V: Next-Generation 5G-Enabled IoT for Industrial Automation

Chapter 20 covers healthcare applications based on 5G technology. It presents new challenges and techniques in the healthcare area. This chapter also proposes a wearable biotechnology platform based on 5G networks to show the bio-information methods and bio-sensing platforms. This chapter also discusses a relative comparison of healthcare system environments like user environment, bio-information gathering type and method, etc.

Chapter 21 gives a comparative analysis of curated survey papers with specific parameters to understand the subject coverage and to discover the research gaps. Then, the research issues, implementation challenges, and future trends are highlighted. A case study of a world-class tool manufacturing company is presented, and the chapter concludes with a holistic view of IoT applications.

Chapter 22 discusses real-time monitoring of the patient's condition by clinicians, and sharing of medical records by patients to avail second referral on their medical condition with high-speed 5G network and enhanced services provided employing blockchain and IoT. The major concerns for acceptance of this technology are data misuse during sharing with third parties or indirect user identification through pseudonymous identifiers. This research focuses on the use of technologies for existing patients and normal users and improves the services of the healthcare industry.

Chapter 23 discusses the substantial development of the latest mobile and satellite communication; the multiple frequency antennas with high isolation and low mutual coupling are of particular interests. Additionally, low cross-polarization, high gain, and maximum front-to-back ratio are obtained and how these are used in 5G-enabled IoT applications is discussed.

The editor is very thankful to all the members of Springer, especially Ms. Mary James and Mr. Aninda Bose, for the opportunity to edit this book.

Ahmadabad, Gujarat, India Sudeep Tanwar

Contents

About the Editor

Sudeep Tanwar is an associate professor in the Computer Science and Engineering Department at the Institute of Technology, Nirma University, Ahmedabad, Gujarat, India. He is visiting Professor in Jan Wyzykowski University in Polkowice, Poland, and University of Pitesti in Pitesti, Romania. He received his B.Tech in 2002 from Kurukshetra University, India, M.Tech (Honors) in 2009 from Guru Gobind Singh Indraprastha University, Delhi, India, and Ph.D. in 2016 with specialization in Wireless Sensor Network. He has authored or coauthored more than 200 technical research papers published in leading journals and conferences from the IEEE, Elsevier, Springer, Wiley, etc. Some of his research findings are published in top-cited journals such as *IEEE TNSE*, *IEEE TVT*, *IEEE TII*, *Transactions on Emerging*, *Telecommunications Technologies*, *IEEE WCM*, *IEEE Networks*, *IEEE Systems Journal*, *IEEE Access*, *IET Software*, *IET Networks*, *JISA*, *Computer Communication*, *Applied Soft Computing*, *JPDC*, *JNCA*, *PMC*, *SUSCOM*, *CEE*, *IJCS*, *Software: Practice and Experience*, *MTAP*, and *Telecommunication System*. He has also edited/authored 13 books with national/international publishers like IET and Springer. One of his edited textbooks, *Multimedia Big Data Computing for IoT Applications: Concepts, Paradigms, and Solutions*, published by Springer in 2019 has been downloaded 3.7 million times until March 13, 2021. This text book attracts attention of researchers across the globe. (https://link.springer.com/book/10.1007/978-981-13-8759-3). He has guided many students leading to M.E./M.Tech and Ph.D. He is currently serving the editorial boards of Physical Communication, Computer Communications, International Journal of Communication System, and Security and Privacy. His current interest includes wireless sensor networks, fog computing, smart grid, IoT, and blockchain technology. He initiated the research field of blockchain technology adoption in various verticals in 2017. He was invited as guest editor/editorial board member of many international journals, as keynote speaker at many international conferences held in Asia and as program chair, publications chair, publicity chair, and session chair at many international conferences held in North America, Europe, Asia, and Africa. He has been awarded best research paper awards from IEEE GLOBECOM 2018, IEEE ICC 2019, and Springer ICRIC-2019. He is a Senior Member of IEEE, CSI, IAENG, ISTE, CSTA, and the member of Technical Committee on Tactile Internet of IEEE Communication Society.

Contributors

Samir Abdelrazek Information Systems Department, Faculty of Computers, and Information, Mansoura University, Mansoura, Egypt

Puneet Kumar Aggarwal ABES Engineering College, Ghaziabad, Uttar Pradesh, India

P. Ajitha Sathyabama Institute of Science and Technology, Chennai, Tamil Nadu, India

Nurshod Akhmedov Tashkent University of Information Technologies named after Muhammad al-Khwarizmi, Tashkent, Uzbekistan

Darpan Anand Chandigarh University, Mohali, India

Shalini Bhaskar Bajaj Amity University, Noida, India

G. Nidhi Bhat Department of Information Science & Engineering, Ramaiah Institute of Technology, Bangalore, India

Dinesh Bhatia North Eastern Hill University, Shillong, Meghalaya, India

Jitendra Bhatia Department of Computer Engineering, Vishwakarma Government Engineering College, Ahmedabad, Gujarat, India

Madhuri D. Bhavsar Department of Computer Science and Engineering, Institute of Technology, Nirma University, Ahmedabad, Gujarat, India

Mohammed Husain Bohara Devang Patel Institute of Advance Technology & Research (DEPSTAR), Faculty of Technology and Engineering (FTE), Charotar University of Science & Technology (CHARUSAT), Changa, Gujarat, India

Poorvi Chaudhary HMRITM, Delhi, India

Prakrut Chauhan Institute of Technology, Nirma University, Ahmedabad, India

D. Deepa Sathyabama Institute of Science and Technology, Chennai, Tamil Nadu, India

Ibrahim Elhenawy Faculty of Computer Science and Information Systems, Zagazig University, Zagazig, Egypt

Amit Ganatra Devang Patel Institute of Advance Technology & Research (DEPSTAR), Faculty of Technology and Engineering (FTE), Charotar University of Science & Technology (CHARUSAT), Changa, Gujarat, India

Xiao-Zhi Gao University of Eastern Finland, Kuopio, Finland

Riya Garg HMRITM, Delhi, India

Aishwarya Gupta Amity University, Lucknow Campus, Lucknow, Uttar Pradesh, India

U. Hariharan Department of Information Technology, Galgotias College of Engineering and Technology, Noida, Uttar Pradesh, India

Ahmed Ismail Nordson, Munich, GermanyReDI School of Digital Integration Munich, Munich, Germany

J. Jabez Sathyabama Institute of Science and Technology, Chennai, Tamil Nadu, India

Parita Jain KIET Group of Institutions, Ghaziabad, Uttar Pradesh, India

Swati Jain Institute of Technology, Nirma University, Ahmedabad, India

Upinder Kaur Department of Computer Science, Akal University, Talwandi Saboo, Punjab, India

Shweta Kaushik ABES Engineering College, Ghaziabad, Uttar Pradesh, India

Pooja Khanna Amity University, Lucknow Campus, Lucknow, Uttar Pradesh, India

Doston Khasanov Tashkent University of Information Technologies named after Muhammad al-Khwarizmi, Tashkent, Uzbekistan

Vineeta Khemchandani J.S.S. Academy of Technical Education, Noida, India

Khalimjon Khujamatov Tashkent University of Information Technologies named after Muhammad al-Khwarizmi, Tashkent, Uzbekistan

R. V. Kulkarni CSIBER, Kolhapur, Maharashtra, India

Sachin Kumar Amity University, Lucknow Campus, Lucknow, Uttar Pradesh, India

Malaram Kumhar Department of Computer Science and Engineering, Institute of Technology, Nirma University, Ahmedabad, Gujarat, India

L. Lakshmanan Sathyabama Institute of Science and Technology, Chennai, Tamil Nadu, India

Satyasundara Mahapatra Pranveer Singh Institute of Technology, Kanpur, Uttar Pradesh, India

Kshirja Makar HMRITM, Delhi, India

S. R. Mani Sekhar Department of Information Science & Engineering, Ramaiah Institute of Technology, Bangalore, India

Jaya Mehta HMRITM, Delhi, India

Animesh Mishra North Eastern Indira Gandhi Regional Institute of Health and Medical Sciences, Shillong, Meghalaya, India

Aznida Hayati Zakaria Mohamad Faculty of Informatics and Computing, Universiti Sultan Zainal Abidin, Besut, Terengganu, Malaysia

Zarina Mohamad Faculty of Informatics and Computing, Universiti Sultan Zainal Abidin, Besut, Terengganu, Malaysia

Nazeeruddin Mohammad Cybersecurity Center, Prince Mohammad Bin Fahd University, Al Khobar, Saudi Arabia

Parveen Mor Amity University, Noida, India

Arjuna Muduli KL Education Foundation, Guntur, Andhra Pradesh, India

Moumita Mukherjee Adamas University, Kolkata, West Bengal, India

Sharif Nawaz Jaypee University of Engineering & Technology, Guna, Madhya Pradesh, India

Padmini Nigam DIT University, Dehradun, India

Wan Nor Shuhadah Nik Faculty of Informatics and Computing, Universiti Sultan Zainal Abidin, Besut, Terengganu, Malaysia

Amrindra Pal DIT University, Dehradun, India

Prateek Pandey Jaypee University of Engineering & Technology, Guna, Madhya Pradesh, India

Hiren Patel Vidush Somany Institute of Technology and Research, Kadi, Gujarat, India

Khushi Patel Devang Patel Institute of Advance Technology & Research (DEPSTAR), Faculty of Technology and Engineering (FTE), Charotar University of Science & Technology (CHARUSAT), Changa, Gujarat, India

Vivek Kumar Prasad Department of Computer Science and Engineering, Institute of Technology, Nirma University, Ahmedabad, Gujarat, India

Rohit Rai Indian Naval Ship (INS) Valsura, Jamnagar, Gujarat, India

K. Rajkumar Department of Information Technology, Galgotias College of Engineering and Technology, Noida, Uttar Pradesh, India

Ernazar Reypnazarov Tashkent University of Information Technologies named after Muhammad al-Khwarizmi, Tashkent, Uzbekistan

Atufaali Saiyed Devang Patel Institute of Advance Technology & Research (DEPSTAR), Faculty of Technology and Engineering (FTE), Charotar University of Science & Technology (CHARUSAT), Changa, Gujarat, India

Suneeta Satpathy College of Engineering Bhubaneswar, Bhubaneswar, Odisha, India

Eman Shaikh Department of Computer Science and Engineering, American University of Sharjah, Sharjah, UAE

Shalu Department of Computer Science, Baba Farid College, Bathinda, Punjab, India

Sandeep Sharma Center for Reliability Sciences & Technologies, Chang Gung University, Taoyuan City, TaiwanOMKARR Tech, New Delhi, India

Bela Shrimali Department of Computer Engineering, LDRP Institute of Technology and Research, Gandhinagar, Gujarat, India

G. M. Siddesh Department of Information Science & Engineering, Ramaiah Institute of Technology, Bangalore, India

Anupam Singh University of Petroleum and Energy Studies, Dehradun, Uttarakhand, India

Shubham Kumar Singh Jaypee University of Engineering & Technology, Guna, Madhya Pradesh, India

A. Sivasangari Sathyabama Institute of Science and Technology, Chennai, Tamil Nadu, India

D. K. Sreekantha NMAM Institute of Technology, Nitte, Karnataka, India

Shivangi Surati Department of Computer Engineering, LDRP Institute of Technology and Research, Gandhinagar, Gujarat, India

Sudeep Tanwar Department of Computer Science and Engineering, Institute of Technology, Nirma University, Ahmedabad, Gujarat, India

S. Vaishnavi Department of Information Science & Engineering, Ramaiah Institute of Technology, Bangalore, India

Arpit Verma Jaypee University of Engineering & Technology, Guna, Madhya Pradesh, India

Jai Prakash Verma Institute of Technology, Nirma University, Ahmedabad, India

Nurnadiah Zamri Faculty of Informatics and Computing, Universiti Sultan Zainal Abidin, Besut, Terengganu, Malaysia

Part I
Background and Preliminaries

Chapter 1
Blockchain and 5G-Enabled Internet of Things: Background and Preliminaries

Shweta Kaushik

1 Introduction

1.1 Blockchain

Blockchain is generally known as the fundamental innovation of the cryptographic money Bitcoin [1]. The centre feature of a blockchain is decentralization. This implies it does not store any of its databases in a pivotal area. Rather, the data is replicated and distributed over a system of members. At whatever point any block is added to the blockchain, each computer on the system refreshes its blockchain to mirror the change. This distributed engineering guarantees a robust and secure procedure on the blockchain with the upsides of alter obstruction and no single-point weaknesses. Specifically, blockchain can be available for everybody and is not constrained by any system element. This is enabled by an instrument called accord, which is a set of rules to guarantee the understanding between all members on the position of the blockchain top. The overall idea on how blockchain works is shown in Fig. 1.1. Blockchain, a circulated record innovation empowers clients to connect and execute (store and recover information) with guaranteed information legitimacy, changelessness and non-disavowal. The appropriated idea of blockchain permits the mechanical elements and different 5G/IoT gadgets to trade information, to and from their friends, removing the necessity for concentrated operation.

The blockchain-assisted 5G biological system is suitable for building up responsibility, information provenance, and non-denial for each client. The principal hinder in a blockchain is alluded to as the beginning block, which does not contain any exchange. Each block from that point contains various approved exchanges

S. Kaushik (✉)
ABES Engineering College, Ghaziabad, Uttar Pradesh, India
e-mail: shweta.kaushik@abes.ac.in

S. Tanwar (ed.), *Blockchain for 5G-Enabled IoT*,
https://doi.org/10.1007/978-3-030-67490-8_1

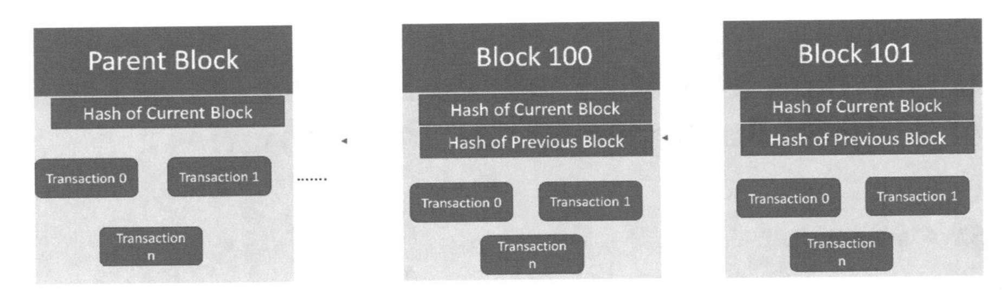

Fig. 1.1 How blockchain works

and is cryptographically connected with past blocks. When all is said and done, blockchains can be named either an open (permission-less) or a private (permissioned) blockchain [2]. An open blockchain is available for everybody and anybody can join and cause exchanges but only some can decide the agreement procedure. The most popular open blockchain applications are used in Bitcoin and Ethereum. Private blockchains have an access control system overseen by a focal element. A member must be permissioned to utilize an approval component. To understand the capability of blockchain in 5G systems, it is important to comprehend the activity idea, principle components of blockchain, and how blockchain can bring opportunities to 5G-enabled applications. Moreover, blockchain is unique in relation to other disseminated frameworks dependent on agreement and the following properties [3], as shown in Fig. 1.2:

- Trust-less: The elements associated with the system are obscure to one another. However, they can convey, coordinate and work together without realizing one another, which implies there is no prerequisite of guaranteed advanced character to carry out any exchange between the substances.
- Permission-less: There is no limitation of who can or cannot work inside the system, that is, there is no sort of consent.
- Restriction safe: Being a system without supervisors, anybody can communicate or execute on the blockchain. Additionally, any affirmed exchange cannot be altered or blue-pencilled. Notwithstanding the previously mentioned legitimacies, blockchain innovation has four fundamental segments [4], which are referred to as:
- Consensus: The PoW convention is mindful to check each activity in the system which is fundamental to avoid a solitary excavator hub from commanding the whole blockchain system and to control the exchanges history.
- Register or Ledger: It is a common and conveyed database that encompasses data exchanges performed inside the system. It is commonly changeless, where data once put away cannot be erased using any means. It ensures that each exchange is checked and afterwards acknowledged as a legitimate one, by the greater part of the customers required at a specific moment [5].

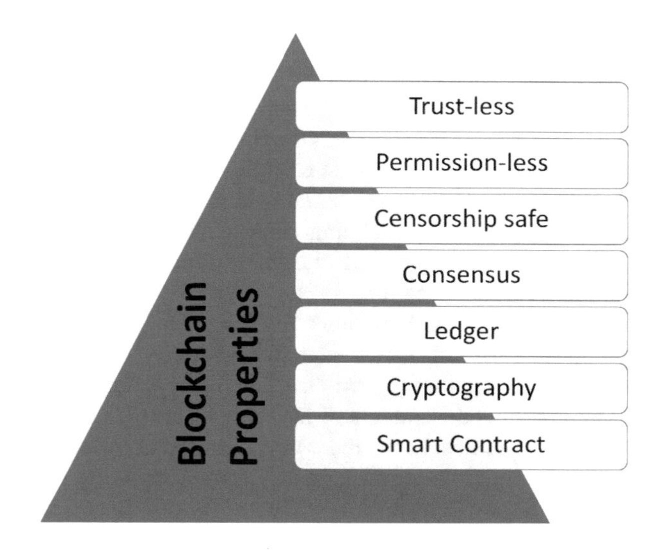

Fig. 1.2 Blockchain properties

- Cryptographic operations: It guarantees that all the information in the system is verified with a solid encryption process. Only approved clients are permitted to unscramble the data.
- Smart Contract: It is utilized to approve and confirm the members of the system.

1.2 5G Technology

In the course of recent decades, the world has seen a consistent improvement of correspondence systems, beginning with the original and moving towards the fourth era. The worldwide correspondence traffic has demonstrated an extraordinary increment as of late and is continuing to proceed, which has triggered the emergence of the prospective age of media transmission systems, to be specific 5G that aims to address the constraints of past cell guidelines and scope with ever-expanding system limits. The 5G system can surpass previous adaptations of remote correspondence innovation and offer assorted assistance capacities as well as support full systems administration among nations all-inclusive. Likewise, 5G presents answers related to the effective and financially smart dispatch of a huge number of new administrations, customized for various vertical markets with a wide scope of administration prerequisites. Specifically, the advances in 5G correspondence are imagined as inaugural of new submissions in different areas through extraordinary effects on almost all parts of our lives, for example, IoT, smart medical services [11], vehicular systems [12], smart frameworks [13] and smart cities [14]. Blockchain innovation promises the possibility for various specialized advantages regarding 5G systems and administrations. We sum up the potential applications that blockchain can provide to 5G in Table 1.1.

The 5G organization engineering must help to organize the security systems and capacities (for example, virtual security firewalls) at whatever point required in any system border. The most conspicuous innovation for streamlining management arrangement is Software-Defined Networking (SDN). SDN isolates the framework control from the data-sending plane. The control plane is configured to administer the entire system and control organization of assets through programmable Application Programming Interfaces (APIs). System Functions Virtualization (NFV) executes Network Functions (NF) for all intents and purposes by decoupling equipment machines (for example, firewalls, entryways) from the capacities that are running on them to give virtualized entryways, virtualized firewalls and, indeed, even virtualized segments of the system, prompting the arrangements of adaptable system capacities. In the meantime, cloud processing/cloud RAN underpins boundless information stockpiling and information preparing to adapt to the developing IoT information traffic in 5G. The variety of 5G empowering innovations guarantee to encourage portable systems with recently accelerating administrations such as smart information investigation and huge information handling. Contrary to past system ages (for example, 3G/4G), 5G promises to offer portable types of assistance with incredibly low inertness of vitality investment funds because of adaptability (for example, arrange cutting edge processing), which will improve quality of service (QoS) for the system and guarantee from top to bottom quality of experience (QoE) aimed at clients.

1.3 IoT

The Internet of Things (IoT) signifies the system of different unmistakable electronic or electrical gadgets that are competent to communicate with one another utilizing an open channel, for example, the Internet. This association is made utilizing remote innovations, for example, sensor systems, radio recurrence ID (RFID), close to handle correspondence (NFC), M2M and ZigBee [8]. The IoT has changed the domain of omnipresent registering with various mechanical applications working with different kinds of sensors. However, constraints exist regarding the use of the IoT, which should be addressed to advance it into a progressively effective framework [6], as shown in Fig. 1.3:

- Security: As the quantity of associated gadgets of the system expands, the odds to exploit weaknesses by outside assaults also increases. This occurs because of the usage of low standard gadgets.
- Privacy: The information gathered from IoT gadgets is sent to a focal distributed storage for investigation and handling, which involves an outsider. This sort of dissemination of information without the assent of the client can additionally cause information spills; thus, trading off the protection of the end clients.

Table 1.1 Blockchain characteristics and their possibilities for 5G

Blockchain characteristics	Description	Application for 5G
Data security and privacy	Blockchain hires uneven steganography for protection with excessive verification, truthfulness and nonrepudiation. Smart contracts to be had at the blockchain can aid records auditability, obtain entry to manage and records provenance for privacy.	Deliver excessive protection for 5G applications concerned with decentralized registers. It enables stability in the 5G networks with the aid of using supplying disbursed trust methods with excessive permissioned entry to authentication, in flip allowing 5G structures to shield themselves and ensure privacy. By loading facts statistics throughout a community of computers, the project of cooperating facts will become substantially more difficult for hackers. In addition, smart contracts, as trust-less third parties, probably assist 5G services, which include facts authentication, person verification and upkeep of 5G useful resources in opposition to attacks.
Immutability	It is very hard to regulate or alternate the records recorded within the blockchain.	Empower unbalanced unchanging nature for 5G administrations. Range sharing, records sharing, virtualized network helpful valuable asset arrangements, useful helpful asset looking for and advancing can be recorded permanently into the best-affixed blockchain. Also, D2D correspondences, pervasive IoT organizing, and wide-ranging human-driven interconnections should be possible through shared organizations of universal blockchain hubs without being adjusted or altered. The radical permanence can be exceptionally helpful for 5G organizations to call up and display bookkeeping entries, for example, logging of meeting information and utilization information for charging, helpful valuable asset use and style investigation.
Transparency	All data during transitions on blockchain (i.e. public ledgers) may be viewable to all public contributors.	Deliver better-localized perceptibility into 5G carrier utilization. The identical replica of information in blockchain banquets throughout a massive community for public verifiability. This permits carrier companies and customers to completely obtain right of entry to, verify and music transaction spots over the community with identical rights. Also, blockchains probably provide obvious ledger answers for open 5G architectures. Blockchain registers additionally assist truthful carrier buying and selling applications (i.e. useful resource buying and selling, payment) beyond the manipulation of all community entities.

(continued)

Table 1.1 (continued)

Blockchain characteristics	Description	Application for 5G
Decentralization	No principal authority and not dependent on a third party to carry out transactions. Users have complete access to their personal records.	Removes the need to rely upon the government in 5G environments, for example, range licenses, band chiefs and information base administrators are the executives; chief cloud/territory transporter manager in cell figuring and D2D networks; UAV control centre in 5G UAV networks; and convoluted cryptographic natives in 5G IoT structures. Decentralizing 5G networks most likely discards single-factor disappointments, ensures realities accessibility and amplifies transporter transport proficiency.

Fig. 1.3 IoT constraints

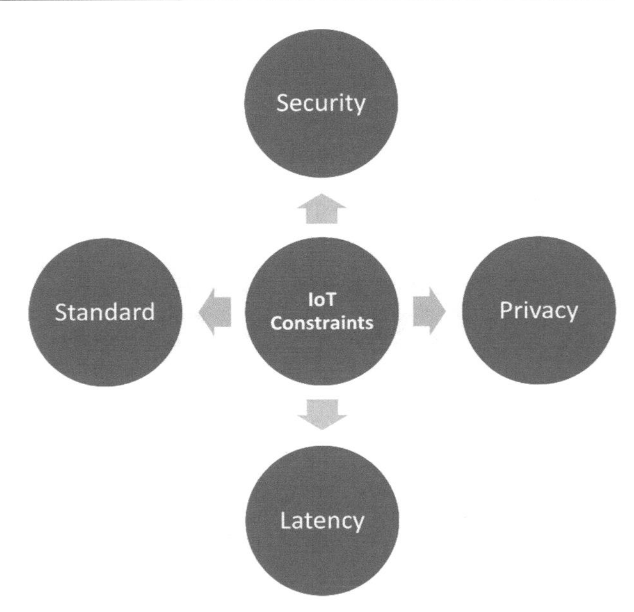

- Standards: Lack of guidelines and standards can cause unfortunate outcomes while managing the designed gadgets.
- Latency: The current correspondence principles utilized for cooperation between numerous IoT gadgets encounter dormancy issues. The continuous increase in the quantity of IoT-empowered gadgets causes a requirement for an innovation that can withstand this tremendous number of information transmissions effectively at an incredibly high data transfer capacity.

Additionally, the gadgets themselves must have the option to deal with these changes in setup, for example, enormous transfer speed limit, improved information rate and low latencies [7]. The appearance of quicker remote advancements, particularly the fifth era remote frameworks (5G), is a driver for the 5G-empowered

IoT applications. It likewise assists with managing an enormous number of IoT-empowered gadgets. The term 5G incorporates Massive-Input Massive-Output (MIMO), which helps to accomplish better arrangement capacities than the current 4G LTE, and "little cells," which permit a denser organization framework [9]. Contrasted with the current 4G innovation, which utilizes frequencies under 6 GHz, 5G systems use much higher frequencies which range from 30 to 300 GHz. 5G also empowers the making of another mechanical application that works outside of the current portable broadband range. This inescapable availability is the venturing stone to accomplish higher accessibility, which has been the focus since the initiation of cell framework [10]; thus, 5G innovation is a key empowering influence for IoT innovation. Along these lines, 5G supplements IoT to give higher information rates, diminished latencies, lower vitality prerequisites and higher versatility. The fast development of IoT innovation and 5G promises to carry substantial advantages to end clients, particularly purchasers and business companies. Purchasers are offered certain administrations dependent on their exercises. For instance, they can travel all the more proficiently by avoiding gridlocks and taking alterative driving routes when informed by the smart IoT-empowered gadget introduced in their vehicle. Furthermore, they can stay healthy by utilizing wearable gadgets that critique their wellbeing after monitoring their physical activity and body parameters for the duration of the day. Organizations can utilize the information of clients to give better administrations and items. Likewise, they can utilize area trackers and remote locking on certain hardware to verify their resources. Government and open specialists can bring about diminished medical services costs with the arrangement of better wellbeing support by remote wellbeing observing, particularly for senior individuals. In addition, street maintenance and smart road lighting can make the residents' life simpler by diminishing the general upkeep cost of the structures.

2 Adoption of Blockchain with IoT Systems

Blockchain can be used to follow the sensor data assessments and hinder duplication with some different malicious information. Positionings of IoT contraptions can be many-sided, and an appropriated record is fitting to give IoT device ID, approval and reliable secure data transfer. Rather than relying an outsider to set up trust, IoT sensors can trade information through a blockchain. All the possible advantages after adoption of blockchain in IoT systems are described in Fig. 1.4.

2.1 Trust Formation

The foundation of trust is one of the most critical prerequisites in the majority of enterprises. For the partners, including the hosts and buyers, of a specific service, such as electronic budgetary biological system or social insurance, management

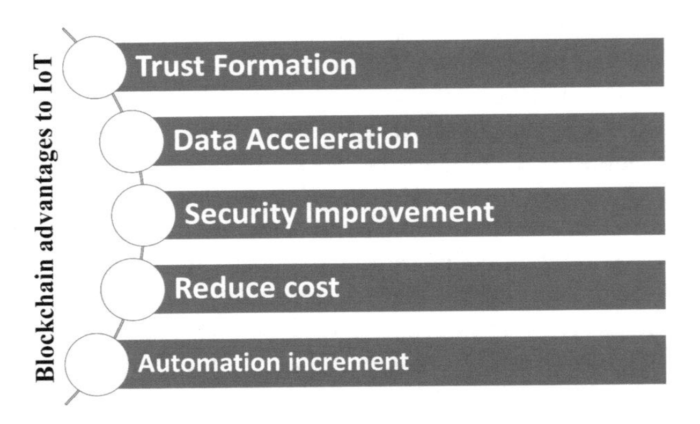

Fig. 1.4 Blockchain advantages for IoT

framework requires trust in various measurements. The measures of the trust are characterized as administrative requirements and internationally shared in the vast majority of the ventures. For example, Payment Card Industry-Information Security Standards (PCI-DSS) in a fund setting, Health Data Portability and Accountability Act (HIPAA) in a clinical setting and General Data Protection Regulation (GDPR) in a close to home information setting, are the instances of the guidelines for the foundation of trust. The smart agreements are recognizable as the trust delegates of the administrative definitions in real life. The smart agreements can be characterized as programming codes implementing the administrative standards and make them transparently available. The smart agreements completely rely upon transparency and integrity of all hubs involved.

The consistency is a fundamental reality for the trust foundation inside the system. Through the transparency of smart agreements, the trust is decentralized without being a "Black Box" in tasks. In the IoT setting, the arrangement of smart agreements makes the hubs reliable and consistent in the particular business biological system. Kuo et al. [15] clarified the advantages of utilizing blockchain-based smart agreements in the medical services space. Dagher et al. [16] clarified the utilization of blockchain for the entrance control of human services information. Yu et al. [17] depicted the establishment of trust in the IoT biological system utilizing blockchain.

2.2 *Data Acceleration*

The quickened information exchange with higher throughput and negligible inactivity is a key necessity to develop IoT biological systems. The presentation of the whole framework relies upon the quickened activity of information exchange in the IoT hubs. The appropriated idea of blockchain and smart contracts change the information exchange scene towards decentralization by lifting the presentation highlights. For example, the incorporated approval of specific information can be supplanted by decentralized approval with the utilization of smart contracts conveyed on the IoT hub itself or close to the IoT hubs, for example, the edge or the fog registering hubs which go about as blockchain organized occupants. In this manner, the solicitation reaction full circle lead-time compared to information approval, control or access can be radically reduced utilizing blockchain. Manzoor et al. [18] proposed a blockchain stage which underpins the robotized IoT information trade. Androulaki et al. [19] introduced the ability to increase up to 3500 exchanges every second with lower inertness in Hyperledger Fabric along with different strategies.

2.3 Security Improvement

Privacy, Integrity and Availability are the head model of data security which is otherwise called the confidentiality integrity and accessibility (CIA) triangle. Keeping the classification aside, the integrity and accessibility are the key highlights in the blockchain-based smart contracts by structure. The blockchain-based smart agreements guarantee the integrity by applying hashing and reaching out to the advanced marks to the individual exchange and maintaining the chain of trust altogether inside the blockchain. In addition, the blockchain innovation uses cryptographic methods, for example, Merkle trees to guarantee the predictable respectability over the system. Yu et al. [20] researched the run of the mill security and protection issues in the IoT setting and further explained by developing a system for the combination of blockchain and IoT for the affirmation of different functionalities, including validation and adaptability. Khan et al. [21] clarified the critical security issues in the IoT and how the blockchain can address these issues. The accessibility is guaranteed by the disseminated operational nature of the blockchain arrangement. For example, Denial-of-Service assaults (DoS) assailants who endeavour to vanquish the blockchain arrangement need to survive computationally troublesome obstacles, for example, commandeering 51% of the mining intensity of the system. Blockchain's solid insurance against information altering prevents a rebel gadget from disturbing the cooperative nature of correspondence frameworks, including home, processing plant or transportation framework by infusing or transferring malicious data. In this manner, the blockchain innovation holds the possibility to safely open the business and operational estimations of 5G systems to bolster basic undertakings, for example, detecting, handling, storing and conveying data. Dwivedi et al. [22] proposed a novel system to empower highlights, for example, maintain control and uphold security for the IoT gadgets and clinical information in the medical services setting. Ramani [23] introduced an expansive clarification of blockchain in various settings, including the authorization of IoT security. Ramani et al. [23].

2.4 Reduce Cost

The operational expenses of an IoT environment can be limited in various view-points when the blockchain and smart contracts are used. The decentralized activity of smart agreements and the record remove the prerequisite of sending costly high quality registering framework, for example, multi-centre distributed computing hubs for simultaneous exchange handling. In addition, the incorporated information stockpiling can be eliminated by using the conveyed record rather than concentrated information bases. The information transmission overheads for the solicitation trips there and back to the incorporated hubs, for example, cloud occasions in the concentrated frameworks can be eliminated in the blockchain-based smart agreements related environments. Effective information use is an imperative necessity in any IoT-based framework, including the arrangements associated with 5G. However, the IoT framework requires some information overhead for the synchronization between the hubs. Clauson et al. [24] clarified a couple of utilization instances of blockchain featuring the cost-cutting advantages.

2.5 Automation Increment

Blockchain combined with IoT is ideal for the mechanization necessities of future industry. The smart agreements execute naturally when the conditions have reached the executable state without mediation of some other gathering. The blockchain and smart agreements sent in the IoT gadgets are equipped for executing the smart agreements and log the occasions in the appropriated record. For example, the temperature changes of the transitory payload can be executed through the smart agreements dependent on the outer temperature. In addition, the area-based traditions obligation count is operable through the smart agreements. Griggs et al. [25] proposed a framework which uses private Ethereum blockchain and a master–slave demonstrated clinical gadget organization model so that IoT-fuelled medication actuators function precisely. Gallo et al. [26] presented BlockSee, which is a blockchain-based video reconnaissance framework to approve and guarantee the permanence of camera settings similar to the observation recordings in the smart urban communities.

3 Challenge of Blockchain and IoT Integration

There is no uncertainty that the Internet of Things (IoT) and blockchain innovation will have a significant effect on the mechanized modern world. Despite the fact that the utilization of IoT is expanding quickly, it is filled with adaptability, security, protection and integrity issues. Although blockchain was originally made for

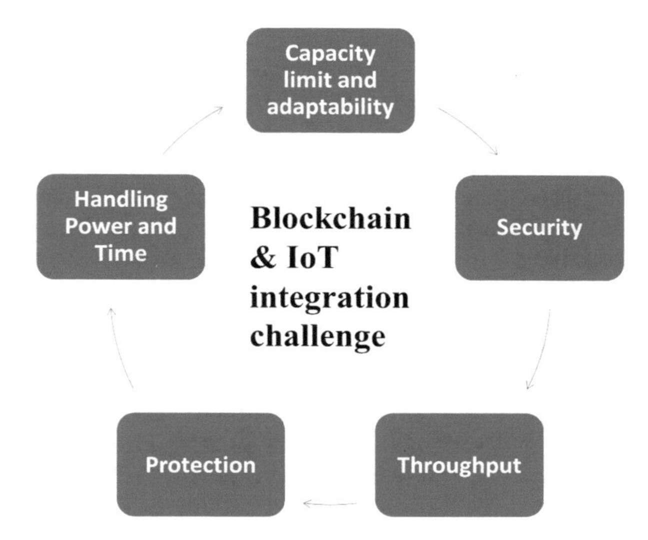

Fig. 1.5 Blockchain & IoT integration challenges

overseeing cryptographic forms of money, its decentralized nature, higher security and trustworthiness has prompted it to be incorporated with the IoT to improve the IoT. There are different difficulties emerging from this reconciliation which expands the complexities. It is important to examine the difficulties engaged with this joining before doing it. The various possible challenges are shown in Fig. 1.5.

A. Capacity Limit and Adaptability

The reliable stockpiling of exchanges and blocks is an essential necessity of the blockchain innovation. Hypothetically, every hub must contain a duplicate of the record which is developing with the exchanges. From an adaptability point of view, the effect on capacity for the IoT environment will influence the usefulness of the whole framework. Particularly, the advancing exchanges with scaling up the framework require noteworthy capacity.

B. Handling Power and Time

There are a couple of computational-asset serious tasks in the blockchain environment. These activities incorporate exchange confirmation and block age, which incorporate few cryptographic tasks. Because of the asset-limited nature of the IoT, there are certain constraints in calculation which will lead to security dangers. Therefore, use of the less asset-escalated choices must be applied explicitly when the blockchain is applied in the IoT setting. The Elliptic Curve Cryptography (ECC)-related advancements are one of the huge options, which bring about less computational overheads to the asset-confined IoT equipment. The cryptographic tasks in the limited equipment will lead to execution restrictions when scaling up the framework.

C. Security

The trustworthiness, accessibility and access control are the essential security worries in any framework. Therefore, the blockchain implements respectability and accessibility characteristically by design. Each exchange is checked with the computerized signature and the blocks of exchanges connected with confirming advanced marks. The exchange confirmation is an asset-escalated activity due to the impediments of IoT registering framework. The exchange confirmation and block age will have adaptability impediments in cryptographic procedure on the blockchain execution. Kumar and Mallick [27] portrayed the security and protection issues in the IoT and examined the noteworthiness of blockchain in this specific situation. Novo et al. [28] proposed a blockchain-based access to management engineering for the IoT. The proposed design eliminates the correspondence overheads and improves versatility. The proposed arrangement joined Software-Defined Networking (SDN), haze processing and blockchain to empower easy and low latency access to the information in a verified way. Hu et al. [29] introduced a deferred open-minded Ethereum blockchain-based instalment plot for country zones.

D. Protection

The enormous volume of IoT gadgets is common in present day sending models. The IoT gadgets uncover more extensive danger surfaces and huge restrictions in protection implementation because of the asset-limited equipment. Particularly, when blockchain is thought of, the information protection is not inbuilt because the exchanges are affixed to the record openly upon confirmation. Security conservation is a critical challenge with broadly utilized encryption strategies. In any case, the lightweight cryptographic components produced for the asset-limited computational foundation will be the perfect answer for authorized information security in the IoT setting. Zhou et al. [30] proposed BeeKeeper, which uses homomorphic calculations on the information without uncovering any bits of knowledge regarding the clients who receive the information. The framework was assessed on the Ethereum blockchain stage.

E. Throughput

Other than the versatility issue of blockchain, the throughput is another task that is difficult to handle. The exchange throughput and inertness experience predictable difficulties, and as the size of exchanges increase, they present the difficult issues that the IoT framework cannot deal with. While hypothetical examination of a stage may give a thought regarding its presentation, only useful execution can give a real use examination. We can investigate the relevance of blockchain frameworks dependent on the objective use by considering the number of exchanges important to be served in an objective time outline. Concerning IoT gadgets, private blockchains might be appropriate, as the quantity of estimations for any single gadget will be small. Regardless, as we scale to bigger IoT-based smart world frameworks serving widely dispersed gadgets, or enormous information frameworks that follow up on an extraordinary amount of information, the capacity to apply blockchain becomes more troublesome.

4 Application of Blockchain in 5G-Enabled Services

Blockchain, when coordinated into the 5G organization, will offer numerous advantages at different levels in the whole 5G environment. Organizations incorporated with blockchain can be redone dependent on the spot and supporter needs and changed progressively to fulfil the flexibly and need. Blockchain can help to improve the interior activity in the centre organization, to lessen expenses and increment adaptability. The various possible application areas are as shown in Fig. 1.6.

4.1 Healthcare

Medical care is one of the significant components for the overall improvement of any country. It could be thought about as an outline of a general public's notable prosperity. With a development regarding people and logical conditions, the weight on contemporary-day medical services structures will also increase. 5G-empowered IoT contemplated a capacity choice to lighten the burden at the medical services devices [42]. One of the appropriate responses is remote wellness monitoring, which incorporates the utilization of IoT sensor contraptions to a degree and examines numerous wellness boundaries of a customer remotely. For instance, Baker et al. [43] perceived an IoT-based medical care device for remote monitoring of the wellness of significantly ill patients. Electronic wellness records (EHR) is the combined virtual model of patients' wellness realities, while non-public wellness

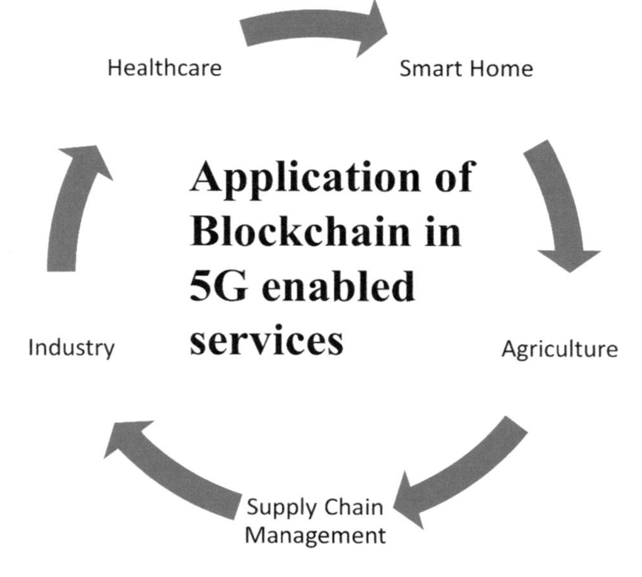

Fig. 1.6 Application of blockchain in 5G enabled services

record (PHR) is related to the virtual report of a character understanding. EHR permits consistent, continuous sharing of logical and cure accounts of patients to ensure a legitimate logical work force [44]. It handles non-public realities and oversees fundamental issues, including confirmation, privacy, responsibility and realities sharing. It empowers logical partners, including public wellness specialists, analysts and clinical specialists, to participate within the blockchain network as "miners" to offer sure motivating forces. Saravanan et al. [45] proposed a medical services worldview named Secured Mobile Enabled Assisting Device (SMEAD) to monitor diabetes. It is an offer up-to-surrender blockchain-based medical care contraption, which continuously monitors diabetic patients. In addition, it changed into fundamentally based at the guarantee that wearable devices have been presently not, at this point, suitable for crisis conditions and have been basically utilized for monitoring purposes. It helps patients who are looking for personal consideration and consistent management from specific clinical specialists.

The main aim of smart wellness applications is to organize wellness in the smart city (or society) in a green and manageable way. Capossele et al. [46] proposed a variant that encouraged the improvement of such s-wellness applications. It is assumed as an overhauled model of the existent e-wellness or m-wellness answers. It calls for realities accumulated from the various EHR and PHR, notwithstanding obtaining permission to the smart urban areas' realities and lays the foundation for the utilization of innovations such as IoT and 5G to flexibly provide appropriate continuous remarks to the residents. However, this strategy has a couple of security issues that need to be addressed. Substantially fewer environmental factors of the stage inferred that there has been a need for a consistent middleware to eliminate any third parties permissions. To address the previously mentioned issue, the creators of [46] proposed a blockchain-based on the absolute s-wellness stage to ensure security, protection, consistency, interoperability and concur with the use of 5G and IoT. Additionally, it allows the relationship of a few IoT devices with low dormancy and exorbitant unwavering quality.

Table 1.2 offers an in-depth contrast of the present procedures in healthcare, with regard to parameters, including utilization of blockchain, wearables, smart fitness, protection, open problems with possible challenges and merits/demerits of the prevailing procedures.

4.2 *Smart Home*

A smart home is an exemplification of a mechanically improved dwelling, which has the objectives to upgrade the ways of life of the populace. It bears the cost of wellbeing, comfort and extravagance to the proprietors, by letting them adjust the settings in accordance with their decisions with the assistance of a smart phone program. Through the IoT, smart home devices organize with one another to robotize home functions in accordance with the clients' inclinations. The literature mentions a few designs for energy efficient green smart houses. A notable framework of a smart

Table 1.2 Comparative analysis of different approaches for healthcare applications

Model	Description	Advantages	Disadvantages	Blockchain usage	Wearable	Smart health	Security	Challenges
Islam et al. (2015)	Overviewed the advances in IoT-based medical services innovations.	Utilizations of IoT in medical care industry examined in detail.	A few utilizations of IoT were not talked about in detail.	No	Yes	No	Yes	Yes
Saravanan et al. (2017)	Proposed a medical care worldview for diabetes monitoring.	Employed prototype in disaster circumstances talked about.	Difficulties of the model, aside from crisis circumstances, not examined.	Yes	Yes	No	Yes	No
Baker et al. (2017)	Proposed a prototype for application in IoT medical services.	Widespread conversation about vesture medical care frameworks.	A noticeable effect of movement on sensors, which may obstruct the motivation behind these wearables.	No	Yes	Yes	Yes	Yes
Capossele et al. (2018)	Proposed a model for inspiring the improvement of wellbeing applications.	Nitty-gritty arrangements of fundamental difficulties in the wellbeing environment.	Security parts of s-wellbeing applications not delineated.	Yes	No	Yes	No	Yes

home incorporates the ensuing assets: network availability (for the most part Wi-Fi), IoT-empowered sensor devices, and cell programming for remote access. Some basic contributions outfitted through smart houses comprise smart lighting, smart entryway lock, smart indoor regulator, video reconnaissance and smart stopping [47]. To offer reasonable quality for the people living inside the home, the particular contributions should continually change depending on the desired outcome. A smart entryway lock device is a basic part of any smart home. Its main objective is to prevent any unapproved guests from entering the house. The data about the populace is kept in an important worker, which allows white-listed individuals to gain admission to the house. Nonetheless, the realities managed through such a device can be projected through an undesired individual who endeavours to avoid the lock contraption to attempt to gain unapproved admission to the device. To adapt to this issue, Han et al. [48] proposed a blockchain-based smart entryway lock device, which bears the cost of security capacities such as verification, realities integrity and non-renouncement. They utilized fixed Passive Infrared (PIR) sensors, ultrasonic sensors and a development sensor to find indoor/outside gatecrashers. The blockchain network blocks keep the information about exchanges which contain open/lock order. The changeless idea of the blockchain network makes it impractical for any interloper to gain unapproved admission to the contraption and make any adjustment to effectively finished exchanges. Notwithstanding, the inactivity of IoT contraptions (sensors) can presumably be an obstacle to find such an interruption. This issue may be addressed by utilizing 5G Wi-Fi innovation, which manages the cost of discernibly low inactivity, fast interruption recognition and block mining of the exchanges in the blockchain. Dorri et al. [49] proposed a blockchain-based smart home model, which incorporates three transparency levels: the smart home, overlay and distributed storage. In this model, IoT devices controlled midway through an excavator were put within the smart home level. The intersection network carried the designated nature to this structure and is essentially similar to the P2P group used in Bitcoin. In a practically identical line, Aung et al. [50] proposed a decentralized strategy of realities control to manage the smart home device wellbeing and prolateness issues. Table 1.3 presents an in-depth contrast of current strategies in smart houses with regard to parameters together with blockchain, conversation standards, domestic automation interfaces, demanding situations and issues, and pros, cons of the prevailing strategies.

4.3 Agriculture

Smart agribusiness uses present day affects, for example, IoT, GPS and big data, to enhance the standard and extent of the resultant plants. Information such as temperature, light, soil tenacity and moisture are often managed in a central system and broken down using certain AI counts [51]. The blend of assorted movements in agribusiness hopes to form an effective watchful cultivating chain with no compromise to quality. Appropriated Ledger Technologies (DLTs) are considered

Table 1.3 Comparative analysis of different approaches for smart home applications

Model	Explanation	Advantages	Disadvantages	Blockchain usage	Way of communication	Automation interface	Challenges
Lazaroiu et al. (2017)	A smart region model has been proposed, which is essential for constructing a smart city.	Discussed home automation interface more efficiently.	Difficulties and problems of the model not talked about.	Yes	KNX protocol	Yes	No
Aung et al. (2017)	An approach for blockchain implementation in a smart home system is presented, to survive with privacy and security problems.	Smart contract policies related to smart home system discussed very well.	Exchange season of 20s, not reasonable for time delicate circumstances.	Yes	Model and survey	No	Yes
Dorri et al. [49]	Blockchain-based smart home outline presented. Summarizes three main levels of smart homes.	Substantial security and privacy assistances.	Added energy and time overheads	Yes	IPv6 over low power WAN	No	No

to have the most potential to deliver efficiency and simplicity in these common productivity chains [52]. The foremost advantage that DLTs provide is improved prominence. They will ceaselessly follow any trades that happen throughout the productivity chain. The employment of blockchain relies upon the food productivity chain because agribusiness and the food chain are correlative centres, where the end products of cultivating are not any vulnerability used as obligations to varied multi-purpose scattered productivity chains. In such food productivity chains, the client is routinely the last customer. Hua et al. [53] proposed creating a blockchain-based provenance structure, which plans to focus on the trust issues in the productivity chain industry. It records all information associated with the creation of the productivity chain; therefore, everything taken under consideration will be seen by the included individuals. To address the complexities of storing information on the blockchain, they coordinated two related structures:

- Basic Planting Information: Information associated with a selected course of action of the productivity chain, for example, creation, gathering and various techniques, is managed.
- Provenance Record: Information associated with a specific creating improvement is managed.

Caro et al. [54] proposed a blockchain-based decentralized distinguishable quality structure for an agri-food productivity chain management, called AgriBlockIoT. It ensures transparency and auditable asset obviousness to store data from the IoT devices along the whole supply chain within the key blockchain. It uses present day devices as focal points of the layered blockchain to improve the facility of the structure. The vital modules of AgriBlockIoT were API, Controller and blockchain. Another essential part, aside from the agrarian productivity chain, is the smart water system, which provides a more efficient utilization of water. The variable access to open freshwater resources encourages the planning of a system to utilize water resources sensibly, given such advancement in science and headways, for example, IoT, spread enlisting and big data. Robotization of water framework structures together with warm imaging has been a probable response for staggering water structures, which evaluates the water levels within the earth and controls the actuators to flood. It is an improvement to the back-and-forth movement of previous water frameworks, therefore causing a more controlled use of water. Sushanth et al. [55] proposed a smart farming framework, in view of the ideas of IoT and distributed computing. It empowers a rancher to devise a productive, doable water system plan for their homestead dependent on their inclinations. According to the rancher's information sources, a computerized smart water system framework was created, which gave the appropriate timetable to them. At that point, with the assistance of significant sensors and actuators, a particular methodology was executed to control the water amount delivered. Table 1.4 gives the point-by-point examination of existing methodologies in horticulture, with reference to boundaries, for example, utilization of blockchain, smart agribusiness, food recognizability, calculation and professionals, and cons of the current methodologies.

Table 1.4 Comparative analysis of different approaches for agriculture applications

Model	Description	Advantages	Disadvantages	Blockchain usage	Smart Agriculture	Food Traceability	Algorithm
Lin et al. (2018)	A food recognizability framework identified with blockchain and IoT is introduced.	Talked about the information handling stream and structure.	Hard for law-agents to discover and handle issues in the framework.	Yes	Yes	Yes	No
Caro et al. (2018)	A blockchain-based discernibility answer for agri-food supply chain is talked about well overall.	"Ranch to-people" use case.	Utilizing a solitary language for actualizing smart agreements, may meddle while growing more refined business rationale.	Yes	No	Yes	No
Sushanth et al. (2018)	Brilliant farming dependent on IoT and WSN is portrayed.	Start to finish calculation for smart cultivating framework.	The necessity of consistent web availability, which may not be consistently accessible.	No	Yes	No	Yes
Hua et al. (2018)	Proposed a rural provenance framework dependent on blockchain.	Nitty-gritty utilization of information hubs clarified.	The information transferred by taking an interest in organizations will be obvious to all members, which implies there is an absence of access control.	Yes	No	Yes	No
Tripoli et al. (2018)	Investigated the chances of use of blockchain in the agri-food industry.	An exhaustive conversation about DLT in the rural areas.	Real models were excluded.	Yes	No	Yes	No

4.4 Industry

Currently, the total mechanization of industry and business is becoming a reality. Enormous improvements in innovation and their presentation into industry have brought about the development of a next generation of industry, known as Industry 4.0. It plans to join the ability of different innovative areas, for example, IoT, blockchain and Cyber-Physical Systems (CPS) [56]. In Industry 4.0, IoT is relied upon to offer promising ground-breaking answers for existing mechanical frameworks. Therefore, it is being viewed as a key empowering agent for the up and coming age of cutting-edge modern mechanization [57]. Because of the profoundly serious market, organizations plan to pick up business points of interest at any expense. This powers business processes management (BPM) frameworks in Industry 4.0 to digitize and mechanize business procedures to build their benefits. In any case, by adding independent specialists to these business forms, the exchange expenses and dangers related to them also increase. A potential answer for handling these dangers is that every operator should discuss transparently with one another. It tackled the issue of exchange costs for self-ruling operators. However, there emerges an issue of trust between those taking an interest specialist. To handle all previously mentioned issues, Kapitonov et al. [58] recommended the utilization of decentralized frameworks (blockchain innovation) for productive and secure correspondence between the self-sufficient operators in a multi-specialist system. Unlike other dispersed records such as Ethereum and Bitcoin, which experience high deferrals, being founded on the PoW, the QoS blockchain requires constant data updates. In this situation, the opportune execution of a smart agreement makes the anchoring of another block to the primary blockchain conceivable in real time. Moreover, customers that have additional registering force can get UNET tokens as remunerations, in the event that they distribute those unused assets into an unordered arrangement. The job of the "QoS chain" is to check the quality, throughput and dependability of the system suppliers. It improves administration quality, making unchained fit for wide reception. Table 1.5 gives the point-by-point examination of existing methodologies in Industry 4.0 regarding boundaries, for example, utilization of blockchain, BPM, QoS, smart agreements, utilization of AIRA convention, difficulties and issues, and experts, and cons of the current methodologies.

4.5 Supply Chain Management

A supply chain is the system of people, associations, assets and exercises that are engaged with the existence pattern of an item. It begins from item creation to its purchase, from the conveyance of crude materials from provider to producer, directly dependent upon its conveyance to the end client. The standard stream in a supply chain starts with the provider, followed by the producer, distributer,

Table 1.5 Comparative analysis of different approaches for Industry applications

Model	Description	Advantages	Disadvantages	Blockchain usage	BPM	QoS	Smart Con-tracts	AIRA protocol	Challenges
Viryasitavat et al. (2018)	Proposed an answer for coordinate blockchain with mechanized BPM frameworks.	Administration choice and arrangement in industry 4.0	The proposed QoS blockchain is unequipped for identifying exchange cheats.	Yes	Yes	Yes	Yes	No	Yes
Kapitonov et al. (2018)	Proposed a structure to sort out monetary communications between specialists utilizing a P2P network dependent on blockchain.	Proposed a structure to sort out monetary communications between specialists utilizing a P2P network dependent on blockchain.	Issues and challenges of the protocol were not discussed.	Yes	No	No	Yes	Yes	No
Xu et al. (2018)	Study of the cutting edge in industry 4.0 as it identifies with businesses.	Digital physical systems examined in detail.	Security angles identified with industry 4.0 were not investigated.	No	Yes	No	No	No	Yes

retailer and the buyer. Supply Chain Management (SCM) is the technique to oversee materials, data and funds as they travel through a procedure in the supply chain. Given the importance of supply chains, it likewise faces difficulties, some of which are as follows [59]:

1. Logistic blunder
2. Lack of perceivability and resources
3. Improper treatment of information
4. Inefficient treatment of stock
5. Ineffective hazard management

Kothari et al. [59] examined the effect of two advances on supply chains; blockchain and 5G-empowered IoT. 5G-empowered IoT expands the transfer speed limit to verify transmission of product-related information. Blockchain gives a changeless, conveyed record which enables secure capacity of information. Additionally, it may be utilized as a device to forestall malicious IoT gadgets from entering the system. In addition to the prudent effect of blockchain innovation on organizations as far as operational cost, it can conceivably assist with relieving lawful charges emerging from questions. The primary segment of blockchain innovation is smart agreements which can empower programmed instalment of products upon their receipt, and thus eliminate the requirement for an outsider affirmation. Another significant angle is to resolve debates with respect to whether a wholesaler is qualified for a volume motivation refund. This can be addressed by utilizing smart agreements combined with 5G to follow a shipment. Casado-Vara et al. [60] proposed a model of supply chain, where the purpose of blockchain is to give security to the data of organizations associated with the agrarian supply chain alongside multi-operator frameworks for viable coordination of inside exercises. It shows the theoretical design of a supply chain and the executives with blockchain. This model empowers another market model called circular economy. It utilizes the "Make–Use–Recycle" model, as opposed to the current "Take–Make–Dispose" model. It permits the economy to act naturally sufficient. Unlike physical resources, utilities do not have a stock of their basic digital resources. Also, they do not have the capacity to follow the various exercises related to programming and equipment, for example, their turn of events, shipment and establishment, which sometimes make the frameworks vulnerable to outside digital assaults. The utilization of blockchain for this situation helps to review and track the subtleties of the product and equipment supply chain. Table 1.6 gives the point-by-point examination of existing methodologies in supply chains regarding boundaries, for example, utilization of blockchain, sort of industry, difficulties and issues, and professionals, of the current methodologies.

Table 1.6 Comparative analysis of different approaches for supply chain management

Model	Description	Advantages	Disadvantages	Blockchain usage	Industry	Challenges
Dewey et al. [9]	Explores the uses of blockchain, IoT, and 5G technology in supply chains and trade finance.	Detailed discussion on using blockchain with 5G/IoT.	Practical experiences not discussed in detail	Yes	General	Yes
Holland et al. [32]	Describes the use of DRM in additive manufacturing methods.	Discussion of business development by SAMPL ecosystem.	Implementation part was not explained in detail.	Yes	3D print	Yes
Mylrea et al. [61]	Software patch and configuration management using blockchain.	Detailed diagrammatic explanation of the research area.	Applied use case of the concept was not discussed.	Yes	Software development	Yes
Kothari et al. [59]	Explores how IoT addresses the challenges of current supply chain.	Conceptual model of IoT in SCM.	Challenges related to IoT were not discussed.	No	General	No
Casado-Vara et al. [60]	Presents the concept of circular economy.	Thorough comparison of current and blockchain-based supply chain.	Use case of circular economy was ignored	Yes	Alimentary	Yes

5 Future Research Direction

5.1 Blockchain with Big Data & 5G

In the time of information overload, large information is a hot topic in 5G [34]. A lot of interactive media information created from omnipresent 5G IoT gadgets can be abused to empower information related applications, for instance, information investigation and information extraction engaged by computerized reasoning programs. Distributed computing administrations can offer high stockpiling capacities to adapt to the increased volume of and decent variety of advanced IoT information. However, large information innovations can confront different difficulties, extending from information protection leaks, and they must maintain control over security weaknesses to prevent exceptionally complex information theft [35]. Further, huge information investigations on cloud/edge processing are profoundly helpless against cyberattacks in the complex operational business situations.

In such settings, blockchain shows up as the perfect up-and-comer to comprehend huge information related issues. To be sure, the decentralized executives related to validation and unwavering quality of blockchain can give high-security certifications to large information assets. In particular, blockchain can offer transparency and dependability for the sharing of enormous amounts of information among administration suppliers and information proprietors. By taking out the dread of security bottlenecks, blockchain can empower all-inclusive information trade which engages the wide range of 5G large information organizations. As of late, some huge information models empowered by blockchain have been proposed, for example, information offering smart agreements, maintaining control for huge information security [36], or protection safeguarding for huge information investigations [37]. Such fundamental outcomes show that blockchain can acquire different points of interest in terms of security and execution improvement to large information applications in the time of 5G.

5.2 Blockchain with Machine Learning in 5G

The fast advancements in blockchain innovation are opening new doors for computerized reasoning applications. The unrest of AI or ML innovation changes current 5G administrations by empowering its capacity to gain from information and give information driven bits of knowledge, choice help and forecasts. These focal points of AI would change the way information examinations are performed to help smart administrations in the time of 5G. For instance, ML has the capacity to associate with the remote condition to encourage executives and client correspondence [31]. ML also shows incredible potential regarding information highlight disclosure to foresee information use conduct to create control calculations, such as information traffic estimation to organize clog shirking or client monitoring for security protection [33]. Currently, there is a developing pattern of coordinating AI with blockchain in 5G applications. For instance, deep fortification learning [23] has been examined and joined with blockchain to empower secure and insightful assets for management and organization in 5G systems.

5.3 *Blockchain for 6G*

Past the fifth-age (B5G) systems, or purported 6G, will develop to give prevalent execution to 5G and meet the inexorably high prerequisites of future versatile administrations and applications during the 2030s. The key drivers of 6G will be the union of all the past highlights, for example, arrangement densification, high throughput, high dependability, low vitality utilization and monstrous network. As per [38], 6G remote systems are common to help huge client availability and multi-gigabits information transmissions with super-high throughput, incredibly low inactivity interchanges (roughly 10 s), and backing submerged and space correspondences. The 6G systems are additionally imagined to make new human-driven qualities [39] empowered by various imaginative administrations with the expansion of new innovations. The new administrations may incorporate smart wearables, inserts, autonomous vehicles, processing reality gadgets, 3D planning, smart living, space travel, Internet of Nano-Things, remote ocean touring and space travel [40]. To fulfil such applications for the 2030 smart data society, 6G should meet various rigid specialized necessities. Following this method of reasoning, high security and versatility are the significant highlights of 6G, which will be given exceptional consideration from the remote exploration network. With the promising security capacity, blockchain is expected to assume a significant role in future 6G systems. Blockchain conceivably gives a wide range of security administrations, from decentralization, protection, transparency to security and recognizability without requiring any outsiders, which will not only upgrade the security of 6G arranges but also guarantee to advance the change of future versatile administrations. The Federal Communications Commission (FCC) likewise proposes that blockchain will be a key innovation for 6G administrations. For instance, it is accepted that blockchain-based remote sharing [41] is a promising innovation for 6G to give secure, more intelligent, minimal effort and profoundly productive decentralized remote sharing. Blockchain can likewise empower security and protection of quantum correspondences, such as processing, atomic interchanges and the Internet of Nano-Things, through secure decentralized records. All the possible future research directions are shown in Fig. 1.7.

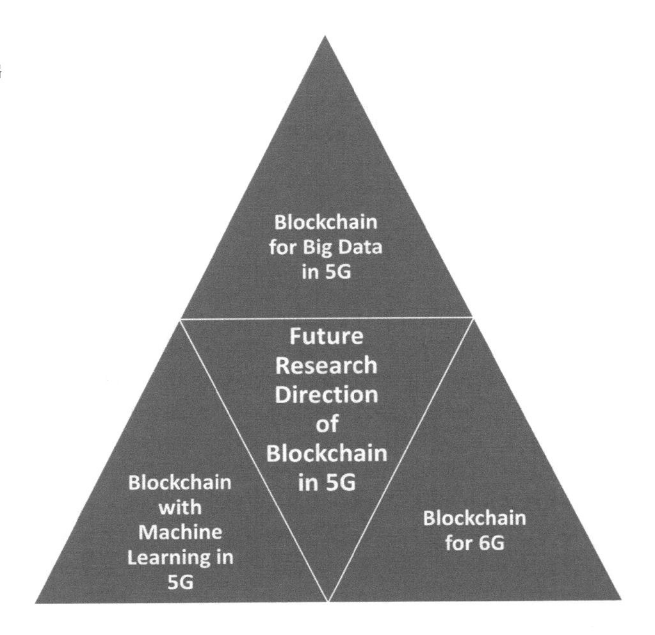

Fig. 1.7 Future research direction of blockchain in 5G

6 Conclusion

Blockchain has moved past the domain of digital currency and is currently reforming a few businesses. The intrigue goes past the promotion as a few ventures have begun embracing the blockchain-based answer for improved business measures. Similarly, 5G organization and past 5G networks are no special case as a few examinations have been directed to carry the advantages of blockchain to 5G organizations. As future 5G networks are required to be exceptionally appropriated and decentralized in nature, network management and security issues become more common and difficult contrasted with prior times. Blockchain, because of its protected plan ideas, addresses centre security issues, for example, integrity, verification, trust and accessibility, in a dispersed style. Likewise, the smart agreements can empower start to finish asset portion/sharing, network management and coordination conveying wanted administrations imagined by 5G. Moreover, blockchain will empower a few new plans of action, diminish the issue related to collaboration among network administrators and consistently handle a few cycles.

References

1. S. Nakamoto et al., Bitcoin: A Peer-to-Peer Electronic Cash System, 2008
2. W. Wang, D.T. Hoang, P. Hu, Z. Xiong, D. Niyato, P. Wang, Y. Wen, D.I. Kim, A survey on consensus mechanisms and mining strategy management in blockchain networks. IEEE Access **7**, 22328–22370 (2019)

3. A. Panarello et al., Blockchain and IoT integration: A systematic survey. Sensors **18**(8), 2575 (2018)
4. M. Singh, A. Singh, S. Kim, Blockchain: A game changer for securing IoT data, in *2018 IEEE 4th World Forum on Internet of Things (WF-IoT)*, pp. 51–55
5. S.Z. Khan, S.R. Kamble, A.R. Bhuyar, A review on BIoT: Blockchain IoT. IJREAM **04**(03), 808–812 (2018)
6. E.P. Yadav, E.A. Mittal, H. Yadav, IoT: Challenges and issues in indian perspective, in *2018 3rd International Conference On Internet of Things: Smart Innovation and Usages (IoT-SIU)*, IEEE, pp. 1–5
7. J.M. Khurpade, D. Rao, P.D. Sanghavi, A survey on IOT and 5G network, in *2018 International Conference on Smart City and Emerging Technology (ICSCET)*, pp. 1–3
8. M.H. Miraz, M. Ali, Blockchain enabled enhanced IoT ecosystem security, in *International Conference for Emerging Technologies in Computing*, (Springer, Cham, 2018), pp. 38–46
9. J.N. Dewey, R. Hill, R. Plasencia, Blockchain and 5G-enabled internet of things (IOT) will redefine supply chains and trade finance, in *Proc. Secured Lender*, 2018, pp. 43–45
10. A.Y. Ding, M. Janssen, Opportunities for applications using 5G networks: Requirements, challenges, and outlook, in *Proceedings of the Seventh International Conference on Telecommunications and Remote Sensing*, ACM, 2018, pp. 27–34.
11. A. Ahad, M. Tahir, K.-L.A. Yau, 5G-based smart healthcare network: Architecture, taxonomy, challenges and future research directions. IEEE Access **7**, 100747–100762 (2019)
12. D. Garcia-Roger, S. Roger, D. Martin-Sacristan, J.F. Monserrat, A. Kousaridas, P. Spapis, C. Zhou, 5G functional architecture and signalling enhancements to support path management for ev2x. IEEE Access **7**, 20484–20498 (2019)
13. T. Dragičević, P. Siano, S. Prabaharan, et al., Future generation 5G wireless networks for smart grid: A comprehensive review. Energies **12**(11), 2140 (2019)
14. M. Usman, M.R. Asghar, F. Granelli, 5G and d2d communications at the service of smart cities, in *Transportation and Power Grid in Smart Cities: Communication Networks and Services*, (2018), pp. 147–169
15. T.-T. Kuo, H.-E. Kim, L. Ohno-Machado, Blockchain distributed ledger technologies for biomedical and health care applications. J. Am. Med. Inform. Assoc. **24**(6), 1211–1220 (2017)
16. G.G. Dagher, J. Mohler, M. Milojkovic, P.B. Marella, Ancile: Privacy-preserving framework for access control and interoperability of electronic health records using Blockchain technology. Sustain. Cities Soc. **39**, 283–297 (2018)
17. B. Yu, J. Wright, S. Nepal, L. Zhu, J. Liu, R. Ranjan, IoTChain: Establishing trust in the internet of things ecosystem using Blockchain. IEEE Cloud Comput. **5**(4), 12–23 (2018)
18. A. Manzoor, M. Liyanage, A. Braeke, S.S. Kanhere, M. Ylianttila et al., Blockchain based proxy re-encryption scheme for secure IoT data sharing, in *2019 IEEE International Conference on Blockchain and Cryptocurrency (ICBC)*, IEEE, 2019, pp. 99–103
19. E. Androulaki, A. Barger, V. Bortnikov, C. Cachin, K. Christidis, A. De Caro, D. Enyeart, C. Ferris, G. Laventman, Y. Manevich et al., Hyperledger fabric: A distributed operating system for permissioned blockchains, *in Proceedings of the Thirteenth EuroSys Conference*, ACM, 2018, p. 30
20. Y. Yu, Y. Li, J. Tian, J. Liu, Blockchain-based solutions to security and privacy issues in the internet of things. IEEE Wirel. Commun. **25**(6), 12–18 (2018)
21. M.A. Khan, K. Salah, IoT security: Review, blockchain solutions, and open challenges. Futur. Gener. Comput. Syst. **82**, 395–411 (2018)
22. D. Dwivedi, G. Srivastava, S. Dhar, R. Singh, A decentralized privacy-preserving healthcare blockchain for IoT. Sensors **19**(2), 326 (2019)
23. V. Ramani, T. Kumar, A. Bracken, M. Liyanage, M. Ylianttila, Secure and efficient data accessibility in blockchain based healthcare systems, in *2018 IEEE Global Communications Conference (GLOBECOM)*, IEEE, 2018, pp. 206–212
24. K.A. Clauson, E.A. Breeden, C. Davidson, T.K. Mackey, Leveraging blockchain technology to enhance supply chain management in healthcare. Blockchain Healthc. Today **1**, 1–12 (2018)

25. K.N. Griggs, O. Ossipova, C.P. Kohlios, A.N. Baccarini, E.A. Howson, T. Hayajneh, Healthcare Blockchain system using smart contracts for secure automated remote patient monitoring. J. Med. Syst. **42**(7), 130 (2018)
26. P. Gallo, S. Pongnumkul, U.Q. Nguyen, BlockSee: Blockchain for IoT video surveillance in smart cities, in *2018 IEEE International Conference on Environment and Electrical Engineering and 2018 IEEE Industrial and Commercial Power Systems Europe (EEEIC/I&CPS Europe)*, IEEE, 2018, pp. 1–6
27. N.M. Kumar, P.K. Mallick, Blockchain technology for security issues and challenges in IoT. Procedia Comput. Sci. **132**, 1815–1823 (2018)
28. O. Novo, Blockchain meets IoT: An architecture for scalable access management in IoT. IEEE Internet Things J. **5**(2), 1184–1195 (2018)
29. Y. Hu, A. Manzoor, P. Ekparinya, M. Liyanage, K. Thilakarathna, G. Jourjon, A. Seneviratne, A delay-tolerant payment scheme based on the Ethereum Blockchain. IEEE Access **7**, 33159–33172 (2019)
30. L. Zhou, L. Wang, Y. Sun, P. Lv, Beekeeper: A Blockchain-based IoT system with secure storage and homomorphic computation. IEEE Access **6**, 43472–43488 (2018)
31. D.C. Nguyen, P.N. Pathirana, M. Ding, A. Seneviratne, Privacy preserved task offloading in mobile blockchain with deep reinforcement learning. arXiv preprint arXiv:1908.07467 (2019)
32. M. Holland, J. Stjepandić, C. Nigischer, Intellectual property protection of 3D print supply chain with blockchain technology, in *2018 IEEE International conference on engineering, technology and innovation (ICE/ITMC)*, 2018, pp. 1–8
33. Y. Dai, D. Xu, S. Maharjan, Z. Chen, Q. He, Y. Zhang, Blockchain and deep reinforcement learning empowered intelligent 5G beyond. IEEE Netw. **33**(3), 10–17 (2019)
34. M.G. Kibria, K. Nguyen, G.P. Villardi, O. Zhao, K. Ishizu, F. Kojima, Big data analytics, machine learning, and artificial intelligence in next-generation wireless networks. IEEE Access **6**, 32328–32338 (2018)
35. K. Sultan, H. Ali, Z. Zhang, Big data perspective and challenges in next generation networks. Future Internet **10**(7), 56 (2018)
36. U.U. Uchibeke, K.A. Schneider, S.H. Kassani, R. Deters, Blockchain access control ecosystem for big data security, in *2018 IEEE International Conference on Internet of Things (iThings) and IEEE Green Computing and Communications (GreenCom) and IEEE Cyber, Physical and Social Computing (CPSCom) and IEEE Smart Data (SmartData)*, 2018, pp. 1373–1378
37. K. Lampropoulos, G. Georgakakos, S. Ioannidis, Using blockchains to enable big data analysis of private information, in *2019 IEEE 24th International Workshop on Computer Aided Modeling and Design of Communication Links and Networks (CAMAD)*, 2019, pp. 1–6
38. W. Saad, M. Bennis, M. Chen, A vision of 6G wireless systems: Applications, trends, technologies, and open research problems. arXiv preprint arXiv:1902.10265 (2019)
39. S. Dang, O. Amin, B. Shihada, M.-S. Alouini, From a human-centric perspective: What might 6G be? arXiv preprint arXiv:1906.00741 (2019)
40. M.Z. Chowdhury et al., 6G wireless communication systems: Applications, requirements, technologies, challenges, and research directions. arXiv preprint arXiv:1909.11315 (2019)
41. Z. Zhang, Y. Xiao, Z. Ma, M. Xiao, Z. Ding, X. Lei, G.K. Karagiannidis, P. Fan, 6G wireless networks: Vision, requirements, architecture, and key technologies. IEEE Veh. Technol. Mag. **14**(3), 28–41 (2019)
42. S.M. Riazul Islam et al., The internet of things for health care: A comprehensive survey. IEEE Access **3**, 678–708 (2015)
43. S.B. Baker, W. Xiang, I. Atkinson, Internet of things for smart healthcare: Technologies, challenges, and opportunities. IEEE Access **5**, 26521–26544 (2017)
44. What is an electronic health record (EHR)? URL: https://www.healthit.gov/faq/what-electronic-health-record-ehr
45. M. Saravanan et al., SMEAD: A secured mobile enabled assisting device for diabetics monitoring, in *2017 IEEE International Conference on Advanced Networks and Telecommunications Systems (ANTS)*, pp. 1–6

46. A. Capossele et al., Leveraging blockchain to enable smart-health applications, in *IEEE 4th International Forum on Research and Technology for Society and Industry (RTSI)*, 2018, pp. 1–6
47. C. Lazaroiu, M. Roscia, Smart district through IoT and blockchain, in *2017 IEEE 6th International Conference on Renewable Energy Research and Applications (ICRERA)*, pp. 454–461
48. D. Han, H. Kim, J. Jang, Blockchain based smart door lock system, in *2017 International Conference on Information and Communication Technology Convergence (ICTC)*, pp. 1165–1167
49. A. Dorri, S.S. Kanhere, R. Jurdak, P. Gauravaram, Blockchain for IoT security and privacy: The case study of a smart home, *in 2017 IEEE international conference on pervasive computing and communications workshops (PerCom workshops)*, pp. 618–623
50. Y.N. Aung, T. Tantidham, Review of ethereum: Smart home case study, in *2017 2nd International Conference on Information Technology (INCIT)*, pp. 1–4
51. J. Lin et al., Blockchain and IoT based food traceability for smart agriculture, in *Proceedings of the 3rd International Conference on Crowd Science and Engineering*, ACM, 2018, p. 3
52. M. Tripoli, J. Schmidhuber, *Emerging Opportunities for the Application of Blockchain in the Agri-Food Industry* (FAO and ICTSD, Rome and Geneva. Licence: CC BY-NC-SA 3, 2018)
53. J. Hua et al., Blockchain based provenance for agricultural products: A distributed platform with duplicated and shared bookkeeping, in *2018 IEEE Intelligent Vehicles Symposium (IV)*, pp. 97–101
54. M.P. Caro et al., Blockchain-based traceability in Agri-Food supply chain management: A practical implementation, in *IoT Vertical and Topical Summit on Agriculture-Tuscany (IOT Tuscany)*, 2018, pp. 1–4
55. G. Sushanth, S. Sujatha, IOT based smart agriculture system, in *2018 International Conference on Wireless Communications, Signal Processing and Networking (WiSPNET)*, IEEE, 2018, pp. 1–4
56. W. Viryasitavat et al., Blockchain-based business process management (BPM) framework for service composition in industry 4.0. J. Intell. Manuf **31**, 1737–1748 (2018)
57. L. Da Xu, E.L. Xu, L. Li, Industry 4.0: State of the art and future trends. Int. J. Prod. Res. **56**(8), 2941–2962 (2018)
58. A. Kapitonov et al., Blockchain based protocol for economical communication in Industry 4.0, in *2018 Crypto Valley Conference on Blockchain Technology (CVCBT)*, pp. 41–44
59. S.S. Kothari, S.V. Jain, A. Venkteshwar, The impact of IOT in supply chain management. Int Res. J. Eng. Technol. **5**(08), 257–259 (2018)
60. R. Casado-Vara et al., How blockchain improves the supply chain: Case study alimentary supply chain. Proc. Comput. Sci. **134**, 393–398 (2018)
61. M. Mylrea, S.N.G. Gourisetti, Blockchain for supply chain cybersecurity, optimization and compliance, in *2018 Resilience Week (RWS)*, 2018, pp. 70–76

Chapter 2
Background and Research Challenges for Blockchain-Driven 5G IoT-Enabled Industrial Automation

U. Hariharan and K. Rajkumar

1 Introduction

The IoT and its use in different manufacturing areas have developed at an exponential pace over the past several years. Based on Gartner's survey, the miscellaneous collection of IoT-enabled products was expected to reach 24 billion by 2020 [1]. Overwhelming amounts of information have been produced from the gadgets, not to mention a requirement of adequate storage space and processing methods to manage it. This method comes with an increase in the information relays from machine-to-machine (M2M) as well as Device-to-Device (D2D) [2–4] interactions. To manage this particular immense information proliferation, a robust IoT process stack is needed, which works with each problem relevant to information transmission and processing at various phases. Using standardized protocols and levels, a structure could be created to perform the appropriate providers regarding IoT products [5]. The enumerations will be utilized within the automotive business to fulfill the computer users' needs and realize their business goals. Recently, there is a requirement for competitive by nature, quality products that are good with decreased merchandise expenses. The IoT has transformed the scenarios that use 5G infrastructure, including the real-time interaction involving devices, information, and a human who is able to plug in the above problems.

Nowadays, the vast majority of IoT-based methods were planned to make use of the centralized client-server, cloud servers, powerful data source, and on the internet [6–8]. The authors noticed some limitations on the IoT's centralized infrastructure:

U. Hariharan (✉) · K. Rajkumar
Department of Information Technology, Galgotias College of Engineering and Technology, Greater Noida, Uttar Pradesh, India
e-mail: u.hariharan@galgotiacollege.edu; k.rajkumar@galgotiacollege.edu

S. Tanwar (ed.), *Blockchain for 5G-Enabled IoT*,
https://doi.org/10.1007/978-3-030-67490-8_2

- Single point disaster that could topple the whole system.
- Absence of trust between the entities concerned within the device.

BC-based decentralized architectures could be utilized for P2P data transmission between sensor nodes to overcome the mentioned limits. Nevertheless, the methods consider the number of securities and privacy applied to avoid opening up the way for hackers to release various attacks.

A large number of examples can be found utilizing IoT product programs for the improvement of the system, including smart homes, smart cities/communities, industry 4.0, healthcare 4.0, and secured VANET systems [9–12]. A long-term goal relies on the fact that 5G-enabled IoT deployment to protect IoT products could utilize BC. The IoT system is secure, when it comes to the circumstance in which the new works are rapidly and efficiently processed well over the unreliable networks [13, 14]. Consequently, 5G-enabled IoT products can use identical networks within real-time. Protection might not be the only focus regarding the utilization of IoT products. Likewise, it can undoubtedly participate in information division because it operates within a quicker method. Moreover, as writers' expertise, unbiased mathematical evidence, only for the quick fixes, remains unavailable. As soon as the accessibility of mathematical evidence is available an ideal atmosphere is likely to be precisely where rapid and reliable nodes are already linked with the networking and reap the benefits of 5G by utilizing the particular cloud or maybe fog levels.

Therefore, the BC-based decentralized device is among the remedies. Among the positive aspects of utilizing BC and IoT solutions could be to keep the information within an immutable fashion, which calls for merely having no centralized data source. Furthermore, it likewise offers a means to be observed and perform transactions along with different individuals in a reliable situation. In addition to the use of effective encryption with public, private crucial pairs, IoT and BC offer vast amounts of protection to its participants.

A few BC-decentralized applications (D-Apps) are accessible within the marketplace, using BC and IoT. Utilizing IoT infrastructure and data sharing between the products could be accomplished using wireless smart devices and a high amount of network connectivity. The ubiquitous network integration is extremely hard to attain in the contemporary era, and it could be accomplished because of 5G. These solutions reduce the latency by a hundred occasions as opposed to 4G. Additionally, incorporating IoT and BC allows maintaining an immutable ledger of transactions for the conversation shared between the devices. Using the particular decentralized P2P fashion, the man in the middle attack could be removed, enabling the subscribers to communicate while not including a reliable third party [15].

Driven through the above conversation, with this chapter, an introduction to the connectivity of blockchain for the 5G-enabled IoT for manufacturing product operations is developed. We then discuss a few sensory challenges and issues that might impede BC's expansion for 5G-enabled IoT technology integrations.

1.1 Contribution of the Study

Existing surveys investigated several areas of 5G-enabled IoT for BC [16–23]. Mistry et al. [18] discussed the protection requirements for BC utilization. Vora et al. [19] acknowledged BC's usage for information suppliers to the IoT apps. Miraz et al. [20] evaluated a setup for BC using IoT data protection. Dorri et al. [16] discussed the lightweight as well as secure structure for IoT built over BC for 5G-enabled IoT technologies. Atlam et al. [21] described a base foundation for integrating BC for IoT to showcase the challenges and benefits. Singh et al. [59] additionally concentrated on the protection elements of BC dependent IoT products. Christidis et al. [17] inspected the possibilities for BC storage within the IoT products. Hwang et al. [22] recommended the powerful entry management way of dynamic interaction between products.

Nevertheless, to the best of our knowledge, these were centered on how you can make use of BC with IoT for both decentralized storage or security purposes only. Thus, this particular extensive investigation was manipulated for 7 years primarily. A view on existing mechanisms available along with advantages and disadvantages is shown in Table 2.1.

A few investigation proposals occur within the literature covering the IoT and BC, but not many had regarded the possibility of high-speed connectivity within the IoT products utilizing BC. This chapter tried to investigate the job for a

Table 2.1 Survey on existing mechanisms available

Reference no.	Advantages	Disadvantages
[15]	Easier to interface BC and IoT	Data management in UAV's is not easier
[16]	With the help of network overlaying, we could reduce the operational complexity for each frame/block	Data protection can be obtained for some attacks, not for all attacks
[19]	Architecture for IoT applications	Cannot integrate with hybrid IoT
[21]	Use of BC for 5G-enabled IoT	Cannot use D2D communication
[22]	DAC for BC	Insufficient authentication, due to modified standard
[24]	Secured application for IoT devices	Interfacing BC with IoT has high risk
[25]	The framework provides security for EHR	Restrictions on reviewing and confidentiality
[26]	Automated car parking has reduced human errors	The system is not that secured
[27]	Provides better security for IoT application in healthcare 4.0	Restrictions on reviewing and confidentiality
[28]	CPS system provides an interface between BC and 5G-enabled IoT	High security risk involved
[29]	More accessible communication between the user using BC-based P2P networks	High risk in interfacing BC with 5G-enabled IoT

manufacturing operation with BC for 5G-enabled IoT units. Below are the primary investigation efforts of this particular chapter:

- We show a systematic and comprehensive overview of 5G-enabled IoT and discuss its future industrial applications.
- Likewise, different uses for the integration of 5G-enabled IoT are discussed.
- The end of the chapter bridges the gap between scalability, interoperability, and additional studies issues for decentralized apps with 5G-enabled IoT to develop industrial automation.

The chapter is organized as follows: Section 2 discusses the essential information regarding the BC for 5G-enabled IoT, characteristics, brief working, 5G-enabled IoT usage, and challenges while integrating BC with 5G-enabled IoT. Section 3 explains BC for 5G-enabled smart industrial automation and applications in use. Section 4 illustrates different receptive challenges and issues for the incorporation of industrial automation. Finally, the end of the chapter is realized in Sect. 5.

2 Survey on BC Using 5G-Enabled Internet of Things

Specifically, this segment tracks the record for BC using 5G-enabled IoT. This section is split into three sub-sections. Initially, we discuss the fundamental information regarding BC followed by basic ideas of the IoT, limitations, features, and how 5G transforms the IoT standard regarding 5G-enabled IoT and how it is changing the contemporary era. Finally, we discuss a likely benefit for merging these two frameworks that provides the basis of the chapter.

2.1 Blockchain

With the emergence of cryptocurrencies, specifically Bitcoin, blockchain (BC) was developed with disorganized engineering [15] that works in the center of the changeover, coming through the centralized client-server system to a decentralized cryptographic protected interconnected structure. Furthermore, BC holds a decentralized and immutable ledger that keeps all the information captured in economic transactions. It contains a sequentially connected chain to the time-frame blocks, collectively utilizing cryptographic hashes [5]. This enables an end-user to obtain a distributed peer-to-peer system, in which mistrust users could swap info through one another, without a reliable third party [17].

Trust is crucial in BC, which is attained through requiring the resulting hash of the prior block to produce the subsequent block. To obtain an opinion to discover the nodes responsible for confirming the ensuing hashes, it was implemented to discover the hash of the subsequent block. A handful of transactions are bundled up collectively by blocks utilizing the hash tree, and only the hash tree root includes the block. This process is referred to as a proof-of-work (POW) node, which compensates for the effort done on the system [19]. This kind of motivator encourages the miner nodes to get involved within the system to swap computational energy supplied by them for mining blocks.

There are various kinds of BC, which could be categorized according to the foundation of details, handled information, usefulness, and accessibility to management. There are private/public and permissioned/permission-less blockchains. The main change is based on authentication principles, showing that they can use the BC (private vs. public) and permission that shows precisely what the individuals can perform (permissioned vs. permission-less).

2.2 *Industrial Automation Using 5G-Enabled IoT*

The internet of things represents the system of different unique electronic devices. They are able to interact with one another while using any receptive channel such as the internet. The links are evident in utilizing wireless devices and know-how, including RFID, sensors, Near Field Communication (NFC), machine-to-machine, and ZigBee protocol [20, 23]. The IoT has transformed the world of pervasive operation with several modern business uses constructed with different receptors. Nevertheless, at this time, there are specific limits to using the IoT that need to be addressed to develop it within a more effective program [30, 31]:

- Security: Because the variety of connected products of the system increases, the likelihood of malicious actors taking advantage of vulnerabilities also grows. It could occur because of the utilization of lower regular units.
- Privacy: The information gathered from IoT products is transferred to a cloud storage space in support of processing and analysis, including third parties. This particular distribution of details ignores the consent of the end-user and could result in information leaks. As a result, the secrecy of the end-user is compromised.
- Criteria: Not enough regulations and low standards can lead to unwanted effects while working with setup products.
- Latency: The present correspondence requirements employed for interaction between several IoT products experience latency problems.

An increase in the variety of IoT-enabled gadgets induces a requirement for such a technology that can help support this particular massive quantity of information transmissions effectively at an incredibly high bandwidth. Furthermore, the equipment themselves should deal with the modifications in setup such as large bandwidth capability, enhanced data rate, and minimal delay [32]. The creation of quicker wireless solutions, particularly the wireless model for 5G-enabled IoT uses, can also help to handle a vast variety of IoT-enabled gadgets [33]. 5G consists of Multiple-Input Multiple-Output that helps to attain better network features compared to the current 4G technology, as well as “small cells,” allowing a far denser network structure [34]. As compared to the current 4G, which employs wavelengths below 6 GHz, 5G networks support an incredibly high frequency and range from 30 to 300 GHz.

Moreover, 5G allows producing new manufacturing uses that work outside of the present movable broadband internet span. These particular omnipresent connections are a stepping-stone in attaining substantial accessibility, which has been the focus from the beginning of the wireless technologies [35]. The development of the 5G technological innovation is a vital enabler for IoT know-how. As a result, IoT is complemented by offering substantial details, lowered latencies, reduced power needs, and greater scalability [36].

The fast development of IoT engineering raised expectations to improve quality of service for businesses and consumers [37]. Individuals are available for particular solutions depending on the activities of the device. For instance, they could move around more proficiently by avoiding traffic jams and using another route, when informed by the smart IoT-enabled unit set up on the transportation vehicle. Additionally, they can stay healthy using wearable devices that offer responses associated with users' health after monitoring the physical activity of the patient and body parameters during the day. Companies can utilize the information about the owners to offer more tailored products and services. In addition, they can easily employ spot followers and remote future on specific tools to balance the effects. National, as well as public companies, can incur decreased healthcare bills together with the provision of more excellent wellness assistance by remote health and wellbeing monitoring, particularly for the elderly. Additionally, smart street and road safety can make the citizens' daily life more transparent, with a minimized general expense to maintain the components.

2.3 *Benefits of Integrating BC for 5G-Enabled IoT*

Through all of the improvements of smart programs (i.e., smart applications) for enhancing the people's caliber, IoT displays a crucial role in digitizing the services. With the fast development of IoT, one of the many entry factors to share and access information over the system (the internet) comes up. Centralized information storage space methods that use cloud computing may contribute substantially toward the improvement of IoT. However, it appears to act as a tan package in which the individuals are ignorant of the use of the information they discuss on the system. Also, these kinds of centralized frameworks might not be successful in offering information transparency [37]. To improve privacy and security, utilization of decentralized storage engineering (BC) reveals the advantages of utilizing the BC for 5G-enabled IoT for manufacturing operations, as shown in Fig. 2.1.

BC can transform the IoT, which may have an opened, trusted, and adaptable sharing dais, in which any kind of data exchanged is traceable and reliable. Several of the advantages of particular integration areas use [16, 21, 26]:

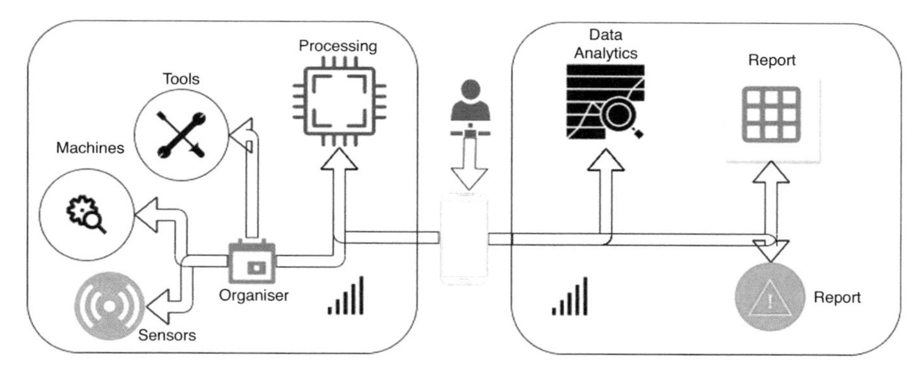

Fig. 2.1 The advantages of utilizing BC for 5G-enabled IoT for industrial automation

- Scalability as well as decentralization: The paradigm changes from a centralized to distributed cloud, it can eliminate any personal reasons for the disappointment that improves the fault tolerance. Additionally, it stops the control of online resources, in which several impressive companies might manage the compilation and process the information of many computer users.
- Identity: The utilization of one common BC and IoT enables improved identification of every unit. Becoming immutable can also have the ability to trace the foundation of just about any needed information. Furthermore, it can additionally supply a reliable way for authorization and authentication of IoT products.
- Autonomy: Using IoT and BC, products can connect without having the participation of just about any intermediary. It can pave the way for creating a device with unbelievable IoT-based manufacturing uses.
- Security: With the aid of smart providers, data switches are viewed as being a transaction; it supplies secure interservice reception.
- Reliability: Users are enabled by this integration to confirm any transaction's authenticity with confidence and respond responsibly.
- Protected code Organization: Being an inflexible log, a maker can trace the update past-history very easily. Furthermore, it allows them to upgrade IoT devices correctly [38].

2.4 *Challenges in Integrating BC for 5G-Enabled IoT for Industrial Automation*

- Space Complexity and extensibility: The standard storage space of transactions plus blocks is a primary necessity of the BC know-how. In theory, every node should have a text on the ledger to develop together with the transactions. According to the scalability point of view, your IoT environment's effect on storage space would affect the whole system's performance. Remarkably, the changing transactions increase the storage required by the system.

- Time Complexity and Energy Management: There are some computational resource intensive activities along with the BC environment, such as business transaction verification and block development, including a couple of cryptographic activities. Because of the IoT's limited source dynamics, there are specific limits in deep computation that could have direct protection consequences. Thus, using substantially fewer learning resources, comprehensive options need to be utilized, especially once the BC is used in the IoT. The Elliptic Curve Cryptography (ECC) associated solutions are among the substantial options because they incur much lower computational overheads on the source restricted IoT hardware. The cryptographic activities within the limited hardware could lead to general performance limits when scaling the product.
- Security & Privacy Policy: Integrity, accessibility, and gaining access to management tend to be the primary protections within every product. Nevertheless, the BC enforces availability and integrity inherently by design and style. Each transaction confirms using the electronic signature and the blocks of transactions are connected with confirming electronic signatures. The transaction verification is a source of comprehensive functioning because of the limits of the IoT computing infrastructure. The transaction verification and block development can have scalability limits in cryptographic activities on the BC setup. Viriyasitavat et al. [39] described the protection as well as secrecy problems in IoT and the significance of BC within this context. Da et al. [28] unveiled a prominent style of a transmitted, secure data storage program created for the IoT with moderated overhead within the product. The system enforces fine-grained entry management and sharing of time-series sensor information within the IoT uses. Kapitonov et al. [29] recommended a BC-based entry managing structure for IoT. The suggested structure removes the correspondence overheads and increases scalability. Kapitonov et al. [40] proposed a BC-based transmitted cloud structure to deal with the problems such as excessive accessibility, low latency, resilience, and real-time data delivery. The suggested answer integrated Software-Defined Networking (SDN), fog computing, and BC are to allow low and low-cost latency entry regarding the information in a secured fashion. Kothari et al. [41] discuss Ethereum BC-based transaction patterns for remote places.

The substantial amount of IoT devices is common in contemporary deployment versions, and it leads to a wider exposure of the IoT devices, which have significant threat limitations and surfaces in deep privacy enforcement because of the resource-restricted hardware. Significantly, once the BC is recognized, the information secrecy is not built-in because the transactions are added publicly to the ledger after verification. Privacy maintenance is a tremendous struggle with popular encryption methods. Nevertheless, the limited cryptographic systems created for the resource-restricted computational infrastructure are the perfect strategy for implementing information privacy within the IoT context. Casado-Vara et al. [42] proposed BeeKeeper that uses homomorphic computation on the information without revealing insights with the people who access information.

The device should be examined regarding the Ethereum BC wedge. Mylrea et al. [43] proposed BC-connected gateways, and they serve as mediators in between the IoT equipment and the people.

- Productiveness/Throughput

  Aside from BC's scalability issue, the throughput is yet another problematic condition to contend with. The transaction throughput and latency are uniform complications; thus, when the dimensions of transactions increase, usually, so do they, and those are the challenging issues that IoT devices cannot manage. While a wedge's theoretical evaluation might explain its performance, only a functional setup can supply real-world evaluation use. We can evaluate BC methods' applicability depending on the goal utilized by thinking about the number of transactions required to serve within a goal time frame. Within the situation of IoT systems, personal BC's might be ideal, as the number of dimensions for almost any unit is going to be too small.

  Nevertheless, as we level to more effective IoT-based smart world methods that primarily assist transmitting devices or maybe extensive detailed methods that act on an unprecedented number of information products, the capability to utilize BC becomes more challenging. Bocek et al. [70] determined the overall performance bottlenecks within the opinion systems and suggested architectural alterations that decrease other and computational overheads to improve the throughput to as many as 20,000 transactions per minute. In [42], several intriguing brand-new structures and analytical scientific studies suggested improving the throughput and transaction latency on the Bitcoin BC. However, considerable analysis has been performed to migrate that principle to the IoT/5G domains.

3 Deployment Using BC for 5G-Enabled IIoT (Automation)

Understanding the taxonomy of BC for 5G-enabled IoT-based manufacturing uses regarding 5G-enabled IoT is mentioned in Fig. 2.2; this includes present aspects of attention for software such as Smart network, Industry 4.0, Healthcare 4.0, Autonomous automobiles, Agriculture and Supply Chain Management (SCM). When it comes to the applications, 5G and BC are accustomed to enhancing the protection, boosting the bandwidth, and decreasing the complete functionality and capital spending. The comprehensive explanation of the apps is discussed in the next subsections.

3.1 Smart City/Smart Community

The growing trend of city migration and the connection procedure for urbanization pose several complicated obstacles regarding the cities' general infrastructure and a

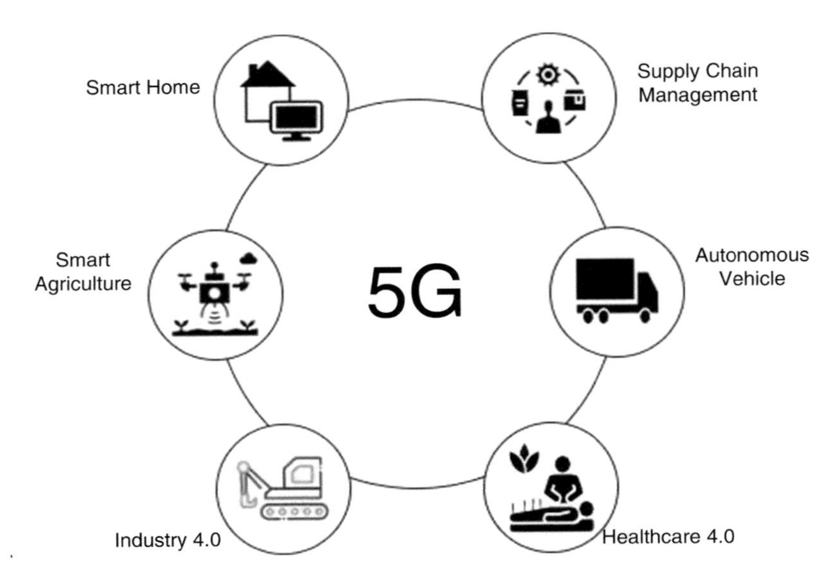

Fig. 2.2 Widely used applications of 5G-enabled IoT

device's ability to offer people the fundamental essentials of water, transportation, energy, and healthcare. This unique urbanized development results from local weather shifts, increased publicity, and the scarcity of online resources. "Smart Cities," stand out as a hands-on reaction to the issues; they guarantee an efficient and optimal utilization of available resources using the IoT and cloud computing solutions. The target of just about any smart network is providing a much better quality of solutions to the people while lowering public administration's all-around functional expense [44]. A survey of the IoT-based smart network highlights the disadvantages, benefits, and various applications, as suggested by Talari et al. [24], who also incorporated a few moderate cases of smart cities, particularly the situation of the "Padova Smart City" [25]. Figure 2.3 reveals the conceptual framework of the smart network suggested. Table 2.2 provides a comprehensive comparison of pre-existing methods in a smart city.

An essential element of the smart network is a smart auto parking process, which helps improve automobile traffic control methods to reduce the price incurred by selecting a related group. For instance, Pham et al. [26] proposed an algorithm that enhanced cloud-based smart parking methods' effectiveness dependent on IoT technology. The objective was to reduce the number of situations in which computer users neglect to locate an auto parking area while concurrently reducing the typical waiting period of owners for auto parking. Nevertheless, it lacks specific protection attributes such as extensive waiting periods for real-time deployment. The protection of these smart methods could be accomplished with the assistance of distributed ledger technologies. For instance, Lazaroiu et al. [45] recommended a smart auto parking method design that thinks as two entities, collectively within a sensor and

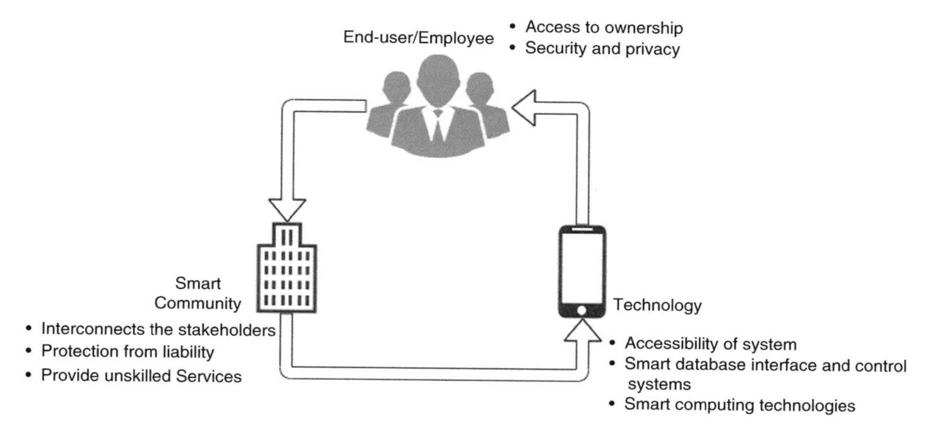

Fig. 2.3 Widely used applications of 5G-enabled IoT

Table 2.2 Survey on existing mechanisms available in smart city/smart community

Reference no.	Advantages	Disadvantages
[16]	Enhance privacy and security	High computational and energy consumption
[45]	Provides better interface smart city automation	High cost involved
[44]	A simple platform to interface IoT components for smart cities	High risk in interfacing BC with 5G-enabled IoT
[24]	Secured application for IoT devices	Interfacing BC with IoT has high risk
[26]	Automated car parking has reduced human errors	The system is not that secure

BC-based parking program. Entity A presents visitors who settled the auto parking charges to entity B system authorized user. The transaction data is in blockchain-based decentralized storage, which was represented on the internet as a block that contains data regarding the block quantity, hash of prior block, and then POW. When the vast majority of the nodes confirm the transaction's originality, the block is added to the storage unit's particular transaction. Lastly, the delivery associated with a predecided smart agreement initiates the fund transaction out of the pocketbook of entity P to entity Q, the aim is to address a variety of protection risks, as follows:

- Accessibility threat: It is concerned about the unapproved, validating of online resources.
- Integrity threats: It is concerned about unauthorized changes in the information, such as data corruption or manipulation.
- Confidentiality threats: It is focused on disclosing private information by any unauthorized entity.
- Authenticity threats: It is concerned about accessing vulnerable details with no appropriate authorization.
- Accountability threats: It is concerned about the repudiation of reception or transmission of communications by an entity.

3.2 Industry 4.0

Within the current era, the entire operation of manufacturing and industry procedures turn into a simple fact. Considerable innovations in deep technologies and the introduction of the sector have led to the growth of a new generation method, called Industry 4.0. It is designed to mix different technical domains, such as the IoT and Cyber–Physical Systems (CPS) [39]. When it comes to Industry 4.0, IoT anticipates providing promising transformational ways to pre-existing manufacturing methods. As a result, the IoT is essentially regarded as a vital enabler for the coming model of innovation in the field of manufacturing and industrial automation [28].

Because of the extremely cutthroat sector, businesses wish to obtain company benefits at virtually any price. This drives Business Process Management (BPM) devices in Industry 4.0 to digitize and automate company procedures to increase production profits. Nevertheless, by having independent representatives of the company procedures, the transaction charges related to them also improved significantly. A probable strategy to deal with the changes is that every representative can talk directly with one another. It solved the issue of transaction expenses for independent elements. However, at this time, there arises a query of loyalty between these participating elements.

To deal with almost all the problems mentioned above, Kapitonov et al. [29, 40] recommended using BC systems for secure and efficient interaction between the independent elements in a multi-agent phone system. Additionally, they improved the Autonomous Intelligent Robot Agent (project AIRA) [46], which implemented a standard financial interaction format between agent–agent and human–agent. Along with the equivalent type, Viryasitavat et al. [39] explored the chance to carry out the hands-free operation inside BPM methods utilizing D-Apps storage technologies. Using security policies in the company procedure guarantees providers' interoperation with security and trust among the integrated people. The advantages of utilizing policies know-how of BPM are as follows:

- Construct Trust: To create loyalty involving devices and parties to lessen the danger of collision and tampering.
- Cost Cutting: To bring down expenses and get rid of overhead linked with intermediaries and mediators.
- Increase and Initiate transactions: To minimize the settlement period of days or weeks to close to instantaneous.
- Quality and efficiency: Increase the effectiveness by decreasing costs and time, automated enhancements with no need for a primary representative.
- Compliance and agility: Automated conformity checking out boosts the agility of contemporary business owners.
- Networking and integration: Automation that incorporates cross-organizational business procedures by eliminating activities completed by mediators.

Real-time QoS monitoring is additionally an essential portion of the new company procedure. Having an increased variety of solutions as a result of IoT

Table 2.3 Survey on existing mechanisms available in industry 4.0

Reference no.	Advantages	Disadvantages
[39]	Easy to configure and assist the system	Cannot detect transaction deceit
[28]	CPS system provides an interface between BC and 5G-enabled IoT	High security risk involved
[29]	More accessible communication between the user using BC-based P2P networks.	High risk in interfacing BC with 5G-enabled IoT
[40]	Secured application for IoT devices	Interfacing BC with IoT has high risk

and cloud solutions, which proves challenging to pick the most suitable among the solutions, presents an effective workflow for internet business procedures.

As opposed to some other dispersed BC such as Bitcoin and Ethereum, which encounter very high waiting times, essentially based upon the PoW, the QoS demands real-time data updating. With this situation, the regular delivery associated with a smart agreement helps make the chaining of a new block to the primary feasible only in real-time. For instance, UNCHAINET [47] is a heterogeneous cloud infrastructure driven by the IoT that links underutilized detailed materials with all customers that require them. Additionally, customers have additional computing energy to grab UNET tokens as incentives, in case they set aside all those rarely used sources within the UNCHAINET network. The "QoS chain" job verifies the product quality, throughput, and then the system suppliers' dependability. It improves program quality, producing a UNCHAINET match for large-scale adoption. Table 2.3 shows the comparison of existing methods in industry 4.0.

3.3 Healthcare 4.0

Healthcare is among the majority of crucial facets for the general advancement of every country. It may be considered as a sign of a society's all-round wellbeing. With an increase in both population and medical conditions, concern regarding contemporary healthcare methods has similarly increased. 5G-enabled IoT is viewed as a possible resolution to ease the pressures along with the healthcare method [27, 48, 49]. Among the remedies is remote wellbeing monitoring, which consists of using IoT sensor devices to calculate and assess a variety of health and fitness details of a person remotely. For instance, Baker et al. [50] determined the primary key component of an end-to-end IoT-based healthcare process to remotely monitor significantly ill patients.

Electric health report (EHR) records [51] could be the electronic models of affected persons' wellness data. Personal Health Record (PHR) history is connected to the electronic history of a private long-term illness patient. EHR allows protected, real-time sharing of healthcare and therapy track records of individuals to particular authorized health-related personnel [52]. Ekblaw et al. [53] recommended a BC-based decentralized report managing process called MedRec to deal with EHRs utilizing BC storage technologies. It manages public & private information and controls essential aspects, including authentication, accountability, confidentiality, and information sharing. It encourages medical-related stakeholders, such as public health and fitness authorities, scientists, and medical doctors, to get involved within the system as a miner by offering particular perks.

Saravanan et al. [54] discussed a paradigm for healthcare 4.0 called Secured Mobile-Enabled Assisting Device (SMEAD) for diabetic monitoring. It is a one-to-one based healthcare process; it works in real-time, monitoring diabetic patients. Furthermore, the mechanism depending on wearable gadgets was not ideal for crisis scenarios and had only been utilized to check functions. Patients that looks for constant supervision and special care of specialized physicians are assisted by it.

Solanas et al. [55] projected smart-health and fitness (i.e., s-health) as the counterpart of mobile-health (i.e., m-Health), and these are the subsets of e-health, within the background of the smart network. The main objective of s-health and healthcare applications is prioritizing overall health in the smart network (or maybe modern society in general) in a sustainable and efficient fashion. Later on, Capossele et al. [56] proposed a unit that promoted these e-health uses' advancement. It is meant as a more advanced edition of existent e-health fixes [55, 57]. It takes information gathered as a result of the different PHR and EHR, and the entry on the smart cities' information beyond infrastructure with solutions such as the IoT and 5G to provide related real-time comments to the people.

Nevertheless, this method has several protection problems that need to be resolved. The wedge's trustless setting implied that we had a dependence on a protected middleware to eliminate almost any third party entry. To address the mentioned problem, Capossele et al. [56] proposed a healthy wedge to guarantee safety measures, interoperability, consistency, privacy, and loyalty by using BC for 5G-enabled IoT. Furthermore, this enables the integration of several IoT devices with high reliability and low latency.

Table 2.4 supplies the comprehensive comparability of pre-existing healthcare methods, with a guide to details such as BC, wearables, s-health & fitness, safety measures, open challenges & issues, advantages, and disadvantages.

Table 2.4 Survey on existing mechanisms available in healthcare 4.0

Reference no.	Advantages	Disadvantages
[25]	The framework provides security for EHR	Restrictions on reviewing and confidentiality
[27]	Provides better security for IoT application in healthcare 4.0	Restrictions on reviewing and confidentiality
[54]	The framework helps to monitor diabetes	It is not able to manage emergencies
[55]	S-health	It cannot be integrated with hybrid IoT

3.4 Autonomous Automobile Carrier

Autonomous Automobile Carrier has quick technical innovations, which realize interaction, computation, and evaluation. Expansion of Intelligent Transportation Systems (ITS) has become challenging. It has empowered smarter, less hazardous, plus more hassle-free transportation amenities as well as providers. It has the propensity to centralism. It might trigger a central establishment, and the layout printed momentarily because of particular outside malicious attacks.

Additionally, insufficient loyalty among the agents must be addressed. To overcome the mentioned difficulties, Yuan et al. [58] described an IoT-based protected, reliable, decentralized, and independent IoT environment called B2ITS. It is a stepping-stone for the Parallel Transportation Management System (PTMS) outline, typically intended to enhance the real-world commuter route devices using parallel interactions with the counterparts. Table 2.5 supplies the comprehensive comparability of existing methods in autonomous automobile carriers used in the industry.

Among the B2ITS framework uses, it provides a real-time ride allocation, known as La′zooz [61], designed to build an open-source, decentralized ride-sharing network to challenge the revolutionary developed personal commuter route methods. The benefits of this kind of decentralized use are eliminating the excess risks and decisions used through the primary telephone system, such as surge rates and privacy leaks. Corresponding to some novel undertakings, Huckle et al. [15] recommended an instant transaction process, called AutoPay, which is a program that delivers trust and security through “embodying” customers using smart contracts between device and user. The vehicle is allowed to synchronize instantly with the user’s AutoPay program, which could utilize outside transaction-connected providers such as fuel transactions.

Singh et al. [59] proposed an intelligent vehicle (IV) data sharing framework for a smart automobile corresponding framework based on 5G-enabled IoT technologies. It is used to verify a dependable setting to discuss the particulars involving the automobiles, with Proof of Driving (PoD). Furthermore, it is centered on a fast and secure correspondence process among the smart automobiles (i.e., “self-steering cars” [60]). Three fundamental elements are as follows:

Table 2.5 Survey on existing mechanisms available in an autonomous automobile carrier

Reference no.	Advantages	Disadvantages
[15]	Easier to interface BC and IoT	Data management in UAVs is not easier
[58]	Utilization of BC-based ITS	We are having security and privacy issues
[59]	Less time complexity for the proposed framework	Issues in updating live traffic data
[60]	Secured data management using cloud	Not easier to configure the communication protocol for UAVs

- BC: It guarantees safety and authentic communication among Intelligent Vehicles (IVs)
- Network-enabled associated system: Device using Internet connectivity, which typically enables communication within the VANET. The current situation proves that the system is an intelligent vehicle.
- Vehicular Cloud-Computing (VCC): Enhancing road traffic and highway security helps track various other intelligent vehicles.

3.5 Agriculture

Modern technologies are used by smart agriculture, including IoT, GPS, and Big Data, to enhance product quality and the amount of resulting farming products and solutions. Details such as heat, soil moisture, light, and humidity are usually kept inside the primary command structure and then examined utilizing particular AI algorithms [62]. The amalgamation of different solutions to develop smart farming is intended to improve the economics of the food supply chain without compromising the quality of the end products. Distributed Ledger Technology (DLT) is considered the best chance to increase the usefulness and transparency throughout these food supply chains [63].

Probably the utmost crucial facet that DLT offers is to improve traceability. DLT is able to observe some transactions, all of which happen through the source chain in real-time. The regulatory command becomes more relaxed; it may remain copied through each motion within the food supply chain. The outline of smart agriculture FSC using DLT is shown in Fig. 2.4.

The utilization of BC in smart farming that focuses on SCM remains a supporting area in which a definitive solution for farming is utilized to contribute to different distributed agent resource chains. For the food supply chain, customers are typically the last stop [64]. In a case study, ICT_4Ag was recognized with executing the planet's initial settlement sale made from 21 tons of feed using BC in 2017. This achievement by ICT_4Ag demonstrated that it remains a great idea to some businesses to consider a possible supply chain utilization within the farming source chain. Nevertheless, food quality must be monitored to ensure it is not jeopardized, which calls for a good food's traceability process, overseeing the product quality and security throughout the whole farming source chain. In response to this concern, Lin et al. [62] recommend a reliable nourishment discoverability system, dependent on an IoT inside the surroundings. It depends upon the standard Enterprise Resource Planning (ERP) history process, which comprises a new innovative IoT feature. A customer can use any node (for SCM) to use the smartphone to access the decentralized storage information. Additionally, Zhang et al. [65] recommended a 5G-enabled IoT, according to the discoverability process used for farming and nourishment resources, particularly for the Asian sector that dealt with the food security problems. Table 2.6 explains the existing mechanisms and models/frameworks in agriculture.

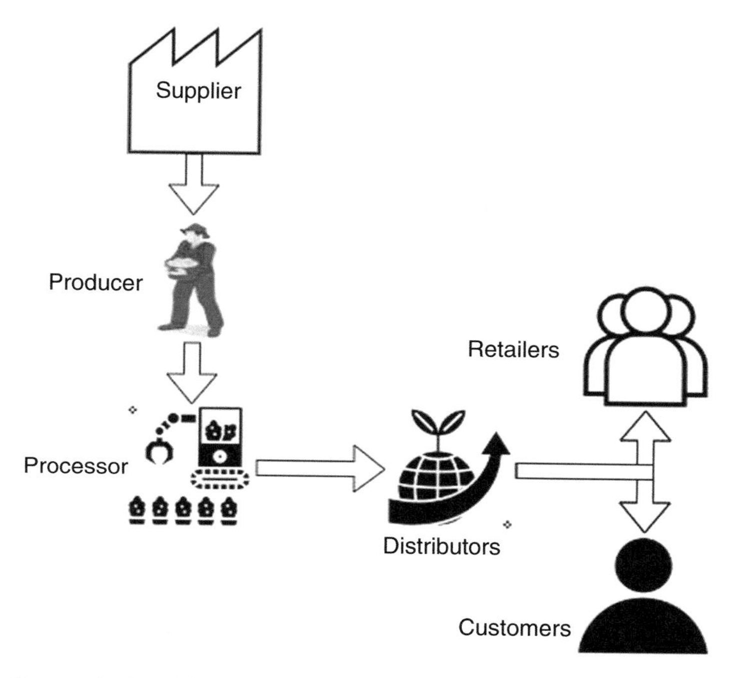

Fig. 2.4 Smart agriculture FSC using DLT

Table 2.6 Survey on existing mechanisms available in agriculture

Reference no.	Advantages	Disadvantages
[62]	More comfortable process flow and data management	Some issues in progress identification
[63]	Distributed ledger technology helps to increase efficiency and transparency	Issues with interfacing smart tracking
[64]	More accessible data management using BC for agriculture	Issues in integrating SCM with agriculture
[66]	The framework allows communicating with each level/stage in the Agri industry	Lack of access control due to high transparency
[67]	Proposed end-to-end algorithm helps for more accessible communication	Continuous connectivity is mandatory

Hua et al. [66] deliberated a framework for the farming province process typically designed to deal with the source chain sector's loyalty problems. It records each data associated with the generation, source chain, which can indeed be administered by the integrated third parties. To avoid needless complexities to maintain information in storage, they created two associated structures:

- Fundamental Planting Information: Information associated with a particular practice on the source chain such as production, storage space, along with various other procedures, is save.
- Provenance Record: Information associated with specific farming functioning is saved.

The main modules of AgriBlockIoT are API and Controller. Another essential element of smart farming, other than the farming source chain, is the smart sprinkler system that is designed for more efficient utilization of irrigation water. The accessible freshwater amounts around the world help the subscribers to develop particular ways to make use of irrigation water prudently, considering the developments in science and solutions for IoT, cloud computing, and Big Data. Hands-free operation of sprinkler system methods accompanied by winter imaging is a possible option for an intelligent sprinkler system that measures warm water amounts within the ground and regulates the actuators to irrigate. It enhances the regular electric sprinkler system, thus designing a much more controlled water consumption utilization. Sushanth et al. [67] recommended a general smart farming process using IoT and cloud computing concepts. A farmer is enabled by it to develop a useful, achievable sprinkler system routine for their farm based on users' preferences. In line with the agriculturalist's involvement, an automated smart sprinkler device is installed, which supplies the appropriate agenda for farm fields. Next, the pertinent receptors and actuators performed a specific process to manage the warm water amount while the sprinkler system was complete. A comparable item was suggested by Pallavi et al. [68], where they recommended a management process for greenhouse farming using IoT products to remotely realize pertinent details. The objective of the producer was to encouraging natural agriculture while enhancing the yield. The variables considered incorporated skin tightening and emission, soil moisture, temperature, and lighting.

3.6 Supply Chain Management

The source chain could be the people, resources, organizations, and pursuits engaged within the life span cycle of merchandise. It starts from merchandise development to its sale, from shipping and delivery of raw substances at the provider to supply the correct product and the delivery of it to the conclusion PC handler. The regular movement inside the supply chain starts with a provider, accompanied by a farmer/producer, retailer, wholesaler, and customer. Supply Chain Management (SCM) could be a process for controlling the funds, data, and materials that are shifted via a procedure within the resource chain [69]. Considering the significance of supply chains, they also face difficulties, including [41]:

- Logistic mishandling
- Insufficient assets and visibility

- Incorrect data management
- Ineffective management of the stock
- Inadequate supervision

Dewey et al. [34] communicated about the effects of two solutions in the supply chain using BC and IoT. It enhances the transmission capability to protect the communication of stock information. Decentralized storage offers an unchallengeable distribution for BC that allows protected depository space for information. Furthermore, it may be employed to stop the incidence of harmful products of IoT in the system. Aside from the affordable effect of storage engineering on businesses within functional price terminology, it can help mitigate authorized charges arising from conflicts. The primary element is an intelligent agreement that can allow automated transactions of the products on their receipt, eliminating the demand for a third-party confirmation. Another essential part is eliminating conflicts concerning if a distributor permits an amount of motivator rebate. It could be managed by utilizing smart contracts accompanied by BC for 5G-enabled IoT to observe the track of delivery.

Casado-Vara et al. [42] recommended a particular style for SCM, in which the IoT is protecting the data regarding businesses concerned within the farming stock through multi-agent devices that are useful to control the inner pursuits. Figure 2.5 reveals a theoretical structure for SCM managing through IoT devices. The particular design provides a novel industrial strategy known as a circular economy. It applies the "Make–Use–Reuse" version, instead of the existing "Take—Make–Dispose." The economy is enabled by it to become self-sufficient.

Unlike actual physical property, utilities do not have an accounting of the user's critical cyber assets. Additionally, they cannot observe the various pursuits linked to software programs and hardware for their advancement, shipment, and set up, making the devices susceptible to outside cyber attacks. The utilization of supply chain management within this situation assists in auditing and monitoring the software program's specifics and the hardware resources chain [43]. Bocek et al. [70] discuss information concerning the business enterprise start-ups work using BC for 5G-enabled IoT products in medical SCM. Table 2.7 describes the summary of the existing model or framework used in SCM.

A similar start-up was initiated in Switzerland (https://modum.io/), which allows effective supply management and regulatory conformity of the commuter routes of healthcare items to monitor each product's heat in the course of shipment. A smart agreement evaluates the information to look for anomalies while maintaining the IoT infrastructure. Utilizing decentralized storage for medical SCM reduces the number of mediators within an organization's course of action, therefore lowering the functional expenses and chances relevant to product or service tampering.

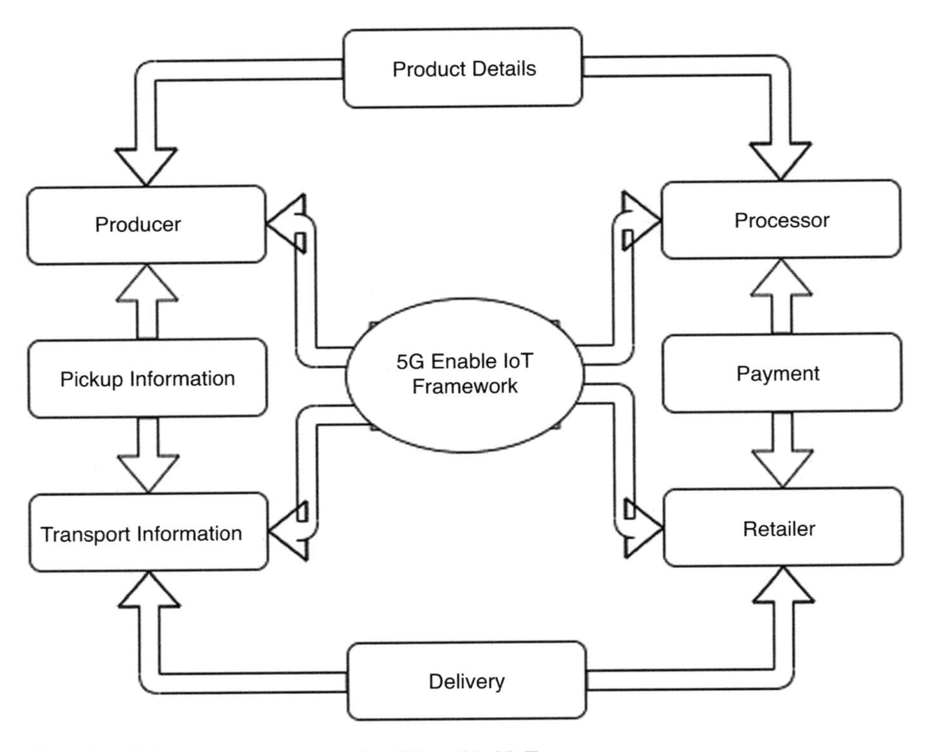

Fig. 2.5 SCM operation structure using 5G-enabled IoT

Table 2.7 Survey on existing mechanisms available in supply chain management

Reference no.	Advantages	Disadvantages
[41]	Capable of handling data and mismanagement of logistics	Lack of risk management
[42]	Provides data security to the company involved in SCM and easy to track the products	Continuous connectivity is mandatory
[43]	Easier to track the logistics using hardware and software	The system may be vulnerable to external attacks
[70]	A smart contract checks the data for anomalies	Lack of logistics security, due to tampering
[65]	High information security could be achieved because of DLT	Interfacing BC and the food supply chain is not easier

4 Challenges and Issues for Incorporation of Industrial Automation

Even though the amalgamation of the IoT and 5G has received a lot of attention from industry and academia, the cryptocurrency style based on the PoW idea possesses attributes that make them poorly suited for many IoT scenarios. In line with the many literature evaluations that discussed the problems regarding 5G-enabled IoT programs, and problems delineated, as displayed in Fig. 2.6, we have determined huge investigation problems relevant to 5G-enabled IoT. Thus, before taking into consideration applying the idea inside pre-existing programs, the problems must be further investigated:

- Because many IoT products remain battery-operated, its useful toward source restrictions due to the power constraints are restricted. Nevertheless, using BC that launches D-Apps for decentralized storage, block mining is a computationally intensive job. Within the scenarios, the power needs and the processing period of products have to be investigated further.
- 5G was introduced to supplant the present client/server methods. Nevertheless, this data is kept within the nodes that are generally the IoT products. These units have low computational energy and minimal storage space capability. Therefore, this confirms a huge hurdle within the adoption of the technologies.
- IoT scales poorly as the number of sensors within the system increases, which should be tackled quickly.
- Generally, there are a few unanswered queries about 5G-enabled IoT being able to eliminate some of the particular vulnerabilities such as DoS attacks together with the infamous 42% occurrence concerning setting up dispersed loyalty.
- Not enough, no interoperability, and proper standardization signify diverse ledgers cannot exclusively communicate with one another. Nevertheless, they call for the integrated stakeholders to excessively compromise (ranging from

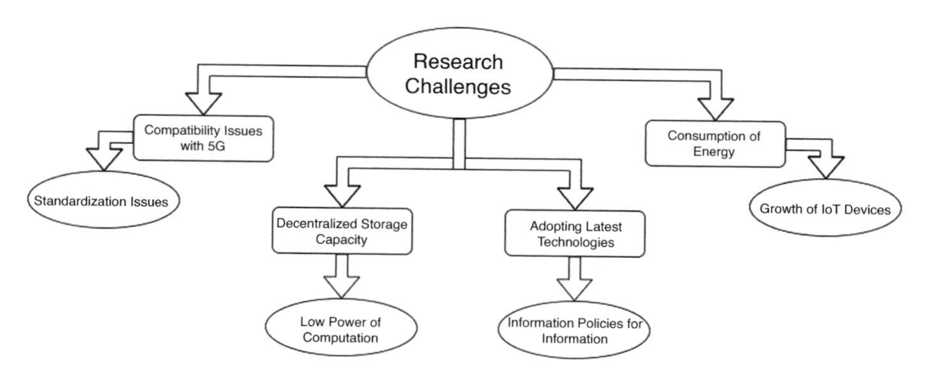

Fig. 2.6 Investigation problems on 5G-enabled IoT

using complete details on the policies) to attain complete interoperability. It takes overseas policies for collective information and trust protection.
- Insufficient reliable company instances because of the substantial amount of concerns remain regarding know-how, and individuals might stay in a difficult situation to predict and accept the changes made due to technological advancement in the industries.
- Additionally, having the capability to link various independent users coming through various areas (possibly countries) without having the demand for any kind of type of authorized conformity to watch. It is equally challenging for program suppliers and companies and might be a big screen for switching many company instances.
- With the arrival of 5G solutions, IoT products should adapt to be suitable for high-speed network connectivity.

Table 2.8 shows the comparison of the various use case summary of BC using 5G-enabled IoT for industrial automation.

5 Conclusion

The user probably is experiencing the 5G consequent integration of BC through IoT products, which became revolutionary. This chapter provided knowledge concerning BC's industrial uses for 5G-enabled IoT products split into four sections. First, we discussed the basics of IoT and BC. Second, we briefly discussed BC using 5G-enabled IoT and then respective industrial applications and uses. In the third part, the deployment using BC for 5G-enabled automation systems was investigated. Finally, sensory challenges and issues were discussed primarily for industrial applications and uses. The requirement of top hardware and the absence of integration to higher network interconnection requires a common framework to keep the solutions tackled within this function. Currently, the future utilization remains to be developed. The majority of the industrial applications and uses discussed how BC for 5G-enabled IoT could protect and provide quicker information flow. Nevertheless, over and above the small-scale innovations and deployments of particular programs, an excellent technical analysis level is necessary to deal with the particular challenges relating to the solutions' effort. Finally, relative evaluation of current BC-based industrial applications and uses needs to be carried out regarding the foundation of specific details.

Table 2.8 Use case summary of BC using 5G-enabled IoT for industrial automation

Application	Description	Applicability of blockchain	Challenges
Smart City	Smart urban centers are innovative contemporary methods within the urbanization to enhance man's everyday living.	BC is perfect for allowing expert services, including payments, security, e-governance, and surveillance of smart urban areas.	The computational energy necessity for mining is a tremendous downside within the blockchain for IoT within the smart urban areas.
Industry 4.0	IIoT can revolutionize how manufacturing production, resource management, supervision, etc., are completed.	A plethora of manufacturing uses 5G IoT, which includes smart production, warehousing, automated maintenance, etc., could be built safe with blockchain.	The scalability, information secrecy, and information storage space overheads are essential issues within the manufacturing 5G IoT context.
Healthcare 4.0	All kinds of healthcare (preventive, rehabilitation, diagnostic, etc.) require overseeing detecting signs bringing about premature analysis, lumber privately health background, revealing health-related papers, etc., properly.	BC can allow safe and trusted healthcare hands-free operation methods for checking, remedies, and healthcare information entry management.	A malicious assailant can dominate almost all mining nodes, which is catastrophic.
Agriculture	To boost the farming efficiency while reducing the price requires constant background parameter acquisition, smart aggregation, and an immense amount of particular data.	Using BC for smart farming allows lifecycle transparency, independent managing, and creating regulatory demands of agri-food items.	Information integrity is an extremely critical facet of the Agri products that call for a unique interest over the ledger.
Autonomous vehicle	Completely automated (level 5) automobiles have to have correct processing and sense with ultra-minimal latency aimed at as the promise of 5G IoT.	BC can perform a considerable part for various start using instances, including protected and personalized background in electric powered autos, automated cost computation, and other providers.	The access and security management systems need to eliminate hacks and ensure specific people's information privacy.

References

1. Gartner, Leading the IoT, https://www.gartner.com/en/newsroom/press-releases/2017-02-07-gartner-says-8-billion-connected-things-will-be-in-use-in-2017-up-31-percent-from-2016
2. I. Budhiraja, S. Tyagi, S. Tanwar, N. Kumar, M. Guizani, CR-NOMA based interference mitigation scheme for 5G femtocells users, in *2018 IEEE Global Communications Conference (GLOBECOM)*, Abu Dhabi, United Arab Emirates, pp. 1–6 (2018)
3. I. Budhiraja, S. Tyagi, S. Tanwar, N. Kumar, M. Guizani, Cross layer NOMA interference mitigation for femtocell users in 5G environment. IEEE Trans. Veh. Technol. **68**(5), 4721–4733 (2019)
4. I. Budhiraja, S. Tyagi, N. Kumar, J.J.P.C. Rodrigues, DIYA: Tactile internet driven delay assessment NOMA-based scheme for D2D communication. IEEE Trans. Industr. Inform. **15**(12), 6354–6366 (2019)
5. S. Tanwar et al., Human arthritis analysis in fog computing environment using Bayesian network classifier and thread protocol. IEEE Consum. Electron. Mag. **9**(1), 88–94 (2020)
6. J. Vora, S. Tyagi, N. Kumar, M.S. Obaidat, A systematic review on security issues in VANET. Secur. Priv. J. Wiley **1**(5), 1–27 (2018)
7. H.-N. Dai, H. Wang, G. Xu, J. Wan, M. Imran, Big data analytics for manufacturing internet of things: Opportunities, challenges and enabling technologies. Enterp. Inf. Syst. **14**(6), 1–25 (2019)
8. J. Bhatia, Y. Modi, M. Bhavsar, Software defined vehicular networks: A comprehensive review. Int. J. Commun. Syst. **32**(12), e4005 (2019)
9. S. Tanwar, P. Patel, K. Patel, S. Tyagi, N. Kumar, M.S. Obaidat, An advanced internet of thing based security alert system for smart home, in *2017 International Conference on Computer, Information and Telecommunication Systems (CITS)*, IEEE, pp. 25–29 (2017)
10. J. Wan, J. Li, M. Imran, D. Li, A blockchain-based solution for enhancing security and privacy in smart factory. IEEE Trans. Ind. Inform. **15**(6), 3652–3660 (2019)
11. S. Tanwar, S. Tyagi, S. Kumar, The role of internet of things and smart grid for the development of a smart city, in *Intelligent Communication and Computational Technologies*, (Springer, Singapore, 2018), pp. 23–33
12. J. Vora, S. Tyagi, N. Kumar, M.S. Obaidat, A systematic review on security issues in vehicular ad hoc network. Secur. Priv. **1**(5), 1–26 (2018)
13. S. Moin, A. Karim, Z. Safdar, K. Safdar, E. Ahmed, M. Imran, Securing IoTs in distributed blockchain: Analysis, requirements and open issues. Futur. Gener. Comput. Syst. **100**, 325–343 (2019)
14. H.A. Khattak, M.A. Shah, S. Khan, I. Ali, M. Imran, Perception layer security in internet of things. Futur. Gener. Comput. Syst. **100**, 144–164 (2019)
15. S. Huckle, R. Bhattacharya, M. White, N. Beloff, Internet of things, blockchain and shared economy applications. Procedia Comput. Sci. **98**, 461–466 (2016)
16. A. Dorri, S.S. Kanhere, R. Jurdak, Blockchain in internet of things: Challenges and solutions. arXiv preprint arXiv:1608.05187 (2016)
17. K. Christidis, M. Devetsikiotis, Blockchains and smart contracts for the internet of things. IEEE Access **4**, 2292–2303 (2016)
18. I. Mistry, S. Tanwar, S. Tyagi, N. Kumar, Blockchain for 5G-enabled IoT for industrial automation: A systematic review, solutions, and challenges. Mech. Syst. Signal Process. **135**, 106382 (2020)
19. J. Vora, S. Kaneriya, S. Tanwar, S. Tyagi, N. Kumar, M.S. Obaidat, Tilaa: Tactile internet-based ambient assistant living in fog environment. Futur. Gener. Comput. Syst. **98**, 635–649 (2019)
20. M.H. Miraz, M. Ali, Blockchain enabled enhanced IoT ecosystem security, in *International Conference for Emerging Technologies in Computing*, (Springer, Cham, 2018), pp. 38–46
21. H.F. Atlam, A. Alenezi, M.O. Alassafi, G. Wills, Blockchain with internet of things: Benefits, challenges, and future directions. Int. J. Intell. Syst. Appl. **10**(6), 40–48 (2018)

22. D. Hwang, J.Y. Choi, K-H. Kim, Dynamic access control scheme for iot devices using blockchain, in *2018 International Conference on Information and Communication Technology Convergence (ICTC)*, IEEE, pp. 713–715 (2018)
23. S.H. Shah, I. Yaqoob, A survey: Internet of Things (IOT) technologies, applications and challenges, in *2016 IEEE Smart Energy Grid Engineering (SEGE)*, IEEE, pp. 381–385 (2016)
24. S. Talari, M. Shafie-Khah, P. Siano, V. Loia, A. Tommasetti, J.P.S. Catalão, A review of smart cities based on the internet of things concept. Energies **10**(4), 1–23 (2017)
25. A. Cenedese, A. Zanella, L. Vangelista, M. Zorzi, Padova smart city: An urban internet of things experimentation, in *Proceeding of IEEE International Symposium on a World of Wireless, Mobile and Multimedia Networks*, IEEE, pp. 1–6 (2014)
26. T.N. Pham, M.-F. Tsai, D.B. Nguyen, C.-R. Dow, D.-J. Deng, A cloud-based smart-parking system based on internet-of-things technologies. IEEE Access **3**, 1581–1591 (2015)
27. S.M.R. Islam, D. Kwak, M.D.H. Kabir, M. Hossain, K.-S. Kwak, The internet of things for health care: A comprehensive survey. IEEE Access **3**, 678–708 (2015)
28. L. Da Xu, E.L. Xu, L. Li, Industry 4.0: State of the art and future trends. Int. J. Prod. Res. **56**(8), 2941–2962 (2018)
29. A. Kapitonov, I. Berman, S. Lonshakov, A. Krupenkin. BC based protocol for economical communication in industry 4.0, in *2018 Crypto valley conference on BC technology (CVCBT)*, IEEE, pp. 41–44 (2018)
30. E.P. Yadav, E.A. Mittal, H. Yadav, IoT: Challenges and issues in indian perspective, in *2018 3rd International Conference On Internet of Things: Smart Innovation and Usages (IoT-SIU)*, IEEE, pp. 1–5 (2018)
31. R. Gupta, S. Tanwar, F. Al-Turjman, P. Italiya, A. Nauman, S.W. Kim, Smart contract privacy protection using AI in cyber-physical systems: Tools, techniques and challenges. IEEE Access **8**, 24746–24772 (2020)
32. J.M. Khurpade, D. Rao, D.P. Sanghavi, A survey on IOT and 5G network, in *2018 International Conference on Smart City and Emerging Technology (ICSCET)*, IEEE, pp. 1–3 (2018)
33. W. Ejaz, A. Anpalagan, M.A. Imran, M. Jo, M. Naeem, S.B. Qaisar, W. Wang, Internet of things (IoT) in 5G wireless communications. IEEE Access **4**, 10310–10314 (2016)
34. J.N. Dewey, R. Plasencia, Blockchain and 5G-enabled Internet of Things (IoT) will redefine supply chains and trade finance, in *Proc. Secured Lender*, pp. 43–45 (2018)
35. A.Y. Ding, M. Janssen, Opportunities for applications using 5G networks: Requirements, challenges, and outlook, in *Proceedings of the Seventh International Conference on Telecommunications and Remote Sensing*, pp. 27–34 (2018)
36. M.M. Alsulami, N. Akkari, The role of 5G wireless networks in the internet-of-things (IoT), in *2018 1st International Conference on Computer Applications & Information Security (ICCAIS)*, IEEE, pp. 1–8 (2018)
37. A. Reyna, C. Martín, J. Chen, E. Soler, M. Díaz, On BC and its integration with IoT. Challenges and opportunities. Futur. Gener. Comput. Syst. **88**, 173–190 (2018)
38. A. Boudguiga, N. Bouzerna, L. Granboulan, A. Olivereau, F. Quesnel, A. Roger, R. Sirdey, Towards better availability and accountability for iot updates utilizing a BC, in *2017 IEEE European Symposium on Security and Privacy Workshops (EuroS&PW)*, IEEE, pp. 50–58 (2017)
39. W. Viriyasitavat, L. Da Xu, Z. Bi, A. Sapsomboon, BC-based business process management (BPM) framework for service composition in industry 4.0. J. Intell. Manuf. **31**, 1737–1748 (2018)
40. A. Kapitonov, S. Lonshakov, A. Krupenkin, I. Berman, BC based protocol of autonomous business activity for multi-agent systems consisting of UAVs, in *2017 Workshop on Research, Education and Development of Unmanned Aerial Systems (RED-UAS)*, IEEE, pp. 84–89 (2017)
41. S.S. Kothari, S.V. Jain, A. Venkteshwar, The impact of IOT in supply chain management. Int. Res. J. Eng. Technol. **5**(08), 257–259 (2018)
42. R. Casado-Vara, J. Prieto, F. De la Prieta, J.M. Corchado, How BC improves the supply chain: Case study alimentary supply chain. Procedia Comput. Sci. **134**, 393–398 (2018)

43. M. Mylrea, S.N.G. Gourisetti, BC for supply chain cybersecurity, optimization and compliance, in *2018 Resilience Week (RWS)*, IEEE, pp. 70–76 (2018)
44. A. Zanella, N. Bui, A. Castellani, L. Vangelista, M. Zorzi, Internet of things for smart cities. IEEE Internet Things J. **1**(1), 22–32 (2014)
45. C. Lazaroiu, M. Roscia, Smart district through iot and BC, in *2017 IEEE 6th International Conference on Renewable Energy Research and Applications (ICRERA)*, IEEE, pp. 454–461 (2017)
46. AIRA (Open source software for smart cities and Industry 4.0 projects), https://aira.life
47. UNCHAINET (Decentralized cloud platform), https:// https://unchained-capital.com/
48. A. Kumari, S. Tyagi, N. Kumar, Fog computing for healthcare 4.0 environment: Opportunities and challenges. Comput. Electr. Eng. **72**, 1–13 (2018)
49. V. Tripathi, F. Shakeel, Monitoring health care system using internet of thingsan immaculate pairing, in *2017 International Conference on Next Generation Computing and Information Systems (ICNGCIS)*, IEEE, pp. 153–158 (2017)
50. S.B. Baker, W. Xiang, I. Atkinson, Internet of things for smart healthcare: Technologies, challenges, and opportunities. IEEE Access **5**, 26521–26544 (2017)
51. J.J. Hathaliya, S. Tanwar, S. Tyagi, N. Kumar, Securing electronics healthcare records in healthcare 4.0: A biometric-based approach. Comput. Electr. Eng. **76**, 398–410 (2019)
52. What is an electronic health record (EHR), https://www.healthit.gov/faq/what-electronic-health-record-ehr/
53. A. Ekblaw, A. Azaria, J.D. Halamka, A. Lippman, A case study for BC in healthcare: MedRec prototype for electronic health records and medical research data, in *Proceedings of IEEE Open & Big Data Conference*, vol. 13, p. 13 (2016)
54. M. Saravanan, R. Shubha, A.M. Marks, V. Iyer, SMEAD: A secured mobile enabled assisting device for diabetics monitoring, in *2017 IEEE International Conference on Advanced Networks and Telecommunications Systems (ANTS)*, IEEE, pp. 1–6 (2017)
55. A. Solanas, C. Patsakis, M. Conti, I.S. Vlachos, V. Ramos, F. Falcone, O. Postolache, et al., Smart health: A context-aware health paradigm within smart cities. IEEE Commun. Mag. **52**(8), 74–81 (2014)
56. A. Capossele, A. Gaglione, M. Nati, M. Conti, R. Lazzeretti, P. Missier, Leveraging BC to enable smart-health applications, in *2018 IEEE 4th International Forum on Research and Technology for Society and Industry (RTSI)*, IEEE, pp. 1–6 (2018)
57. J. Vora, A. Nayyar, S. Tanwar, S. Tyagi, N. Kumar, M.S. Obaidat, J.J.P.C Rodrigues, BHEEM: A BC-based framework for securing electronic health records, in *2018 IEEE Globecom Workshops (GC Wkshps)*, IEEE, pp. 1–6 (2018)
58. Y. Yuan, F-Y. Wang, Towards BC-based intelligent transportation systems, in *2016 IEEE 19th International Conference on Intelligent Transportation Systems (ITSC)*, IEEE, pp. 2663–2668 (2016)
59. M. Singh, S. Kim, Blockchain based intelligent vehicle data sharing framework. arXiv preprint arXiv:1708.09721, 1–4 (2017)
60. M. Singh, D. Singh, A. Jara, Secure cloud networks for connected & automated vehicles, in *2015 International Conference on Connected Vehicles and Expo (ICCVE)*, IEEE, pp. 330–335 (2015)
61. La'zooz, http://lazooz.org/
62. J. Lin, Z. Shen, A. Zhang, Y. Chai, BC and IoT based food traceability for smart agriculture, in *Proceedings of the 3rd International Conference on Crowd Science and Engineering*, pp. 1–6 (2018)
63. M. Tripoli, J. Schmidhuber, *Emerging Opportunities for the Application of BC in the Agri-food Industry*. FAO and ICTSD: Rome and Geneva. Licence: CC BY-NC-SA, 3, pp. 1–40 (2018)
64. A. Kamilaris, A. Fonts, F.X. Prenafeta-Boldύ, The rise of BC technology in agriculture and food supply chains. Trends Food Sci. Technol. **91**, 640–652 (2019)
65. D. Tse, B. Zhang, Y. Yang, C. Cheng, M. Haoran, BC application in food supply information security, in *2017 IEEE International Conference on Industrial Engineering and Engineering Management (IEEM)*, IEEE, pp. 1357–1361 (2017)

66. J. Hua, X. Wang, M. Kang, H. Wang, F-Y. Wang, BC based provenance for agricultural products: A distributed platform with duplicated and shared bookkeeping, in *2018 IEEE Intelligent Vehicles Symposium (IV)*, IEEE, pp. 97–101 (2018)
67. G. Sushanth, S. Sujatha, IOT based smart agriculture system, in *2018 International Conference on Wireless Communications, Signal Processing and Networking (WiSPNET)*, IEEE, pp. 1–4 (2018)
68. S. Pallavi, J.D. Mallapur, K.Y. Bendigeri, Remote sensing and controlling of greenhouse agriculture parameters based on IoT, in *2017 International Conference on Big Data, IoT and Data Science (BID)*, IEEE, pp. 44–48 (2017)
69. What is supply chain (SC), https://whatis.techtarget.com/definition/supply-chain
70. T. Bocek, B.B. Rodrigues, T. Strasser, B. Stiller, BCs everywherea use-case of BCs in the pharma supply-chain, in *2017 IFIP/IEEE Symposium on Integrated Network and Service Management (IM)*, IEEE, pp. 772–777 (2017)

Chapter 3
Fusion of Blockchain Technology with 5G: A Symmetric Beginning

Suneeta Satpathy, Satyasundara Mahapatra, and Anupam Singh

1 Introduction

Blockchain technology is becoming popularized as digital currency in the form of Bitcoin all over the world. It is termed as a distributed database for carrying out transactional operations online and has justified its efficiency and benefits in terms of its key attributes, federalization, secrecy, tenaciousness and controllable features for translating the conventional industrial system [1–3]. On the same hand, 5G is becoming more popularized in mobile technological industries because of its ability to interconnect heterogeneous devices with its broadband, remission services, machine-like communication [4] and enhanced qualitative throughput. 5G has revolutionized the communicational network system with a new set of attributes that have improved the criteria like network security, reliability and ability with smaller latency [5, 6]. Such a communicational network has brought a complete makeover in the industrial organizations in terms of high speed, virtualization among the business sectors and establishing the connection between Internet-operated devices, applications as well as objects. The 5G network [7] has also created many new opportunities for customers as well as business organizers and industries by providing a facility to interconnect communicating devices that can control and connect all spheres of human lifestyle and services. There are several

S. Satpathy (✉)
College of Engineering Bhubaneswar, Bhubaneswar, Odisha, India
e-mail: suneeta@koustuvgroup.ac.in

S. Mahapatra
Pranveer Singh Institute of Technology, Kanpur, Uttar Pradesh, India
e-mail: satyasundara@psit.ac.in

A. Singh
University of Petroleum and Energy Studies, Dehradun, Uttarakhand, India
e-mail: anupam.singh@ddn.upes.ac.in

S. Tanwar (ed.), *Blockchain for 5G-Enabled IoT*,
https://doi.org/10.1007/978-3-030-67490-8_3

technical supports like software-defined networking (SDN), cloud computing, network functions virtualization (NFV), edge computing, network slicing and D2D communication that have strengthened the power of 5G network [8, 9]. So 5G network empowered mobile communicating devices are required to be assisted with online digital payment platforms which in turn can be provided by blockchain technology. Again along with the power of 5G communicating technology, many challenges need to be handled like trustworthiness of network, permanency of data, isolation and secrecy of data [10]. On the other hand, Blockchain being popularized in the digital era can efficiently handle the challenges that have been put forth by 5G networks. So blockchain technology with an intent to make the most of cryptocurrency applications [11] can be associated with a 5G network to more securely carry out digital online transactions to prove the positive potential of it. In the recent smartphone being manufactured, the concept of hardware wallets that makes use of hardware cryptocurrency and empowers the blockchain transaction to be conducted safely over 5G communication networks taking the real benefit of 5G networks has been developed. So the features of blockchain can be thought of as a supporting hand for all sorts of future network technologies [12].

So the major focus of this chapter is to lay out the technological backdrop for 5G communications as well as to review how blockchain technology can tune-up with it as a fused component with an objective of considering it as one of the driving factors for the development of next-generation 6G network services. Further, the chapter flows with the description of the concepts of blockchain and its smart-supportive features for the 5G communicational network in Sect. 2 followed by Sect. 3 that briefs the potential features of 5G communication networks. Section 4 narrates the fusion of blockchain technology with 5G-enabled smart automated applications. Then several challenges and issues that are assisted by the technological revolution [13, 14] are addressed in Sect. 5 to outline the future research direction followed by conclusions in Sect. 6.

2 Blockchain and Its Related Concepts

Blockchain technology has been adopted in various market segments because of its potentiality as well as several benefits. Such distributed technology has been adopted in various applications starting from cryptocurrency, IoT and finance-related transactions to various social and risk-oriented tasks and thus is expected to carry out day-to-day activities [15]. Initially, this technology has been used in terms of digital money named Bitcoin which in turn is described as a protocol in the digital communication network. In a better way, blockchain can be thought of as a decentralized composition of digital transactions which is not under the control of individual or company. Again, the technology is named so as because old blocks are not altered or tampered by anyone and new blocks get linked with the existing blocks resulting in a formation of a chain. Blockchain has a stringent set of rules and structure that makes sure that data can only be inserted into the database without doing any alteration to the existing ones which in turn is a long and complicated process of back tracing the entire history of transactional data. Moreover, we can say that blockchain is a group of shared and linked transactional emergent data that are stored digitally in the form of a ledger. The security aspects in blockchain are maintained by cryptographic techniques, digital authenticating signature and distribution agreement that allow freeness of mind among the people to accumulate, swap over and observe the information in a digital platform securely. Blockchain, with a growing transactional data characterized by date and time stamp, is decentralized and dispersed around the communicational networks with the security rules enforced that all the interaction done through it would be visible to all the users, but require authentication verification before augmenting any information into it. On the same hand, users would be able to update the existing data block to which they have been granted access, and the same would be reflected in the entire network. Prior application of blockchain was named through Bitcoin which is regarded as a digital coin to make business. The successful application of Bitcoin has enabled the utilization of blockchain technology in different fields like healthcare facility, IoT, finance-related services, official document management and tracking, insurance-related services, supply chain management, tourism services as well as handling cyberthreats and criminals.

So blockchain technology can be summarized as a mixture of different disciplinary concepts like mathematics, cryptography, networking, economic modelling and distributed consensus algorithms [16] that have made the inclusion of various features such as decentralized, maintaining secrecy and trust, self-sufficient and automated and visible, secure and verifiable as briefed below to make the digital transactions protected and tamper-proof.

Decentralized Blockchain technology implements distributive transactional operations where data can be stocked and updated.

Maintaining secrecy and trust Blockchain technology maintains the user's trust by allowing anonymous data transfer. It allows to send only their blockchain address and not the original identity during transactional operations.

Self-sufficient and automated The blockchain users make a set of rules on the basis of which blockchain technology works. It is not ruled by any single central authorized person; rather, it has one of the components called smart contract which is a computer program with auto executed actions when the contract conditions are fulfilled.

Visible, secure and verifiable Blockchain technology works on the principle of decentralization which means data is not stored in a single place; rather, it is scattered all over. Again transactional data is visible to all the users. Even when any updation occurs, it can be visualized by all, justifying the transparency for each node. Further, the transactional data remains restricted for any change by the users unless otherwise authorized for it.

Blockchain data storage is distributive in nature and thus maintains the security by not being easily accessible to the hackers for taking any illicit attempts. The security is also enforced by encrypting the transactional data and linking it to the existing blocks only when every node user gives their nod of the validated updations if made to it. So blockchain technology has maintained a trust and security factor for various business organizations where data is the most critical asset and possible to be tampered by the intruders.

2.1 Blockchain Architecture

The blockchain can be seen as consisting of blocks which are sequentially connected representing a complete transaction record. The blocks are connected with each other through a hash reference of the preceding block known as parent block. The first block does not have a parent, hence known as the genesis block, and every block has the following information:

1. Header
2. Body

The header part of the block contains information like version describing various validation protocols to be adopted, the 256 bit hash value of the parent block, the hash value of the Merkle tree root block indicating hash values of all transactions, date and time stamp of every transaction present in the block, a 4 byte nonce starting with zero that amplifies for each hash value calculation and n-bits representing the present hashing value in a compacted manner. The body part of the block contains transaction counters [17]. The capacity of the block and the size of the transaction decide the total number of transactions that can be present in the block. Figure 3.1 shows the functioning of the blockchain technology.

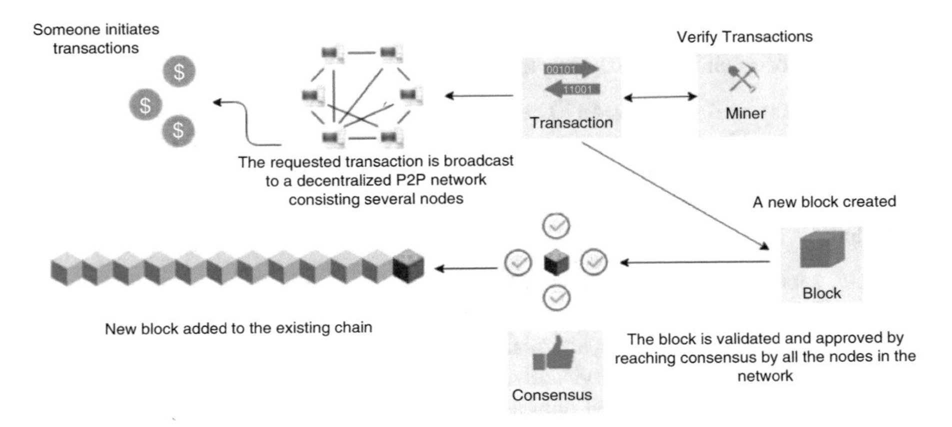

Fig. 3.1 Functional diagram of Blockchain network [1]

The legitimacy of the blockchain technology is maintained with the help of asymmetric cryptography that enables digitally authenticated signatures. The legitimacy is also maintained through two phases named as signing with a private key phase and verifying it with the public key phase. Each blockchain user owns a private and public key. The user uses the private key to put a sign on a transaction that is to be distributed around the network. All others who can see the transaction in the network can access it by using their public key.

Blockchain technology is also augmented with a consensus algorithm [18] that makes it more secure and is distributed across the network. The algorithmic function makes sure that the transactional block, whether the updated one or the new block, has been placed into the existing chain properly or not. It also ensures that the block that is added is the one visible to all in the network and is protected from various cyberthreats.

The algorithm works on two principles:

1. Proof of work
2. Proof of stake

The proof of work is able to generate valid proof in a randomized process which is also known as the mining process. In this, each block is associated with a random value designated as the nonce in the block header. The proof of work has to produce a value that can compare the nonce hash value to be smaller than a value already set up as a targeted value. The comparison with the targeted value is done in terms of the time required for generating it. Such a process of the generation that would make a complete coverage of all sorts of data in the block by proof of work decides the acceptance of the block by the users of the network. As an addition to the proof of work, the security protection from different types of cyberthreats is given by the proof of stake.

2.2 Catalogue of Blockchain Architectures

The flavours of blockchain architectures differ in their design layout and architectural description. The architectures can be discussed under the following names:

1. Public blockchain
2. Private blockchain
3. Consortium blockchain

1. Public blockchain

This type of blockchain architecture (Fig. 3.2) defends itself to be completely transparent as every user in the network is provided with the total history of blockchain and each of them is allowed to check and verify the transaction. The user connected to the network with a computer and Internet connection is treated as a node which is allowed to take part and obtain the consensus. The main advantage of such architecture lies in hiding the user credentials. A peer-to-peer transaction is

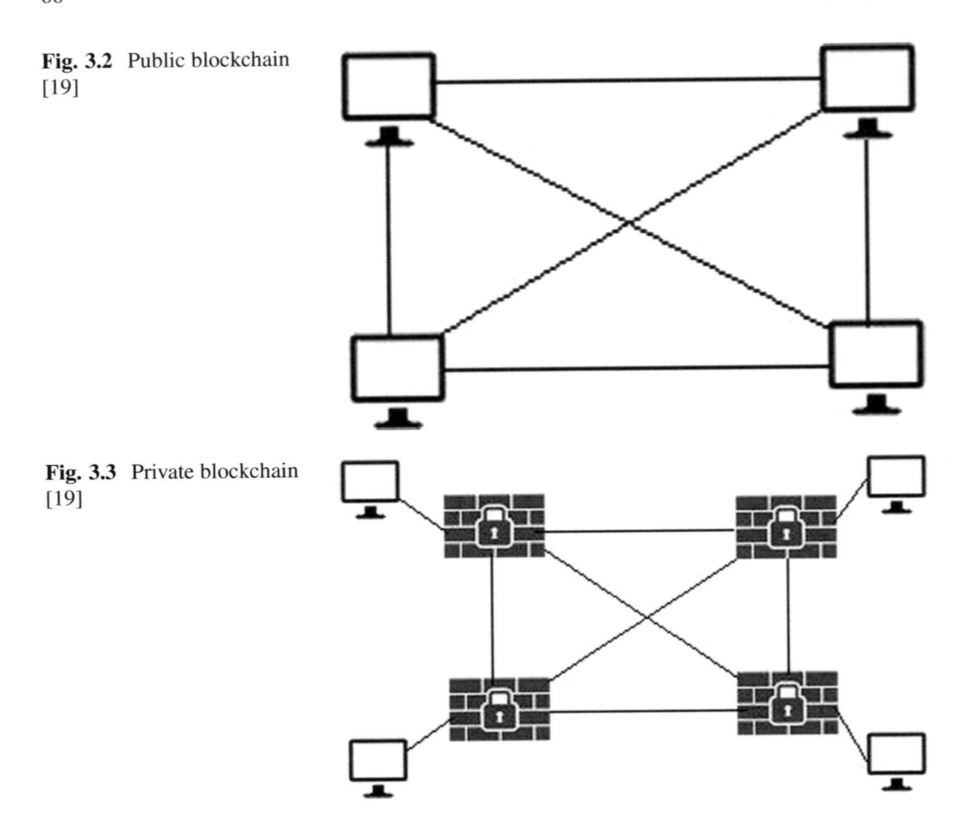

Fig. 3.2 Public blockchain [19]

Fig. 3.3 Private blockchain [19]

built up to emphasize decentralization. Such architecture also makes sure that every transaction is linked with blockchain prior to its updation in the system and thus gets synced with each and every node in the network of blockchain.

2. Private blockchain

In comparison to the public blockchain architecture, private blockchain as shown in Fig. 3.3 has restricted settings to access the data in the network. The participation restrictions are applied on the nodes, i.e. only authorized nodes are allowed to participate. The transaction can only be validated and verified by authorized nodes being initiated by a company or organization. Such a feature in the private blockchain architecture enhances the verification and validation process effectiveness. In comparison to the public blockchain, private blockchain keeps the users' information more private by sanctioning the access privileges for them. Such architecture is more inclined towards conventional modelling and e-governance. The private blockchain is more adopted by the private and government sectors because of the security provided by the central authority as well as for its enhanced efficiency and faster transactions.

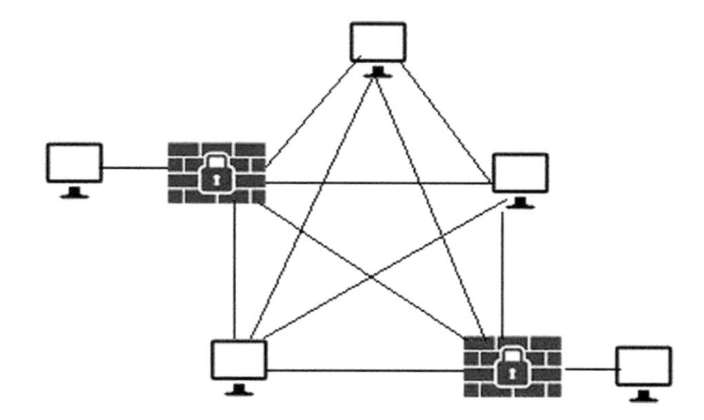

Fig. 3.4 Consortium blockchain [19]

3. Consortium blockchain

This type of blockchain architecture is a mixture of public and private blockchain architectures as shown in Fig. 3.4. That means such architecture makes the blockchain free to the public with a restriction that all the users won't be able to avail the total data. Protocols are defined for access privileges and validation process. In comparison to private blockchain consortium, architecture is partially decentralized. From the total number of nodes, few nodes are treated as trusted nodes that have the responsibility of controlling the consensus process. These nodes also decide the addition of a block to the chain once the consensus is obtained from the validation process. Such architecture is more preferred by the corporate business house in comparison to private blockchain architecture because of its partial decentralization.

3 5G and Its Features

The network communication technology has undergone a transition from the first generation to the fourth generation. The exponential increase in network traffic globally has prompted the development of even more trustworthy, speeder and efficient communication technology named as 5G networks that can fulfil all the limitations and gaps of existing mobile networks [20]. 5G communication technology is also known as global communication technology as it is consumed in various application areas across the globe. It is adopted because of its specific features like broadband connection, low-latency reliability as well as machine-to-machine communication. The adaptation of 5G technology has made a paradigm shift in different industrial business applications. Also the interconnection of the devices creates an ample amount of opportunities for business collaboration. So, 5G communication has made possible to connect all pieces of human life to the network to avail the promising user-oriented services like smart healthcare systems, vehicular networks [21, 22], smart grid and IoT [23, 24] by enabling different talented technologies like cloud computing, SDN, D2D services, network slicing and virtualization features as briefed below.

Cloud computing: The main objective of cloud computing technology is to manage the storage space and resources, thus achieving the challenges and demands of the 5G network. Various 5G network services starting from mobile network management to remote sensor-based services are well handled by virtual computing capability of cloud computing. In addition to the well-powered features of cloud technology, edge computing has also added extra power to 5G networks by providing storage and computing platform with a close relation to low value for transmission [25].

Software-defined networking (SDN): With this feature, it is possible to accomplish different network functions and services with software rather than using hardware. Such adaptability is also enhanced with the flexibility of a 5G network [26] by providing a separation between the control plane and the data plane.

Network functions virtualization (NFV): The main objective of NFV is to bring transformation in the delivery of network services. In addition to the feature of SDN that allows the services to run on the software, NFV encourages the services to run in standardized hardware platforms rather than the exclusive hardware made for them. So, with the features and functionalities of NFV, 5G network services would be enhanced in terms of efficiency, speed and faster services leading to better revenue generation and at the same time simplification of network functionalities.

Network slicing: Due to the versatility nature of the 5G network, it can be used by different applications. Based on the application requirement, there will be a demand for different types of networks leading to network slicing. Network virtualization as well as SDN feature can configure a variety of networks and fulfil the demands of user applications. Thus, it will allow different types of software to be run on the same hardware platform and fulfil different users' demand by providing low latency level for one and other types of the level depending on network performances for other users so that every user gets a taste of a slice of the network.

Device-to-device (D2D) communication: Such a feature allows communication between the IoT devices easily in comparison to conventional signal communication and transmission. Such type of communication makes data transfer between the mobile devices placed in short distance range at a high speed as well as guarantees the latency value at a lower range. It also empowers the 5G communication with flexibility in traffic offloading and enhances efficiency as well as reduces the energy that is supposed to be lost in lengthy data transmission.

Millimetre-wave (mmWave) communication: This communication technology augments novel facilities to 5G communication networks with demanded mobile data dynamically. Such feature also adds on new benefits to 5G networks like high bandwidth with the qualitative transmission, reduced beam, enhanced data accessing ability by ignoring the mobile network traffic, huge number of connected devices and machines with a variety of use cases.

So all these technologies are the augmented features in the 5G network to fulfil the diversified demands of the user applications along with the potential design of the network as summarized below.

3.1 Design Concepts of 5G Network

5G communication network is popular because of its properties as being flexible, scalable and configurable. Such beneficial features are obtained through a variety of software that runs on the hardware platform as per user demand and network virtualization as well as network slicing. Network slicing frees the user from the control panels and introduces dynamic network applications and services leading to ubiquitous platform and infrastructure. The main aim of 5G network is to transform all conventional manual services into automated qualitative services with high data transmission rates. On the same hand, it also aims to overcome the limitations of 4G networks. SDN feature of the 5G network is responsible for the simplification of operations and services by separating the control panel from the data panel. The centrally placed control has the responsibility of controlling the network resources with the help of API that is programmed. Also, network functions virtualization is meant for providing flexibility in carrying out network functions by making an impression of the virtual detachment of hardware platforms from various software-oriented services which results into virtualized gateways, firewalls as well as a network that is flexible enough to fulfil the demands of user applications. In addition to that, cloud computing platform solves the limitation of data storage to keep at par with the growing network traffic created due to the connection of IoT devices [27]. So the main focus of the 5G network is to render novel application-based services as well as amplified mobile network services with low latency value, the flexibility that could improve the quality in comparison to prior generation's network like 3G/4G. But all the benefits of the 5G network should be supported with well-defined security mechanisms which take the form of challenges being faced by the network and thus highlighted in the subsequent section.

4 Challenges Faced by 5G Network and Motivational Gains for Fusing Blockchain Technology with 5G

Any communicational network has to be assisted with authentication mechanisms. It plays a vital role in maintaining the security of the user applications, the identity as well as the underlying network. The authentication technique in the 5G network is handled by a server which takes lots of time even for preliminary authentication. As 5G network gives the assurance of zero latency, maintaining authentication for devices in the network and UEs, sustaining data confidentiality and data integrity, making the network available and having smaller computational complexity value and low cost for communication would be very crucial tasks. Further, it would require enhanced security mechanisms to make it more secure and beneficial. Though there exist security mechanisms for 5G networks to handle data transfer and connections, they are not adequate. So there is a need to have more protocols that can enforce security restrictions, awareness, storage, data transmission and user-demanded software that can form a threat-free valid network. Blockchain technology has the potentiality to handle the challenges put forth in the 5G network. So the integration of both technologies can show promising opportunities for the user to avail the benefits from both [28–30].

The 5G technology follows a centralized architecture which results in dreadful conditions for both network and computing performances. For example, the cloud computing or edge computing feature of 5G communications adopts the Amazon cloud server which is centralized and can have severe security problems. In addition to that, cloud services are mainly popular in fulfilling services as per user demand or choice, but the same cannot be achieved if there is a failure at any single point. IoT-enabled services with multiple user demands are also not possible for a centralized system, if there would be a cyber malware attack. Similarly, NFV provides a virtual environment with multiple cloud services that give the benefit of function chaining in the 5G network, but it can suffer from leakage of data by a few cloud objects. Since the virtual environment is a sharable resource, there is a high probability of threats that can hamper the lucidity and liability of cloud service providers. On the same note, virtualized servers for NFV run through virtual machines that are offering a variety of operating system environment with the help of orchestration protocol, leading to a security concern for the transmission that is established between a VM manager and an orchestrator. The mobile bandwidth data-oriented services like video streaming and a large volume of data processing may require an efficient resource management strategy in 5G communication so that resource usage can be consistent with the avoidance of scarce issues and degradation in performance. Such things can only be achieved by sharing through centralization but that would create the possibility of more cyber-attacks and decreased quality of service, or an authorized person would be devoid of resources. The IoT-enabled services availed through the 5G network making everything smart will also face complex security concerns as a large amount of diversified data would be generated leading to misidentification of malicious objects causing threats [31]. But there is an absolute requirement by the users to get efficient sharing of data along with proper network management assisted with high-throughput and elevated security features. So the limitations and security constraints faced in the 5G communicational network can be overcome to some extent by the fusion of blockchain technology with 5G network services [32].

4.1 Promoting Fusion of Blockchain with 5G

As we know, blockchain technology is considered to be a distributed public ledger associated with properties like being decentralized, maintaining secrecy and trust, being self-sufficient and automated and being visible, secure and verifiable and, hence, has the capability of handling the challenges faced by 5G technology concerning security and management of the network [33]. Moreover, the benefits that blockchain technology can augment to 5G networks can be grouped under enhanced system performance, enforcement of better security features and well-handled and managed network services discussed as follows.

1. Enhanced system performance
The main objective of the 5G network is to maintain qualitative services with low latency. Keeping this into account, blockchain technology is blessed with decentralized nodes with smart contracts that can handle all resource requests intelligently without availing centralized authorization. The integration of blockchain with 5G communication [34] can surely enhance the system performance with better storage and application service administration along with lower value for latency in comparison to conventional SQL-based database platforms [35]. The integration of the technology can provide a flexible data transmission model with a reduced cost supposed to be incurred against management by making direct contact between users and 5G network services. Such a model can still be considered as secure since blockchain makes D2D communication among the users and avail the computational power from its users to manage the network rather than availing from the central authority [36]. Thus, it can achieve lower latency value for communication, for transactional work as well for data accessibility leading to amplified system performance. In addition to that, even if there is a failure at any point due to malicious attacks, the network can still remain useful by the consensus on the publicly distributed ledger.

2. Enforcement of better security features
Blockchain guarantees the enforcement of additional security features such as being decentralized, maintaining secrecy, trust, being self-sufficient and automated and being visible, secure and verifiable. Though cloud computing has a centralized cloud server, blockchain-oriented cloud services facilitate decentralization to enforce an unbiased agreement with the consensus of blockchain technology. It further ignores the failure that can happen at any node and wins the trust of the system as well as the user. Since blockchain technology employs peer-to-peer communication that in turn transfers each device in the network, as a blockchain node, it can have a copy of the ledger to keep an eye on the transactional data to make everything transparent and dependable. In comparison to traditional DBMS, the smart contract feature of blockchain technology makes use of the computing power of authenticated nodes to allow decentralized transaction validation of each node's access [37]. Such a smart contract feature can empower the 5G network with all its services not to be tampered or modified. Further, the immutability characteristics provide security and protection against various cyber-attacks. Protocols for user access along with logic codes in blockchain technology are also capable of providing enhanced authentication mechanisms with contracts that can automate the authorization process for its users without revealing their private identity information as well as discard the intruding ones for 5G networks. The data also remain highly protected in the blockchain platform as those are authenticated by hash values and then get added to the blocks. Along with data protection, the private data when shared in an untrusted communication medium remains in the total control of the blockchain technology that even prevents the users of the node to trace their own data [38].

3. Well-handled and managed network services
Due to the decentralization, blockchain promises the simplification of 5G network services. Practically, the worries of employing a central controlling authority to operate the mobile network services are eliminated with the augmentation of blockchain principles. Both the mobile network service providers and users can avail the 5G services like trading in terms of resources, payments, responses and data access on the distributed decentralized public ledger. Thus, the adoption of blockchain technology can significantly simplify the network and its associated costs. It can enable the sharing of services for both data and spectrum for the 5G network by providing the rights to the nodes for network maintenance and management. The node participants can also be empowered to explore the internal network resources to facilitate better user experience and network services on the mobile platform by fusing blockchain technology with 5G network services [39].

So with the innovative characteristic features of blockchain technology fused with the 5G network services, the challenges in terms of security, privacy, confidentiality and consistency can be definitely get rid of and can become a more powerful.

4.2 The Fusion of Blockchain with 5G Network Leading to Smart Applications and Automations

The potential of 5G network services integrated with blockchain technology has made possible IoT-enabled devices to get connected and set up an environment to render a variety of automated services like auto-sensing, communicating and processing without any human operator across the domain. The features of 5G technology like SDN, NFV, cloud or edge computing integrated with enhanced and inbuilt security mechanisms and with distributed decentralized ledger technology enabled the progression of IoT services across the globe. Many IoT applications are availed by joining together blockchain and 5G network services to suffice the basic needs of human being like smart healthcare, home, city, vehicle, industry, education, grid, trading, etc. which are portrayed as follows.

1. Smart Healthcare
Healthcare services are the industries that cater to the needs of the people by providing medical services and medical insurance facility as well as smoothen the progress of medical facility delivery to patients at the time the need arises. The prospective features of 5G communication technology can modernize the existing healthcare systems to smart medical facilities with reliable trustworthy services [40]. The blockchain technology highlighted with its relevant features like decentralized distributed ledger, improved secure and private platform, qualitative service and simplified network management with low cost can be fused with the most promising 5G networks to provide better assistance to increase performances in healthcare sectors [41]. Various features of blockchain technology like on-demand software can do various network functions through virtualization and enhanced storage through

cloud computing to promote the delivery of services at a faster rate so that a survey can be made on patient's health conditions at a very primary stage. Since blockchain promotes peer-to-peer communication, its fusion with 5G network will result in a healthcare database that can encourage the storage of validated transactional data as well as all patient communications in the distributed public ledger. Since the data would be stored in a secured publicly distributed ledger, all the healthcare workers can exchange and share data during the treatment process. In addition, home-based portable medical services with diversified analysis can also be availed through SDN as well as cloud-based modelling. Such portable services are highly secure to maintain communication between patients and doctors privately. The smart contract feature enables accessing privileges to ensure that healthcare data remains secure from malicious threats. The D2D communication enables feature extraction from patient's private data on a high scale which in turn is tagged with hash values and securely placed in the public decentralized ledger to get rid of data seeping out and tampering issues. Blockchain technology is also proudly implemented in a telemedicine service in mobile edge computing that allows the users to use their mobile network and access the services. The protocols defined through consensus also ensure that the patient is taken care of and is the top priority and optimization is achieved for resource allocation and security is maintained for their sensitive information. Also other applications for disease detection are able to take a capture of all medical test data and store it in an offline medium under the supervision of blockchain technology that gives the permission to the patients to access their own medical clinical test-related data authenticated with cryptographic hash values to maintain integrity checks and transparency of the process. Various cloud-based healthcare solutions also adopt the blockchain technology for enhancing the security of electronic health-related data. IoT-enabled healthcare devices can maintain the communication between cloud-based servers with the help of a protocol meant for communication defined under blockchain technology. The mobile app-based cloud blockchain medium is also used to record electronic health data accessible between patients and doctors. In such type of architecture, data is handled by the smart contract technology. So when blockchain is integrated to a cloud computing platform to enable data sharing, lower latency value is achieved with efficient data management and security in contrast to a centralized cloud-based architecture.

2. Smart City

The potential features of 5G technology have made a revolutionary change in transforming the conventional systems providing services into a completely digitalized and cost-effective one. One of such revolution includes smart city formation which may consist of many IoT devices scattered all throughout and connected through different networks with powerful cloud computing servers for carrying out processing tasks. Since the security issues stay with 5G network connecting IoT smart devices, blockchain technology can be useful in providing the same [42]. Such technology for a smart city application can segregate the city into a number of blocks being administered by a block administrator. It can comprise of IoT-required devices like sensors and cameras augmented with secured private blockchain database for sharing the information from IoT devices. Such a system employs fog computing mechanism to deal with data from mobile and IoT devices and machine learning techniques to carry out the data analysis and storage in secure blockchain ledgers. Blockchain technology is also adopted to make an interconnection between smart cities and IoT devices with utmost security. In smart cities as cameras and sensors would be present everywhere to capture an ample amount of data and to analyse those in case of a malicious attack, detecting the object of threat would be a cumbersome task. Such challenges can be well handled by the concept of edge computing done in a distributive manner. Also secure blockchain can interconnect IoT devices, nodes and users to communicate with each other by setting up a decentralized secure platform. As blockchain technology is a distributed ledger with a central cloud-based server, it provides more benefits in comparison to centralized architecture. In a smart city, the transport providers and the travellers communicate through mobile in a platform known as Mobility-as-a-Service (MaaS) which is more prone to malicious attacks and data leakage. Such services can be enhanced if integrated with the blockchain platform to improve the security of the services offered through a mobile platform like smart contract that can enable secure and trustworthy payment forum [18]. The large amount of data generated by IoT devices in a smart city can be efficiently and securely handled by cloud-based servers assisted by blockchain technology to handle auditing issues leading to assemble a data auditing blockchain (DAB) that in turn uses Practical Byzantine Fault Tolerance (pBFT) protocol by consensus algorithm. In such type of systems, each cloud server is treated as a node in blockchain, and the respective happenings can be stored in ledgers that ultimately reduce the risk of malicious attacks and failure at any single node.

3. Smart Transportation
The development of 5G technology has made a significant impact from traditional transport facilities into intelligent transportation systems (ITS) called smart transportation with smart vehicles, thereby providing better services to people. Smart transportation is an end product of IoT-based communication with vehicles in transportation. But due to the centralized architectural system used in such types of services, there is a high risk of security threat in a vehicle-to-vehicle transmission. Blockchain technology with its all essential features of distributed ledger, decentralization, transparency and peer-to-peer communication can set up the security protocols for secure vehicle transportation [43]. The technology can help in building up peer-to-peer transmission between vehicles and roadside units as well as can set up decentralized storage to store all transactional data of electrical vehicles. Vehicle-to-grid (V2G) is a smart and new device used as mobile power storage for a more secure energy platform between the power supply and electronic vehicles. The power supply in the smart city gets connected to the public blockchain where the communicational data between the supplier and user are stored, and thus the payment orders as well as charging and discharging information are also circulated by the electronic vehicles. Since authentication plays an important role in smart vehicle systems, smart contract can be a good approach

to authorize and confirm all sorts of transactions related to it [44] with the help of programs. The smart contract also enables authorization of vehicles that are registered without mentioning their full details, thereby reducing the risk of cyber-attacks. Even distributed SDN features of blockchain technology can be used to make a secure VANET [45] that can deal with heterogeneous traffic needs. The important requirement of VANET is to maintain security among EVs and V2G for power transmission and trading. So the need of the hour is to set up a good versatile trading model that can deal with different services related to vehicles. Such demand is fulfilled by blockchain technology by setting up a decentralized energy platform so that decentralized ledger is used for secure storage. Also smart contract is integrated with EV and V2G to form a combined trading platform for authorized low cost and efficient communication and authentication.

4. Smart Home
Modern human life has become smart enough because of the IoT-enabled connected devices like home appliances and smart watch, healthcare devices and other wearables. All smart home-based devices can be further enhanced with the features of blockchain technology. The research work also shows a smart power outlet system for a smart home which is further enhanced with auto-monitoring and controlling remotely with the application of blockchain technology [46]. Blockchain technology is also employed to maintain communication between smart electricity supply and smart home for energy trading. A smart home is also equipped with automatic locking of the door with the application of blockchain technology for authorization check, payment and event recording. The application of blockchain technology is also optimized to maintain security in a smart home case study. IoT-enabled device data is provided with an additional level of security with encryption which is narrated with the employment of consortium blockchain application.

5. Smart Industry
Industry 4.0 with IoT has brought the revolutionary transformation with cyber-physical systems like smart manufacturing, smart sensing, smart supply chain management, etc. These smart industrial applications can further be boosted with blockchain technology. For example, the research work by [8, 9] narrated the augmented blockchain technological applications in industrial IoT-enabled manufacturing, supply chain management, diagnostic operations, machine to machine transactional work as well as product certification, etc. The work also developed a smart contract-based prototype that can diagnose and sense the faulty part of the system and sends a report to the user about the necessity of part replacement in the machine. Smart contract technology of blockchain applications also provides the facility of buying electric power from an energy house [47, 48]. The technology is also successfully implemented in supply chain management for manufacturing and allocation of materials [49] and in credit-based trust systems.

6. Smart Agriculture
The application of IoT has also modernized the agricultural system with smart sensing of the area to be cultivated. IoT-enabled devices are able to monitor

temperature, humidity, insect and plant diseases that contribute greatly towards the prediction of crop cultivation and production dynamically. Blockchain technology has also made its role in a smart agricultural system so that there is an exchange of ideas and knowledge among the farms and government sectors like setting up irrigation canals. Further, blockchain applications have enhanced smart agriculture by adding transparency and backtracking of crops [50]. The research study also proves the application of blockchain with agricultural IoT for storing transactional data for the supply chain [51]. The sensitive and essential information regarding food safety like production, storage, processing and selling against its resources is also publicly exposed with an integrated application of blockchain technology with IoT-enabled smart agriculture. The backtracking of products used in agriculture in the supply chain is also made possible with the combined application of blockchain with RFID.

7. Smart Grid
Smart grid is a smart automated management of electricity through a network connection and IoT devices. IoT devices like sensors and metres can collect the data related to the power supply, consumption and load which can further be used for effective management of electricity resources. But such type of smart management of energy can suffer from some shortcomings like convincing the customers for reliable electricity metre reading, handling energy system complexity, etc. Such limitations can be handled by blockchain technology applications. The research is also done to make a replacement of the local grid by trans-active microgrids which is an application of blockchain for power supply transactions. In such an application, grid nodes are capable of maintaining the privacy of electricity trading individually [52]. The transparency of power consumption is also demonstrated with the help of smart contract for smart energy. Even the close monitoring and recording of consumption of energy is done with the application of blockchain technology. The power trading as well as its price has been optimized with the application of blockchain consortium. Smart grid application of blockchain technology is also used for reliable power trading. Even blockchain has been applied to prevent malicious threats and protect the sensitive data for smart energy applications.

8. Smart Trading and Supply Chain Management
The application of blockchain technology with a 5G network enabling the connection of IoT devices has also brought a revolutionary change to supply chain management [53] to support the effective recording of the product-related information when it is transferred from the manufacturer to the customer. The technological application also enables the monitoring of the quality of products and raw materials [48]. Smart contract features on blockchain technology can be used to create business collaborations which can then be used to carry out transactions in an automated manner without waiting for the traditional process of confirmation. The research work also shows the automated filing of taxes with smart contract feature of blockchain technology.

9. Smart Education
The educational domain has also undergone a paradigm shift from a paper-based world to a paperless environment with the development of smart devices like tablets, smartphones, computers and laptops. All such smart devices have revolutionized the way course context is designed, home assignments are uploaded and examination is conducted. All such smart devices can track learner's behavioural patterns as well as intelligence level automatically leading to smart education. The augmentation of blockchain technology into smart education will add on extra advantages of transparency in the learning process, examination grading and evaluation system as well as in the management of certificates so that the educational system becomes more fair, reliable and trustworthy [34].

5 Revealing Future Challenges for Fusion of 5G Network with Blockchain Technology

Though fusion of blockchain technology with the 5G network brings potential benefits, on the other hand, it suffers from several issues and challenges in terms of security from all perspectives and data privacy. The challenges are briefed as follows.

(a) Expandability with trustworthiness
The major goal of the 5G network is to achieve a low latency rate and, currently, for data and payload application to happen in less than 1 millisecond [54]. To achieve such a latency rate, there is requirement of tight protocols and configuration setup to carry out the transactions with high throughput. The public blockchain applications in terms of Ethereum and Bitcoin are only able to handle transactions up to a range of 10–14 per second, and private blockchain applications can handle 3000–20,000 transactions per second. So there is a need for further research in the future for its expansion and optimization starting from the architecture, increasing the size of each block as well as the smart contracts to meet the goal of achieving higher throughput. Along with such achievement of goals, trustworthiness plays a major factor that is achieved through decentralized distributed ledger with security. Such reliability is maintained for information like images and symbols, so it requires enhancement for different information options. Blockchain architecture itself enables trust and transparency factor, so more collaboration works are supposed to evolve with competition, ignoring intermediate competitors.

(b) Upgradation of smart contracts and resolution of vulnerability issues
Public blockchain applications are equipped with ten million smart contracts, and these smart contracts are to be used for 5G networks which enable interconnection of IoT devices. The smart contract of blockchain technology is vulnerable to cyber-attacks and needs to be securely coded for the 5G network [55, 56]. Moreover, the smart contract once coded is not further upgradable or modifiable in place of

any malicious attack by the hackers. So the smart contract code has to be further researched for upgradation and reporting of vulnerability issues.

(c) Data privacy and malicious threats
Privacy and security are the two faces of a single coin. It is an essential requirement of every individual, organization as well as government. It has to be maintained with 5G network communication too as it deals with customer's sensitive data. The integration of blockchain technology with a 5G network enables privacy with an issue that the data that gets stored in the blocks are not able to be erased because of its immutable property. Also, blockchain only identifies the users' address and never stores their private information. But Bitcoin applications are found to record the personal information of the user to make an exchange of the identity information among the users with the public and private key. Even the smart contract application and Ethereum are also indulged in the exchange of identity information which is a major issue and needs to be dealt with. The cryptography security maintained through the public key and the private key is supposed to be prone to a man in the middle attack [57] where a third person can come in the middle carrying an un-genuine public key and decrypt the sensitive information. Similarly, blockchain technology can be threatened with DDoS attacks which can smash away any platform or network and its resources [58]. On the same note, Bitcoin technology is prone to selfish mining threat that threatens the genuineness of the technology which requires urgent attention.

(d) The cost associated with transaction and cloud server setup
There are costs involved to set up the nodes of blockchain, maintain the blockchain consortium and set up the cloud server. The cost investments would be more if the number is not optimized [59]. In applications like Ethereum, costs are calculated in terms of gas units for each transaction, and gas units are the energy consumption units for smart contracts. So the costs are also calculated concerning the code that is being executed for each transaction. These fees would be required more in order to execute complex coding, if involved in the transaction, so it needs to be taken care of.

6 Conclusion

This chapter has presented the outline description for blockchain technology and its associated potential features as well as 5G network basics to mention the need for fusion of both the technologies. The study also highlighted the smart application areas of the 5G network with interconnected smart devices that have become more effective and secure with the integration of blockchain technology with it. Though there are enough beneficiary gains that are obtained through the fusion of both technologies, the present study also put forth various challenges that come into existence and needed to be sorted out with further research to make the technology accomplish all its desired and directed targets.

References

1. A.A. Monrat, O. Schelén, K. Andersson, A survey of blockchain from the perspectives of applications, challenges, and opportunities. IEEE Access **7**, 117134–117151 (2019). https://doi.org/10.1109/ACCESS.2019.2936094
2. F. Casino, T.K. Dasaklis, C. Patsakis, A systematic literature review of blockchain-based applications: Current status, classification and open issues, in *Telematics and Informatics* (Elsevier Ltd., 2019). https://doi.org/10.1016/j.tele.2018.11.006
3. K. Christidis, M. Devetsik, Blockchains and smart contracts for the internet of things. IEEE Access **4**, 2292–2303 (2016)
4. A. Gupta, R.K. Jha, A survey of 5G network: Architecture and emerging technologies. IEEE Access **3**, 1206–1232 (2015)
5. I. Jovovi'c, S. Husnjak, I. Forenbacher, S. Maˇcek, Innovative application of 5G and blockchain technology in industry 4.0. EAI Endorsed Trans. Ind. Netw. Intell. Syst. **6**(18), e4 (2019)
6. M. Agiwal, A. Roy, N. Saxena, Next generation 5G wireless networks: A comprehensive survey. IEEE Commun. Surv. Tutor. **18**(3), 1617–1655 (2016)
7. N. Panwar, S. Sharma, A.K. Singh, A survey on 5G : The next generation of mobile communication. Phys. Commun. **18**, 64–84 (2016)
8. R. Yang, F.R. Yu, P. Si, Z. Yang, Y. Zhang, Integrated blockchain and edge computing systems: A survey, some research issues and challenges. IEEE Commun. Surv. Tutor. **21**(2), 1508–1532 (2019)
9. D. Sukheja, L. Indira, P. Sharma, S. Chirgaiya, Blockchain technology: A comprehensive survey. J. Adv. Res. Dynam. Control Syst. **11**, 1187–1203 (2019). https://doi.org/10.5373/JARDCS/V11/20192690
10. M. Liyanage, I. Ahmad, A.B. Abro, A. Gurtov, M. Ylianttila, *A Comprehensive Guide to 5G Security* (John Wiley & Sons, Hoboken, 2018)
11. F. Tschorsch, B. Scheuermann, Bitcoin and beyond: A technical survey on decentralized digital currencies. IEEE Commun. Surv. Tutor. **18**(3), 2084–2123 (2016)
12. I. Bhudiraja, S. Tyagi, S. Tanwar, N. Kumar, J.J.P.C. Rodrigues, Tactile internet for smart communities in 5G: An insight for NOMA-based solutions. IEEE Trans. Ind. Inf. **15**(5), 3104–3112 (2019)
13. S. Rouhani, R. Deters, Security, performance, and applications of smart contracts: A systematic survey. IEEE Access **7**, 50759–50779 (2019)
14. S. Tanwar, Q. Bhatia, P. Patel, A. Kumari, P.K. Singh, W.C. Hong, Machine learning adoption in blockchain-based smart applications: The challenges, and a way forward. IEEE Access **8**, 474–488 (2020)
15. X. Wang, X. Zha, W. Ni, R.P. Liu, Y.J. Guo, X. Niu, K. Zheng, Survey on blockchain for internet of things. Comput. Commun. **136**, 10–29 (2019)
16. W. Wang, D.T. Hoang, P. Hu, Z. Xiong, D. Niyato, P. Wang, Y. Wen, D.I. Kim, A survey on consensus mechanisms and mining strategy management in blockchain networks. IEEE Access **7**, 22328–22370 (2019)
17. Z. Zheng, S. Xie, H. Dai, X. Chen, H. Wang, An overview of blockchain technology: Architecture, consensus, and future trends, in *2017 IEEE International Congress on Big Data (BigData Congress)*, pp. 557–564, (2017)
18. P. Mehta, R. Gupta, S. Tanwar, Blockchain envisioned UAV networks: Challenges, solutions, and comparisons. Comput. Commun. **151**, 518–538 (2020)
19. S. Kumar, A. Kumar, V. Verma, A survey paper on blockchain technology, challenges and opportunities. Int. J. Comput. Trends Technol. (IJCTT) **67**(4), 16 (2019). ISSN: 2231-2803, http://www.ijcttjournal.org
20. J.G. Andrews, S. Buzzi, W. Choi, S.V. Hanly, A. Lozano, A.C.K. Soong, J.C. Zhang, What will 5G be? IEEE J. Sel. Areas Commun. **32**(6), 1065–1082 (2014)
21. R. Gupta, S. Tanwar, S. Tyagi, N. Kumar, Tactile Internet and its applications in 5G Era: A comprehensive review. Int. J. Commun. Syst. **32**(14), 1–49 (2019)

22. T.T.A. Dinh, R. Liu, M. Zhang, G. Chen, B.C. Ooi, J. Wang, Untangling blockchain: A data processing view of blockchain systems. IEEE Trans. Knowl. Data Eng. **30**(7), 1366–1385 (2018)
23. A. Singh, S. Mahapatra, Network-based applications of multimedia big data computing in IoT environment, in *Multimedia Big Data Computing for IoT Applications*, Intelligent Systems Reference Library 163, https://doi.org/10.1007/978-981-13-8759-3_17, (2020)
24. Internet of Things (IoT). [Online]. Available: https://www.cisco.com/c/en/us/solutions/internet-of-things/
25. J.M. Khurpade, D. Rao, P.D. Sanghavi, A survey on IOT and 5G network, in *International Conference on Smart City and Emerging Technology (ICSCET)*, pp. 1–3, (2018)
26. S. Zhang, X. Xu, Y. Wu et al., 5G: Towards energy-efficient, low-latency and high-reliable communications networks, in *Proceedings of the IEEE ICCS*, pp. 197–201, (2014)
27. I. Mistry, S. Tanwar, S. Tyagi, N. Kumar, Blockchain for 5G-enabled IoT for industrial automation: A systematic review, solutions, and challenges. Mech. Syst. Signal Process. **135**, 106382 (2020)
28. W. Al-Saqaf, N. Seidler, Blockchain technology for social impact: Opportunities and challenges ahead. J. Cyber Policy **2**(3), 338–354 (2017) The Road to the Next Wave of Tech: 5G +Blockchain. [Online]. Available: https://www.asiablockchainreview.com/the-road to-the-nextwave-of-tech-5G blockchain/
29. W.H. Chin, Z. Fan, R. Haines, Emerging technologies and research challenges for 5G wireless networks. IEEE Wirel. Commun. **21**(2), 106–112 (2014)
30. M.T. Hammi, B. Hammi, P. Bellot, A. Serhrouchni, Bubbles of trust: A decentralized blockchain-based authentication system for IoT. Comput. Secur. **78**, 126–142 (2018)
31. J. Liu, Z. Liu, A survey on security verification of blockchain smart contracts. IEEE Access **7**, 77894–77904 (2019). https://doi.org/10.1109/ACCESS.2019.2921624
32. M.S. Ali, M. Vecchio, M. Pincheira, K. Dolui, F. Antonelli, M.H. Rehmani, Applications of blockchains in the internet of things: A comprehensive survey. IEEE Commun. Surv. Tutor. **21**(2), 1676–1717 (2018)
33. M. Chaudhry, Joint IEEE spectrum and comsoc talk, test and measurement virtualization and blockchain: Enablers for 5G networks, Nov 13, (2018)
34. Y. Yu, Y. Li, J. Tian, J. Liu, Blockchain-based solutions to security and privacy issues in the internet of things. IEEE Wirel. Commun. **25**(6), 12–18 (2018)
35. J. Wan et al., A blockchain-based solution for enhancing security and privacy in smart factory. IEEE Trans. Ind. Inf. **15**(6), 3652–3660 (2019)
36. T. Salman, M. Zolanvari, A. Erbad, R. Jain, M. Samaka, Security services using blockchains: A state of the art survey. IEEE Commun. Surv. Tutor. **21**(1), 858–880 (2018)
37. R. Gupta, S. Tanwar, F. Al-Turjman, P. Italiya, A. Nauman, S.W. Kim, Smart contract privacy protection using AI in cyber-physical systems: Tools, techniques and challenges. IEEE Access **8**, 24746–24772 (2020)
38. Y. Dai, D. Xu, S. Maharjan, Z. Chen, Q. He, Y. Zhang, Blockchain and deep reinforcement learning empowered intelligent 5G beyond. IEEE Netw. **33**(3), 10–17 (2019)
39. R. Gupta, S. Tanwar, S. Tyagi, N. Kumar, M.S. Obaidat, B. Sadoun, HaBiTs: Blockchain-based telesurgery framework for healthcare 4.0, in *International Conference on Computer, Information and Telecommunication Systems (IEEE CITS-2019)*, Beijing, China, August 28–31, pp. 6–10, (2019)
40. R. Gupta, S. Tanwar, S. Tyagi, N. Kumar, Tactile-Internet-based Telesurgery System for Healthcare 4.0: An architecture, research challenges, and future directions. IEEE Netw. **33**(6), 22–29 (2019)
41. J. Xie, H. Tang, T. Huang, F.R. Yu, R. Xie, J. Liu, Y. Liu, A survey of blockchain technology applied to smart cities: Research issues and challenges. IEEE Commun. Surv. Tutor. **21**(3), 2794–2830 (2019)
42. S. Talari et al., A review of smart cities based on the internet of things concept. Energies **10**(4), 421 (2017)

43. T. Jiang, H. Fang, H. Wang, Blockchain-based internet of vehicles: Distributed network architecture and performance analysis. IEEE Internet Things J. **6**(3), 4640–4649 (2018)
44. J. Vora, S. Tyagi, N. Kumar, M.S. Obaidat, A systematic review on security issues in VANET. Secur. Priv. J. Wiley **1**(5), 1–27 (2018)
45. A. Dorri, S.S. Kanhere, R. Jurdak, P. Gauravaram, Blockchain for IoT security and privacy: The case study of a smart home, in *2017 IEEE international conference on pervasive computing and communications workshops (PerCom workshops)*. IEEE, pp. 618–623, (2017)
46. J. Al-Jaroodi, N. Mohamed, Blockchain in industries: A survey. IEEE Access **7**, 36500–36515 (2019)
47. K. Rabah, Overview of blockchain as the engine of the 4th industrial revolution. Mara Res. J. Bus. Manag. (ISSN: 2519–1381) **1**(1), 125–135 (2017)
48. M. Tahir, M.H. Habaebi, M. Dabbagh, A. Mughees, A. Ahad, K.I. Ahmed, A review on application of blockchain in 5G and beyond networks: Taxonomy, field-trials, challenges and opportunities. IEEE Access **8**, 115876–115904 (2020). https://doi.org/10.1109/ACCESS.2020.3003020
49. A. Litke, D. Anagnostopoulos, T. Varvarigou, Blockchains for supply chain management: Architectural elements and challenges towards a global scale deployment. Logistics **3**(1), 5 (2019)
50. M. Kouhizadeh, J. Sarkis, Blockchain practices, potentials, and perspectives in greening supply chains. Sustainability **10**(10), 3652 (2018)
51. H. Malik, A. Manzoor, M. Ylianttila, M. Liyanage, Performance analysis of blockchain based smart grids with Ethereum and Hyperledger implementations, in *IEEE International Conference on Advanced Networks and Telecommunications Systems 2019*. IEEE, pp. 1–5, (2019)
52. S. Saberi, M. Kouhizadeh, J. Sarkis, L. Shen, Blockchain technology and its relationships to sustainable supply chain management. Int. J. Prod. Res. **57**(7), 2117–2135 (2019)
53. M. Petersen, N. Hackius, B. von See, Mapping the Sea of opportunities: Blockchain in supply chain and logistics. it-Inf. Technol. **60**(5–6), 263–271 (2018)
54. B. Rodrigues, T. Bocek, A. Lareida, D. Hausheer, S. Rafati, B. Stiller, A blockchain-based Architecture for Collaborative DDoS Mitigation with Smart Contracts, in *IFIP International Conference on Autonomous Infrastructure, Management and Security*, (Springer, Cham, 2017), pp. 16–29
55. K.N. Griggs, O. Ossipova, C.P. Kohlios, A.N. Baccarini, E.A. Howson, T. Hayajneh, Healthcare blockchain system using smart contracts for secure automated remote patient monitoring. J. Med. Syst. **42**(7), 130 (2018)
56. L. Zhou, L. Wang, Y. Sun, P. Lv, Beekeeper: A blockchain-based IoT system with secure storage and homomorphic computation. IEEE Access **6**, 43472–43488 (2018)
57. M.A. Khan, K. Salah, IoT security: Review, blockchain solutions, and open challenges. Futur. Gener. Comput. Syst. **82**, 395–411 (2018)
58. Market Pulse Report, Internet of Things (IoT). URL: https://growthenabler.com/flipbook/pdf/IOT%20Report.pdf. Accessed 30 Nov 2019
59. P. Daugherty, B. Berthon, *Winning with the Industrial Internet of Things: How to Accelerate the Journey to Productivity and Growth* (Accenture, Dubl'ın, 2015)

Chapter 4
Introduction of Blockchain and 5G-Enabled IoT Devices

Shivangi Surati, Bela Shrimali, and Hiren Patel

1 Introduction

The advancements in the industry have various stages beginning from industry 1.0 to industry 5.0 as per the requirements and the challenges of the respective age. The industry 5.0 aims at the co-operation between human and machine to improve industrial productions and standards with the skilled manpower. This goal can effectively be achieved by utilizing the IoT paradigm that is revolutionized from the decades to handle the diversity of applications, *viz. smart homes, smart cities, smart grids, healthcare, agriculture, smart industries, and many more* efficiently. These IoT-based applications can preliminary be distinguished as per consumer IoT and industrial IoT [1]. The consumer IoT focuses on improving people's life that offers quality living, reduces human efforts, and saves time and money. The cIoT applications include smart homes, smart cities, and offices. On the other side, industrial IoT combines Operational Technology and Information Technology together. It concentrates on improving the devices and services related to the industries such as smart machines, sensor devices, food processing, data integration and analytics for business purposes. Thus, iIoT contributes in the tremendous growth of industries that are now driven by distinct manufacturing, transportation, logistics, and functionalities. However, both the types of IoT require common functionalities such as scalability, security, privacy, huge data storage, and lesser response time.

S. Surati (✉) · B. Shrimali
Department of Computer Engineering, LDRP Institute of Technology and Research, Gandhinagar, Gujarat, India

H. Patel
Vidush Somany Institute of Technology and Research, Kadi, Gujarat, India

S. Tanwar (ed.), *Blockchain for 5G-Enabled IoT*,
https://doi.org/10.1007/978-3-030-67490-8_4

IoT connects billions of devices to each other using Internet technologies such as 2G/3G/4G/LTE/5G or using M2M (Machine-to-Machine) technologies such as Bluetooth, Wi-Fi, or ZigBee [2]. The total count of devices connected together and the applications are increased with each enhancement of Internet technology, however, 5G networks accelerate IoT and especially iIoT due to following characteristics [2–4]:

- A higher transmission speed
- A lower latency (Improved competency for remote applications)
- An increased number of devices connected together
- The additional feature of possible implementation of virtual networks (network slicing)
- Reliability and Integrity
- Scalability (Increased number of connected devices)
- Optimized networks
- Lower latency and response in real-time
- Less human intervention
- Reduced network traffic or network related issues

These features improved the performance of time-sensitive industrial applications, *viz. healthcare, robotics, drones, video streaming* tremendously. Thus, industrial applications of 5G-IoT can be deployed with gigantic number of devices without worrying about network capabilities. However, 5G-enabled IoT devices still suffer from the issues of diverse and heterogeneous data and devices, decentralization, interoperability, security and privacy of sensitive data.

In 5G-IoT, if the generation, storage, and access of the data are made centralized, then the system will become more vulnerable to the attackers. The security threats even increase with the accommodation of billions of devices and gigantic amount of data generated by them. The previous methods applied for security in 2G/3G/4G networks are not powerful enough to handle the security and privacy demands of 5G networks due to new technologies, models, and advancements provided in the key features of 5G networks. Hence, to resolve the aforementioned issues, Blockchain stands out as a promising technology as it offers secure, immutable, transparent, reliable, and tempered-proof environment for 5G-IoT.

Apart from this, 5G enabled networks are working at high frequencies, *i.e., millimeter Wave (mmWave)* that operates above 30 GHz, resulting in signals that travel shorter distances. This leads to the requirement of a larger number of cellular stations or sites to cover the same 4G landscape. Deployment of a larger number of cellular station costs high. To reduce infrastructure deployment cost, options like (1) infrastructure crowdsourcing, (2) roaming sharing, and (3) spectrum sharing can be used. However, massive communication between the devices and cellular tower is another key challenge in 5G-IoT. Such a massive number of distributed and heterogeneous IoT and mobile devices instigate new challenges with respect to authentication and scalability of 5G networks. Again, the Blockchain can be promising technology that plays a salient role in addressing these challenges and providing trusted, decentralized, tempered-proof, and cost-efficient solutions [5].

The Blockchain is a distributed ledger technology that became popular and successful with the crypto currency called Bitcoin [6]. It follows decentralized and peer-to-peer network architecture instead of following the centralized storage. Data is stored and shared among the participating nodes in the network and hence, tempering of data at any node needs to be reflected at every other node that results in the tempered-proof environment. Due to this, Blockchain revolutionized as an astonished, promising, and secure technology not only for data storage but also to standstill the entrance of malicious IoT devices in the network. The integration of Blockchain with 5G-IoT will give secured data as per time, format, and location requirements.

Thus, the combination of Blockchain with 5G-IoT aims at security and performance improvement in cIoT as well as iIoT applications (Fig. 4.1). Blockchain in 5G-IoT and Blockchain in iIoT 4.0 are compared based on different parameters [5, 7–9] as shown in Table 4.1.

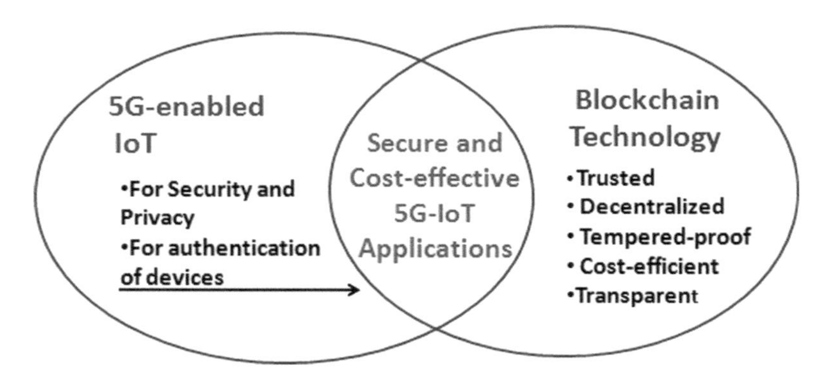

Fig. 4.1 The convergence of 5G-enabled IoT with Blockchain Technology

Table 4.1 Comparison of Blockchain in 5G-enabled IoT with Blockchain in iIoT 4.0

Parameter	Blockchain in 5G-IoT	Blockchain in IoT 4.0
Latency	Very high	High
Security	High	High
Privacy	High	High
Throughput	Very high	High
Transparency	Yes	Yes
Ubiquity	Ensure	May get disturbed
Usage/Adoption	Even at cellular broadcast level	At the industries
Threat	High	Comparatively low

1.1 Motivation

Industrial IoT is an emerging industry experiencing significant challenges, such as the necessity of an increased amount of data transmission with security, transparency, anonymity, and credibility. The fifth Generation Mobile Networks and Blockchain are contemporary emerging technologies that can facilitate these needs. 5G will enable extremely large data channel capacities and less data latency, while Blockchain's innovative data-sharing mode of operation provides an improved high level of security, transparency, and credibility of stored data. Therefore, the strengths of these two technologies motivated us to study the applications and the advantages of them when combined together. The aim of this chapter is to introduce 5G-enabled IoT merged with the Blockchain technology and their applicability in Industrial IoT. The chapter will be helpful to the researchers and beginners to dive into the area of 5G-IoT and Blockchain.

1.2 Our Contributions

After introducing the revolution of 5G networks in IoT and how Blockchain technology can be utilized with it in Sect. 1, our contributions in this chapter are as follows:

- The background knowledge required to understand both the technologies is discussed in detail (Sect. 2).
- The integration of 5G-IoT with Blockchain technology is explored along with the discussion of the security requirements to manage 5G-IoT devices (Sect. 3).
- The case studies of 5G-IoT and Blockchain in industrial applications are presented (Sect. 4).
- Furthermore, the issues, challenges and limitations and the opportunities and research directions of Blockchain-based 5G-enabled iIoT are explored (Sects. 5 and 6, respectively).

At the end, the chapter contents are concluded in Sect. 7.

2 Background

The description of 5G-enabled IoT, its key features and the other integrated technologies are explored in this section. In addition, a general overview of Blockchain including its definition, features, components, and categories are discussed.

2.1 5G-Enabled IoT

Over the last few decades, the communication networks are progressively revolutionized from the first generation networks and then moving towards the fourth generation networks. This evolution aims to satisfy the increasing communication traffic of data across the world. The eternal demand of the increased traffic motivated the upcoming telecommunication networks, *viz. 5G*. These networks take benefits from various underlying technologies as discussed below for the improvement in the features of existing networks [1, 3]:

- **Software-Defined Networking (SDN):** SDN defines two separate planes for routing: the Data plane handles the routing of data and the Control plane makes decisions about routing. It can be considered as an enabling key technology for 5G networks. By separating software and hardware requirements in two separate planes, SDN offers velocity and flexibility to 5G networks. In addition, Vehicular Ad hoc Networks (VANETs) in combination with SDN provides programmability, security, flexibility, and other networking advantages leading to software-defined vehicular networks (SDVNs) [10].
- **Network Functions Virtualization (NFV):** NFV deploys a set of network functions into a wide array of software packages and stored into servers. They are then combined and chained to implement the services that are same as offered by the old traditional networks, but with the improved scalability and flexibility for mobile devices.
- **Device-to-Device Communications (D2D):** It is the key feature added to 5G network with an aim to permit the direct communication between two devices with each other without intervening base station or with just a partial aid of base station. It includes four types of communications between devices: Device relaying with operator controlled link establishment, Direct D2D communication with operator controlled link establishment, Device relaying with device controlled link establishment, Direct D2D communication with device controlled link establishment.
- **Millimeter Wave Communication:** As the capacity requirements in the mobile communication networks are increasing rapidly, the mmWave frequencies can be utilized to improve radio wireless spectrum bands beyond 4G networks.
- **Cloud/Edge Computing:** The increasing demands of data storage, analysis, and transfer between various devices in 5G networks are satisfied by the use of resources from Cloud computing. The computing services can also be provided at the edge of the devices using edge computing that reduces the transmission delays.

On the other side, global traffic is increasing enormously with the inclusion of applications based on Internet of Things in day-to-day life and it is expected to continue. The 5G networks can overcome the limitations of the previous standards and protocols by increasing the capabilities of the networks as well as ensure full network coverage at the global level [3]. Thus, the industrial IoT applications can be enhanced efficiently with the support of 5G networks that is developed as 5G-enabled IoT.

2.1.1 Key Features of 5G-Enabled IoT

The development of 5G communication networks is beneficial for the IoT devices due to the following features of 5G-IoT [2–4]:

1. **High-Speed Data Transmission:** The IoT devices can communicate with each other at rapid speed using 5G networks. The data transfer speed is increased tremendously after evolution of 5G networks as it is availed with more spectrum, advanced modulation schemes for wireless communication and advanced radio technology. The expected maximum and average speed of 5G networks are 1–10 Gbps and 50 Mbps and up, respectively. Thus, the required data will reach to the other IoT devices and to the IoT application users very quickly. This enhances the number of static and mobile IoT devices connected with each other for various applications.
2. **Lower Latency:** Due to the increased data rate and bandwidths in 5G, the latency time is reduced considerably as compared to the previous communication networks. This ultra-low latency time (1 ms or even lesser) resolved the major challenge of real-time applications, *viz. latency time*. Thus, 5G-enabled IoT has improved not only applications like smart homes, but it has also enabled new services in time-critical industrial IoT applications, *viz. healthcare, remote controlled vehicular as well as industrial applications, video streaming*, etc.
3. **Network Slicing Capabilities:** 5G networks support virtual networking architecture by using SDN and NFV. The network slicing in 5G enables the network operators to deploy the procedures and modules required for specific customers of application segments. This concept of using only a part of the procedures results in savings of execution time and space as compared to the deployment of full functionality of the devices. Thus, the network slicing offers greater flexibility and elasticity to the diverse use cases and IoT devices that work with the constrained resources.
4. **Greater Network Reliability:** Apart from the increasing speed and ultra-low latency time, 5G networks also work more reliably that makes the connections more stable. Thus, the connected devices in IoT can work efficiently in such a reliable and stable network conditions. Especially the consumers using the connected devices like CCTV cameras for security and monitoring, IoT locks, and the other applications will be benefited due to more reliability for real-time updates.
5. **Support for Optimized Distributed Network Applications:** The optimization algorithms are applied in these distributed networks that enhance the performance of IoT applications.

Thus, 5G-enabled IoT is capable to fulfill the increasing demands of reliable data transmission between static and mobile devices with higher throughput.

2.1.2 Integrated Technologies

The 5G-enabled IoT can be integrated with the other technologies [2] as follows:

- **Artificial Intelligence (AI):** 5G-enabled IoT provided the infrastructure and huge amount of data as required by AI [11]. It could give contextual awareness to AI assistants for better understanding and learning and hence, reduces the complexity of 5G networks [12].
- **Machine Learning (ML) and Big Data Analytics:** Due to the enhancements in the Machine Learning techniques, *viz. Deep Neural Networks (DNN), Convolutional Neural Networks (CNN), and Recurrent Neural Networks (RNN)*, the mathematical complexities in data and network management are resolved. Apart from that, big data analytics and algorithms are capable to analyze the data and generate the expected outcomes at much faster rates using 5G-IoT.
- **Optimization Techniques:** 5G-enabled IoT utilizes various optimization techniques, *viz. heuristic approach, genetic algorithms, evolutionary algorithms, bio-inspired heuristic algorithms, fuzzy logic, stochastic algorithms*, to optimize the network related problems in IoT. This helps in proper monitoring and reduction of the network traffic generated by IoT devices.
- **Blockchain:** As discussed earlier, the decentralized Blockchain technology can also be integrated with 5G-IoT to provide security and privacy of data.

Thus, 5G-enabled IoT improves the performance and throughput of the IoT networks, but it also necessitates the security and privacy of the data provided by the devices. Thus, as per the existing literature survey, the Blockchain technology is observed as the most promising technology for the same.

2.2 Blockchain

The popularity and success of crypto currency called Bitcoin [6] bring out this promising technology in the attention. It has replaced the traditional client server architecture by a pure decentralized environment wherein the participants having no trust with each other can communicate and their transactions are stored and maintained in the shared ledger without any third party intervention. Thus, Blockchain is called as a peer-to-peer decentralized environment that maintains a publicly shared distributed ledger between the participating nodes. It is a cryptographically linked chain of time-stamped blocks. Each of these blocks has their unique identity in the form of hash value. In addition to the transactions, each block also stores its own hash value as well as the hash value of the previous block. Hash of the block is calculated based on the data stored in it and hence, change in a data of any single block changes the hash of that particular block and further, this hash change needs to be reflected in all the successive nodes.

Immutability is one of the important features of Blockchain. Once the transaction/data is stored in a block, it cannot be changed. Transaction/data can be added to the block by the common consensus between the participating nodes. Node has to prove its authenticity to introduce new block in the network/chain. Moreover, every participating node keeps their own copy of distributed shared ledger. In case, intruder wants to attack on any single block, he needs to attack on every copy of the distributed shared ledger to maintain the change. Otherwise, the network discards the odd copy. This methodology strengthens the Blockchain. Along with the immutability, the other important features/benefits of this technology are as follows [13]:

- **Transparency:** The Blockchain stores the transactions that are transparent to all the participant nodes. Any changes made in the data of block or a block itself introduced to the network is transparent to every participant node.
- **Security:** Blockchain offers more security as compared to the other record keeping systems. Transactions/data are included after the consensus by all the participant nodes. After the successful verification and validation of the transaction, it is encrypted and stored in the block and the block gets linked with the previous block. Secured hashing mechanisms are applied to secure blocks[14].
- **Fast and Efficient:** Journey from traditional pen-paper based system to automated computerized system increased the speed of process execution. Blockchain along with the computerized system takes out the third party or human intervention and hence, it is fast and efficient.

Apart from the above-mentioned features, the four major components of Blockchain technology are briefly described as follows:

1. **Block [15]:** A block is divided into two components:

   (a) Block header
   (b) List of transactions

   As shown in Fig. 4.2, along with transaction, a block also stores the imperative information, *viz. its own address/identity called hash, previous block's hash, other mining statistics including time stamp, nonce*, etc. The transaction cannot be removed or altered once it is included in the block. Blocks are chained together to form a Blockchain. It is introduced by authentic miner in the network. Block will be accepted in the chain only if it is authentic and it has proof of work. Once the Block is verified by all the participating nodes, it will be added to the longest chain. The block that is not included to the chain is called orphan block [15] and the small chains other than the longest accepted chain are called fork. Orphan blocks and the other shorter chains are discarded by the network later on.
2. **Distributed shared ledger:** A distributed shared ledger is basically a database asset that can be mutually shared across a multiple network, institutions, and over geographical locations [16]. All the participants that are included in a network can have their own identical copy of the ledger. Any type of changes/updates to the ledger is also reflected in all the copies within the predefined time interval. The ledger is kept secured and accurate through the use of cryptographic

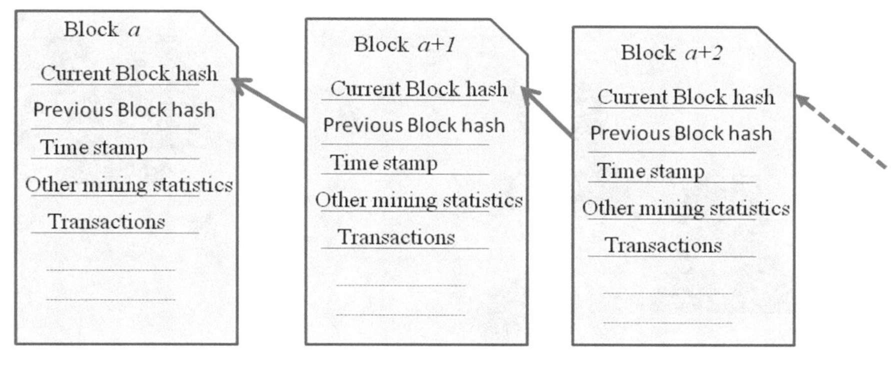

Fig. 4.2 Blockchain and Block concept

algorithms such as digital signatures and hash functions. The control over the ledger for modification or creation is defined by the mutual agreement called consensus. This will define different role of usage in a different way of the shared ledger.

3. **Smart contract [17]:** Smart contract can be defined as a code in the form of contract in which the terms of agreement between the multiple stakeholders are written in the line of code that is executed according to predefined processes [18]. In simple term, smart contract is a line of code that is self-executable and it is implemented/maintained/regulated on terms and agreements made between two or more parties. Distributed shared ledgers are applied and executed through smart contracts.
4. **Cryptographic techniques:** Security is a prime concern for online applications. Blockchain provides security through users', blocks', and transactional data's validation and verifications. Cryptographic techniques like public key cryptography are used for authentication and prevention of non-repudiation. While hashing algorithm like Secured Hash Algorithm (SHA)-256 ensures the security of blocks.

Furthermore, based on different parameters, *viz. user requirements, access control, and availability*, Blockchain technology can be categorized into three types [19].

1. **Public or Permission-less Blockchain [20]:** As shown in Fig. 4.3, no permission is required to join this network. It is a Blockchain environment that anyone can join and utilize. It is open for all. It is transparent and strongly resistant to tempering of data, e.g., Bitcoin Blockchain.
2. **Private or Permissioned Blockchain [19]:** As shown in Fig. 4.4, only an identified user can participate and join the network. It is the Blockchain environment with the private or closed networks that defines and decides the participants

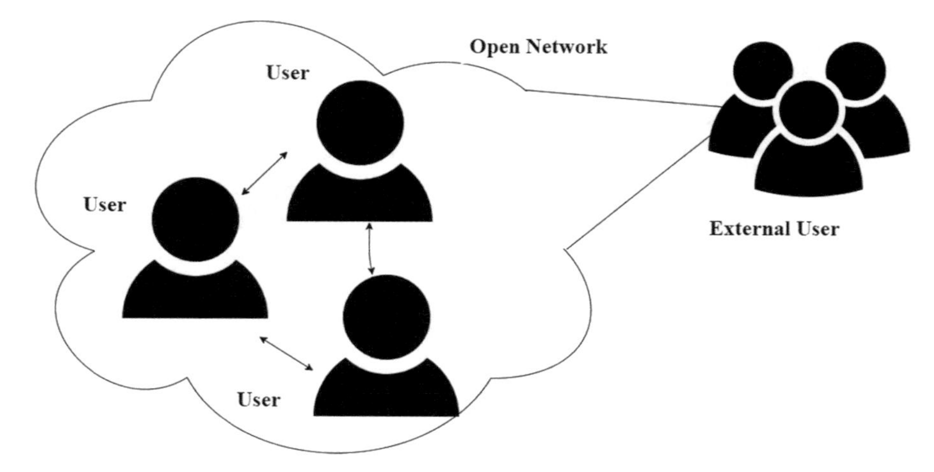

Fig. 4.3 Public Blockchain environment

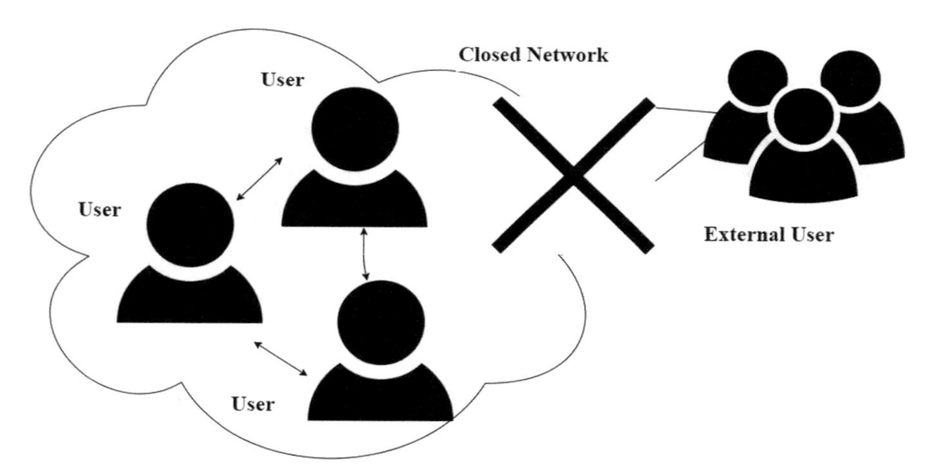

Fig. 4.4 Private Blockchain environment

and their access control. Generally, this environment is used and built by the industries/enterprises. It is speedy as compared to the public Blockchain.

3. **Hybrid Blockchain [21]:** It is a suitable combination of public and private Blockchain in which the distributed shared ledger is available to every user with predefined access control. It is the most suitable type of Blockchain environment for the industries.

Thus, Blockchain is peer-to-peer, decentralized, reliable, and tempered-proof technology that mainly aims at providing security and transparency of data. In the current era of 5G-IoT, the devices generate tremendous amount of data that also requires confidentiality. Hence, the architecture of 5G-IoT is required to be modified accordingly to include Blockchain in the existing IoT architecture. The opportunities and the modified architecture of 5G-enabled IoT utilizing Blockchain for security purpose are discussed in the next section.

3 Integrating Blockchain Technology in 5G-Enabled IoT

Due to the rapid and continuous development in the embedded technology, network, and IoT over a decade, the lifestyle of citizens is drastically changed. Usage and applications of IoT in last few years bring the industry in pure digital era. Data generated by IoT devices are used to be stored in Cloud. Thus, Cloud computing technology plays a vital role to store, analyze, and process the huge data generated and being used by IoT devices [22]. Cloud technology is built upon the centralized architecture and hence, there are few limitations reflected by default to the usage of IoT, *viz. trust, transparency, and security*. The upcoming technology called Blockchain overcomes this black-box type limitations by replacing the centralized architecture to distributed and decentralized one. Initially, the use of Blockchain was limited to crypto currency like bitcoin. But after the success of bitcoin, this technology is accepted and adopted widely. Evolution of Blockchain is shown in Fig. 4.5. The next generation Blockchain talks about a distributed Applications (dApp) to facilitate anonymity, privacy, security, and immutability to the different industries. Thus, the integration of Blockchain with 5G-IoT brings out new opportunities and architectures when applied to real-world applications as discussed in the subsections.

3.1 Opportunities in 5G-Enabled IoT Through Blockchain

The opportunities through Blockchain to enhance the efficiency in the 5G-enabled IoT in industry are as follows:

- **Transparency in supply chain management:** Many of the industries deal with the operations like supply chain in real-time. Once the data is stored on the chain, it cannot be altered. However, all the transactions are visible to all the legitimate stakeholders of the system. Transactional data may contain product details like its origin, shipment timestamp, price, etc. Thus, every stakeholder transparently tracks the location and quality of the product that is also known as public verifiability.
- **Appropriate prices of products and online payment facilities to stakeholders:** Every stakeholder transparently views the original price of the crop. So, producers, retailers, and wholesalers get the fair price of their products. Prize alteration is not possible once it is included on Blockchain. Stipulated price of the products cannot be changed resulting in a rare possibility of any fraud.

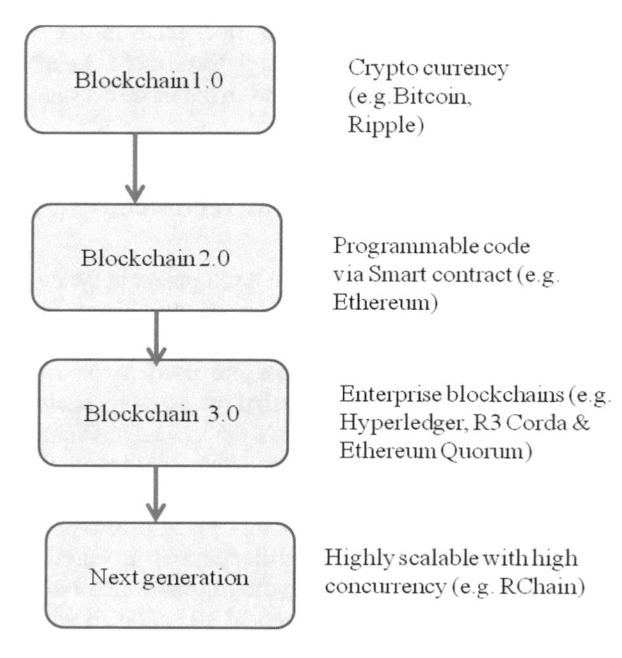

Fig. 4.5 Blockchain evolution

- **Easy traceability and auditability:** Starting from production place (i.e., industry) to every single intermediate place of stakeholders, all the required information of the products is digitally linked to within the Blockchain; every stakeholder including producers to end-customers/consumers can explore everything by exploring the entire supply chain cycle. The supply chain based on the Blockchain for any industry can help different stakeholders to access information related to the products' quality at every stage. Thus, Blockchain brings transparency in the supply chain network.
- **Reduction in operational cost:** In case of the loss due to any types of crisis, *viz. financial crisis or natural crisis*, industry can directly apply for the insurance claim amount through the Blockchain. The transparent and immutable property of the Blockchain enables the insurance companies and the other authorized parties the access of the data provided by industry easily. Different operations on Blockchain can be performed internally by using smart contracts. After the claim request is approved, industries will automatically get the requested amount in their respective wallets. In addition, industry can trace the activity of approval of subsidy during the whole process. Thus, industries can get the compensation stainlessly and quickly through Blockchain.
- **Building the trust between parties:** Only verified and authenticated devices are allowed to communicate with each other in the network using Blockchain technology. Even the miners will first verify each and every block and then only the block is allowed to enter in Blockchain.

3.2 Architecture of Blockchain-Based 5G-Enabled IoT

The architecture of 5G-enabled IoT is extended to include the Blockchain technology that is known as Blockchain-based 5G-enabled IoT (B5G-IoT) (Fig. 4.6). The Blockchain layer is merged as an intermediary between IoT and industrial applications. The aim of this layered architecture is to show and provide the pure abstraction from lower layers in IoT and to provide Blockchain view to users. In particular, the physical device layer deals with the heterogeneity of physical devices and their communication with upper layer. The network layer deals with the communication methodologies. On the other side, the Blockchain layer between the network layer and an application layer provides various Blockchain-based services. These services are basically distributed Application programming User Interfaces (dAppUIs) that are implemented to support different industrial applications. This introduction of Blockchain layer provides an abstraction that decreases the complexity of developing industrial applications [23].

A possible deployment scenario of B5G-IoT is depicted in Fig. 4.7. The advantage of using this deployment is that the whole Blockchain (or partial Blockchain) data can be stored on the Cloud servers while the IoT devices may only save the partial Blockchain data as shown in the figure. The deployment of B5G-IoT represents the possible communications between IoT and Blockchain as follows [24]:

- Firstly, the edge servers are deployed with the IoT gateways, Macro Base Stations (MBS) or small BS and the Blockchain data is stored in these edge servers. The IoT applications can directly communicate with the Blockchain and access the stored data.
- Secondly, IoT nodes can communicate directly and they can directly share partial Blockchain data.
- Lastly, hybrid communication is offered between cloud/edge servers and IoT devices. In this case, IoT devices can interact with Blockchain data through edge/cloud servers.

3.3 Applications of 5G-Enabled IoT Using Blockchain

Existing areas of interest for different industries, viz. Healthcare, Agriculture, Autonomous transportation industry, etc., with blockchain-based 5G-enabled IoT are shown in Fig. 4.8. In all these applications, blockchain-based 5G enabled iIoT are used to amend the security, to enhance the bandwidth, and to reduce the overall capital as well as operational expenditure, respectively [25]. The detailed description of these industries is discussed in the following:

- Healthcare: Healthcare is one of the principal aspects of the prosperity of any nation. With the increase in the population all over the world, their health

APPLICATION LAYER
(Industrial applications, dAppEnvironment, GUI etc)
BLOCKCHAIN COMPOSITE LAYER
(Encryption mechanisms, consensus algorithms, hashing algorithm,digital signature etc)
NETWORK LAYER
(IoT gateway,Wifi, Bluetooth etc)
PHYSICAL DEVICE LAYER
(Sensors, Readers, actuators, computer device etc)

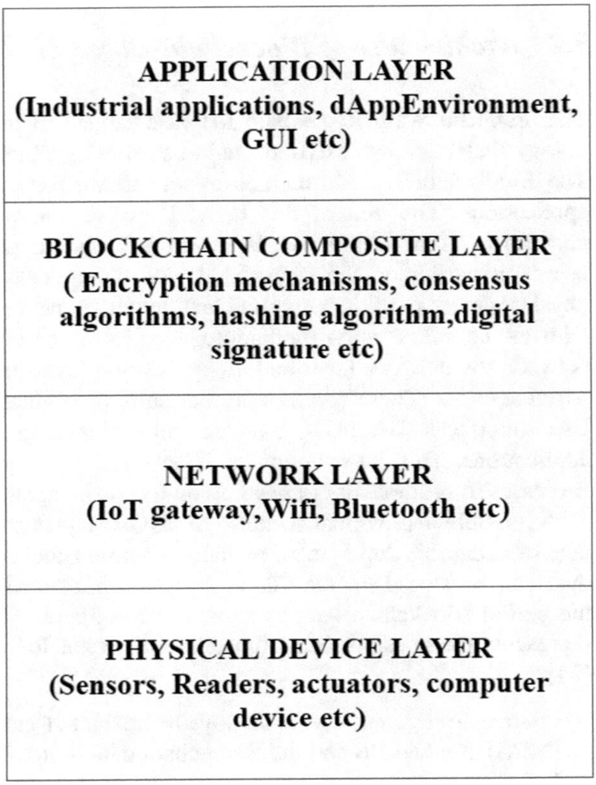

Fig. 4.6 Layered architecture of 5G-enabled iIoT

BLOCKCHAIN
BLOCKCHAIN
BLOCKCHAIN
INTERNET
CLOUD

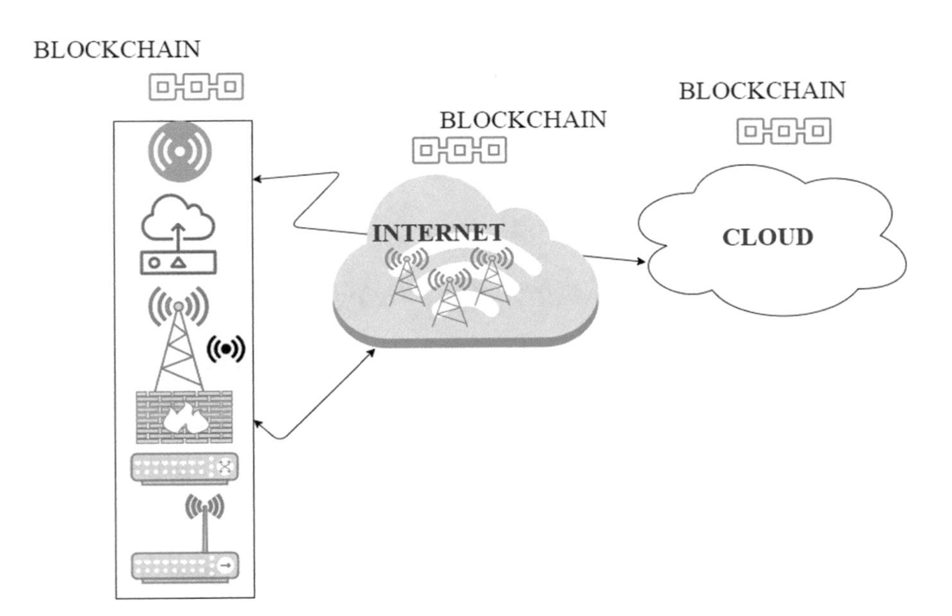

Fig. 4.7 Development scenario of B5GIoT

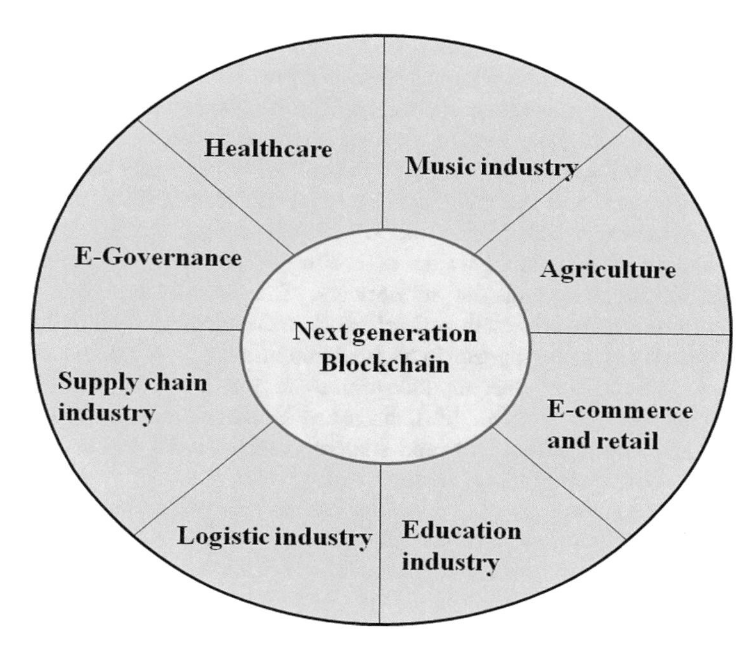

Fig. 4.8 Scope of Blockchain in iIOT

management becomes crucial. There are numerous opportunities brought by IoT in healthcare. It helps real-time, continuous monitoring, and error-free diagnosis. Electronic Health Record (EHR) repository is used by IoT in healthcare 4.0 [26]. Wherein data is stored centrally so that doctors can access it from anywhere. Privacy and security of this real-time huge data is a major concern and here, the Blockchain can be used. Its distributed, decentralized, and tempered-proof behavior can resolve the privacy and security concerns. Dwivedi et al. [27] proposed a novel model of blockchain for IoT-based healthcare system. They implemented distributed architecture and included security properties in the network. To facilitate the anonymous transactions by authentic users, they introduced a privacy-preserving ring signature scheme.

- Agriculture: Smart agriculture comes out as a vital IoT application for the agri-products exporting countries in agri-products exporting countries [28]. Use of IoT in agriculture that covers different operations like supply chain, precisions, and smart farming increases drastically. Ferrag et al. [28] analyzed and discussed the privacy-oriented blockchain-based solutions for IoT applications and their implementations for green IoT-based agriculture. They also discussed the security and privacy issues and solutions for green IoT-based agriculture.
- E-commerce and retail: Retailers and industries worldwide are very enthusiastic to use IoT to improve their business and customer experience. Inventory tracking and management become more easy and effective with IoT. By doing so, human errors can be reduced. The use of IoT also helps to minimize waste, control costs,

and reduce the shortage. For example, sensors for temperature-monitoring can be used to maintain optimal temperatures for perishable products and generate an alarm in case of predefined conditions [29]. Blockchain along with IoT in E-commerce will introduce a self-executing secure code called smart contract to manage various E-commerce activities. Also, it will open many directions to use IoT for smart contracts and B2B business more dexterously [29].

- Education industry: Introducing the Blockchain across the IoT platform for the education system could be an interesting methodology to enhance communication efficiency with the 5G network. The use of intelligent, powerful, and high-speed networks with several extant technologies, such as low power consumption, and so on, appear to be available in today's world to connect with each other. The need of a secure education system invites the use of Blockchain along with IoT. Alam et al. [30] discussed Blockchain in IoT to create an efficient interaction system between students, teachers, employers, developers, facilitators, and accreditors on the Internet.
- Music industry: Because of the Revolution in internet technology and easy accessibility to numerous streaming facilities over the internet, the music industry has become one of the industries that can benefit from Blockchain. The music industry involves many stakeholders, viz. producers, publishers, lyricists, artists, labels, and streaming service providers. Due to the growth of internet technology and its easy access, maintenance of music ownership has become more crucial. Transparency in payments and intellectual property rights is very essential for all the stakeholders of this industry. The integration of the Blockchain and the music industry can resolve several issues of ownership payment and transparency. The Blockchain can work as the best solution to create an accurate, immutable, and transparent distributed database to secure information of musical rights in a shared distributed ledger. Also, smart contracts can be used to execute a digital and secure contract/self-executing code for the music industry [31].
- Transportation industry: With the rapid development in sensing, communicating, and computing devices, the use of the Intelligent Transport System (ITS) is started and gets popular. It introduced a comfortable safe lifestyle with smarter transport facilities and more convenient transport services. The major issue with the traditional ITS is its centralized approach to store the data generated by IoT devices including sensors and vehicles. The usage of Blockchain for transportation can lead to secure, private, and safe methods of transportation service. Yuan et al. [32] discussed the Blockchain technology and its latent applications in the transportation industry. They also proposed a novel framework called Blockchain-Based ITS (B2 ITS) as an emerging decentralized architecture and distributed computing paradigm.

The applications of 5G-IoT and Blockchain are explored through various case studies in the different domains of society. The interesting case studies are discussed in the next section that covers the real-world scenarios of the combinations of these technologies.

4 Blockchain with 5G-Enabled iIoT: Case Studies

In this section, we present the few interesting proposals and implementation of Blockchain in 5G-enabled iIoT.

4.1 Smart Factory/Smart Industry

The two essential parameters of industrial IoT- system security and transaction efficiency are achieved using a credit-based proof of work (PoW) mechanism for IoT devices [33]. A data authority management method is designed that controls the access of data coming from different sensors. The blockchains developed are based on directed acyclic graph structure to improve the performance efficiency. A case study of smart factory is conducted for evaluation, demonstration, and analysis of the credit-based PoW mechanism in industrial IoT. The privacy of data is also sustained using the authority management that can be extended for the other iIoT scenarios also.

In [34], the authors have discussed two case studies for two different industry segments, viz. telecommunications-oriented 5G network slice brokering and the energy industry-related internal electricity allocation in a housing society. In telecommunications system, the 5G network with underlying Blockchain use case is proposed that utilizes a 5G network slice broker. This broker facilitates the manufacturing equipment so that it can obtain the required slice autonomously and dynamically. Moreover, the manufacturing device can lease independently the required network slice as per demand, it can approve service-level agreement (SLA) and pay for the service according to actual usage using network slicing in a Blockchain. Thus, the first use case enables autonomous 5G network resource leasing, brokering, payout, and billing in the Blockchain. The second case study is a renewable energy-oriented smart grids use case, in which entities within a housing society can balance electricity production and consumption among themselves without the control of an external intermediator that can save delay and energy.

In another case study [35], the authors have discussed and highlighted the impact of Blockchain management and 5G on supply chain management and trade Finance. The case study shows how Blockchain and IoT will offer opportunities to various organizations in integration with 5G networks. Illustration is given through a shipment tracking device that can easily track a shipment so that the manufacturer as well as the distributor knows transparently, promptly, and accurately where they stand with respect to incentive and wages.

4.2 Smart Agriculture

An Emergent Routing Scheme for Smart Agriculture is proposed in [36] to reduce energy consumption and to enhance throughput and network's life of IoT network. The proposed protocol modifies the IoT devices and heterogeneous IoT networks. It includes smart contracts to discover a route to the Base Station. The redundant data is removed from the combined data and attacks on network are blocked by adding smart contracts. The comparison of energy consumption, stability period, and network lifetime is done with that of IoT-based agriculture and with LEACH in Agriculture. In IoT with Blockchain, the energy consumption is reduced, stability period is improved, and network lifetime is improved as compared to the other two approaches

4.3 Smart Healthcare

An architecture based on Blockchain called Healthcare Data Gateway Application (HBD App) is proposed in [37] that enhances the Healthcare domain by an additional intelligence. The security and privacy are maintained for patients' data and they can share, manage, own, and control their personal data effortlessly. They proposed a purpose-centric access model that ensures the management of patient's healthcare data.

Another case study in Healthcare domain is Medical delivery drones discussed in [38]. As per their research study, medical delivery drones are gradually improving human society by providing medical facilities and delivery via drone-based aspects by emerging 5G-IoT. This study includes 5G-IoT architecture that combines Blockchain and medical delivery drones (via unmanned aerial vehicles) with extraterrestrial communication support from satellites.

5 Issues and Limitations of Blockchain and 5G-IoT

The issues and limitations of Blockchain and 5G-enabled IoT are as follows [39]:

- **Storage capacity and scalability:** The fundamental and basic requirement of the Blockchain technology is a consistent storage of transactions and blocks. As discussed in the previous section, every node must contain a copy of the Blockchain which is growing rapidly with every transaction. The overall performance of the entire system is improved by using this larger storage for the IoT ecosystem. Especially, when the transactions are evolved with the highly scalable application, the sufficient storage, and highly execution machines are required.

- **Processing Power and Time:** The Blockchain environment requires a few self-executive operations such as transaction and block verification. These operations need cryptographic operations. In IoT environment, there are certain limitations in computation which will lead to security risks. Therefore, less resource-intensive substitutes should be applied especially while using Blockchain with respect to IoT context.
- **Throughput:** On the one side, the Blockchain is facing scalability issue, while on the other side, throughput can be considered as another major problem of Blockchain technology. The continuous increase of transactions and its size is hard to tackle for IoT. Latency and transaction throughput are other consistent challenges.
- **Security:** Though Blockchain technology is a tempered-proof and highly secured, it is possible to modify its immutable transactions by covering its 51% of stack. This control over a network can reverse the transactions and cause double-spends by gaining majority control of a Blockchain's hash rate is called 51% attack. This attack can be prevented by continuous monitoring of mining pools and mining activities. Mining complexity and higher hash rate can control the attack activities.

6 Research Directions and Opportunities

5G-enabled IoT devices and networks required huge amount of data transfer with enhancement in existing security, transparency, and trust model which can be addressed through Blockchain technology. Unlike centralized architecture having single point of failure, distributed nature of Blockchain technology makes it a perfect option to the decentralized architecture of 5G-enabled network of IoT devices. Though the research directions and opportunities of Blockchain technology are very wide, the major facets of the same are discussed in this section (Table 4.2). Broadly, we classify the research tracks into following categories:

- Integration of Blockchain with other existing technologies: Researchers work in the direction of integrating Blockchain with distributed database [40], IoT [22], Cloud, AI, and many others.
- Use cases and application subdomains which make use of Blockchain: Though there are many different use cases and application subdomains wherein Blockchain can be utilized, the major focus is on intelligent vehicle management [41, 42], manufacturing [43], healthcare [44], election/voting mechanism [45], agriculture [46], supply change management [47], logistic management [48], and smart contracts [48].
- Security and Privacy concerns: Being a distributed ledger available publicly to all, the concerns such as security (e.g., attacks such as Double spending, DoS, 51%, etc.) and privacy [49, 50], public verifiability [51], and fault tolerance [52] are vital to address.

Table 4.2 Blockchain opportunities in 5G-enabled IoT devices and networking

Sr. No	Blockchain coverage area
1	Smart infrastructure and utilities
2	Integration of Blockchain with other technologies (Big data, IoT, AI, Cloud, etc.)
3	Industry 5.0
4	Application sectors (Healthcare, agriculture, supply chain management, financial sector, manufacturing)
5	IPR and digital right management
6	Security and privacy
7	Consensus mechanism

- Intellectual Property Rights: Researchers also work on protecting Intellectual Property Rights (IPR) [53] and digital rights management [54].
- Consensus mechanism: Being a public network, common consensus among the interacting nodes is an important factor in decision making. Major consensus algorithms in permission-less network are Proof of Work (PoW) [6], Proof of Stack (PoS) [55], Proof of Activity (PoA) [56], Proof of Location (PoL) [57], Proof of ELapsed time (PoEL) [58], etc. These algorithms have their pros and cons and hence; researchers work upon to derive a common consensus mechanism which should be universally adopted.

Hence, it is pragmatic that the Blockchain technology has potential to offer much to the existing and forthcoming industry requirements as opportunities. Due to compute-intensive mining process, energy requirements have increased drastically. Further, IoT devices are low-powered low-configured apparatus. Hence, devices with performance and energy prerequisite need to be reconnoitered. Scalability, standardization (in terms of consensus mechanisms and protocols), interoperability (among different types of devices), and legal compliance (different acts in different regions/countries) are also huge opportunities for research community.

7 Conclusion

Today the world is witnessing the initial dispersal of 5G networks that introduced extreme mobile wireless communications, fastest services, low delays, and very pervasive connectivity via mobile devices. It is worth mentioning that the important paradigm that benefited from 5G is really the usage of the Internet of Things for the industry. However, the extreme use of 5G technology also generates important concerns in terms of security and privacy due to the facts (a) the continuous wireless connection to the network, (b) the reliability of the involved devices is tough to verify. Particularly, the 5G-enabled iIoT integrates the emerging 5G techniques into future industrial IoT applications that require more security and privacy attention.

To address the same, the Blockchain technology could be a prominent option to get integrated with 5G-IoT in order to maintain privacy and security. Through this chapter, we contribute in the direction of exploring the opportunities and challenges while integrating Blockchain technology with 5G-IoT. Further, we investigate the architecture of Blockchain-based 5G-IoT and its application, issues, and limitations. The chapter also helps understand the researchers wishing to plunge into the domain of 5G-enabled IoT and Blockchain through various case studies.

References

1. M.R. Palattella, M. Dohler, A. Grieco, G. Rizzo, J. Torsner, T. Engel, L. Ladid, Internet of Things in the 5G era: enablers, architecture, and business models. IEEE J. Sel. Areas Commun. **34**(3), 510–527 (2016)
2. N.N. Srinidhi, S.M. Dilip Kumar, K.R. Venugopal, Network optimizations in the Internet of Things: A review. Eng. Sci. Technol. Int. J. **22**(1), 1–21 (2019)
3. D.C. Nguyen, P.N. Pathirana, M. Ding, A. Seneviratne, Blockchain for 5G and beyond networks: A state of the art survey. J. Netw. Comput. Appl. **166**, 102693 (2020)
4. J. Ni, X. Lin, X.S. Shen, Efficient and secure service-oriented authentication supporting network slicing for 5G-enabled IoT. IEEE J. Sel. Areas Commun. **36**(3), 644–657 (2018)
5. A. Chaer, K. Salah, C. Lima, P.P. Ray, T. Sheltami, Blockchain for 5g: opportunities and challenges, in *2019 IEEE Globecom Workshops (GC Wkshps)* (IEEE, Piscataway, 2019), pp. 1–6
6. S. Nakamoto, Bitcoin: A peer-to-peer electronic cash system. Technical report, Manubot, 2019
7. N. Mohamed, J. Al-Jaroodi, Applying blockchain in industry 4.0 applications, in *2019 IEEE 9th Annual Computing and Communication Workshop and Conference (CCWC)* (IEEE, Piscataway, 2019), pp. 0852–0858
8. U. Bodkhe, S. Tanwar, K. Parekh, P. Khanpara, S. Tyagi, N. Kumar, M. Alazab, Blockchain for industry 4.0: A comprehensive review. IEEE Access **8**, 79764–79800 (2020)
9. I. Jovović, S. Husnjak, I. Forenbacher, S. Maček, 5G, Blockchain and IPFS: a general survey with possible innovative applications in industry 4.0, in *3rd EAI International Conference on Management of Manufacturing Systems-MMS* (2018)
10. J. Bhatia, Y. Modi, S. Tanwar, M. Bhavsar, Software defined vehicular networks: a comprehensive review. Int. J. Commun. Syst. **32**(12), 1–26 (2019)
11. K. Sheth, K. Patel, H. Shah, S. Tanwar, R. Gupta, N. Kumar, A taxonomy of AI techniques for 6G communication networks. Comput. Commun. **161**, 279–303 (2020)
12. A. Kumari, R. Gupta, S. Tanwar, N. Kumar, Blockchain and AI amalgamation for energy cloud management: challenges, solutions, and future directions. J. Parallel Distrib. Comput. **143**, 148–166 (2020)
13. L. Hughes, Y.K. Dwivedi, S.K. Misra, N.P. Rana, V. Raghavan, V. Akella, Blockchain research, practice and policy: applications, benefits, limitations, emerging research themes and research agenda. Int. J. Inf. Manag. **49**, 114–129 (2019)
14. X. Li, P. Jiang, T. Chen, X. Luo, Q. Wen, A survey on the security of blockchain systems. Future Gen. Comput. Syst. **107**, 841–853 (2020)
15. Z. Zheng, S. Xie, H. Dai, X. Chen, H. Wang, An overview of blockchain technology: architecture, consensus, and future trends, in *2017 IEEE International Congress on Big Data (BigData Congress)* (IEEE, Piscataway, 2017)), pp. 557–564
16. M. Walport, Distributed ledger technology: beyond block chain (a report by the UK government chief scientific adviser), UK Government (2016)

17. V. Buterin, et al., A next-generation smart contract and decentralized application platform. Ethereum white paper **3**(37) (2014)
18. SCALA Blockchain, Accessed: 2019 [Online]. Available: https://www.scalablockchain.com/
19. R. Andreev, P. Andreeva, L. Krotov, E. Krotova, Review of blockchain technology: Types of blockchain and their application. Intellekt. Sist. Proizv. **16**(1), 11–14 (2018)
20. M. Vukolić, Rethinking permissioned blockchains, in *Proceedings of the ACM Workshop on Blockchain, Cryptocurrencies and Contracts* (2017), pp. 3–7
21. Z. Li, J. Kang, R. Yu, D. Ye, Q. Deng, Y. Zhang, Consortium blockchain for secure energy trading in industrial internet of things. IEEE Trans. Ind. Inf. **14**(8), 3690–3700 (2017)
22. A. Reyna, C. Martín, J. Chen, E. Soler, M. Díaz, On blockchain and its integration with IoT challenges and opportunities. Future Gen. Comput. Syst. **88**, 173–190 (2018)
23. H.N. Dai, Z. Zheng, Y. Zhang, Blockchain for internet of things: A survey. IEEE Internet Things J. **6**(5), 8076–8094 (2019)
24. T. Alam, Design a blockchain-based middleware layer in the internet of things architecture. JOIV Int. J. Inf. Vis. **4**(1), 28–31 (2020)
25. T. Alladi, V. Chamola, R.M. Parizi, K.K.R. Choo, Blockchain applications for industry 4.0 and industrial IoT: A review. IEEE Access **7**, 176935–176951 (2019)
26. J.J. Hathaliya, S. Tanwar, S. Tyagi, N. Kumar, Securing electronics healthcare records in healthcare 4.0: a biometric-based approach. Comput. Electr. Eng. **76**, 398–410 (2019)
27. A.D. Dwivedi, G. Srivastava, S. Dhar, R. Singh, A decentralized privacy-preserving healthcare blockchain for IoT. Sensors **19**(2), 326 (2019)
28. M.A. Ferrag, L. Shu, X. Yang, A. Derhab, L. Maglaras, Security and privacy for green IoT-based agriculture: review, blockchain solutions, and challenges. IEEE Access **8**, 32031–32053 (2020)
29. N. Kshetri, 5G in e-commerce activities. IT Prof **20**(4), 73–77 (2018)
30. T. Alam, M. Benaida, Blockchain and internet of things in higher education. Univers. J. Educ. Res. **8**(5), 2164–2174 (2020)
31. H.F. Atlam, G.B. Wills, Technical aspects of blockchain and IoT. Adv. Comput. **115**, 1–39 (2019)
32. Y. Yuan, F.Y. Wang, Towards blockchain-based intelligent transportation systems, in *2016 IEEE 19th International Conference on Intelligent Transportation Systems (ITSC)* (IEEE, Piscataway, 2016), pp. 2663–2668
33. J. Huang, L. Kong, G. Chen, M. Wu, X. Liu, P. Zeng, Towards secure industrial IoT: Blockchain system with credit-based consensus mechanism. IEEE Trans. Ind. Inf. **15**(6), 3680–3689 (2019)
34. K. Valtanen, J. Backman, S. Yrjölä, Blockchain-powered value creation in the 5G and smart grid use cases. IEEE Access **7**, 25690–25707 (2019)
35. S. Kesharwani, M.P. Sarkar, S. Oberoi, Impact of blockchain technology and 5G/IoT on supply chain management and trade finance. Cybernomics **1**(1),18–20 (2019)
36. S.H. Awan, S. Ahmed, A. Nawaz, S.S. Maghdid, K. Zaman, M.Y. Ali Khan, Z. Najam, S. Imran, BlockChain with IoT, an emergent routing scheme for smart agriculture. Int. J. Adv. Comput. Sci. Appl. **11**(4), 420–429 (2020)
37. X. Yue, H. Wang, D. Jin, M. Li, W. Jiang, Healthcare data gateways: found healthcare intelligence on blockchain with novel privacy risk control. J. Med. Syst. **40**, 218 (2016)
38. P.P. Ray, K. Nguyen, A review on blockchain for medical delivery drones in 5G-IoT era: progress and challenges, in *2020 IEEE/CIC International Conference on Communications in China (ICCC Workshops)* (2020), pp. 29–34
39. T. Hewa, A. Kalla, A. Nag, M. Ylianttila, M. Liyanage, Blockchain for 5G and IoT: opportunities and challenges, in *The 8th IEEE International Conference on Communications and Networking (IEEE ComNet'2020)*, (2020)
40. M. Muzammal, Q. Qu, B. Nasrulin, Renovating blockchain with distributed databases: an open source system. Future Gen. Comput. Syst. **90**, 105–117 (2019)
41. S. Kim, Blockchain for a trust network among intelligent vehicles. Adv. Comput. **111**, 43–68 (2018)

42. B. Leiding, P. Memarmoshrefi, D. Hogrefe, Self-managed and blockchain-based vehicular ad-hoc networks, in *Proceedings of the 2016 ACM International Joint Conference on Pervasive and Ubiquitous Computing: Adjunct* (2016), pp. 137–140
43. Z. Li, L. Liu, A.V. Barenji, W. Wang, Cloud-based manufacturing blockchain: secure knowledge sharing for injection mould redesign. Procedia CIRP **72**(1), 961–966 (2018)
44. P. Zhang, D.C. Schmidt, J. White, G. Lenz, Blockchain technology use cases in healthcare. Adv. Comput. **111**, 1–41 (2018)
45. B. Wang, J. Sun,Y. He, D. Pang, N. Lu, Large-scale election based on blockchain. Procedia Comput. Sci. **129**, 234–237 (2018)
46. K. Leng, Y. Bi, L. Jing, H.C. Fu, I. Van Nieuwenhuyse, Research on agricultural supply chain system with double chain architecture based on blockchain technology. Future Gen. Comput. Syst. **86**, 641–649 (2018)
47. J.N. Dewey, R. Plasencia, Blockchain and 5g-enabled internet of things (IoT) will redefine supply chains and trade finance, in *Proceedings of Secured Lender* (2018), pp. 43–45
48. N. Álvarez-Díaz, J. Herrera-Joancomartí, P. Caballero-Gil, Smart contracts based on blockchain for logistics management, in *Proceedings of the 1st International Conference on Internet of Things and Machine Learning* (2017), pp. 1–8
49. K. Ikeda, Security and privacy of blockchain and quantum computation. Adv. Comput. **111**, 199–228 (2018)
50. M. Milojkovic, Privacy-preserving framework for access control and interoperability of electronic health records using blockchain technology. Sustain. Cities Soc. **39**, 283–297 (2018)
51. C. Yang, X. Chen, Y. Xiang, Blockchain-based publicly verifiable data deletion scheme for cloud storage. J. Netw. Comput. Appl. **103**, 185–193 (2018)
52. J. Sousa, A. Bessani, M. Vukolic, A byzantine fault-tolerant ordering service for the hyperledger fabric blockchain platform, in *2018 48th Annual IEEE/IFIP International Conference on Dependable Systems and Networks (DSN)* (IEEE, Piscataway, 2018), pp. 51–58
53. G. Gürkaynak, İ. Yılmaz, B. Yeşilaltay, B. Bengi, Intellectual property law and practice in the blockchain realm. Comput. Law Secur. Rev. **34**(4), 847–862 (2018)
54. Z. Ma, M. Jiang, H. Gao, Z. Wang, Blockchain for digital rights management. Future Gen. Comput. Syst. **89**, 746–764 (2018)
55. P. Vasin, Blackcoin's proof-of-stake protocol v2, Accessed: 2020 [Online]. Available: https://blackcoin.co/blackcoin-pos-protocol-v2-whitepaper.pdf (2014)
56. I. Bentov, C. Lee, A. Mizrahi, M. Rosenfeld, Proof of activity: Extending bitcoin's proof of work via proof of stake. ACM SIGMETRICS Perform. Eval. Rev. **42**(3), 34–37 (2014)
57. M. Amoretti, G. Brambilla, F. Medioli, F. Zanichelli, Blockchain-based proof of location, in *2018 IEEE International Conference on Software Quality, Reliability and Security Companion (QRS-C)* (2018), pp. 146–153
58. L. Chen, L. Xu, N. Shah, Z. Gao, Y. Lu, W. Shi, On security analysis of proof-of-elapsed-time (poet), in *International Symposium on Stabilization, Safety, and Security of Distributed Systems* (Springer, Cham, 2017), pp. 282–297

Chapter 5
The Influence of 5G, IoT, and Blockchain Technologies in Industrial Automation

Eman Shaikh and Nazeeruddin Mohammad

1 Introduction

Since the dawn of mankind, human beings used to manufacture goods by hands or with the help of working animals. These methods were sufficient only for a short period. As time passed by and the global population grew rapidly, the demand for goods also increased subsequently. Existing manufacturing methods posed to be time-consuming, tedious, and inefficient to facilitate the production of goods on a large scale. Therefore, these challenges led to the advancement of an efficient technique called automation.

Automation in simple terms can be defined as a technique that employs machines and the latest technologies to make a process operate without the requirement of manpower. Earlier human input was used to ensure the functioning of these machines. However, with the employment of automation, the requirement of human intervention is negligible. In its initial stage, automation was a simple assembly line of workers that consistently performed repetitive tasks daily. The problem with this approach was that the majority of these tasks posed to be monotonous, dangerous, and unsanitary. Modern automation has become much more advanced. In terms of industrialization, it refers to the use of control devices like information technologies, robots, computers, etc., that handles and manages various processes and machines of the industry. However, to take full advantage of these benefits, the utilization of the Internet of Things (IoT) was crucial.

E. Shaikh (✉) · N. Mohammad
Cybersecurity Center, Prince Mohammad Bin Fahd University, Al Khobar, Saudi Arabia
e-mail: 201501096@pmu.edu.sa; nmohammad@pmu.edu.sa

S. Tanwar (ed.), *Blockchain for 5G-Enabled IoT*,
https://doi.org/10.1007/978-3-030-67490-8_5

IoT is an interconnected network of industrial devices, objects, processes, and humans through which participating entities can operate, communicate, and utilize the collected data. This is done to further boost the manufacturing operations and productivity. With the rapid increase in the number of IoT devices every day, larger network bandwidth is required. This can be fulfilled by the latest 5G technology. Along with network bandwidth, 5G technology also provides other benefits like low latency, ubiquitous connectivity, and efficient utilization of energy. However, there still exists a lack of security between the communication of the industrial objects of the network [22]. Therefore, blockchain technology is employed to provide security to the network.

Blockchain is a public ledger that is decentralized and immutable by nature. This ensures the transparency and authenticity of the received data, that is, the data is not altered or modified without the consent of anyone present in the network. Any change in the transaction of the network is first verified by all the nodes present in the network and then only it is recorded. Along with this, these nodes present in the network are also granted permission to acquire and transmit the respective transaction. Thus, the primary features of the blockchain (such as security, privacy, and trust) induce a positive influence on further development and innovation of industrialization.

Modern industries are benefited from IoT, Blockchain, and 5G communications. Figure 5.1 shows concisely how each technology is shaping modern industries. Automation in industries is known for a long time. However, with the advent of industrial IoT (IIoT), automation has advanced considerably and contributed to bigger industrial philosophy commonly known as industry 4.0. Figure 5.2 further illustrates this relationship. This figure is plotted using Google Ngram viewer [15] that shows how these terms (industry 4.0, smart manufacturing, IIoT, 5G, Blockchain, industrial automation) evolved during the years 2005–2019. Figure 5.2a

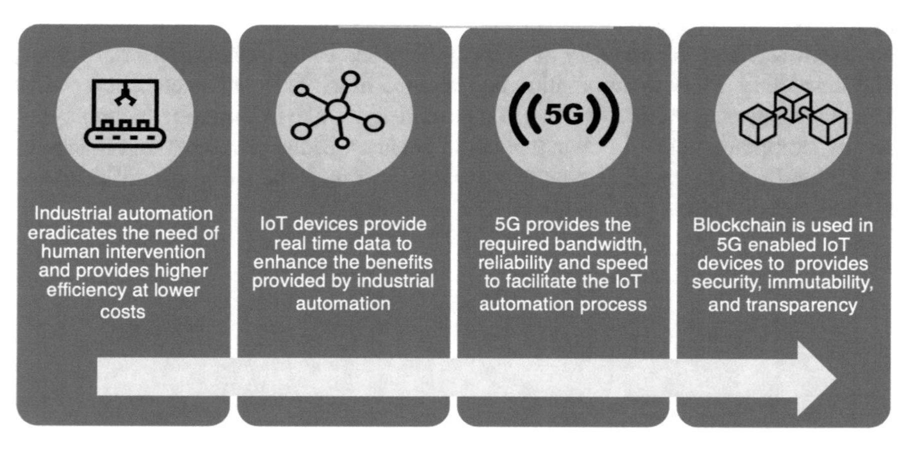

Fig. 5.1 Evolution to Modern Industrial Automation

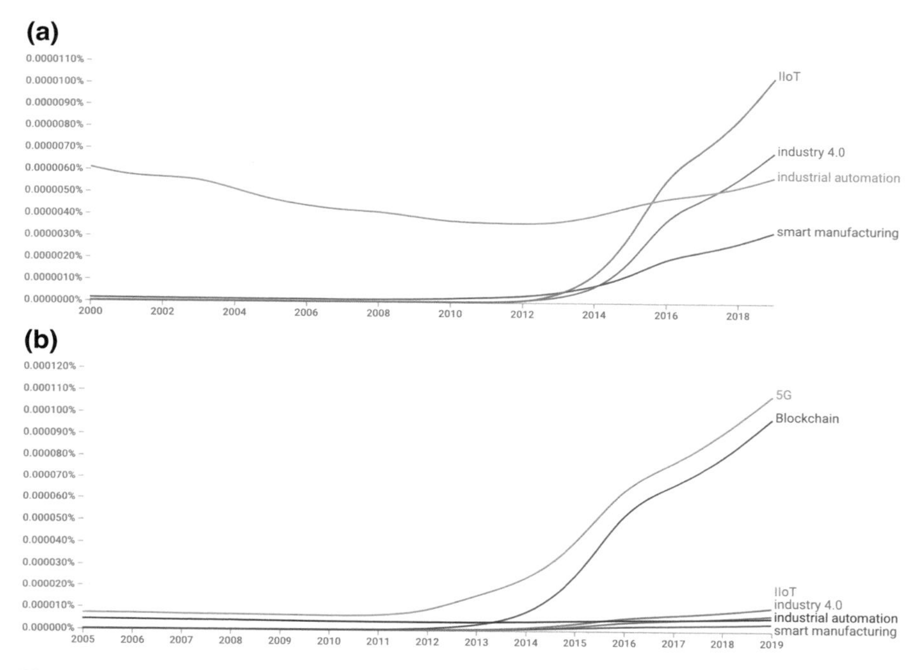

Fig. 5.2 Trends of various contributing technologies for Industry 4.0

shows that industrial automation is commonly used even before the term "industry 4.0" was coined. The trend of "industry 4.0" looks similar to IIoT because it is one of the main enabling technology. Figure 5.2b has additional terms 5G and blockchain, which are more frequently used as they are employed in various sectors in addition to industrial automation.

1.1 Motivation

With the advancement towards industry 4.0, a subsequent rise and need for industrial automation have emerged. The latest technologies, such as IoT and blockchain provide promising solutions to facilitate industrial process automation. Furthermore, the integration of the 5G network features has led to the further enhancement of these processes. The novelty of this chapter is that till now, various papers have been published that talk about these technologies, but none of them has addressed the influence achieved by the integration of these technologies. Therefore, this chapter provides an in-depth introduction to enabling technologies for industrial automation. It discusses the fundamentals of each trend separately and elaborates its impact on industrial automation. It also explains how 5G technology is helping in eliminating bottlenecks in closed-loop automation. It also gives an insight into the applications and influence of these technologies in industrial automation. Lastly, it provides a use case of the integration of these technologies in the banking and finance sector.

1.2 Contributions

In this chapter, we have provided a brief review of the emerging technologies in the various industrial sectors. Mentioned below are the following major contributions of this chapter:

1. Provides a brief introduction of blockchain, IoT, and 5G networks.
2. Discusses the impact of IoT, 5G and blockchain technologies in industrial automation, along with each of their applications towards different industrial sectors.
3. Finally, provides a use case of these integrated technologies in the banking and finance sector.

1.3 Organization

The structure of this chapter is as follows: Sect. 2 talks about the rise of industrial automation, Sect. 3 talks about IoT, Sect. 4 talks about the emergence of 5G wireless networks, Sect. 5 talks about blockchain technology, Sect. 6 demonstrates the use case of blockchain and 5G-enabled IoT in the banking and finance sector. Finally, Sect. 7 concludes the paper.

2 The Rise of Industrial Automation

Currently, there are four stages of industrial automation as illustrated in Fig. 5.3. The first industrial revolution is known as industry 1.0 started during the late eighteenth century. Steam and water-powered machines were developed to replace human labor so that efficient production could be achieved. This soon got replaced by industry 2.0, in which machines that operated on electrical energy were developed to massively further improve the work rate and reduce the expenses. This is because unlike the machines powered by water and steam, electrical machines have proven to work more efficiently. During this period, the first assembly lines were also designed to promote mass production. Later with industry 3.0, the use of electronics and computers first started to automate the process of manufacturing. This helped immensely to enhance speed, accuracy, and productivity. This was soon replaced by the current era we live in—industry 4.0. Industrial automation is the cornerstone

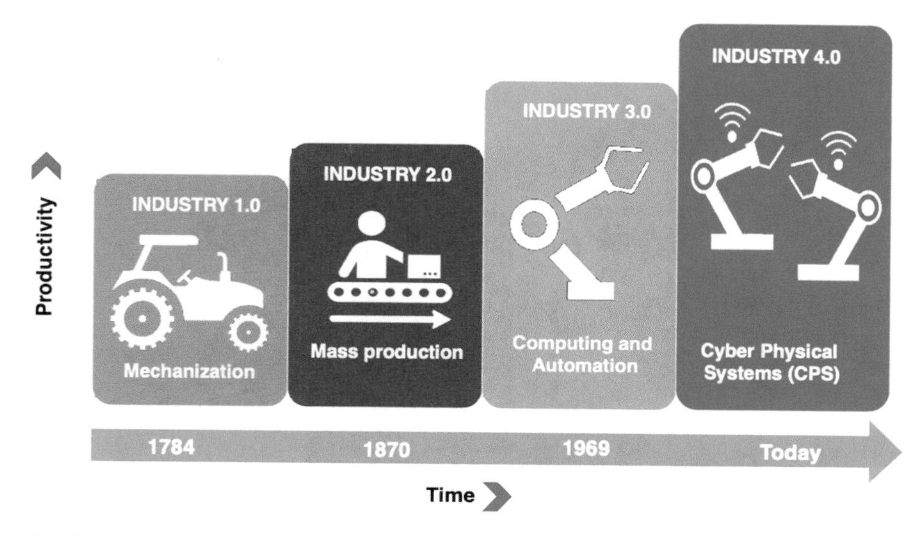

Fig. 5.3 Timeline of the industrial revolution

of industry 4.0, which majorly compromises of various Cyber-Physical Systems (CPS) such as manufacturing facilities, storage, systems, smart grids, and smart machines. This involvement of CPS helped in essentially creating a fully automated environment to ensure that no human intervention is necessary to execute the required tasks.

2.1 Benefits

The beginning of industrial automation brought forward rapid advancements towards various industries as illustrated in Fig. 5.4. Through it, the following major benefits were achieved:

2.1.1 Productivity

In industries, tasks performed by manual labor exhibit certain drawbacks. For instance, humans often get distracted and tired while executing the same repetitive tasks. Moreover, humans often require vacations and cannot operate 24 h a day and 7 days a week. However, with the involvement of industrial automation, continuous mass production can be achieved with ease. The tasks that earlier required multiple workers to complete can now be performed by a single machine. Thus, the productivity levels of the industry can be massively improved.

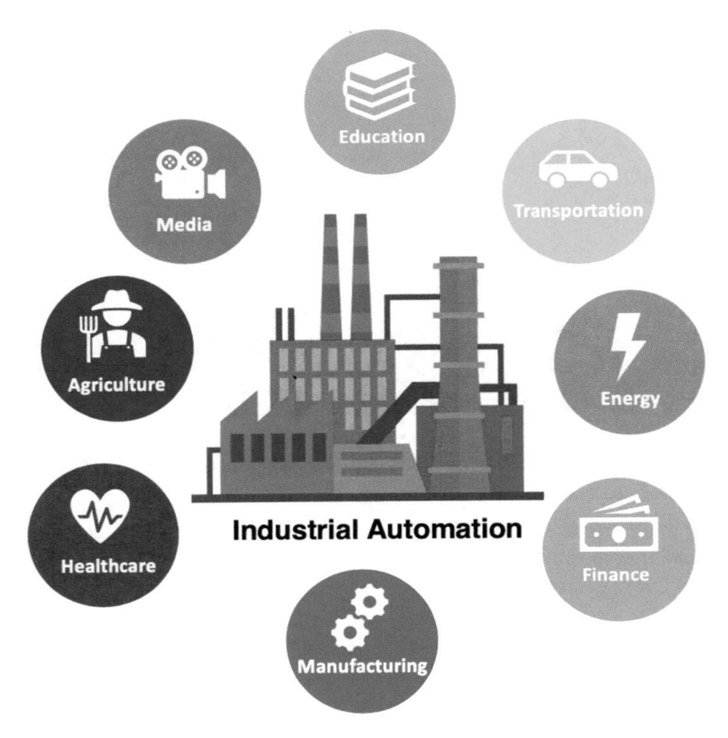

Fig. 5.4 Common industry sectors influenced by industrial automation

2.1.2 Accuracy

Human beings are prone to error and often make mistakes when they are fatigued. Any mistakes made could cause damage to raw materials, components, and final manufactured products. Industrial automation is employed to alleviate human errors and thus improve the quality of products produced.

2.1.3 Safety

As compared to machines, humans are fragile. They cannot be exposed to a dangerous work environment that deals with hazardous materials, extreme temperatures, polluted air, broken equipment, etc. Therefore, industrial automation is implemented to remove humans from hazardous conditions and replaced robots instead.

2.1.4 Costs

Implementing machine and robots pose to be cheaper than recruiting human workers to perform tasks. After the initial capital expenses, the only other cost is in the maintenance of the machines. This is way cheaper than combining the annual salaries of human laborers. Therefore, these lower production costs and higher productivity lead to an increase in financial gain.

To completely take advantage of these benefits, modern automation integrates the latest emerging technologies like IoT, 5G, and Blockchain. The next sections talk in brief about the impact of these technologies for industrial automation.

3 Internet of Things (IoT)

Today we live in an age where we see the Internet of Things (IoT) present everywhere around us. It has made our lives easier and more convenient to live in. Just like how the discovery of the internet changed the way we communicate to one another, IoT has further taken this connectivity to a whole new level. Through it, multiple devices are connected to the internet to facilitate man to machine and machine to machine communications. These devices could be anything from a simple light bulb to huge industrial machines. However, just like any technology, there is a certain process that needs to be executed to complete the desired tasks. Mentioned below are the following steps that a typical IoT system executes (Fig. 5.5):

1. Sensors are used to gather data from the environment. The nature of this data depends on the device it is attached to. It could be anything from a temperature sensor to heat sensors. Moreover, a device could even have multiple sensors attached that can collect a variety of different data.
2. The received data needs to be sent to a processing server (e.g. a private cloud); however, it requires a medium to be sent. Therefore, the sensors are connected to the cloud via a communication technology such as Bluetooth, Wi-Fi, Ethernet, etc. The type of communication technology employed depends upon the IoT application, cost, range, and power consumption.
3. Once the data is delivered to the cloud, a software application is used to perform the necessary actions on the acquired data. This action can be anything like checking the energy consumption of the devices to air quality. If the situation requires the need for user interaction, then an alert is sent to users via their phones, smartwatches, laptop, etc.

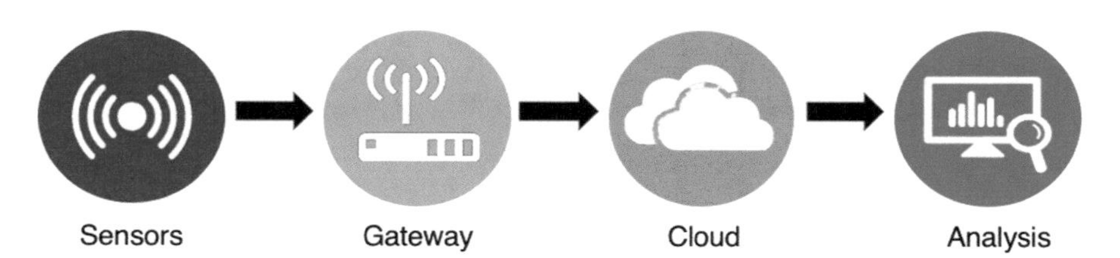

Fig. 5.5 Process flow of IoT components

4. Historical data can also be used to obtain valuable insights that can help in the smooth operation of the devices. For instance, predicting the possible future malfunction of machinery.

3.1 Impact on Industrial Automation

The implementation of IoT in industrial automation has completely revolutionized the way industries operate. It helped to create an environment of intelligent machines that could operate tasks more effectively than human beings. Mentioned below are the major benefits that are provided with the implementation of IoT in industrial automation:

3.1.1 Improvement in Energy Efficiency

Consumption of a large amounts of electricity leads to higher costs. IoT is used to significantly reduce massive energy consumption. From the data collected through IoT, companies can identify the devices that consume a lot of energy. Furthermore, the data collected can help to create operation profiles that instruct the machines when to operate and when not to conserve energy. Systems that drain energy or consume more energy than required could mean that the system itself might be faulty. IoT connectivity helps to address such systems. IoT connectivity can also further help to create energy profiles of an individual or several facilities. This information would help to provide an overview of how to smoothly execute operations while keeping in mind the impact on the energy consumption of such operations.

3.1.2 Predictive Maintenance

Predictive maintenance is a technique that helps to identify the condition of equipment in order to estimate when maintenance is required. This technique is essential to ensure the smooth operations of the machines present in the industries. However, it requires analysis of a large amounts of data and executing complicated algorithms. Therefore, an IoT based solution is employed to eradicate these challenges. With IoT, large terabytes of data can be stored and various machine algorithms can be executed in parallel on different computers to forecast any potential damages and identify when industrial equipment could most likely fail to operate.

3.1.3 Reduction in Operational Cost

Advancements brought towards IoT benefit the industries to reduce operational cost and maximize profits. Therefore, industries that maximize the utilization of IoT will thereby obtain maximum profits. This is because, with the help of IoT devices, real-time information can be obtained instantly. This information can help to facilitate a variety of operations. For instance, in the manufacturing industry, the IoT devices can constantly monitor the equipment for any faults, that is predictive maintenance can be achieved as explained earlier. The implementation of such techniques can also reduce power consumption, this in turn helps to further reduce the overall costs of the industries.

3.2 Industrial Applications

The execution of the IoT process helps to ensure the smooth operations of the industrial processes to provide a closely connected and intelligent environment. This would bring benefits across diverse industries. Mentioned below are some of the major industrial applications:

3.2.1 Agriculture

Maintaining the quality of the soil is crucial to ensure that high-quality goods can be obtained from it. The utilization of IoT helps to ensure this. With the help of the IoT sensors, farmers can get the detailed information regarding the current state of the soil like temperature, moisture content, nutrient deficiency, acidity level, presence of diseases in plants, etc. This information lets farmers implement necessary actions that can help to enhance the present condition of the soil. Moreover, the farmers can deduce the optimal time to plant seeds and identify the presence of diseases in plants/crops.

3.2.2 Healthcare

Before the emergence of IoT, the interaction between patients and healthcare professionals was carried out via face to face visits, phone calls, or text messages. These methods did not provide healthcare professionals with the ability to constantly monitor the status of their patients. However, with the advent of IoT enabled wearable healthcare devices, continuous monitoring of patients is possible. With remote monitoring, the interactions between doctor and patient have become much easier and efficient. Furthermore, it also helps to immensely reduce hospital stay and re-admission of the patients. Another application of IoT is the use of smart beds equipped with sensors that help to constantly monitor vital signs of patients like blood pressure, temperature, pulse rate, etc.

3.2.3 Energy

The energy sector has undergone a variety of changes over the past 20 years. It has been estimated that the total energy consumption across the world would increase to 40% in the near future [11]. This means that there is a need to develop smart energy solutions that would help to achieve efficient use of energy resources. The use of IoT can help to achieve this goal. For instance, the installation of smart energy meters can be used to facilitate the management of the electrical network. Furthermore, the establishment of communication between consumer and service providers can help in the acquirement of a large amounts of data to detect a fault, repair it, and enable decision making. Consumers can benefit from IoT as the data collected can help them to gain insights about their consumption history and the optimal way to reduce their energy consumption if possible.

4 The Emergence of the Fifth-Generation (5G) Wireless Network

The first generation (1G) of the wireless network had first emerged during the 1980s. Within just 40 years, the wireless communication network has completely transformed itself. It now plays a key role in the development of modern infrastructure. The evolution of wireless network took place almost every 10 years. With every evolution, along with better speed and connectivity, various services were also provided. Figure 5.6 illustrates a summary of the evolution from 1G to 5G networks.

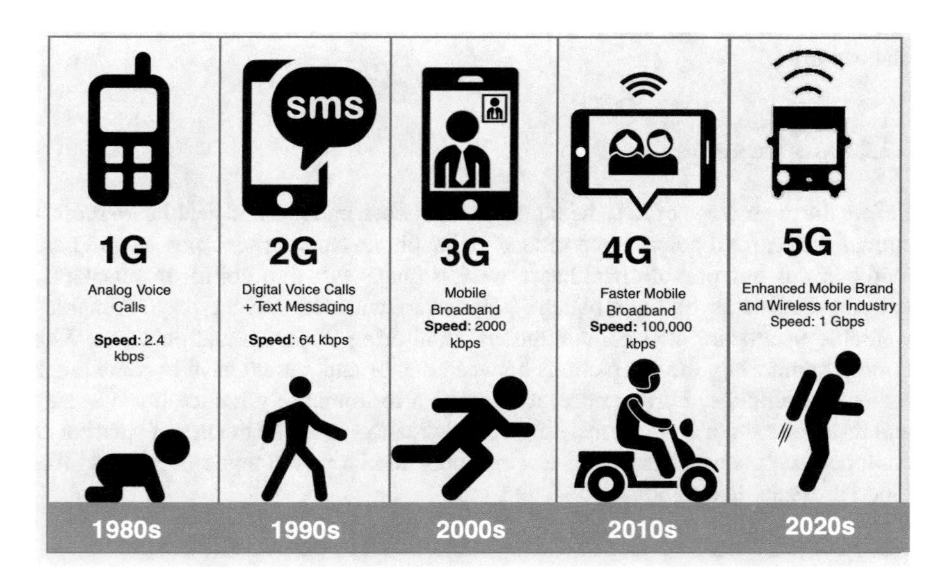

Fig. 5.6 Evolution of mobile communications from 1G to 5G

4.1 *Impact on Industrial Automation*

Industrial automation powered by IoT, because of the various benefits it provides, proved to be successful in enhancing the automation process; the total number of IoT devices employed in different industries rose rapidly. As illustrated in Fig. 5.7, it has been estimated that the total number of IoT devices is set to increase to 75.44 million by 2025 [10]. This massive rise in IoT devices means that a rise in the amount of data will also be transmitted and processed.

Existing technologies are inadequate to handle massive volumes of data with expected reliability or latency requirements. Therefore, 5G implementation can help to facilitate the handling of a large number of IoT devices. Apart from this, the introduction of 5G network helps to facilitate IoT processes in the following ways:

4.1.1 Improved Reliability

5G networks provide a more reliable and stable network, which is extremely important for the various applications of connected IoT devices. This feature is important in applications such as the implementation of security locks, cameras, and other kinds of systems that provide monitoring services and require real-time data as input.

4.1.2 Faster Data Rate

The success of IoT majorly depends upon the speed with which communication can take place. With the introduction of the 5G network, data transfer speed rose rapidly.

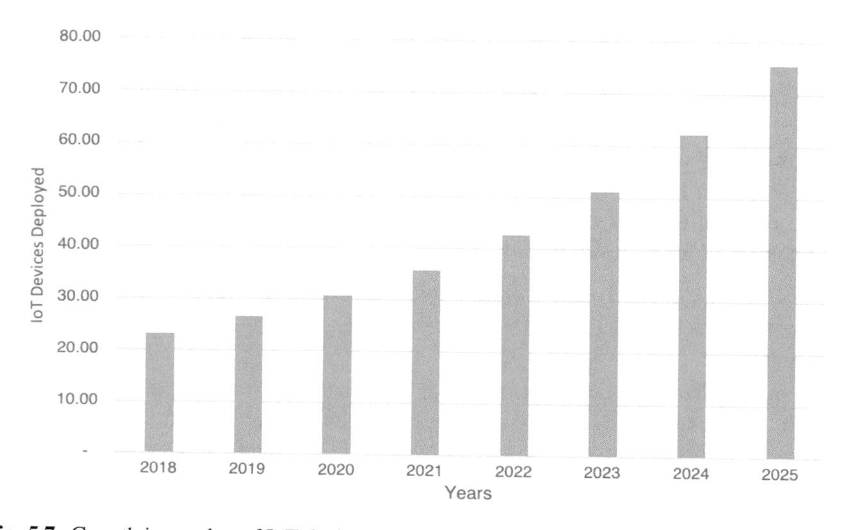

Fig. 5.7 Growth in number of IoT devices

Table 5.1 Comparison of mobile networks in terms of their influence on the industrial applications [23]

Generation	Industrial impact
1G	No impact
2G	Provides remote control and facilitates text messages to and from remote machines
3G	Provides video monitoring, remote access to machines as well as remote condition monitoring
4G	Facilitates live remote access
5G	Facilitates autonomous logistics, machines, edge computing

The increase in data rates allows IoT devices to communicate and share data faster than ever. This feature would benefit various IoT based applications such as the operations in smart healthcare, smart manufacturing, autonomous vehicles, etc.

4.1.3 Lower Latency

5G provides lower latency that is about ten times faster than the traditional 4G network. This low latency will immensely help to facilitate actions that can occur in industrial plants, remote transport, remote surgery, autonomous driving, etc. (Table 5.1).

4.2 Industrial Applications

Now we are quickly shifting and preparing ourselves to the next generation of mobile networks: the fifth generation of mobile networks, also known as 5G. The introduction of the 5G network has completely transformed the way we perform our everyday tasks. It is expected to be the next big thing in terms of mobile connectivity. And has extensively helped in the operations of major industries as illustrated in Table 5.2. Apart from just being a successor of previous networks, the 5G network aims to foster the era of digitization. This is mainly because of the following key features it provides over 4G networks [17]:

1. Ultra-Reliable Low Latency (URLL): In this feature, the communication can happen with a latency of less than one millisecond, which is about 50 times faster than 4G.
2. Enhanced Mobile Broadband (eMBB): This feature allows the data transfer rate up to 10 Gbps which is about 100 times more than what 4G can provide.
3. Massive Machine-Type Communication (mMTC): This feature provides scalable connectivity for a large number of devices, which is 100 times greater than what 4G can provide.

Table 5.2 5G application scenarios across industries [1]

Industry	5G service	Few example cases
Manufacturing	5G smart factory application	Flexible 5G slice, smart toolboxes, production monitoring, remote maintenance/inspection, VR transparent factory
Energy	5G smart power application	Remote control of power distribution systems, advanced tele-metering, robot inspection, electric vehicles (EVs)
Healthcare	5G mobile remote medical care application	Ward inspection with the remote robot, mobile medical vehicles, remote surgery/tests, connecting massive numbers of devices in one hospital ward
Transportation	5G smart transportation application	In-vehicle entertainment, V2V, V2I, and V2P are for automatic driving, vehicle formation, collision avoidance
Security	5G video integrated application	HD video IPTV, remote surveillance, VR/AR live broadcast
Municipal administration	5G public services and society government application	City monitoring for safety, security, environment, cleanliness, smart citizenship services
Education	5G smart park application	Holographic projection, virtual innovation teaching, intelligent recognition

Thus with the implementation of 5G network we will be able to attain a future where not only are mobiles, computers, and laptops connected to the internet, but also other objects like industrial equipment, grocery products, and city assets. This will help to foster the business growth for the organization and completely change the way we communicate, operate in business, and live as a society.

5 Blockchain Technology: The Next Best Thing

For centuries, humans have experimented with different ways to obtain goods and services. At first, the barter system was implemented and used for years. In this system, people exchange goods and services in return for other goods and services. However, this system soon got replaced with the use of coins and money. To further enhance the speed and efficiency of transactions, credit card and electronic payment methods were introduced. Internet and mobile phones have also played a major role in facilitating the process of electronic payment. Nevertheless, these methods still experienced certain challenges such as dependence on a third-party validation, extra payment for transaction costs and service fees, vulnerability to fraudulent activities, etc. Therefore, challenges have led to advancement towards a new form of digital currency cryptocurrency.

Employment of cryptocurrency brought forward several features like faster transactions, anonymity, and the absence of third-party intermediaries. Currently,

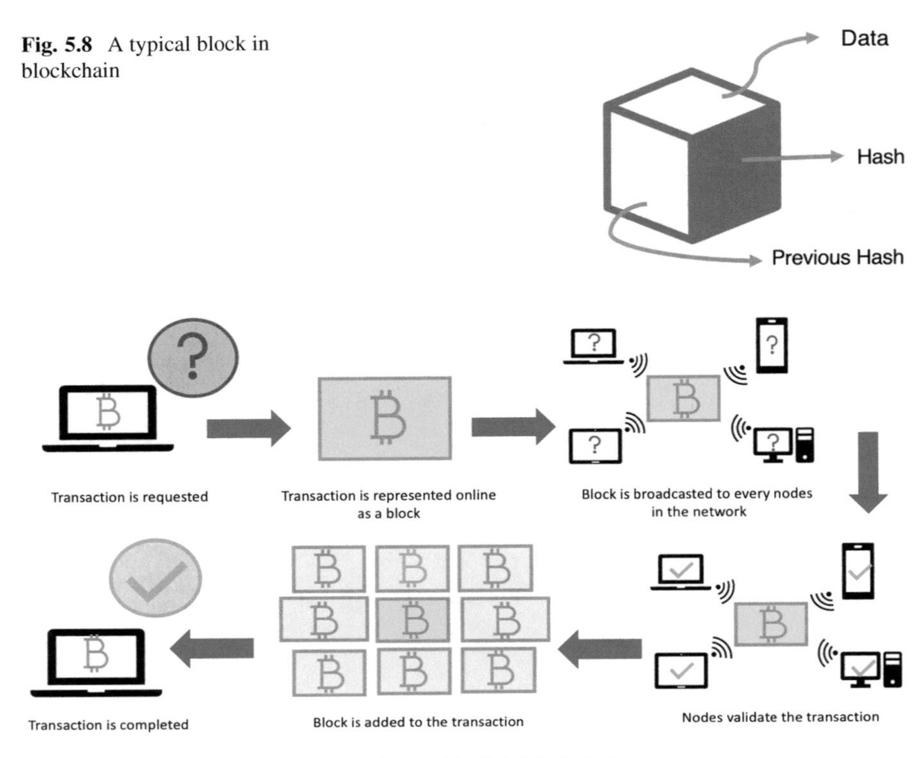

Fig. 5.8 A typical block in blockchain

Fig. 5.9 Process depicting inclusion of a new block in blockchain

several cryptocurrencies are present in the market. Bitcoin, Ethereum, Ripple, Litecoin, etc. are some of the known cryptocurrencies. Among them, Bitcoin is the most widely used cryptocurrency that was founded by Satoshi Nakamoto in 2008 [18]. To facilitate the features of cryptocurrencies, an underlying technology called blockchain is used [21].

Blockchain in layman terms is simply known as a chain of multiple blocks. It is a distributed, decentralized public ledger that is open to anyone. Figure 5.8 illustrates a typical block that compromises of three basic elements: data, a cryptographic identifier called a hash, and the hash of the previous block. The data stored in the block can be anything and it depends upon the application. For instance, the bitcoin blockchain stores information about the buyer, seller, and the amount. The function of the hash is to uniquely identify the block and its contents. It is just like our fingerprint, unique in nature and specific to each block. However, the hash is created by a hash function that maps the block data of arbitrary size into a single fixed value size. The third element of the block is the hash of the previous block, which helps create the chain of blocks. It is due to this type of connection that blockchain can provide a secure environment. Figure 5.9 depicts the steps that are required to be executed in order to successfully add a block in the blockchain network.

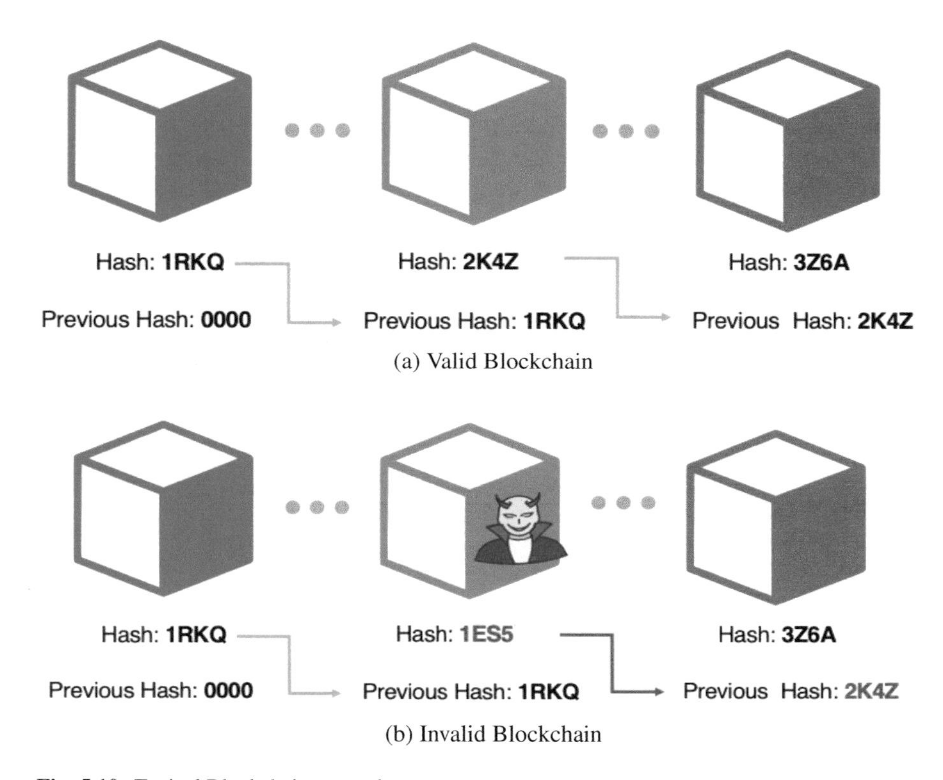

Fig. 5.10 Typical Blockchain network

After the block is created, it is impossible to change the contents of the block without the hash itself getting changed. This is because once a block is created, its hash is always calculated. Any change in the block will change the hash too. That is, a hash is useful to detect whether any changes occur in the block. Changing a single block will make the corresponding blocks invalid. A visual representation of a valid and invalid blockchain network is illustrated in Fig. 5.10.

However, the sole use of hash is not enough to prevent any changes in the block. This is because computers are much more advanced today and can easily compute thousands of hashes per second. This means that the data in the block can be easily tampered as all of the hashes of other blocks can be recalculated to make the blockchain valid. Therefore, blockchain compromises of the proof of work algorithm to overcome this challenge. The proof of work is a mechanism that helps to slow down the process of the block creation process. Another way in which blockchain gains security is by being distributed. That is, rather than employing a central entity to manage the network, blockchain utilizes a peer to peer network. This means that anyone who enters the network gets a complete copy of the blockchain. In addition to this, every node in the network has the right to verify a new block and ensure that it is intact. The verified block is then added to the blockchain network. To successfully tamper with the blockchain network one must satisfy the following conditions:

1. Firstly, every block in the network must be altered.
2. Next, the proof of work for each of these blocks must be re-constructed.
3. Finally, complete control of more than 50% of the network must be achieved.

Only after the execution of all the above conditions, the blockchain network can be altered. With the current technology, it is computationally infeasible to match these conditions.

5.1 Features

Apart from providing security to the system, blockchain also provides other major features. Mentioned below are some of them [3, 24]:

5.1.1 Decentralized

The main role of the blockchain network is to store and copy data (cryptocurrencies, contracts, and other digital assets) across all the nodes present in the network. This means that the network is decentralized in nature, that is, it does not depend upon a central entity, but a group of nodes that maintain the network. Therefore, through the implementation of the decentralized blockchain network, owners will be able to gain direct control over their account by using their private key. Furthermore, this also helps to provide common people the power and rights to perform actions on their assets as they wish for. Stock exchange, transactions are done in real estate, personal identification, etc. are some of the major areas where the decentralization feature can benefit.

5.1.2 Distributed Ledger

A distributed ledger is a database that is sent to all the nodes present in the blockchain network. Whenever a new block is added, a message is broadcasted to make sure that every node has the updated version of the ledger. This feature facilitates the applications beyond digital currencies, as it removes the requirement of dealing with a central authority to record information. As this feature allows the ledger to be stored on several devices that are located in different locations, it, therefore, helps to protect the system in case of data loss caused by devices or servers during downtime.

5.1.3 Immutable

Blockchain is an immutable distributed ledger that is decentralized in nature. It is because of this nature that the information once stored in it cannot be altered easily. This helps to create trust in the transaction record. In case a data is added to the block it first needs to be approved by all the parties present in the network. Without consent from the majority of the nodes, the blocks cannot be added. Furthermore, the user cannot go back and change, edit, delete, or update the appended block. Therefore, this helps to create an environment that is transparent and free of corruption.

5.1.4 Consensus

All participants present in the network must come in terms with a given set of rules that determines the validity of the block. A consensus algorithm is used to put forward a common agreement for all the nodes present in the network. In case, a block violates any of the given rules, then that block is considered as invalid and thus does not get added to the network. Therefore, this feature helps to provide a sense of trust between the nodes present in the network as every node can be assured that every other node follows the ratified rules. At present, there are various types of consensus algorithms that are based upon different principles. Some of the commonly known algorithms are: Proof of Work (PoW), Proof of Stake (PoS), Proof of Burn, Proof of Space, Proof of Activity, etc. Paper [19] talks in brief about each of these algorithms and other common algorithms.

5.2 Impact on 5G-Enabled IoT

The major challenges IoT presently face is security, privacy, compatibility, and centralization [8]. Fortunately, the majority of these challenges are solved via blockchain with 5G enabled IoT. The major feature of blockchain is that its architecture is decentralized in nature. Therefore, its integration with IoT helped to eradicate centralization and make transactions more secure and transparent. Apart from this, the employment of blockchain with IoT provides the following additional benefits:

1. Blockchain uses a distributed ledger, which is unalterable in nature. This helps to remove the need for trust among the involved parties. That is, no single party has control over the massive amount of data produced by the IoT devices.
2. Blockchain helps to store the data collected by the IoT devices. This helps to further enhance the present security levels. This is mainly due to the encryption mechanism that blockchain provides to protect its data. Thereby, making it difficult for hackers to gain access to the network.

3. Blockchain also provides the smooth processing of the transactions of billions of IoT devices. With the increase in the number of devices, the distributed ledger facilitates the processing of the massive transactions obtained from these devices.

5.3 Impact on Industrial Automation

The implementation of blockchain technology in industrial automation processes will bring numerous benefits [13]. Some of the major benefits are mentioned below [9]:

5.3.1 Improved Transparency

With the implementation of blockchain technology, the transaction history can become more transparent. Through it, all the nodes are now able to share the same information that can be updated via consensus. To change a single transaction record, all the transactions present in the network would need to change, which is not possible. This allows anyone in the network to view all the information present in the network. Furthermore, it also helps to ensure that the data on the network is transparent, accurate, and consistent. Therefore, this feature can benefit various industries like real estate, automotive, manufacturing, etc.

5.3.2 Better Security

Blockchain helps to ensure that the information stored is not vulnerable to cyber-attacks. This is because of the way the transactions are stored in the network. That is, first the transaction must be validated by the nodes present in the network, only after it is approved that it is encrypted and linked to the previous transaction. Furthermore, the information stored is the same across all the nodes present in the network. This is the reason why it can be difficult for hackers to alter or change the stored data. This feature would therefore immensely help in industries such as healthcare, government, finance, etc., to protect sensitive data. Thus, blockchain can help to prevent any unauthorized activity that can occur on the stored data.

5.3.3 Enhance Traceability

This feature helps to solve the issue of tracing back the products to their origin. That is, it helps industries whose products have to go through a complex supply chain. This is because blockchain helps to store information related to the present status of the transaction such as how the goods are manufactured, shipment location, how they are managed, etc. As this data is immutable and can be easily shared with the supply chain network, it, therefore helps to provide extensive tracing and tracking abilities. Whenever the transfer of goods is recorded on the blockchain, the entire journey of the blockchain can be seen. This helps to provide validation to the authenticity of the assets in industries like drugs, agriculture, etc.

5.4 *Industrial Applications*

There are many research efforts/proposal/implementations studying the use of blockchains across various industries [2, 16]. For example, blockchain applications are studied for drones [6], finance [20], and autonomous vehicles [7]. This section highlights some of the key applications:

5.4.1 Agriculture

The agriculture sector is another domain that can greatly benefit from the major features that blockchain provides. For instance, it can help to provide an immutable record from origin to the retail store of any product [14]. This would help to create a sense of trust and transparency for the consumers regarding the products they buy. It can also help in enhancing the productivity and efficiency of smart farming. For example, the data gathered by the IoT devices can now be stored in a blockchain and be executed for particular actions. This would, therefore, help to enhance the quality of the crop and the quality of farming.

5.4.2 Energy

Blockchain technology can help improve the energy sector. Through it, the following three major benefits can be achieved: reduced costs, the sustainability of the environment, and enhanced transparency for stakeholders. For instance, it facilitates peer to peer transactions, that is the users can directly trade energy [25]. This feature is useful for energy resources that are renewable like wind and solar energy. This would thus allow prosumers to enter the market and act as suppliers too. Moreover, by employing a decentralized architecture, the consumers can now purchase energy suppliers directly from the utility providers. This would help to reduce costs.

5.4.3 Healthcare

With the rising human population and medical conditions, the need for optimal healthcare facilities is also rising. Fortunately, 5G enabled blockchain has promised to provide a variety of applications in the healthcare domain [2]. For instance, it helps to provide data security to clinical trials done during research and experimentation. Since the data stored in a blockchain is immutable, it also does not allow any tampering of this sensitive data stored. It can also protect the privacy of the patient. That is, sharing of the patient data to a third party like pharmacies can be done with protecting the identity of the patient.

6 Blockchain and 5G-Enabled IoT Use Cases in Finance Sector

As mentioned in the earlier sections, the emergence of blockchain, 5G, and IoT helped to disrupt the way industries operate. The banking and finance sector is one such industry that has observed a wide range of implementation of blockchain technology [5]. The majority of the top banks have started to incorporate blockchain technology to leverage the financial services that they offer. Bank al Etihad is an example of a bank that has already started using blockchain technology to enhance their paperwork and documentation processes [4]. Furthermore, they have also incorporated blockchain to help their customers verify and securely issue confidential documents. Deutsche Bank, HSBC, ING are examples of other banks that have also implemented blockchain technology to enhance their operations and services.

6.1 Characteristics

This is mainly because of the inherent characteristics of blockchain that helps to facilitate the execution of operations and services in this sector. Mentioned below are the following ways on how the characteristics of blockchain technology help the banking and finance sector:

6.1.1 Decentralized Trust

The primary feature of blockchain is that it helps to track and verify transactions. This enables the organizations and the customers to process their respective transactions without the need of a third party or a centralized bank [12]. Various banks have implemented blockchain technology mainly because it consists of a shared infrastructure in which the control is distributed among all the nodes present in a given transaction chain. This helps to immensely reduce any possible counterparty risks.

6.1.2 Enhanced Security

The reason why blockchain is so secure is that once any data is added to its network, it is impossible to tamper with the data. Furthermore, as it is shared by all of the nodes present in the network, it is difficult to hack the network. The decentralized architecture of the blockchain helps to ensure that there is no central point of failure in the network. This allows it to effectively resist any attacks.

6.1.3 Efficient Transactions

The elimination of a centralized entity/third party helped to immensely improve the settlement time and the transaction time of a transaction. Because of this, the transactions can be processed at any time and any day of the week. Furthermore, the transactions can also be done in a faster manner as compared to the traditional methods. This will allow more transactions to be completed at a given time.

6.2 Challenges

Just like any other technology, blockchain also faces few challenges. Therefore, we have mentioned below some of the major challenges that needs to be addressed before implementing it in the banking and finance sector.

6.2.1 Scalability

Although blockchain technology is a prime focus in the financial industry. It however is not capable of handling the large scale of financial transactions that occur daily. This is because multiple nodes are required to validate every transaction. This could lead to a reduction in the transaction speed and an increase in the cost per transaction. Therefore, it is critical to consider before deploying blockchain on a large scale.

6.2.2 Cost

Another challenge faced while deploying blockchain technology is the high cost faced during its initial setup. This makes small companies and banks hesitant to invest in something that does not guarantee a promising success. Therefore, it is important to address this issue before a company thinks about deploying blockchain to facilitate its operations.

6.2.3 Policies and Regulations

One of the major issues faced by banks is the fact that blockchain suffers from the lack of clarity of policies and regulations. Currently, there is no set of standard rules and regulations regarding the transfers done with cryptocurrencies. Unless and until a formal regulatory framework has been established, banks cannot deploy blockchain to facilitate its services.

7 Conclusion

Automation of industrial processes, to meet the conservative business goals, is ever increasing with the help of modern technologies such as IIoT and 5G. IIoT facilitated cooperation among the multiple connected entities for intelligent decisions and subsequent actions. The latest 5G technology has provided the required support for real-time, reliable, and low-latency communications. With so many crucial advanced process control and data analysis applications depending on the IIoT network, there is an increased need for proper access control. Blockchain offers a decentralized tamper-proof robust mechanism for transaction management, which is well-suited for IIoT applications. Blockchain technology in 5G-powered IIoT showed a wide variety of applications and industries started adopting them. This trend is going to continue and benefit various industrial sectors. In parallel, there will be lots of research to improve the scalability, cost efficiency, and standardization of blockchain technology.

Acknowledgments Authors would like to thank the Prince Mohammad Bin Fahd Center for Futuristic Studies (PMFCFS) at Prince Mohammad Bin Fahd University for supporting this research.

References

1. 5G network slicing boosts the digital transformation of vertical industry. Website (2019). https://openlab.zte.com.cn/en/news/2019/5/5G-Network-Slicing-Boosts-the-Digital-Transformation-of-Vertical-Industry
2. T. Alladi, V. Chamola, R.M. Parizi, K.K.R. Choo, Blockchain applications for industry 4.0 and industrial IoT: a review. IEEE Access **7**, 176935–176951 (2019)
3. H. Anwar, 6 key blockchain features you need to know about! Website (2018). https://101blockchains.com/introduction-to-blockchain-features/
4. Bank al Etihad slashes paperwork with blockchain, cloud technology. Website (2020). https://www.cio.com/article/3575990/bank-al-etihad.html?upd=1601752555407
5. P. Chaudhari, Blockchain technology- a silver lining to BFSI industry. Website (2019). https://www.esds.co.in/blog/blockchain-technology-a-silver-lining-to-bfsi-industry/
6. R. Gupta, A. Kumari, S. Tanwar, N. Kumar, Blockchain-envisioned softwarized multi-swarming UAVs to tackle covid-19 situations. IEEE Netw. **PP** (2020). https://doi.org/10.1109/MNET.011.2000439

7. R. Gupta, S. Tanwar, N. Kumar, S. Tyagi, Blockchain-based security attack resilience schemes for autonomous vehicles in industry 4.0: A systematic review. Comput. Electr. Eng. **86**, 106717 (2020). https://doi.org/10.1016/j.compeleceng.2020.106717. http://www.sciencedirect.com/science/article/pii/S0045790620305723
8. R.M. Haris, S. Al-Maadeed, Integrating blockchain technology in 5G enabled IoT: a review, in *2020 IEEE International Conference on Informatics, IoT, and Enabling Technologies (ICIoT)* (2020), pp. 367–371
9. M. Hooper, Top five blockchain benefits transforming your industry. Website (2018). https://www.ibm.com/blogs/blockchain/2018/02/top-five-blockchain-benefits-transforming-your-industry/
10. Internet of things (IoT) connected devices installed base worldwide from 2015 to 2025. Website (2016). https://www.statista.com/statistics/471264/iot-number-of-connected-devices-worldwide/
11. Internet of things in the energy sector. Website. https://www.hiotron.com/iot-energy-sector/
12. N. Kabra, P. Bhattacharya, S. Tanwar, S. Tyagi, Mudrachain: Blockchain-based framework for automated cheque clearance in financial institutions. Future Gen. Comput. Syst. (2019). https://doi.org/10.1016/j.future.2019.08.035
13. O. Lage, Blockchain: From Industry 4.0 to the Machine Economy (2019). https://doi.org/10.5772/intechopen.88694
14. J. Lin, Z. Shen, A. Zhang, Y. Chai, Blockchain and IoT based food traceability for smart agriculture (2018), pp. 1–6. https://doi.org/10.1145/3265689.3265692
15. J.B. Michel, Y.K. Shen, A.P. Aiden, A. Veres, M.K. Gray, J.P. Pickett, D. Hoiberg, D. Clancy, P. Norvig, J. Orwant et al., Quantitative analysis of culture using millions of digitized books. Science **331**(6014), 176–182 (2011)
16. I. Mistry, S. Tanwar, S. Tyagi, N. Kumar, Blockchain for 5G-enabled IoT for industrial automation: A systematic review, solutions, and challenges. Mech. Syst. Signal Process. **135**, 106382 (2020)
17. M.A. Monem, 5G will enrich the telecommunication ecosystem. Website (2017). https://www.netmanias.com/en/post/blog/12440/5g/5g-will-enrich-the-telecommunication-ecosystem
18. S. Nakamoto, Bitcoin: a peer-to-peer electronic cash system. Technical report, Manubot (2019)
19. G.T. Nguyen, K. Kim, A survey about consensus algorithms used in blockchain. J. Inf. Process. Syst. **14**(1), 101–128 (2018)
20. M.M. Patel, S. Tanwar, R. Gupta, N. Kumar, A deep learning-based cryptocurrency price prediction scheme for financial institutions. J. Inf. Secur. Appl. **55**, 102583 (2020). https://doi.org/10.1016/j.jisa.2020.102583. http://www.sciencedirect.com/science/article/pii/S2214212620307535
21. E. Shaikh, N. Mohammad, Applications of blockchain technology for smart cities, in *2020 Fourth International Conference on Inventive Systems and Control (ICISC)* (2020), pp. 186–191
22. E. Shaikh, I. Mohiuddin, A. Manzoor, Internet of things (IoT): Security and privacy threats, in *2019 2nd International Conference on Computer Applications Information Security (ICCAIS)* (2019), pp. 1–6
23. M. Siddiqi, H. Yu, J. Joung, 5G ultra-reliable low-latency communication implementation challenges and operational issues with IoT devices. Electronics **8**, 981 (2019). https://doi.org/10.3390/electronics8090981
24. K. Sultan, U. Ruhi, R. Lakhani, Conceptualizing blockchains: Characteristics & applications, arXiv preprint arXiv:1806.03693 (2018)
25. S. Wang, A.F. Taha, J. Wang, K. Kvaternik, A. Hahn, Energy crowdsourcing and peer-to-peer energy trading in blockchain-enabled smart grids. IEEE Trans. Syst. Man Cybern. Syst. **49**(8), 1612–1623 (2019)

Part II
Enabling Technologies and Architecture for 5G-Enabled IoT

Chapter 6
Emerging Communication Technologies for 5G-Enabled Internet of Things Applications

Malaram Kumhar and Jitendra Bhatia

1 Introduction

The IoT has gained much interest recently due to its ability to connect many real-world objects. It can play a vital role in many domains to improve the quality of life in different areas such as smart homes, medical health, education, industry, agriculture, transports, and smart cities. Now there is a paradigm shift towards the concept of "anytime, anywhere, anyone, connected to anything." In many areas, whether it is our homes or workplace, the devices are becoming essential for our daily lifestyle. IoT has made it possible to connect everyone with everything in the environment [1]. Connecting different devices and adding sensors enhance a level of intelligence to devices, otherwise they would be dumb, making them transfer up-to-date data without any human involvement. The organization of the chapter is shown in Fig. 6.1.

1.1 Contribution and Motivation

Internet of Things (IoT) is very important for the upcoming 5G mobile networks, which will enable several innovative IoT applications which include Self-driving cars, Healthcare, Smart cities, and other enormous IoT use cases set in 5G standards.

M. Kumhar
Department of Computer Science and Engineering, Institute of Technology, Nirma University, Ahmedabad, Gujarat, India
e-mail: malaram.kumhar@nirmauni.ac.in

J. Bhatia (✉)
Department of Computer Engineering, Vishwakarma Government Engineering College, Ahmedabad, Gujarat, India

S. Tanwar (ed.), *Blockchain for 5G-Enabled IoT*,
https://doi.org/10.1007/978-3-030-67490-8_6

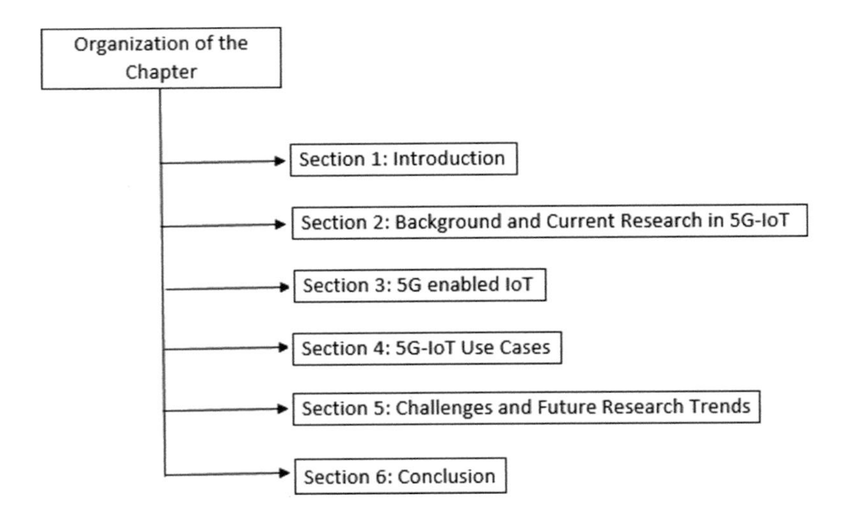

Fig. 6.1 Organization of the chapter

The success of any IoT system depends on its communication speed and proper connectivity with other devices such as smartphones, laptops, and tablets. 5G is faster than other past and existing mobile communication technologies. This high-speed connectivity will allow IoT devices to communicate and share data faster with other smart devices. For 5G enabled IoT, the concept of 5G requirements and its viable technologies should be properly examined. The key motivation behind this contribution is that a large number of devices are expected to come with emerging applications, which need high speed, good connectivity, and high data rates. The key contribution in this chapter includes:

- Various emerging applications in IoT
- Background and current research in the area of 5G enabled IoT
- Key enabling technologies for 5G enabled IoT and Use cases
- Challenges and future research trends
- Case study in 5G enabled IoT

1.2 IoT and Emerging Applications

In IoT, objects located around us can be connected. These objects use Internet services, transfer data, and provide useful information about events that had to happen with the object like changes in their environment or your actions. IoT enables users to bring physical objects into the scope of information technology. This can be accomplished by using different technologies like RFID, NFC, Bar Codes, and QR codes that permit the user to recognize and refer to different physical objects [2, 3]. There are no standards available to connect IoT devices to the Internet, besides their networking protocols.

IoT may be implemented with novel security features [4, 5] for home automation systems, vehicle electronics, telephone networks, control of various domestic useful services, etc. There are numerous innovative applications that came into existence and revealed for a different set of goals and service sectors. Typically, IoT applications are characterized based on the area of deployment. Figure 6.2 depicts the categorization of the multimedia and non-multimedia based various IoT Applications [6], and also, the type of end-user, i.e., system or person. These applications play a major role for making our day-to-day life activities easy, smart, and safe [7].

Smart Surveillance Earlier the surveillance systems were used only in big malls and shopping centers. Nowadays, we find CCTV cameras deployed at many places, in a small departmental store, homes, educational institutes, holy places, etc. This provides a promising solution for public security by paying a minimal cost. Smart cameras are critical sensors in IoT-based surveillance systems that combine sensing, processing, and communication capabilities on an embedded device. These cameras can automatically capture the video of the view of the area where the motion is detected. Modern technologies enable real-time monitoring and reply to alarms immediately. Figure 6.3 shows the numerous components and services of an IoT enabled smart surveillance application.

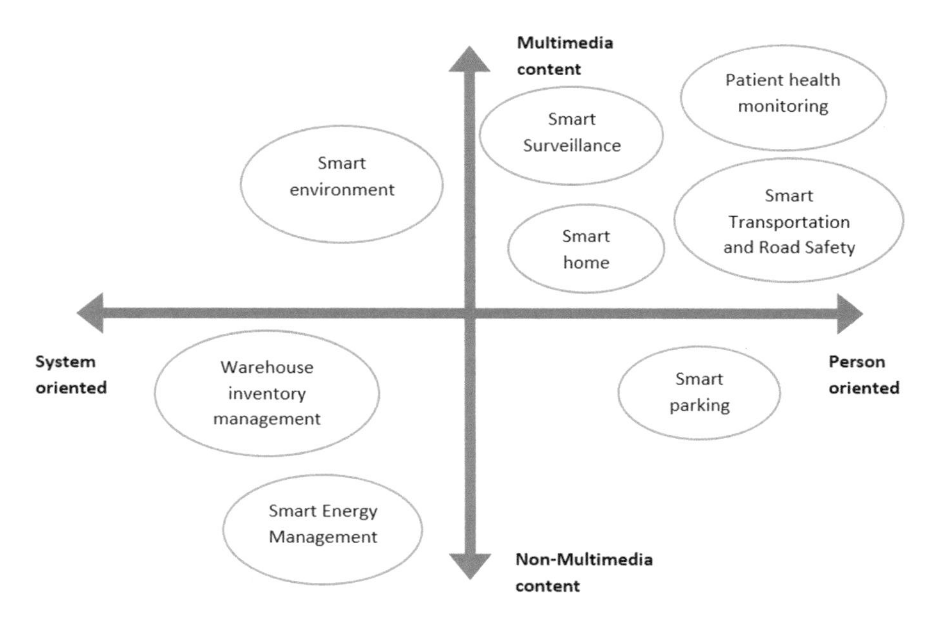

Fig. 6.2 Categorization of IoT applications

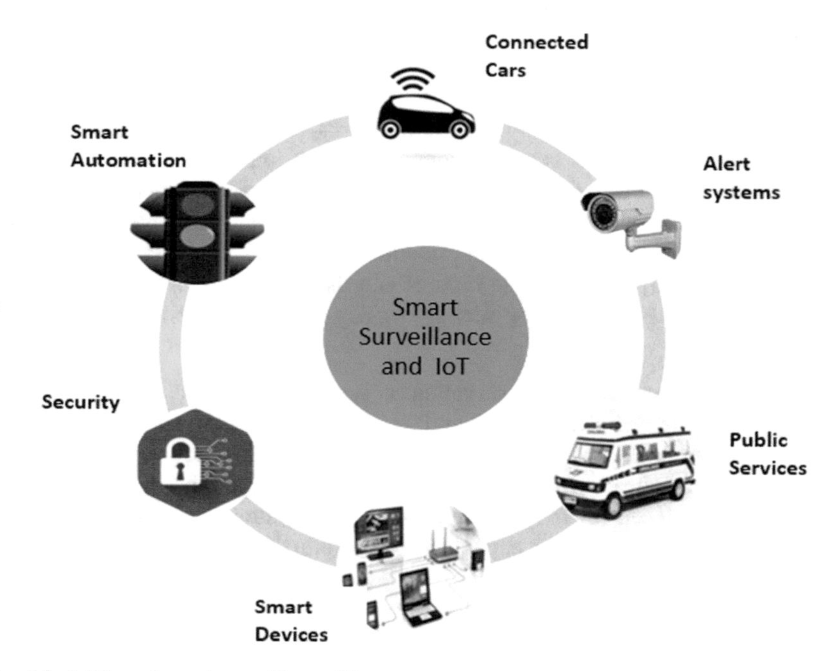

Fig. 6.3 IoT-based smart surveillance [8]

Smart Home The smart home is simply an amalgamation of all household connections controlled remotely via a control unit in a particular room using the computer and Internet. Various devices equipped with the sensors are the main components of IoT, and they sense things from the surrounding, such as pressure, temperature, and send to the control unit via the wireless network. The comfort level at home can be automatically controlled with the help of sensing devices that sense the temperature pressure and humidity of surroundings [9]. The refrigerator can perceive and monitor the items kept in it, and make the user aware about scare products and also order for stock up the required items. Human daily activities can be monitor, and help can be provided, especially to disabled or older people, such as controlling the lights, automatically open the door, or pulling the curtains. Entry in the building can be controlled using RFID Tags, connected with the persons those who are authorized to enter the building [10]. Figure 6.4 shows the devices and sensors used in IoT enabled smart home system.

Patient Heath Monitoring The IoT can be used in the medical field to improve the health condition of many people by monitoring parameters related to health and managing medicines in the record. IoT empowered devices have made remote health monitoring possible in the medical field, allowing them to keep patients safe and healthy and enable the doctors to provide the utmost care. Moreover, remote monitoring helps minimize the time to stay at the hospital for the patients, prevents

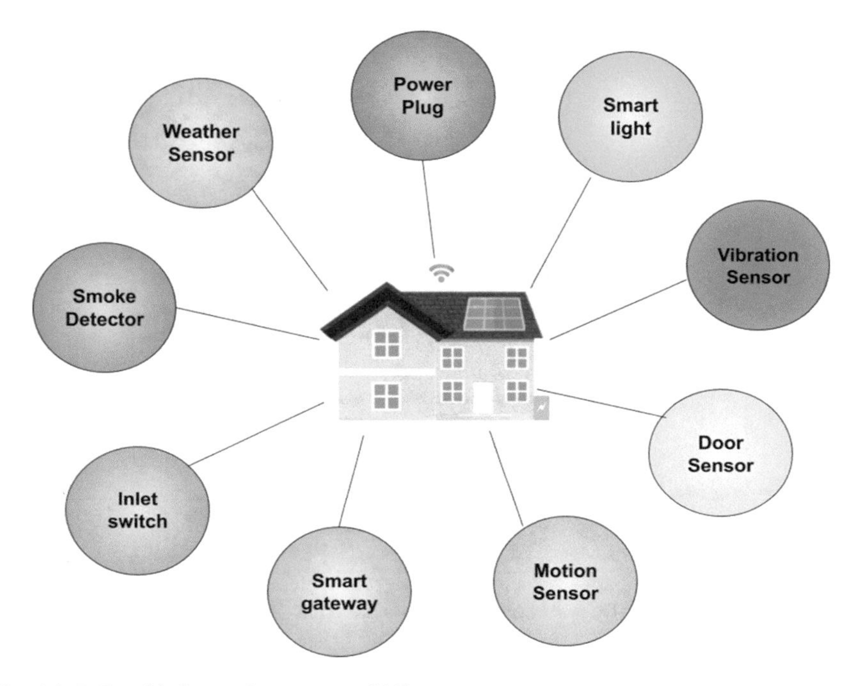

Fig. 6.4 IoT enabled smart home system [11]

re-admissions, and can save lives in case of medical emergencies like heart attack, diabetes, asthma attacks, and many more critical diseases. IoT has a significant impact on reducing healthcare costs considerably and improve treatment outcomes. Figure 6.5 depicts the Components of a remote patient monitoring system.

Smart Environment A healthy environment is an utmost need for daily human life activities. Plants and animals living in the environment may be affected in an unhealthy atmosphere. Various research efforts have been made to answer the environmental issues such as pollution control and waste management [13]. Smart environment monitoring requires intelligent techniques to exchange information between databases, storing, and processing of the data collected from various monitoring devices to avoid many environmental problems and natural disasters. The integration of GPS, mobile and wireless communication technologies, databases, cloud to the internet for building what is known by IoT, the smart environment monitoring will become easier. Figure 6.6 shows the applications where research efforts can be concentrated to solve the environment-related problems.

Smart Transportation and Road Safety Traffic monitoring, adaptive traffic signals, accident avoidance, monitoring the road condition, traffic divert alerts, etc. are the essential application of IoT revolution for the intelligent transportation system (ITS) [15]. Various IoT technologies can work jointly to provide better

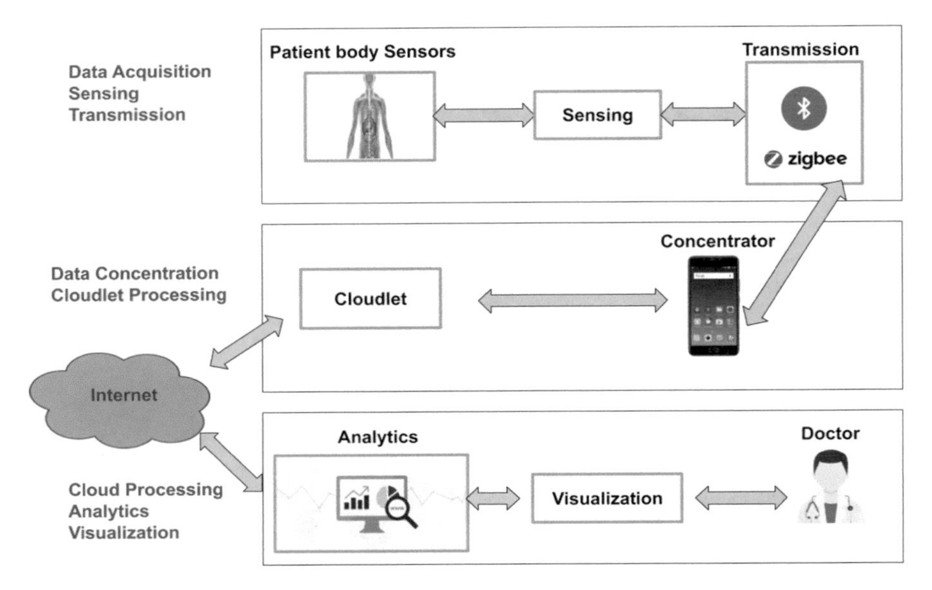

Fig. 6.5 Components of IoT-cloud based remote patient monitoring system [12]

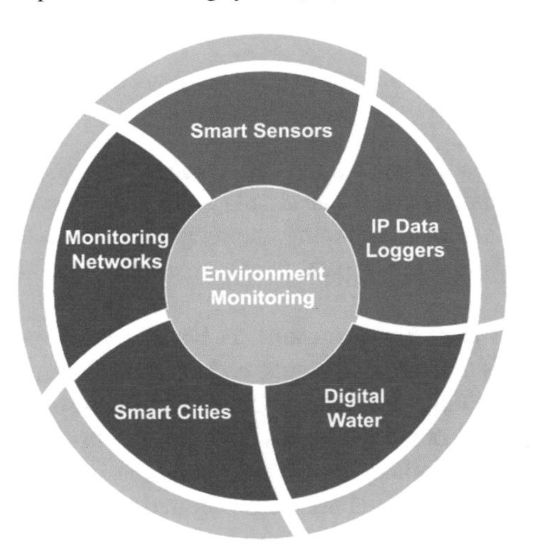

Fig. 6.6 IoT-based environmental monitoring applications [14]

transport services and guarantee people's safety on the roads. History information maintained by RFID can be utilized with the vehicles' part to evade its forging. Tags attached with passenger luggage to quickly identify during transport to save time. Monitoring the level of fuel, pressure in the tires, speed level, brake condition, etc. could be a great help in vehicles to protect from sudden breakdown and road accidents [14]. Figure 6.7 depicts the tasks performed by the IoT-based transport system to facilitate the users.

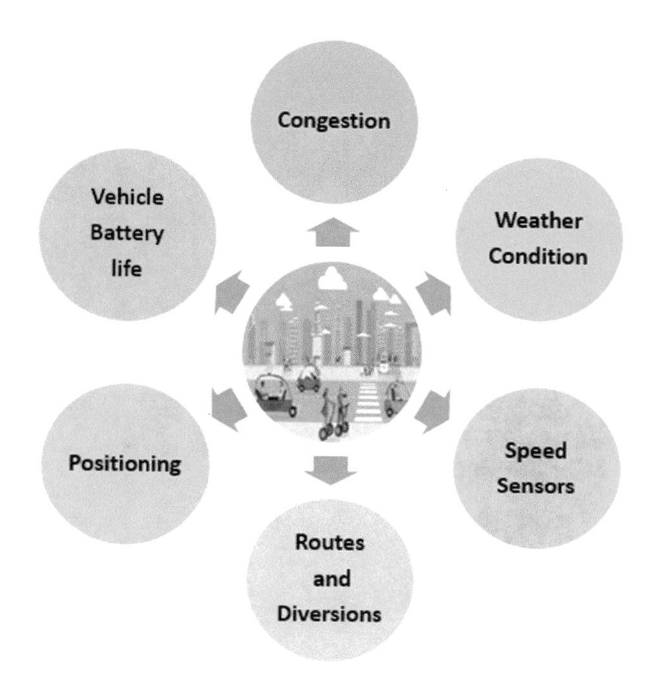

Fig. 6.7 IoT-based smart transport system [16]

Smart Parking The critical issue in metropolitan and congested cities is to get spaces for parking the vehicles. The IoT technology helps design and develop a smart parking system that provides information on available parking spaces and helps find the nearest availability. Furthermore, we can use computer vision techniques to detect a vehicle's number plate to monitor the vehicles in the parking area, which enhances security and help users find their car when they forget about the place, where they parked the vehicle [17]. Figure 6.8 shows the diagram of IoT enabled smart parking system.

Warehouse Inventory Management IoT for warehouse inventory management became more than a promising concept; companies started applying sensors, RFID tags, D2D communication, and other connectivity to manage daily work. RFID tags can store a significantly large volume of data compared to the Barcodes. It is worth using IoT in warehouse inventory management because it improves transparency, analytical maintenance, real-time product tracking, and employee productivity. Figure 6.9 shows the IoT-based warehouse management system proposed in [19].

Smart Energy Management Energy management has become crucial with the increasing demand for energy supply. Also, the increase in the cost of viable energy is making industries to enhance their energy usage [20]. The development of the IoT has transformed energy management systems for proper energy consumption. Smart energy management systems use IoT sensors to collect and analyze the data

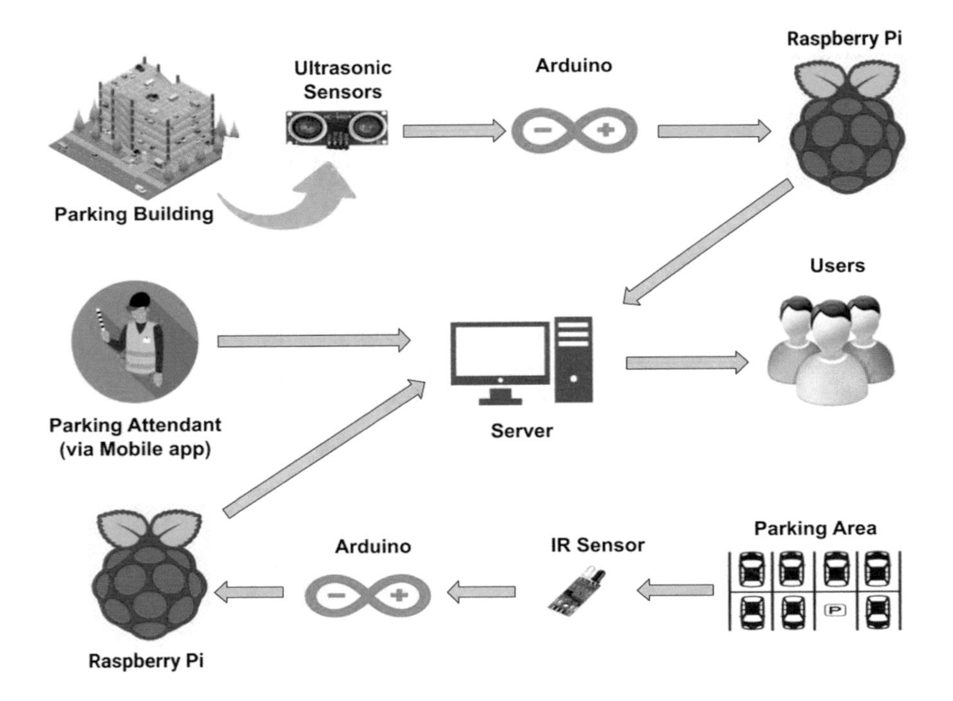

Fig. 6.8 IoT enabled smart parking system [18]

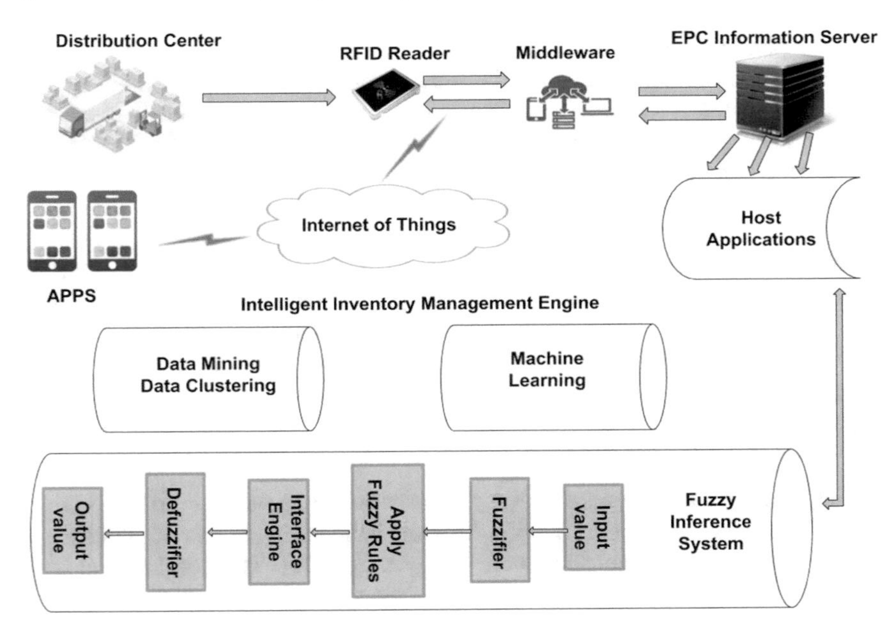

Fig. 6.9 IoT-based warehouse management system [19]

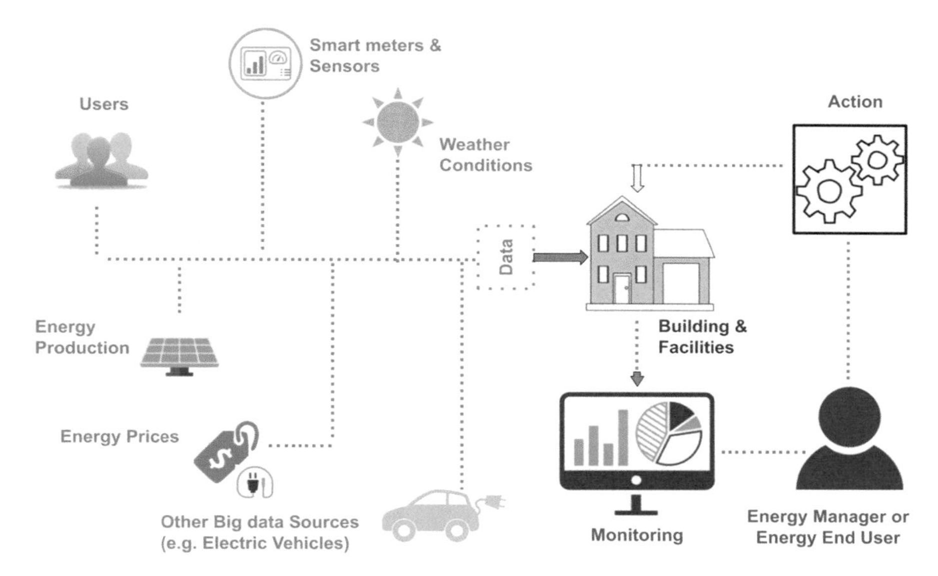

Fig. 6.10 Smart energy management system [21]

for making the necessary decisions to save energy. Figure 6.10 shows the functional diagram of the smart energy management system.

1.3 *Evolution of Wireless Technologies*

The different IoT-based applications need suitable communication technology due to various constraints and requirements. The advancements in wireless technologies have allowed many industries to replace costly and often unreliable, wired communication with wireless communication. The wireless mobile communication system has passed many evolution phases over the past few decades since the inception of the 1G mobile communication network. Because of the massive demand for many networks globally, mobile communication standards improved speedily to support several users. Cellular wireless technologies have accomplished first to fifth generations of technology rising and development in the past few decades, as shown in Fig. 6.11.

Today's market of mobile communications is developing at swift growth. This rapid development is due to the increasing number of devices and users, and it is spanning various generations from 1G to 4G mobile communication technologies. The next generation of mobile internet connectivity, 5G, is going to launch worldwide by 2020 formally. There are different wireless and mobile communication technologies present, such as Universal Mobile Telecommunication

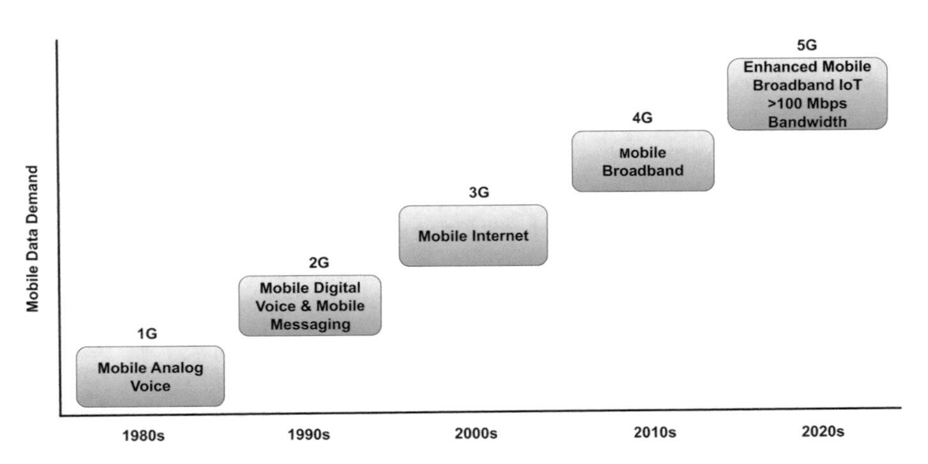

Fig. 6.11 Evolution of mobile technologies [23]

Systems (UMTS), CDMA2000, LTE, WiFi, WiMAX, WSN, or PAN such as Bluetooth and ZigBee. All wireless and mobile networks follow IP standards, i.e., all data and signals are communicated through network layer based IP protocol [22]. Table 6.1 shows a comparison between various generations of mobile communication technologies.

First Generation (1G) 1G has been introduced in the 1980s and comprises an analog system, usually known as cellular phones. It presents communication technologies such as mobile telephone system (MTS), Advanced MTS, and Push to Talk (PTT). It uses analog transmission techniques for transmitting voice signals. For voice calls, it uses Frequency Division Multiple Access (FDMA) technique, and voice calls are modulated at frequency approximately 150MHz. It has issues such as low capacity, unreliable hand-off, reduced voice quality, and lack of security makes voice calls vulnerable to undesirable snooping by third parties [24].

Second Generation (2G) It has presented a new digital technology known as GSM for wireless transmission. It has created the path for the development of new wireless communication standard. This standard was able to support the data rate, which is appropriate for SMS and email services. Next, two generations 2G and 2.5G use packet switching and circuit switching techniques and provide the data rate of up to 144 kbps, i.e., CDMA, GPRS, and EDGE.

Third Generation (3G) 3G wireless communication technology merges several 2G wireless communications systems into a single global communication system and provides the transmission rate up to 2 Mbps. It has started along with the UMTS enabling the video calling facility, first time in mobile devices; therefore, smartphones became popular worldwide. Specific applications to handle email, video calling, social media, and health related services in smartphones were developed. Additional services like global roaming and better voice quality made

Table 6.1 Comparison of various generations

Generation	Deployment	Bandwidth	Technology	Core network	Services
1G	1984	2Kbps	Analog	PSTN	Mobile telephony
2G	1999	14–64 Kbps	Digital	PSTN	Digital voice and short messaging
3G	2002	2 Mbps	Broadbandwith/CDMA/IP technology	Packet network	Unified high quality audio, and video data
4G	2010	200 Mbps	Unified IP and uninterrupted combination of different networks	Internet	Dynamic information access and variable devices
5G	2020	1Gbps and higher	4G+WWWW	Internet	Dynamic information access and variable devices with AI capabilities

3G as an amazing communication technology. The major drawback for 3G is that more power is needed compared to 2G and 3G voice and data plans are more costly than 2G [25].

Fourth Generation (4G) It is an upgraded version of 3G mobile networks with high data rate capability to manage advanced multimedia services. LTE and advanced LTE wireless communication technology used in 4G systems made it possible to transmit voice and data simultaneously, which considerably improve data rate. All services, including voice services, can be communicated over IP packets. 4G provides a downloading speed of 100 Mbps, providing similar features like 3G and additional facilities like multimedia newspapers, watching TV channels with better quality and much faster data transmission speed than earlier generations.

Fifth Generation (5G) It highly supports world wide Wireless web (WWWW) with high capacity and speed. In this technology, IPv6 technology is used, and speed is higher than 1 Gbps. 5G will be using advanced communication technologies to provide high-speed internet and better multimedia data service to the users. The current LTE system will change into modified 5G networks in the coming years. To get a higher data rate, 5G will be using millimeters waves and an unlicensed spectrum for data communication. A multifaceted modulation technique has been developed to provide a high data rate for IoT. 5G will hopefully come with the solutions to all those issues which are experienced in 4G.

2 Background and Current Research in 5G-IoT

Emerging use cases and new business needs in the upcoming IoT systems require new assessment parameters, such as connectivity, reliability, coverage, security, low latency, throughput, etc., for many IoT devices. The growing LTE and 5G will offer better connectivity and interfaces for fulfilling the demands of the emerging applications [26]. A wide variety of research has been done on 5G and IoT systems, from academics to industry sectors, to propose a possibility for the applications, theory and also, the deployment of 5G in IoT [27].

2.1 Communication Technologies and Their Limitations

Various communication technologies are an essential part of any wireless network. The wireless networks consist of limited energy devices that need low power communication technologies. The new communication technologies provide uninterrupted connectivity, which is the prime need for major IoT applications. IoT systems use communication technologies with low energy utilization, poor bandwidth, low processing power, smooth communication with devices because of its feature, computing for everyone, anywhere, in any network and any type of service [28].

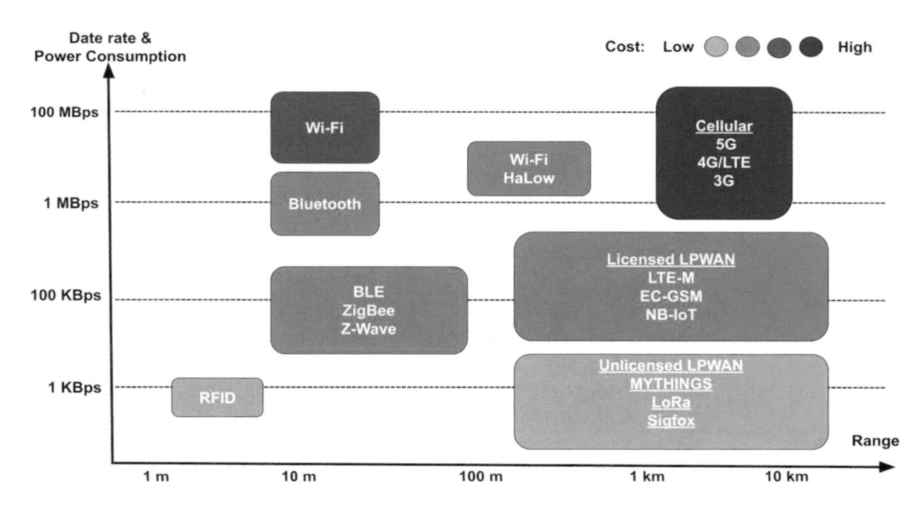

Fig. 6.12 Types of IoT wireless communication technologies [29]

To select the best wireless technology for IoT applications requires a precise examination of power consumption, bandwidth, Quality of service, security, and network management. Figure 6.12 shows the types of wireless communication technologies and comparisons concerning data rate and range. Table 6.2 summarized the advantages and Limitations of various Communication technologies.

Low Power Wide Area Networks (LPWANs) It can connect with many types of IoT sensors, enabling various applications such as smart metering, remote monitoring, worker's safety, and building control. LPWANs can transfer only small size blocks of data at a low rate, so they are not suitable for the time-critical and high bandwidth application requirement.

Cellular Technology Cellular technologies provide consistent broadband communication to enable voice calls and multimedia applications. 5G is likely to provide real-time surveillance for public safety, on time delivery of medical services, and several automation applications in industries soon. However, they enforce very high operative costs and energy consumption. Cellular technologies are not feasible for IoT applications, where battery-operated sensors are used. They are suitable in particular applications such as advanced driver assistance systems, traffic routing, and tracking services [29].

Zigbee It is a low power, short range, wireless technology, typically installed in a mesh topology to improve coverage transmitting data over various sensor nodes. It provides higher data rates, but much less energy efficiency due to mess structure. Due to the short range (< 100m), it is well suited for medium-range IoT applications such as home automation.

Table 6.2 Limitations of communication technologies

Technology	Advantages	Limitations
LPWAN	Long range, Ultra low power, Low bandwidth	Slow bit rate, Not suitable for time-critical applications
Cellular Technology	Poor battery life	High operative costs and energy consumption
Zigbee	Smaller memory, Scalable, Low power than Bluetooth	Less power-efficient, Suited for only medium-range IoT applications
RFID	Reusable, Easy data transmission, Non-line-of-sight communication	High cost, Low data rate, Security and privacy
WiFi	High throughput, long communication range	Low location accuracy, High power consumption
Bluetooth	Low cost, Adequate data rate	Limited QoS, scalability and security, High power consumption
Bluetooth low energy	Low cost, Energy efficient	Limited scalability and QoS, limited design flexibility

RFID RFID is a technology not famous and commonly used compared to other communication technologies but is very useful and used for IoT applications. RFID tags are either embedded or tagged onto devices used in IoT applications. Like other available wireless technologies, RFID has a few security and privacy issues in current IoT applications. The primary concern with IoT applications is to maintain user privacy. To accept RFID enabled IoT technology, strong technical, security, and privacy solutions should be required.

WiFi WiFi plays a pioneer role to give high throughput for both industry and home automation IoT applications. The fundamental limitations of WiFi are coverage, scalability, and energy utilization. It is generally not suitable for vast networks of sensors equipped with batteries, mainly in industrial and smart building IoT applications. Better hardware solutions will make the difference in all WiFi and IoT convergence, in the new phase of intelligent technology.

Bluetooth and BLE Bluetooth and Bluetooth Low Energy(BLE) are used for totally different purposes and requirements. Bluetooth can handle a lot of data, but it quickly drains battery life. BLE is not suitable for applications in which enormous amount of data transfer is exchanged. BLE's energy efficiency has made it a preferred and one of the most scalable options for IoT applications.

The upcoming 5G cellular mobile communications launch will have a significant impact on nearly every aspect of our lives. At the same time, working on the 5G with its unique features makes it a better solution for IoT applications needs to maintain quality of service and effectiveness.

2.2 *Role of 5G in IoT*

IoT is growing rapidly, improving the interconnectivity between devices and people via Internet. Wireless networks have boosted their features to sustain the growth of technologies. Numerous mobile cellular networks generations have been passed until the introduction of 5G cellular networks. The earlier 2G networks cover 90% of the earth's community, 3G includes 65% of the earth's community, 4G and 4G LTE came into existence in 2012 is the fastest and most reliable network yet [27]. The 5G networks' development promises to provide very high data transfer rates, low latency [30], and high integrity. Also, the feature of 5G cellular network provides the heterogeneous connection between devices in the IoT applications.

The 5G will improve IoT security massively, significantly increase speed, enhance cellular operation with improved bandwidth, and solve various network issues faced in the earlier generations of mobile communication networks. Due to the enormous data exchanged among many connected devices in IoT applications, there is a need to provide improved capability, a high data rate, and better connectivity. Therefore, 5G is considered as a key enabler for IoT [31, 32]. The 5G will allow IoT devices to communicate in a smart environment using connected, intelligent sensors. It also increases the range and scale of IoT applications coverage area by providing better communication and capability service. In recent years, the 5G-IoT has got massive attention both in industry and research and first 5G networks are expected by 2020, and the complete 5G network will be available after 2025 [33].

2.3 *Requirements for 5G-IoT*

IoT is growing very fast due to the invention of new technologies. The number of devices is set to increase from million to billion. Many factors contribute to this growth; one of the most important will be the growth of 5G networks. The forthcoming launching of 5G cellular mobile communications is excellent news for the IoT applications because 5G networks will go a long way towards enhancing the performance and reliability of these connected devices. Researchers have made significant contributions in the past few years to address many challenges for the 5G IoT and are expected to see extensive participation in the coming years.

In IoT, various devices and applications demand more advanced networks that can provide high throughput and low latency, energy-efficient techniques, high scalability to manage many devices, and pervasive connectivity for users. This section summaries these different requirements for 5G enabled IoT.

- **High data rate:** One of the requirements of 5G enabled IoT is to provide much higher data rates. This is required to meet the high expectations and demands of customers. The emerging IoT applications have changed our life very much, for example, multimedia streaming, AR/VR, which require a high data rate at around 25 Mbps to deliver adequate service to users [34].

- **Low Latency:** To deliver faster connections and better capacity, an essential requirement of 5G is the quick response time, which is referred to as latency. In 5G-IoT applications, like AR/VR, video games, healthcare, sports, etc., require low latency of approximately 1ms and real-time communications with the user. The delay in communication will cause a reduction in user experience; therefore, latency is a critical factor in 5G-IoT. Low latency communications also open the door for remote medical care and treatment.
- **High Scalability:** Network scalability becomes a vital factor in supporting the growing number of mobile devices connected to the wireless network and communicating with each other in 5G-IoT. High scalability is essential for better performance of emerging IoT applications.
- **Connectivity:** In addition to increasing the speed, 5G-IoT networks will operate more consistently, creating more steady connections. A reliable and stable network connection is significant for any of the IoT applications, but particularly for connected devices like security cameras, locks, and other monitoring systems that depend on real-time event updates.
- **Security:** Security is fundamental and cannot be ignored without risking severe consequences on society. Future of IoT in a 5G era, IoT manufacturers, need to safeguard that their products meet the requirements of the security standards. The 5G-IoT demands enhanced security mechanisms as the future IoT applications differs from the conventional security mechanisms to protect user's confidentiality. The 5G-IoT needs enhanced security strategies on the complete network [20].
- **Mobility:** In the future, there will be a tremendous increase in the density of connected things in 5G-IoT, including high-mobility. An efficient algorithm is needed to manage and control the highly mobile IoT devices. The 5G enabled IoT applications should be capable of providing device to device communication with high movement to support current and future applications.

3 5G Enabled IoT

Many research efforts motivated on advanced research in numerous features of 5G and IoT systems from the perspectives of academics and industry. 5G has the potential to create a more IoT-friendly environment, with massive improvements in the 4G network's current capabilities. The 5G can deliver high data rates, low latency, and enhanced connectivity for M2M communication compared to 4G-LTE to most challenging IoT applications. IoT systems are trying to improve the living standard of life, which includes the connection between devices in applications, such as smart home automation, smart buildings, smart environments, and smart cities [26, 35].

3.1 5G-IoT Architecture

It is expected that 5G-IoT will offer time-critical applications, on-demand, and societal involvements, and for that, 5G enabled IoT system advocates for the architecture that can perform end-to-end synchronization, intelligent, and automatic operation at each level [34]. Shortly, the number of innovative IoT applications and data generated by them will be extremely high. Hence, the amount of data and customer demands will be highly increased. This situation will make it challenging to need a faster, intelligent, reliable, and scalable architecture for IoT applications. Currently available IoT architecture are not responsive and reliable for future IoT applications.

The technologies used to design and develop the possible architectures will not offer seamless connectivity between connected devices due to high service demand and data transfer rates. Therefore, we need a structure based on advanced technologies to handle upcoming challenges in IoT applications [36]. 5G enabled IoT architecture is presented in [26]. The 5G-IoT architectures are likely to deliver logically autonomous networks according to the need of IoT applications. It uses the cloud services [37, 38] to rebuild RAN called Cloud Radio Access Network (CRAN) to support substantial connections of various standards and perform the on-demand implementation of RAN operations needed by 5G.

The conceptual IoT architecture based on next-generation 5G network, named 5G-IoT, is depicted in Fig. 6.13. The proposed architecture leverages the features such as modular, efficient, agile, scalable, and responsive on-demand services. The

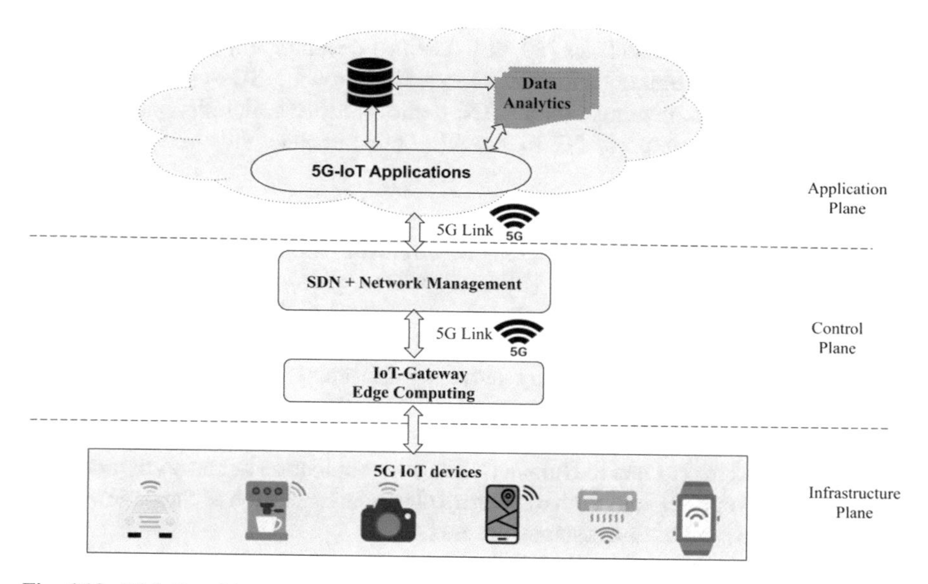

Fig. 6.13 5G-IoT architecture

architecture is comprised of three planes, viz., Infrastructure Plane, Control Plane, and Application Plane with two-ways data transfer. This architecture will be able to achieve the needs of future IoT applications and helping IoT specialists to design and propose practical as well as scalable IoT systems. Infrastructure plane consists of several 5G enabled IoT devices that generate the data. The control plane plays the role of central network control capability and edge computing to minimize the latency. The SDN controller and IoT devices communicate through 5G links with each other. An Application plane comprises several 5G based IoT applications which are powered by SDN policies. The next section discusses about the various technologies involved in 5G enable IoT architecture.

3.2 *Key Enabling Technologies for 5G Enabled IoT*

The 5G-IoT contains many key enabling communication techniques for IoT applications. As 5G technology offers more excellent connectivity, many more IoT applications are likely to come shortly. This section describes the key enabling technologies for 5G-IoT applications.

Cloud Computing The IoT consists of millions of devices that produce a large amount of data. Due to the enormous amounts of data being gathered by IoT technology, the management of large amounts of data at the data center has become critical as the network is responsible for processing and transferring data rapidly. Cloud computing is key to managing it while gaining business insights. Cloud-based solutions offer adequate data storage and data security to manufacturers, as well as for analytics capabilities [39, 40]. The fast development of IoT depends on consistent wireless communication networks. The growing 5G system overcomes these issues by implementing Cloud RAN. It allows better scalability, performance, and flexibility that supports 5G to provide better connectivity between a huge number of IoT devices [41].

Heterogeneous Network (HetNet) Next-generation 5G network runs IoT applications that demand high data rates. This requirement can be fulfilled by deploying a dense network of small cells (Macro-, micro-, pico-, and femto cells). The concurrent operation of these small cells is termed as heterogeneous networks (HetNets). HetNet enables 5G-IoT to transfer information on-demand. The HetNets optimize the spectral efficiency by recycling the spectrum strongly to make it energy efficient. It allows us to manage network traffic and node density to make it appropriate to fulfill the necessities of service-oriented 5G enabled IoT. It was originally aimed for Human to Human (H2H) communication but later characterized to use for M2M applications with distinctive characteristics such as the large volume of devices, QoS, and high-frequency access [26].

Device to Device (D2D) Communication In small range communication (< 200 m) among two devices, suggested a novel method for data communication, which help the 5G based IoT with low energy utilization, load balancing, and improved QoS to end-users. In IoT, many applications need an extended battery and widespread connectivity. Existing communication technologies such as Zigbee, WiFi, low energy BLE, and cellular communication cannot fulfill these requirements. In recent years, many advanced communication technologies have been evolved to support the IoT requirements, such as LPWAN, NB–IoT, LoRa, SigFox, and LTE–M [42]. The D2D communication is going to provide new unique openings for 5G–IoT applications, but this poses many challenges and needs to update the cellular communication architecture.

Software Defined Networking (SDN) Fast growth in communication technology and the emerging usage of Internet of Things (IoT) devices generate a massive amount of data, the 5G mobile network support this growth. Traditional approaches to deploying cellular communication technologies are generally hardware dependent. The hardware-based implementation limits the scalability of the network. SDN is proposed as an essential technology to solve this problem and provide high flexibility, scalability, low cost, and low energy consumption. Because of the growing demands of the huge number of connected devices in IoT, a unified software-based model like SDN is required to maintain QoS [43, 44]. Centralized control for real-time flow management, Network Function Virtualization (NVF), energy management Security and Privacy and the complete network view for better reliability provided by SDN are required to address the issues related to IoT. This will also streamline problems such as node failure, resource allocation, and deployment of new nodes. Therefore, it is considered as a crucial enabler for future 5G enabled IoT applications.

3.3 Impact

The IoT applications are growing very fast, and connected devices are also increasing on a large scale. Many factors are contributing to this growth; one of the most vital will be the growth of 5G networks. The coming 5G is good news for the IoT applications, and that will increase the performance in terms of high bandwidth, high reliability, and low latency and the ability to support higher density of connected devices. As the scope of IoT systems proliferating, their reach will also increase in terms of the impact on industry and customers' lives. 5G will improve IoT networks by managing radio frequency to meet the requirements of both narrowband IoT applications, and those require on-demand high bandwidth.

In the near future, 5G will significantly impact IoT services such as smart home, industrial, healthcare, and autonomous vehicles. Table 6.3 shows that 5G is used in many emerging IoT applications in comparison with its predecessor wireless communication technologies. 5G is going to be the game changer in future IoT applications. The remote medical diagnosis and remote surgery are cases where

Table 6.3 IoT applications with connectivity technologies

IoT applications	LPWAN	BLE	Zigbee	WiFi	LTE	RFID	Sigfox	5G
Smart city	✓							✓
Smart home		✓	✓	✓				✓
Autonomous vehicles		✓		✓	✓			✓
Smart agriculture	✓				✓		✓	
Smart grid	✓		✓				✓	✓
Health monitoring				✓	✓			✓
Smart warehouse		✓	✓	✓	✓	✓		✓
Smart traffic lights	✓	✓	✓					

5G will help improve the AR/VR (Augmented Reality/Virtual Reality) experience, making powerful tools in the medical industry, especially at a time of emergency like the widespread COVID-19 pandemic.

4 5G-IoT Use Cases

5G mobile communication technology is mainly developed for several IoT applications with demanding needs, which include high data rate, low latency, high network reliability, and immense connection density. The integration of 5G and IoT is going to change the scenario of businesses by making real-time decisions. Some recent and future applications that 5G-IoT could promote are shown in Fig. 6.14.

Industrial IoT Communication 5G mobile communications can make better connectivity between various IoT devices. 5G network with machine learning [45] and big data analytics techniques will improve the control of many operations that are being performed manually. Several commercial use cases are Wireless Industrial Control, Smart Factories, and Wearable Technology communication [6, 46].

Vehicle Telematics Autonomous vehicles will revolutionize the transportation industry and will become dominant area to accept 5G technologies in IoT applications. We have already lots of applications, such as vehicle identification and location tracking. These applications are usually installed over existing cellular technologies that provide excellent performance to manage their communication needs. The introduction of 5G offers more opportunity to collect more real-time data about the vehicle's condition and performance and enable the delivery of more intelligent services. In the coming years, IoT analytics believe that vehicle telematics applications will support connectivity with 4G and 5G [47].

Smart Grid Automation The electricity demand is increasing very fast in day-to-day activities. Smart grids and virtual power plants can play a vital role in managing demand. In the energy and utility sector, 5G technology is suitable for real-time management and these solutions would optimize procedures and maintenance by

Fig. 6.14 5G-IoT use cases

rapidly detecting and responding to problems in the grid. 5G is largely adopted in the smart grid because of its higher deployment flexibility and lower cost than wired technology.

Cooperative Intelligent Mobility 5G is expected to have an impact mainly on the architecture and transportation system to share real-time information traffic and road condition. In the last few years, 5G has created an interest in intelligent transport systems (ITS) for private and public transportation sectors in which vehicles cooperatively share the information using wireless communication to improve the efficiency and comfort. In ITS, real-time traffic data gathering from the road infrastructure, and the vehicles on the road are analyzed and timely alert the drivers about the dangerous road conditions, traffic congestions, and helps to avoid the road accidents.

Video Surveillance One more application that is expected to perform well with the 5G environment is video surveillance. Many governments and private sectors worldwide are investing in surveillance systems and that to grow faster shortly to provide better security. Today, many video surveillance systems still depend on wired connectivity. Still, wireless communications such as WiFi and cellular are getting attention due to an easier, quicker, and low-cost deployment. In contrast, LTE networks give an adequate level of performance to access live remotely and recorded videos. The implementation of 5G will allow the performance enhancement needed for more advanced video data analytics in real-time and the placement of a large number of cameras [46]. Table 6.3 shows the IoT applications with connectivity technologies to select the appropriate connectivity technology for the deployment.

5 Challenges and Future Research Trends

The 5G provides many unique features that could fulfill the needs of future IoT applications. Recent developments confirm that the upcoming 5G enabled IoT applications should be able to support the extensive connectivity between the devices and provide reliable QoS. It also makes way for new challenges in designing a modern architecture for 5G enabled IoT. This section describes the possible challenges and upcoming research trends in 5G-IoT.

- Designing architecture for 5G enabled IoT is a big challenge. Researchers with various benefits have proposed many architectures, but these still enforce many problems, such as interoperability and heterogeneity, scalability, network management, security assurance, and privacy concern.
- Providing efficient service is still a big challenge for 5G networking. Though bringing the better scalability, there are still so many technical issues that need to be solved in SDN. To support the underlying network with high flexibility, the scalable SDN, and the separation of data and control plane are challenges for network scalability.
- Though D2D communication offers high throughput for 5G enabled IoT. Two significant challenges are energy and spectral efficiency. It needs the right spectral resources and interference controlling methods to maximize the performance and reliable service among devices.
- Deploying an IoT application is challenging because of its large scale, resource constraints, and diverse environmental conditions. The capability and efficiency of the 5G-IoT use to collect and distribute data is one more challenge.
- In various 5G-IoT applications, security and privacy are critical. In 5G-IoT, new security features are required at both devices and at network levels to emphasize applications that include smart home [5], automation, smart city, etc.
- Many new applications will be offered in the 5G enabled IoT. The regulation of 5G-IoT will make the deployment of applications more accessible. Because of the varied nature of devices used in 5G-IoT applications, there is a lack of reliability and standardization. Research efforts are needed to solve these issues.

6 Case Study: 5G-Enabled IoT for Industrial Automation

5G embodies an important change in communication network architectures. It has a potential to speed up imminent revenue generation with its innovative services provided using 5G-enabled devices, which includes smartphones, laptops, and IoT.

5G will not only offer a reliable communication services but it also provides an exceptionally secure network for industrial IoT by assimilating security into the basic network architecture. The complete industry automation and its business operations become a reality in present time. Enormous growth in technology and their introduction into industry has given rise to a development of a new methodology to production, which is known as Industry 4.0. In Industry 4.0, 5G-IoT is going to be offer promising innovative solutions to present industrial structure [20]. 5G with its characteristics, such as high security, low latency, high data rates, huge number of devices and modified networks, empowers industries to take benefit of the IoT for monitoring and automation of various processes. Key enabling technologies for 5G enabled IoT such as Cloud Computing, Software Defined Networking, Heterogeneous Network, AI and machine learning greatly help to automate the industry processes. Figure 6.15 shows the general industrial automation pyramid divided into several layers each with various set of networks, demands, and properties

In tradition, communication between different devices and systems in industrial automation is provided by wired medium and it has given successful results.

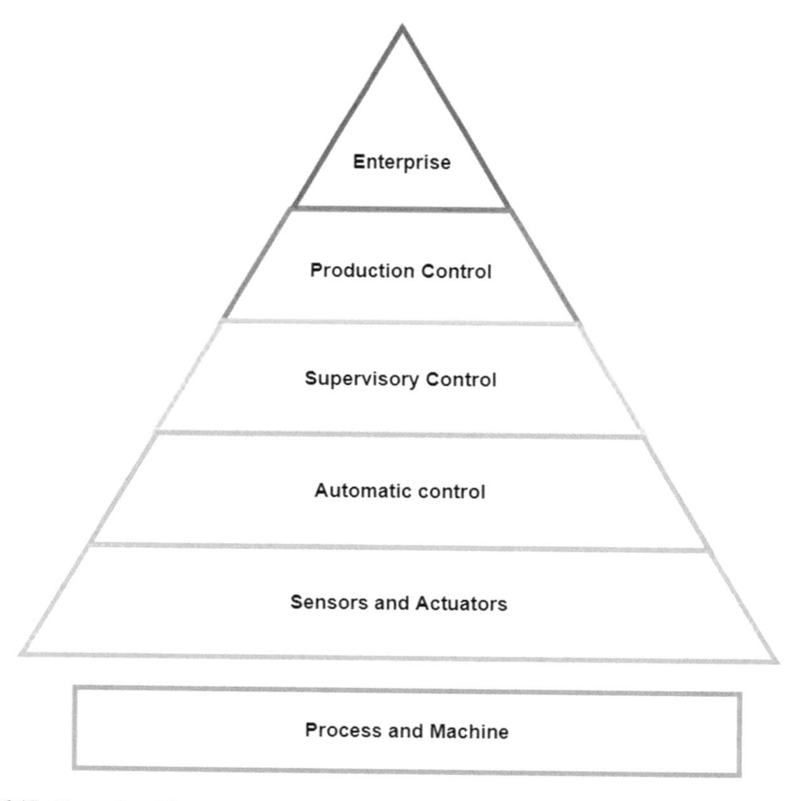

Fig. 6.15 General architecture model for industrial automation [45]

Though, with development of new technologies and upcoming needs on movement, flexibility, cost, and proper maintenance, wireless communication has become a feasible option but still numerous challenges need to be solved before it can be accepted as a complete system [45]. The key research challenges for 5G enabled IoT in industrial automation include:

- Interconnection between 5G and traditional networks (3G, 4G, LTE, and SDN)
- Verifying that QoS metrics latency, jitter, loss, throughput meet the needs in real life
- Meeting security and privacy requirements
- Large scale placements, network and endpoint processes and maintenance
- Deployments of private campus area networks
- Providing effective device to device communication

7 Conclusion

The IoT incorporates the growing 5G techniques to support the IoT applications better. This chapter surveyed an exhaustive 5G wireless communication technology that has grown up as crucial enablers for a ubiquitous deployment of the IoT applications. Also, the evolution of cellular communication technologies in the context of IoT application demands are reviewed. We have presented the state-of-the-art technologies and ongoing research work IoT over 5G. Various 5G based IoT architectures are also discussed, with particular focus on the key enhancements in physical and network layer of 5G networks compared to other cellular wireless technologies. The key enabling technologies in 5G enabled IoT are also discussed and analyzed to improve applications' performance. Finally, research challenges that need to be focus and find the solution in 5G-IoT to support future applications. A case study on importance of 5G enabled IoT in industrial automation also presented along with challenges which can be a motivation for future research trends.

References

1. K. Kaur, A survey on internet of things – architecture, applications, and future trends, in *2018 First International Conference on Secure Cyber Computing and Communication (ICSCCC)* (2018), pp. 581–583
2. S. Nalbandian, A survey on internet of things: Applications and challenges, in *2015 International Congress on Technology, Communication and Knowledge (ICTCK)* (2015), pp. 165–169
3. M. Junaid, M.A. Shah, I.A. Satti, A survey of internet of things, enabling technologies and protocols, in *2017 23rd International Conference on Automation and Computing (ICAC)* (2017), pp. 1–5
4. J. Bhatia, B. Shah, Review on various security threats & solutions and network coding based security approach for VANET. Int. J. Adv. Eng. Technol. **6**(1), 361 (2013)

5. S. Tanwar, P. Patel, K. Patel, S. Tyagi, N. Kumar, M.S. Obaidat, An advanced internet of thing based security alert system for smart home, in 2017 International Conference on Computer, Information and Telecommunication Systems (CITS) (2017), pp. 25–29
6. A. Kumari, S. Tanwar, S. Tyagi, N. Kumar, M. Maasberg, K.-K.R. Choo, Multimedia big data computing and internet of things applications: A taxonomy and process model. J. Netw. Comput. Appl. **124**, 169–195 (2018)
7. A. Floris, L. Atzori, Managing the quality of experience in the multimedia internet of things: a layered-based approach. Sensors **16**(12), 2057 (2016)
8. N. Pahal, A. Mallik, S. Chaudhury, An ontology-based context-aware IoT framework for smart surveillance, in *SCA '18* (2018)
9. J. Vora, S. Tanwar, S. Tyagi, N. Kumar, J.J.P.C. Rodrigues, Home-based exercise system for patients using IoT enabled smart speaker, in *2017 IEEE 19th International Conference on e-Health Networking, Applications and Services (Healthcom)* (IEEE, Piscataway, 2017), pp. 1–6
10. S. Agrawal, M.L. Das, Internet of things – a paradigm shift of future internet applications, in *2011 Nirma University International Conference on Engineering* (2011), pp. 1–7
11. Smart home IoT applications. https://iot5.net/iot-applications/smart-home-iot-applications/. Accessed 18 Aug 2020
12. M. Hassanalieragh, A. Page, T. Soyata, G. Sharma, M. Aktas, G. Mateos, B. Kantarci, S. Andreescu, Health monitoring and management using internet-of-things (IoT) sensing with cloud-based processing: Opportunities and challenges, in *2015 IEEE International Conference on Services Computing* (2015), pp. 285–292
13. M. Wijanarko, A. Djajadi, Ambient environmental quality monitoring using IoT sensor network. Interworking Indonesia J. **134**(1), 1–51 (2019)
14. S.A.A. Elmustafa E.Y. Mujtaba, Internet of things in smart environment: Concept, applications, challenges, and future directions. World Scientific News **8**(1), 41–47 (2016)
15. J. Bhatia, Y. Modi, S. Tanwar, M. Bhavsar, Software defined vehicular networks: A comprehensive review. Int. J. Commun. Syst. **32**, e4005 (2019)
16. Smart transport: Looking to a safer and greener future. https://www.euroscientist.com/smart-transport/. Accessed 12 June 2020
17. R. Lookmuang, K. Nambut, S. Usanavasin, Smart parking using IoT technology, in *2018 5th International Conference on Business and Industrial Research (ICBIR)* (2018), pp. 1–6
18. IoT-based parking system. https://create.arduino.cc/projecthub/102513/iot-based-parking-system-b6a947. Accessed 18 Aug 2020
19. K. Ng, W. Ho, C. Lee, Y. Lv, K. Choy, Design and application of internet of things-based warehouse management system for smart logistics. Int. J. Prod. Res. **56**(8), 2753–2768 (2018)
20. I. Mistry, S. Tanwar, S. Tyagi, N. Kumar, Blockchain for 5G-enabled IoT for industrial automation: A systematic review, solutions, and challenges. Mech. Syst. Signal Process. **135**, 106382 (2020)
21. Intelligent IoT energy management solution. https://vuelogix.blogspot.com/2020/01/iot-energy-management-solution.html. Accessed 18 Aug 2020
22. A. Tudzarov, T. Janevski, Functional architecture for 5G mobile networks. Int. J. Adv. Sci. Technol. **32**, 65 (2011)
23. Goldman sachs 5G: How 100x faster wireless can shape the future. https://knowen-production.s3.amazonaws.com/uploads/attachment/file/3450/Global_%2BTechnology_%2B5G_%2BHow%2B100x%2Bfaster%2Bwireless%2Bcan%2Bshape%2Bthe%2Bfuture.pdf, April (2016). Accessed 21 Aug 2020
24. M.K. Patwa, M.N. Dumbre, M.M. Patwa, 5G wireless technologies-still 4G auction not over, but time to start talking 5G. Int. J. Sci. Eng. Technol. Res. (IJSETR) **2**, 435–440 (2013)
25. K.R. Santhi, V.K. Srivastava, G. SenthilKumaran, A. Butare, Goals of true broad band's wireless next wave (4g-5g), in *2003 IEEE 58th Vehicular Technology Conference. VTC 2003-Fall (IEEE Cat. No.03CH37484)*, vol. 4 (2003), pp. 2317–2321
26. S. Zhao, S. Li, L. Da Xu, 5G internet of things: A survey. J. Ind. Inf. Integr. **10**, 1–9 (2018)

27. M. Simsek, A. Aijaz, M. Dohler, J. Sachs, G. Fettweis, 5G-enabled tactile internet. IEEE J. Sel. Areas Commun. **34**(3), 460–473 (2016)
28. K. Mehmood, A. Baksh, *Communication Technology that Suits IoT – A Critical Review*, vol. 366 (Springer International Publishing, Berlin, 2013), pp. 14–25
29. 6 leading types of IoT wireless tech and their best use cases. https://behrtech.com/blog/6-leading-types-of-iot-wireless-tech-and-their-best-use-cases/. Accessed 12 June 2020
30. P. Patel, J. Bhatia, M. Bhavsar, Effective resource utilization using software defined networking based load balancer for minimizing request service time. Far East J. Electr. Commun. **3**, 755–766 (2016)
31. M.M. Alsulami, N. Akkari, The role of 5G wireless networks in the internet-of- things (IoT), in *2018 1st International Conference on Computer Applications Information Security (ICCAIS)* (2018), pp. 1–8
32. M.R. Palattella, M. Dohler, A. Grieco, G. Rizzo, J. Torsner, T. Engel, L. Ladid, Internet of things in the 5G era: Enablers, architecture, and business models. IEEE J. Sel. Areas Commun. **34**(3), 510–527 (2016)
33. GSA. The road to 5G: Drivers, applications, requirements and technical development. https://www.huawei.com/minisite/5g/img/GSA_the_Road_to_5G.pdf (2015). Accessed 12 June 2020
34. S.-C. Lin, M. Chandrasekaran, I.F. Akyildiz, S. Nie, 5G roadmap: 10 key enabling technologies. Comput. Netw. **106**, 17–48 (2016)
35. J. Vora, S. Kaneriya, S. Tanwar, S. Tyagi, N. Kumar, M.S. Obaidat, TILAA: Tactile internet-based ambient assistant living in fog environment. Future Gen. Comput. Syst. **98**, 635–649 (2019)
36. H. Rahimi, A. Zibaeenejad, A.A. Safavi, A novel IoT architecture based on 5G-IoT and next generation technologies, in *2018 IEEE 9th Annual Information Technology, Electronics and Mobile Communication Conference (IEMCON)* (2018), pp. 81–88
37. J. Bhatia, M. Kumhar, Perspective study on load balancing paradigms in cloud computing. IJCSC **6**(1), 112–120 (2015)
38. J.B. Bhatia, A dynamic model for load balancing in cloud infrastructure. Nirma Univ. J. Eng. Technol. **4**(1), 15 (2015)
39. J. Bhatia, R. Mehta, M. Bhavsar, Variants of software defined network (SDN) based load balancing in cloud computing: a quick review, in *International Conference on Future Internet Technologies and Trends* (Springer, Cham, 2017), pp. 164–173
40. J. Bhatia, R. Govani, M. Bhavsar, Software defined networking: from theory to practice, in *2018 Fifth International Conference on Parallel, Distributed and Grid Computing (PDGC)* (IEEE, Piscataway, 2018), pp. 789–794
41. C. Tsai, M. Moh, Load balancing in 5G cloud radio access networks supporting IoT communications for smart communities, in *2017 IEEE International Symposium on Signal Processing and Information Technology (ISSPIT)* (2017), pp. 259–264
42. M.J. Kaur, R. Ahmed, A.K. Malviya, V.P. Mishra, Comprehensive survey of key technologies enabling 5G-IoT, in *2nd International Conference on Advanced Computing and Software Engineering (ICACSE) 2019* (2019), p. 5
43. K. Shafique, B.A. Khawaja, F. Sabir, S. Qazi, M. Mustaqim, Internet of things (IoT) for next-generation smart systems: A review of current challenges, future trends and prospects for emerging 5G-IoT scenarios. IEEE Access **8**, 23022–23040 (2020)
44. A. Ashishdeep, J. Bhatia, K. Varma, Software process models for mobile application development: A review. Comput. Sci. Electron. J. **7**(1), 150–153 (2016)
45. T. Lennvall, M. Gidlund, J. Åkerberg, Challenges when bringing IoT into industrial automation, in *2017 IEEE AFRICON* (2017), pp. 905–910
46. The leading 5G IoT use cases. https://www.hiotron.com/5g-iot-use-cases//. Accessed 25 June 2020
47. The leading 5G IoT use cases. https://iot-analytics.com/the-leading-5g-iot-use-cases-2019/. Accessed 25 June 2020

Chapter 7
Advance Cloud Data Analytics for 5G Enabled IoT

Vivek Kumar Prasad, Sudeep Tanwar, and Madhuri D. Bhavsar

1 Introduction to 5G, IoT, and Cloud Computing

As an organizational strategy related to the Information and Communication Technologies (ICT) era, the present demands are high-speed, super-low latency, and massive capacity. These demands are fulfilled by the version of the 5G ultra-wide band network and are treated as one of the important features for technological growth [1]. To serve the connectivity needs for the present ICT for faster network with higher capacity, the 5G technology is essential for the IoT. It expands the frequencies from which the digital cellular technologies will transmit the information. This, in turn, will upsurge the overall bandwidth of the cellular linkages and allow other devices to connect into the network. There are many areas where the combinations of the 5G and IoT [2] can prove to be a boon and are discussed here in the chapter. One of the important usages for the same is Virtual Reality and Augmented Reality (AR/VR) [3]. The ultralow latency will enhance the experiences of the AR/VR operations and for the industries, this will be significantly benefited from the ultra-fast information broadcast to the time-sensitive nature of the corresponding outcomes.

Following are the motivation and contribution for the chapter.

Motivation

1. IoT adopters have been unable to handle a massive load of data gathered from multiple end devices due to the rise in connected devices.

V. K. Prasad (✉) · S. Tanwar · M. D. Bhavsar
Department of Computer Science and Engineering, Institute of Technology, Nirma University, Ahmedabad, Gujarat, India
e-mail: vivek.prasad@nirmauni.ac.in; sudeep.tanwar@nirmauni.ac.in; madhuri.bhavsar@nirmauni.ac.in

S. Tanwar (ed.), *Blockchain for 5G-Enabled IoT*,
https://doi.org/10.1007/978-3-030-67490-8_7

2. Businesses now need an active data accumulation system for their edge devices, and remotely positioned IoT end nodes to allow the storing of crucial information.
3. In order to establish high visibility, the IoT and 5G solution should link and facilitate communication between objects, people, and processes, and cloud computing plays a very important role in this collaboration.

Contribution

1. The cloud ecosystem enables easy and rapid integration of IoT devices via 5G and is also useful for the smooth data management from which they are collected.
2. Cloud storage thus allows businesses to explore new potential applications and roll on new operations and processes without thinking about data storage concerns. The case studies discussed here reflects the same.
3. The best data storage method that resonates well with the Internet of Things is cloud storage in the present situation. Because of fragmented data access, the integration of cloud storage for far off IoT devices actually presents speed-related challenges. Even, this problem can be solved once and for all with the rise of the 5G wireless networks.

1.1 Possibilities in 5G and IoT

Before understanding the 5G, let us have a look at the inception of the 1G (the first generation mobile network). The 1G was all about the voice only, then comes the 2G. Through 2G both texting and voice can be done. The 3G was used for data, texting, and voice. In addition to the functions of 3G, the 4G was faster than the 3G. The 5G is even faster than the 4G. This 5G is so fast enough to download an HD movie (full-length) in seconds [4]. The 5G is not only faster, but this also supports high-speed connectivity, fast downloads, lower latency, and supports the ubiquitous coverage for transportation and smart vehicles. The examples for the same can be connected buses, cars, and trucks for the smooth flow of the traffic [5] and to avoid the four ways crash at the intersection pathways.

1.2 Integration of Cloud Computing, 5G, and IoT

CC [6] allows the industries to manage and store information over the services of the Cloud, like Software as a Service (SaaS), Platform as a Service (PaaS), and Infrastructure as a Service (IaaS). This also provides many features such as scalability, security, resource management, etc. [7]. CC also sanctions storage and data transfer through the internet with an uninterrupted transfer of pieces of information between the applications, IoT devices, and Clouds itself. When the capabilities of CC combine with the IoT, these stacks bring added services for the business applications and consumers. The Cloud ecosystem processes and analyzes

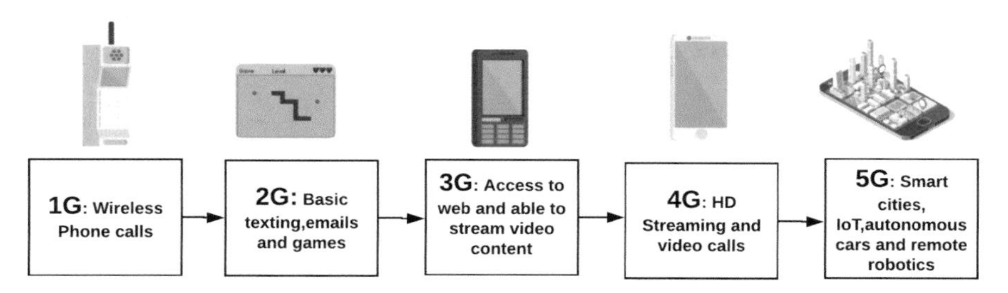

Fig. 7.1 Evolution of wireless mobile network

the data quickly to make an intelligent decision as soon as possible. The IoT developers can store the data remotely and access effortlessly through CC. As CC is booming nowadays and has fuelled the progress of IoT ecosystems. The evolving IoT architecture can be lavish/expensive without the Cloud infrastructure. Hence migrating IoT data into the cloud lets our cost reduce and has other advantages too and is discussed in the other sections.

It is very much likely that the readers are familiar with the concepts and usages of 3G and 4G networks, and the maximum of us are using also. The 5G is the next evolution generation of the wireless mobile network. Figure 7.1 shows the generation wise evolution of the wireless network, where firstly the 1G has come into the picture and was used for wireless phone calls, secondly the 2G has evolved and was used for basic texting, doing emails, and playing games. Then comes the 3G, which was used for accessing the web and have the flexibility for the streaming video contents. After 3G, 4G comes into existence, where this was used for HD streaming and video calls [8]. Finally, it reached 5G and has better functionality and has more functionality like smart city management, IoT data analysis, autonomous cars management, and remote robotics management. [Note every higher step is the having new services as well as the basic feature which was available in the lower generations] As we have discussed before the 5G will impose improvements to the IoT [9]. Any device we interact, will be linked to the internet and is able to transmit and communicate information. These pieces of information are collected and used efficiently and will support real-time decision-making. Let us understand the same using the two scenarios.

Scenario 1: The alarm clock is connected to the coffee maker, and the alarm clock passes the information to the coffee machine as when to start making the coffee. [This is a prospective opening for all of us to learn how IoT is used nowadays for making our job easy and is almost limitless.]

Scenario 2: Smart cities management through IoT devices, where all the things will be connected, transmitting, communicating, and real-time analysis of the data will be happening, such as transport and traffic management, pollution and waste management, to improve and monitor energy utility.

Hence, in scenario 1 and scenario 2, communication plays an important role in quick decision-making. So we recommend, using 5G for bringing real possibilities for the advancements in the technological fields.

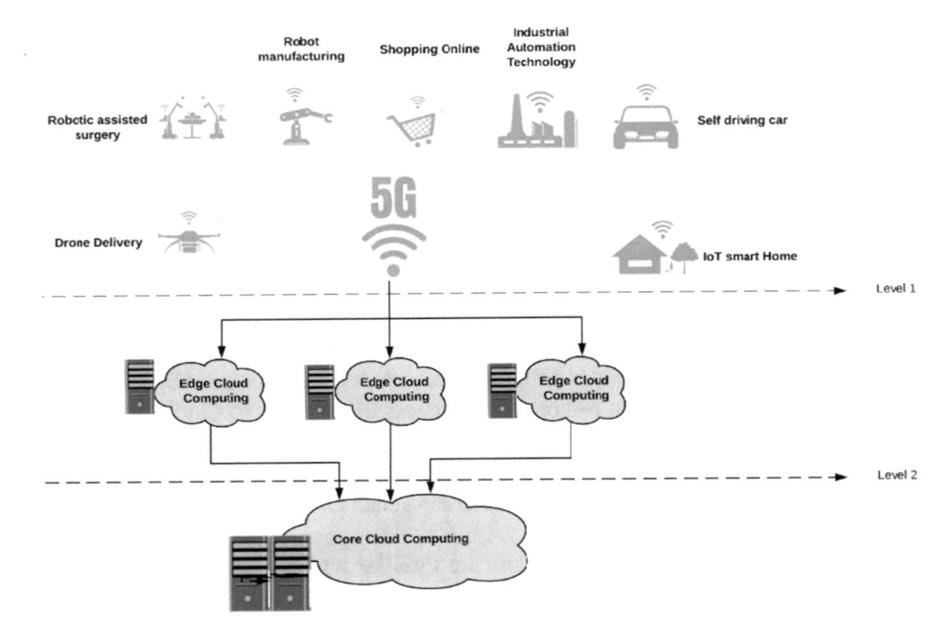

Fig. 7.2 Layered architecture for cloud computing with 5G based IoT

5G and Cloud Computing [10]: The requirement of virtual technology such as CC; demanded a need for the infrastructure and process a huge amount of data that is produced by the IoT and 5G technologies. These data which are stored at CC, are used for making decisions in real-time. The 5G will also transform the existing state of the CC and brings edge computing to the forefront to speed up the transmission of the decisions (like having the physical infrastructure processing data closer to the source/edge of the networks for faster information dispensation).

Figure 7.2 indicates the three layers architecture of the systems where the methodologies such as 5G, IoT, edge, and core cloud computing are used in collaborative ways. The IoT or ubiquitous devices will communicate among themselves with the 5G, in a very fast and convenient way. As many of the IoT devices connect to each other, the coordination among the IoT devices will also increase. The huge data that will be generated via these IoT devices are processed in the edge or core cloud computing [11]. The edge computing is placed near the user's devices, and this will result in decreased latency and response time. The edge computing also offloads the Core cloud computing jobs into itself. The edge computing is placed into the second layer and in the last layer, we have cloud computing. Cloud Computing comes into existence when we have a large number of daily data, and is collected with poor knowledge. To make use of these full data, this information must be mined into the acquaintance/knowledge. The generated knowledge and information is used to make a complex decision-making process in a timely way [12].

These methodologies excited us and motivated us to write a chapter based on the connectivity of the CC to the 5G based IoT ecosystems. The chapter focuses on the existing scenarios and identification of the new applications of the 5G based IoT. The engineers, researchers, and scientists face tremendous challenges in designing the systems for IoT, which can be integrated into the 5G and CC.

2 State of the Art

A recent emerging concept for the future industry is Industry 4.0, which involves many primary enabling technologies such as IoT applications and CC [13]. In this section, the evolutionary stages of Industry 4.0, IoT networking technologies, its benefits, and the leading associated technology supporting like CC of the industrial revolution are briefly described. Later the issues are discussed. Depending on the requirements and problems of the respective era, the industrial revolution with the progression of time has several stages and is shown in Fig. 7.3.

As shown in Fig. 7.3 the industry 1.0, 2.0, and 3.0 were mainly based on Water and steam, electrical energy and assembly lines and automation, electronics and computers. Presently industry 4.0 is booming [14]. Industry 4.0 is a revolution in all fields, including finance, academic, research, industrial, and manufacturing organizations. The industrial revolution had an immense effect on the development

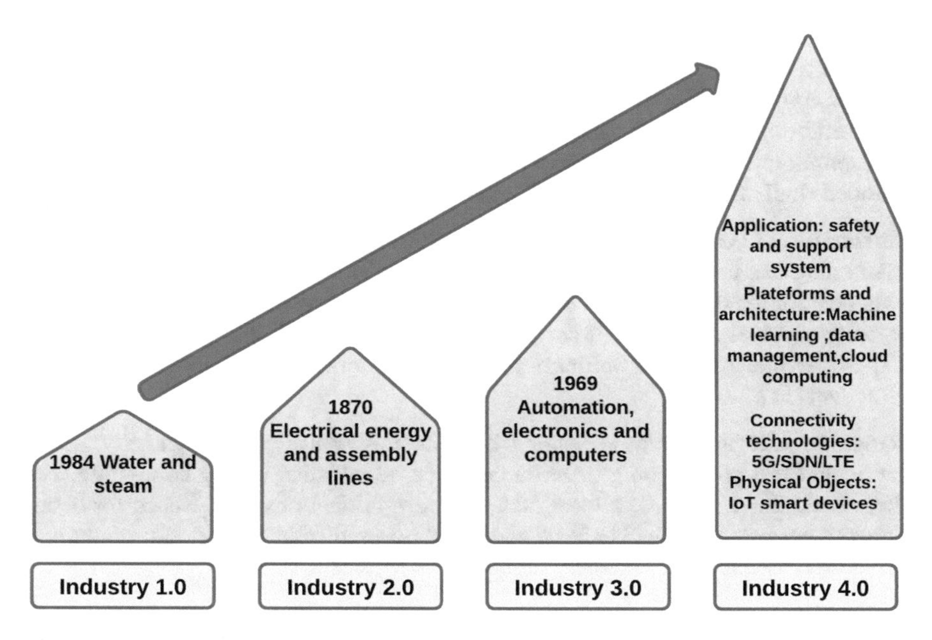

Fig. 7.3 Cloud-based IOT systems and evolution of the Industry 4.0

processes of many leading industries. The introduction of Industry 4.0 demands innovative changes with respect to automation, recognition, machine learning, communications via networks, digital production, manufacturing process, management of production control, decision-making, sensing, and analysis [15]. Researchers are focusing on providing the evolved form of these technologies in terms of versatility and rapid computational process to transform the industrial method. All this automation is very important in the industry and for a country's economic growth [16].

2.1 Open Research Issues for Cloud-Based IoT Systems

The sustainability and performance of the whole system can be improved by well-designed network architecture. The road towards industry 4.0 is on its way, and there are no other criteria, rules, and certifications to observe them so far. As CC is the key enabling technology for IoT architecture to evolve. Its integration into the current architecture of IoT still faces some problems. To achieve an efficient CC network, complex software applications and solutions are required [17]. In terms of main performance metrics, which are bandwidth utilization, energy consumption, low latency, maximum throughput, and resource management, the CC network will be analyzed [18, 19]. Critical technical networking and communication challenges in the context of CC for IoT applications are listed in this section. They all have the potential for future research.

Energy Efficiency (E_{eff}) According to the specifications, many intelligent devices implementing an IoT application can consume a large amount of energy on a different order of magnitude. Ensuring network QoS with minimal energy consumption in an optimized way for smart IoT applications, fog nodes, and CC is an open challenge for any future IoT application to come [20].

Throughput (T_c) Bandwidth, data rate, and throughput or network depend on how much data has been used and where information is stored in a CC network. This data positioning has an impact on cost, bandwidth, delay, and network coverage on fog/edge devices or cloud data servers. One of the main technological challenges of CC-IoT architecture is the optimum positioning of data on cloud servers or fog/edge cloudlets [21].

Resource Allocation (R_{cc}) Features such as multi-tenancy, heterogeneity, scalability, and fast provisioning of resources must be included in Fog computing. For these conditions, resource allocation is the most critical obstacle. Hence for better network performance that has to be addressed by architecture like cloud computing. It has consequences for all other parameters of QoS [22].

Latency (L_t) Real-time networking is a prerequisite for IoT applications. The application of IoT and CC are time-sensitive and require streaming rather than batch processing in real-time. Customizable data center positioning, allocation of resources, system architectures, node energy usage, and node storage capacity have an impact on latency. The number of transmission, propagation, encoding, and queuing delays is the latency for a network. Devices with good channel conditions were among the features that could be used to minimize the latency of the transmission and increase the efficiency of the **transmission link** [23].

Security (S_c) Security is a very important thread, as, over the open network, the cloud is open to the entire planet [24].

An uplink is a path from data communications equipment(transmission link: **T-Link**) to the network center with regard to computer networks. This is often referred to as an upstream link and vice versa for the downlink. Radio waves are used by the cell phone for communication. It does by converting the audio and information into digital signals and sends this as radio waves. It links first via a radio access network (RAN) in order for your mobile phone to connect to a network or the internet [25]. To link you to the cloud, wireless access networks use radio transceivers. C-RAN (Cloud Radio Access Network) is a centralized, cloud computing-based radio access network (RAN) infrastructure that allows for large-scale deployment, support for interactive radio technology, and virtualization capacities in real-time. The "integration of computation with physical phenomena" is a Cyber-Physical System (CPS) that uses sensors and actuators to connect computational systems to the physical realm. This is a new vision of "computing as a physical act" inspired by the current CPS, where the real world is controlled by sensors that relay sensing data to cyberspace, where cyber services and applications use information in real-time to influence the physical environment. The CPS can help individuals understand the physical setting and take optimal action in a wide range of applications: healthcare, transport, energy consumption, production, agriculture, emergency management, critical infrastructure. The features of cloud computing with cyber-physical systems improve the advantages of data processing and management.

A Cyber-Physical Cloud Computing (CPCC) architectural framework is therefore defined as "a system environment that can dynamically create, change and provide cyber-physical systems composed of a collection of the sensor, control, processing and data services based on cloud computing [26]." The digitalization shift of the industry in Industry 4.0 needs research and development in all fields (intelligent cities, D2D connectivity, transportation, healthcare management system, etc.). These are all that belong to 5G based IoT domains, which have the same essential communication and networking issues/challenges. While CC has countless applications in several research areas, few notable applications have been mentioned in Table 7.1 in the respective IoT domain to gain an idea of the diverse usage of CC [27].The primary aim here is to achieve maximum benefits with efficient, optimized QoS based measurements using CC for the industrial/agricultural revolution [28]. In recent years, several writers/researchers have suggested solutions to communication and connectivity problems in order to leverage the advantages of using CC in all IoT domains, and the same is mentioned in Table 7.1.

Table 7.1 Existing methods and their comparison

Ref. No	Area of the IoT application	R_{cc}	L_t	E_{eff}	T_c	S_c	T-link	Architecture
[29]	Smart IoT devices	✓	✓	✓	–	–	Downlink	CC
[30]	VANETS	–	✓	–	–	–	Downlink	CC
[31]	Healthcare	✓	✓	–	–	–	Down and uplink	CC
[31]	Virtualized Passive optical network(VPON)/5G	–	✓	–	–	–	Down and uplink	Radio Access N/w + CC
[32]	5G Network	✓	✓	–	✓	–	Downlink	CC
[33]	IoT and security	✓	–	–	✓		Routing	CC
[34]	Heterogeneous IoT application	–	✓	–	–	–	Downlink	CC
[35]	Heterogeneous IoT application	✓	✓	–	–	–	Downlink	CC
[36]	Security and microgrid	–	–	–	–	–	Downlink	CC
[37]	Microgrid	–	✓	✓	–	–	Downlink	CC
[38]	Smart city	✓	–	–	–	–	Downlink	CC
[39]	Smart City	–	✓	✓	–	–	Routing	Cyber-Physical System+CC
[40]	Secure and time saving multimedia	–	✓	–	–	–	Downlink	CC
[41]	Mobility+VANETs	–	✓	–	–	–	Downlink	CC
[42]	Mobility + Smart City	–	✓	–	–	–	–	Radio Access N/w and CC
[43]	Big data analytics and security	–	–	–	–	–	–	CC
[44]	Smart Home	–	–	–	–	–	–	CC
[45]	Smart city video applications	✓	–	–	–	–	Routing	CC

3 Case Studies for Cloud Computing 5G Based IoT Systems

3.1 5G Patrol Robots Made by G Gosunch Robot Co., Ltd, China for Controlling COVID-19 in Public Areas

For COVID 19 pandemic prevention, its inspection and to assist police officers of China, the scientists have come up with patrol robots using the technologies such as 5G, edge computing, and cloud computing [46]. As the primary symptom of the disease is high temperature. Hence the robots are armed with the high-resolution cameras (a total of five cameras) and infrared red thermometer capable of measuring the temperature of the ten peoples simultaneously within a radius of 5–6 m. Doing these things physically/manually will lead to potential health threats and public safety will be exposed. If the conditions such as high temperature and

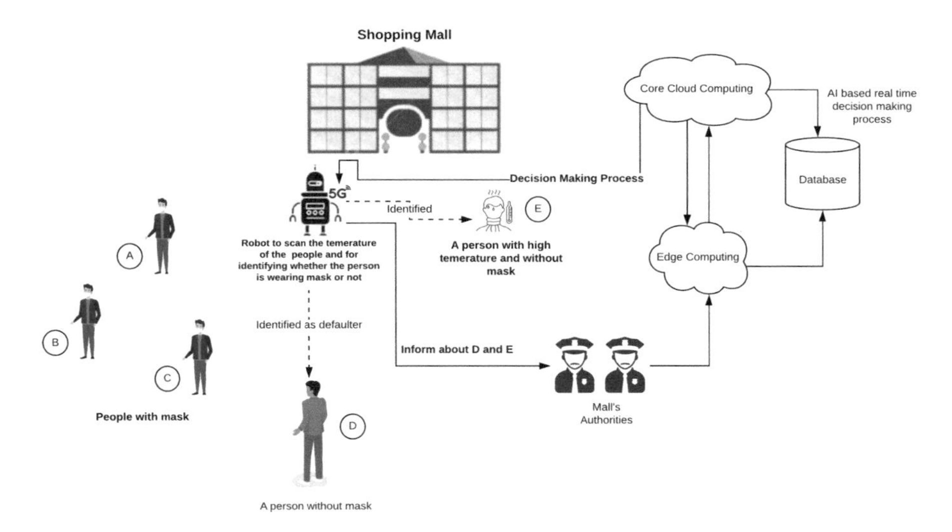

Fig. 7.4 Shopping mall security automation through 5G based robots

a person without the mask are identified, then the robot will send the alert to the concerned authorities (the central control system for making a real-time decision based on the responses). These robots can be handled through remotes and are self-driven and reduce the need for manpower and patrolling responsibilities too. The same is depicted in Fig. 7.4. There are various demands of such 5G based robots in every corner of the world and have already been spotted in areas such as shopping malls and airports. The technologies that these 5G robots will use are Big data, Cloud Computing, edge computing, AI, and IoT. By combining these aforementioned technologies, the robots can sense, do autonomous motion control, dynamic decision-making, and can do behavioral penetration and sensing. To take action at run time the 5G patrol robots are power-driven by industrial edge computing (high performance-MIC-770 and is equipped with eighth Generation Intel Core i and aimed at IoT applications using GPU iModule i.e. MIC-75G20). The industrial-grade edge computer combined with MIC-75G20 GPU iModule delivers a high-performance system for AI inference and training. Its cast aluminum heatsink and ruggedized chassis protect against shock and vibration, the passive thermal solution ensures silent setup.

3.2 5G Based IoT and Smart Cities Healthcare Systems

The impenetrable population of the cities stretches the healthcare services during any pandemic and can speed up the blow-out of the disease, such as COVID-19. How will the technology become a shield for this? The answer is to go with 5G and IoT technology, which will reduce the workforce and avoid the necessity for the human being while reducing its physical presence. These technologies will improve the monitoring capabilities of the health of city populations, and the services will have a reduced response time in emergencies. The following case study will focus on working of this.

The impact of 5G based IoT system on medical healthcare will be of great use [47]. Through 5G and the IoT devices, the healthcare benefactors can gather health-related data from the patients immediately and monitor their conditions for preventive and personalized care. This can be used for training junior doctors for doing surgery through AR/VR with high bandwidth 5G and low latency. The replacement of the wired connections in the operating theatres could be replace with the powerful, secure wireless connections of 5G and low latency. The other usages are to enrich the remote real-time diagnostics by delivering good quality video using 5G. The 5G based network with high bandwidth connectivity and low latency can be used in robotics technologies for dispensing medicines, support for diagnostics, and even for performing surgery. The usages of cloud connectivity will support for doing data analytics across the records of the medical information. This information will contain data about the CT scans and can help with treatment prioritization. Figure 7.5 reflects about the information on various health parameters of the patients is detected using the wireless IoT devices and the information's are collected and passed to the cloud with the help of 5G technology [48]. The doctors and other services will be connected to the cloud for doing their operations based upon the intelligent decision made through data analysis.

3.3 5G Smart Industries Development and Production: Managing Its Utility and Energy

The production environments in the present era are in extreme volatility state; because of the product lifecycle and shorter business, the manufacturing industries around the world are under extreme pressure. As the components increasingly become more complex and varied to produce; which in turn shows that the margins are being squeezed. As the conditions of workforces are being matured, this becomes costlier to maintain [49]. The 5G based IoT and cloud computing can solve these problems to an extent and are being discussed here. The edge computing will enable conveniences to scale the number of deployed platforms, connected devices, and real-time data analytics. The last mile fiber can be replaced with the 5G network, which will result in a cost-effective and flexible approach. For the reduction or avoiding failures, the 5G based microrobots could accomplish the sensor's inspection and share data in real-time for reducing cost and fault prevention. For security enhancement cybersecurity facilities that safeguard the large information and to be implemented in collaboration with the partners of securities. The greater control can be achieved through running private networks to

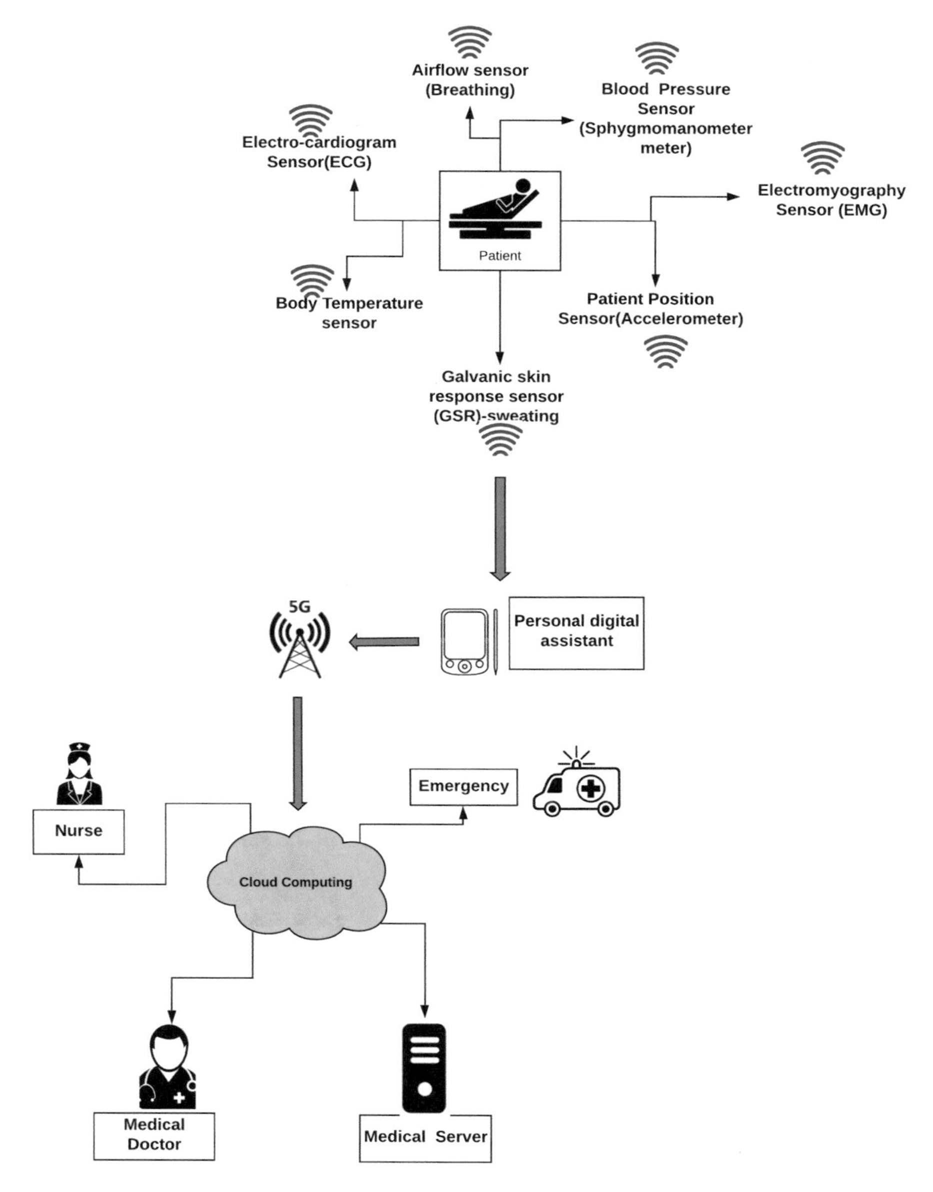

Fig. 7.5 5G-IoT and Healthcare systems

provide better utility and control. The demanding Service level agreements (SLAs) must be incorporated for the adaption of the 5G and trusted data services. For the management of cybersecurity and increasing volume of information (the veracity of data), 5G wireless networks are required. The market-oriented computing works in collaboration with the utilities of the operators to provide platforms for energy administration [50].

3.4 Connected Vehicles and 5G Era

As per the data of Ericsson 2019, there will be 500 million connected cars(vehicles) on the road by 2025, and the connected vehicle services will be worth 81 US $ billion by 2030 as per the data of strategy and PWC 2019 [51]. These data indicate that the vehicle market with IoT integration will grow in the future to provide immense revenue opportunities for the vehicular collaborative organization in the wider ecosystem, automotive manufacturers, and for mobile operators. Below is the discussion of the technologies that are useful for the management of vehicle networks. The other aspect related to the connected vehicles system is the impact of the safety and well-being of the citizens. As per the data released by the Bosch, 2017 by 2025, the connected vehicular systems can save 11,000 lives and may lead to 260K fewer mishaps (in terms of accidents) each year. Avoiding 400K tonnes of emissions produced by CO2 and also saving 280 millions hours spend during the driving of the vehicles. For communications of these devices, the 5G based IoT and CC will play an important role. Let us see how this is done by C-V2X (Cellular Vehicle to Everything). Figure 7.6 shows the basic architecture of the connected vehicles and how does this work with the neighboring environment (such as interaction with the cellular networks, pedestrian and traffic management, and information data) using 5G based IoT systems and Cloud Computing.

The C-V2X allows the vehicle networks to scale to its full potential.C-V2X is enhanced 3GPP (third Generation Partnership Project) LTE standard that describes

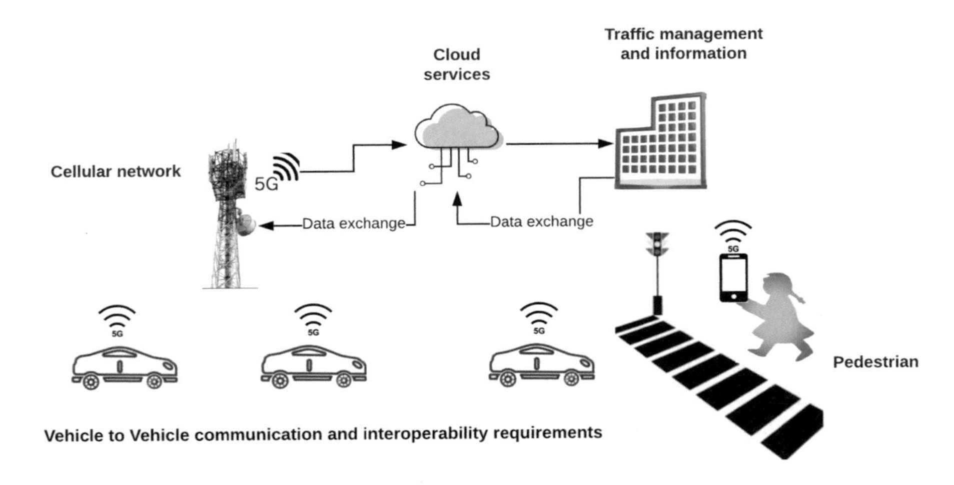

Fig. 7.6 5G-IoT and vehicle network

a set of technologies for supporting the communications between the vehicles, people (mostly pedestrians in our case), and infrastructure. The benefits for the same are some value-added services for the end-users such as safety, enhanced driving management, pollution control, and reduced traffic congestion [52].

Note: The 3GPP has classified the standards of 5G into two releases

- The Release 15: This resembles NR (5G New Radio)-1
- The other one is Release 16: Which resembles Phase-2

In NR (New Radio) phase 1, there are common elements between the NR and LTE and use OFDM (Orthogonal Frequency Division Multiplexing).

The C-V2X is made up of the following communication technologies:

- i.V2V
- ii. V2I
- iii. V2P
- iv. V2N

i. V2V: As shown in Fig. 7.7, the Vehicle to vehicle(V2V) technology permits the vehicle to talk to the other vehicle on the lane for receiving and sending the warning of the collisions, traffic turns, hard emergency braking, etc.

ii. V2I: The Vehicle to infrastructure (V2I) technology as shown in Fig. 7.8 permits the vehicles to communicate with the road infrastructures like toll stations, speed signs, and traffic lights. For example, the drivers can slow down the speed and accelerate through monitoring change times of traffic lights and their status too. This results in a reduction of air pollution, saves fuel, prevents accidents, and improves traffic flow.

iii. V2P: Vehicle to pedestrian (V2P) technology supports to defend vulnerable street operators by a collision warning to the driver and the pedestrian

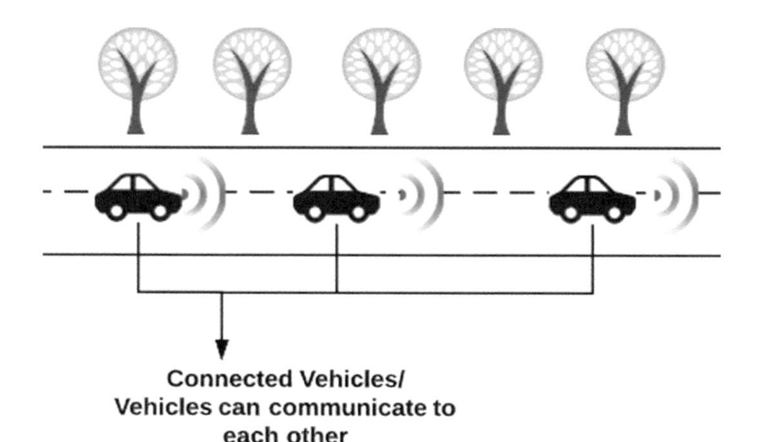

Fig. 7.7 5G-IoT and V2V

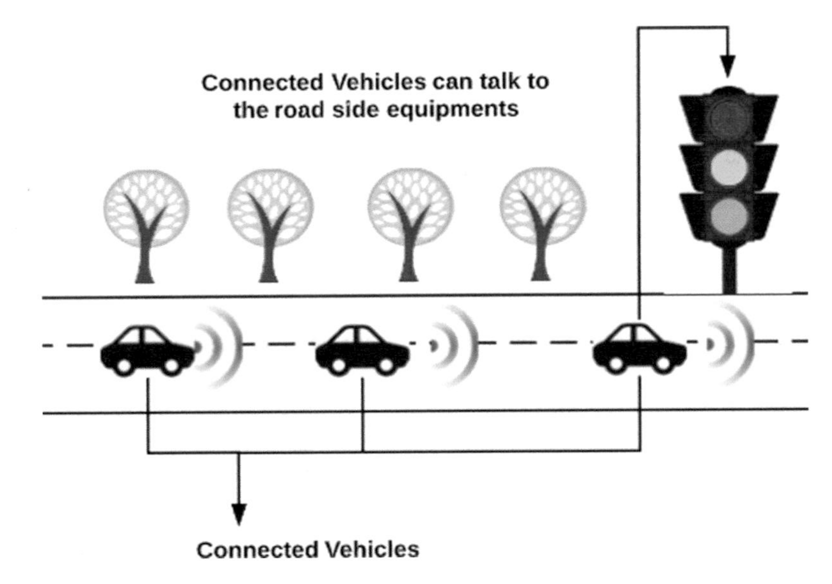

Fig. 7.8 5G-IoT and V2I

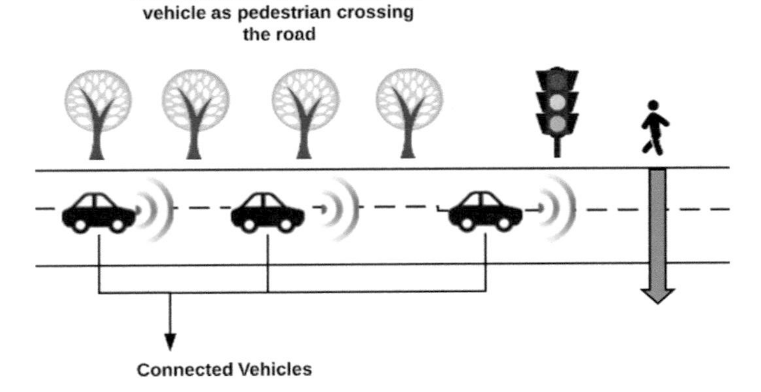

Fig. 7.9 5G-IoT and V2P

or cyclist, thus making the streets safer for the traffic participants and the drivers. This is displayed in the Fig. 7.9.

There are two ways to implement the same:

(a) By making use of the cellular network, where the cyclists or pedestrians are visible to the driver of the corresponding vehicle through an application on their smartphone.

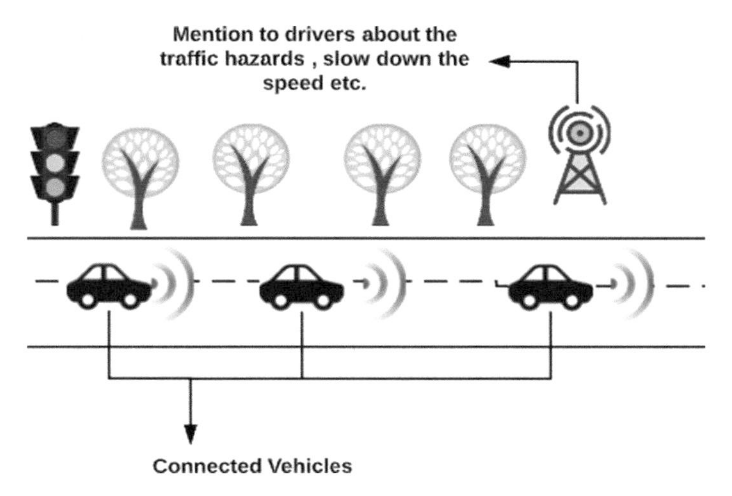

Fig. 7.10 5G-IoT and V2N

(b) Direct communication between vulnerable road users and drivers of the vehicle, in this case, the C-V2X technology is entrenched directly to their smartphones.

iv. V2N: Vehicle to Network technology permits the drivers of the vehicles to communicate with the network, high range communications between vehicles, empowering high reliability, vulnerable road users(pedestrian or a cyclist), roadside infrastructure by the existing cellular network, and is shown in Fig. 7.10. These vehicles link with their surroundings and the existing smart transport solutions through the blind corners and beyond the line of sight. This, as a result, increases safety on the roads and increases traffic efficiency.

As of now we have seen how the Vehicles will communicate with the help of 5G-IoT based System. For the management of the information, Cloud Computing (CC) plays an important role. CC offers services such as SaaS (Software as a Service), PaaS (Platform as a Services), and IaaS (Infrastructure as a Services). These services provide shared resources present over the network. The integrated model of CC in VANETs plays a major role in real-time safety management in vehicular communications and monitoring of the smart traffic systems. The Flow of the system will be as follows:

1. The vehicles moving in a subdivision are linked to each other and exchange significant evidence such as speed, and location, etc.
2. This data is added to the Road Side Units (RSU), which are mounted at consistent gaps of distance on the roads.
3. The information collected by the RSU passes to the central server linked with the CC.

4. The central cloud server segregates and monitors this information so that it can be retrieved/used by the smart traffic monitoring system. The corresponding decisions will be sent to the mobile devices to take appropriate decisions.

4 Research Issues/Challenges in the Field of CC, IoT, and 5G

Till now, this is identified that the future IoT requirements will be fulfilled satisfactorily by using 5G. However, implementing 5G opens new sets of interesting challenges with respect to IoT infrastructure in terms of security issues, architectural issues, and trusted communication between ubiquitous devices issues, etc. and is shown in Fig. 7.11. In the remaining portion of the chapter, we will focus on some of the potential research issues and its future scope. Some of the important technical challenges are mentioned here:

A.

1. The big issue is the **architecture of the 5G based IoT infrastructure**. Even though many architectures have been proposed but still many issues still exist, which includes:
   (a) Network management and scalability [53]: Due to the massive presence of the IoT devices, network scalability is the major issue that needs to be considered.
   (b) Heterogeneity and interoperability [54]: The interconnection among heterogeneous networks is an issue. The enormous number of the various IoT devices is to be coupled with technology to collect, disseminate, and communicate important data with the other applications or smart networks.
   (c) Privacy and security concerns [55]: There must be some mechanism through which we can increase the privacy concern and security and suppress the cyber attacks.

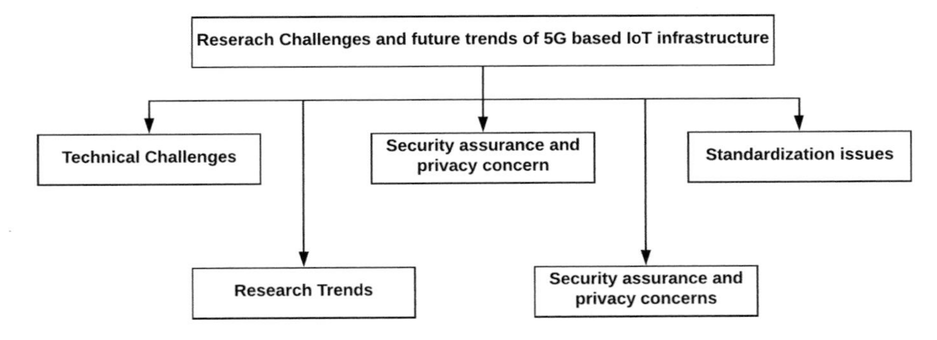

Fig. 7.11 Issues in 5G based IoT infrastructure

2. The 5G based SDN networking is still a challenge [56], such as
   (a) The separation of data plane and control is difficult for most of the SDNs.
   (b) In terms of flexibility of the scalable SD-CN is a challenge for network connectivity expansion.
3. Device to device communication is anticipated to provide better throughput for 5G based IoT. Here the important challenges are spectral efficiencies and energy. To provide high reliability over the communication network in between D2D, we require a good interference management scheme and spectral resource [57].
4. IoT framework deployment: The deployment is difficult for the IoT network because of the heterogeneous environment, limited resources present in the device, and large scalability. Hence the collection and dissemination of the information is a challenging task [58].
5. The other challenges are fixing the dense heterogeneous network deployment in the IoT framework, full-duplex communication at an identical time, and multiple access methods for the 5G and beyond the 5G networks [59].

B. Security Assurance and Privacy Concerns The current situation of the 5G based IoT requires security features to be imposed in the device as well as the network level for complex applications such as smart city, smart traffic management, and others. The designer of the system must consider both local intrusions at the device as well as in the software too [60]. We also need to follow the importance of security w.r.t the following key words: assurance, backward compatibility, storage, mobility, crypto algorithms, authentication, key management, and identity.

C. Issues Related to Standardization Because of the assorted nature of the devices and networks in the 5G based IoT systems, it lacks standardization and consistency for the IoT applications. Again there are still many challenges and hurdles in applying their solutions [61]. These standardizations can be categorized into four parts and are mentioned below:

1. The Killer tasks, which include analysis functions, data collections, and control functions.
2. Business tasks/models, these are expected to satisfy the prerequisite of the consumable market and e-commerce.
3. Connectivity contains protocols and communication networks that connect IoT devices.
4. IoT device platform, design of IoT products, and analytic tools of big data.

With the perspective of cloud computing, the IoT as a Service may be the future scope for the researchers. It not only will support the 5G based IoT but also bridge the gap between the human being and heterogenous environment around.

D. Research Trends for 5G Based IoT The evolution of the 5G is still in the early phase. In addition to this, we have identified some challenges [62] and are mentioned below:

1. The 5G will continue to expand in the future and many devices will be linking to this network. The Named Data Networking (NDN) is required for on-demand feasible network architecture in support of IoT applications for high density. Even NFV (Network function virtualization) is also supportive of the fragmentation of the information generated due to the 5G based IoT.
2. The usage of edge computing will be advantageous for 5G based IoT systems in the following ways: (a) For doing analytics and reducing the communication (b) enhances highly computational tasks such as AR/VR and other information-intensive applications like smart cities and traffic management, etc.
3. The convergence of IoT, data analytics, AI, and 5G will transform the 5G based IoT w.r.t better user's experiences in terms of communications, commerce, digital content, and applications. For example, the AI will enable 5G based IoT in making an intelligent decision on its own for the connected automobiles, connected homes, virtual reality, and wearable devices. These advanced technologies (analytics and machine learning) will support the management of the IoT system framework.
4. It is believed in the future that energy harvesting and spectrum efficient research will play an important role in fast switching between the wireless networks for the 5G enabled IoT.
5. For the present and future scope of IoT, both privacy and security will be a key feature. There is an increasing demand for the security of 5G based IoT w.r.t threat landscape, high privacy concerns, delivery models of the services, and trusted models. These features of the privacy and security cover all the layers of the 5G based IoT systems, for example, this will take care of end to end protection apparatus, protection of privacy location and identity protection from the active attackers.
6. Middleware solution for the context-aware systems: The scalability, heterogeneity, and mobility are expected to speed up as context awareness solutions are also available. This also yields to automatically and autonomously adapt to the dynamic changes w.r.t the context.

E. Privacy Concerns and Security Assurances

Privacy and security concerns include device access control, authorization, authentication, and 5G-IoT based privacy-preserving [62]. Other points related to this are mentioned below:

(a) The challenging research topic is trusted communication over the 5G based networking in the presence of eavesdroppers. The important emerging topic is the device's security of the 5G based IoT systems. Some researchers have tried to solve these issues by using some cryptographic methods.
(b) Need for scalable and flexible architecture for security: The architecture for the 5G based IoT must focus on users' privacy protections, low delay mobility security, security assessment, service-oriented security, identity management, and trust models [63].
(c) Security in terms of energy-efficiency: The 5G based IoT encompasses billions of resource-constrained devices. In many cases, these devices, do not provide

computational security solutions, which in turn, demands the lightweight resources security resolutions over resource-constrained devices are an important issue [64].

Instead of an individual security appliance, it is better to acquire systematic privacy protection and security protection mechanism in the 5G based IoT systems. Other security approaches such as data assurances, trust models, and device security, are to be revisited. The readers can also explore more on cyberattacks or crimes, which leads to attacks that target user devices, core network and access, and home/external networks.

5 Conclusion

There are numerous beneficial range of services, which are provided by the 5G and is not available with other technologies. Which includes flexibility to support high number of mobile and static IoT devices, this have a diverse collection of bandwidth, speed and quality of service necessities. The 5G based IoT systems will integrate the 5G procedures and techniques into the future IoT. The chapter deals with recent research on 5G, IoT, edge computing, and cloud computing. 5G enables to serve wide range of diverse service requirements and different range of devices. The technology such as CC will support and handle the volumes of data encountered due to the IoT. IoT interactions are vital to the growth and development of intelligent communities' progress. The emerging Wireless 5G technologies are designed to provide reliable and scalable communication for fast-growing IoT appliances. This chapter showcased the various case studies where the integration of the CC ecosystem with the IoT Based 5G infrastructure can lead to the smooth management of the data and a secure system. We firstly introduce the background of the 5G, IoT, and Cloud Computing era. Then we have discussed the present applications of the 5G based IoT infrastructure. Afterwards, details about the current research issues and trends were analyzed.

References

1. I. Mistry, S. Tanwar, S. Tyagi, N. Kumar, Blockchain for 5G-enabled IoT for industrial automation: A systematic review, solutions, and challenges. Mech. Syst. Signal Process. **135**, 106382 (2020)
2. V.K. Prasad, M.D. Bhavsar, Monitoring and prediction of SLA for IoT based cloud. Scalable Comput. Pract. Exp. **21**(3), 349–358 (2020)
3. M.T. Vega, C. Liaskos, S. Abadal, E. Papapetrou, A. Jain, B. Mouhouche, G. Kalem et al., Immersive interconnected virtual and augmented reality: a 5G and IoT perspective. J. Netw. Syst. Manag. **28**, 1–31 (2020)
4. A. Kumari, S. Tanwar, S. Tyagi, N. Kumar, M.S. Obaidat, J.J.P.C. Rodrigues, Fog computing for smart grid systems in the 5G environment: challenges and solutions. IEEE Wirel. Commun. **26**(3), 47–53 (2019)

5. A. Khalil, H. Farman, B. Jan, Z. Khan, A. Koubâa, A smart energy-based source location privacy preservation (SESLPP) model for IoT-based VANETs. Trans. Emerg. Telecommun. Technol., **28**, 1–14 (2020)
6. V.K. Prasad, M. Shah, M.D. Bhavsar, Trust management and monitoring at an IaaS level of cloud computing, in *Proceedings of 3rd International Conference on Internet of Things and Connected Technologies (ICIoTCT)* (2018), pp. 26–27
7. V.K. Prasad, M.D. Bhavsar, Exhausting autonomic techniques for meticulous consumption of resources at an IaaS layer of cloud computing, in *International Conference on Future Internet Technologies and Trends* (Springer, Cham, 2017), pp. 37–46
8. L.J. Vora, Evolution of mobile generation technology: 1G to 5G and review of upcoming wireless technology 5G. Int. J. Mod. Trends Eng. Res. **2**(10), 281–290 (2015)
9. D. Wang, D. Chen, B. Song, N. Guizani, X. Yu, X. Du, From IoT to 5G I-IoT: the next generation IoT-based intelligent algorithms and 5G technologies. IEEE Commun. Mag. **56**(10), 114–120 (2018)
10. I. Memon, H. Fazal, R.A. Shaikh, G. Muhammad, Q.A. Arain, T.K. Khatri, Big data, cloud and 5G networks create smart and intelligent world: a survey. Univ. Sindh J. Inf. Commun. Technol. **3**(4), 185–192 (2019)
11. V.K. Prasad, M. Bhavsar, Efficient resource monitoring and prediction techniques in an IaaS level of cloud computing: survey, in *International Conference on Future Internet Technologies and Trends* (Springer, Cham, 2017), pp. 47–55
12. M. Marjani, F. Nasaruddin, A. Gani, A. Karim, I.A.T. Hashem, A. Siddiqa, I. Yaqoob, Big IoT data analytics: architecture, opportunities, and open research challenges. IEEE Access **5**, 5247–5261 (2017)
13. J.H. Kim, A review of cyber-physical system research relevant to the emerging IT trends: industry 4.0, IoT, big data, and cloud computing. J. Ind. Integr. Manag. **2**(03), 1750011 (2017)
14. L.D. Xu, E.L. Xu, L. Li, Industry 4.0: state of the art and future trends. Int. J. Prod. Res. **56**(8), 2941–2962 (2018)
15. H. Ahuett-Garza, T. Kurfess, A brief discussion on the trends of habilitating technologies for industry 4.0 and smart manufacturing. Manuf. Lett. **15**, 60–63 (2018)
16. S.K. Rao, R. Prasad, Impact of 5G technologies on industry 4.0. Wirel. Pers. Commun. **100**(1), 145–159 (2018)
17. V.K. Prasad, M. Shah, N. Patel, M. Bhavsar, Inspection of trust based cloud using security and capacity management at an IaaS level. Procedia Comput. Sci. **132**, 1280–1289 (2018)
18. I.M.A. Jawarneh, P. Bellavista, L. Foschini, G. Martuscelli, R. Montanari, A. Palopoli, F. Bosi, QoS and performance metrics for container-based virtualization in cloud environments, in *Proceedings of the 20th International Conference on Distributed Computing and Networking* (2019), pp. 178–182
19. L. Zhang, G. Zhao, M.A. Imran, Internet of Things and sensors networks in 5G Wireless communications, in *MDPI* (2020)
20. F. Qamar, Mohammad Nour Hindia, Talib Abbas, Kaharudin Bin Dimyati, and Iraj Sadegh Amiri, Investigation of QoS performance evaluation over 5G network for indoor environment at millimeter wave bands. Int. J. Electron. Telecommun. **65**(1), 95–101 (2019)
21. M. Pathan, J. Broberg, R. Buyya, Maximizing utility for content delivery clouds, in *International Conference on Web Information Systems Engineering* (Springer, Berlin, 2009), pp. 13–28
22. H. Mehta, V.K. Prasad, M. Bhavsar, Efficient resource scheduling in cloud computing. Int. J. Adv. Res. Comput. Sci. **8**(3), 809–815 (2017)
23. S. Tanwar, S. Tyagi, I. Budhiraja, N. Kumar, Tactile Internet for autonomous vehicles: latency and reliability analysis. IEEE Wirel. Commun. **26**(4), 66–72 (2019)
24. A. Mewada, R. Gujaran, V.K. Prasad, V. Chudasama, A. Shah, M. Bhavsar, Establishing trust in the cloud using machine learning methods, in *Proceedings of First International Conference on Computing, Communications, and Cyber-Security (IC4S 2019)* (Springer, Singapore, 2020), pp. 791–805

25. S. Tyagi, S. Tanwar, N. Kumar, J.J.P.C. Rodrigues, Cognitive radio-based clustering for opportunistic shared spectrum access to enhance lifetime of wireless sensor network. Pervas. Mobile Comput. **22**, 90–112 (2015)
26. U. Bodkhe, D. Mehta, S. Tanwar, P. Bhattacharya, P.K. Singh, W.-C. Hong, A survey on decentralized consensus mechanisms for cyber physical systems. IEEE Access **8**, 54371–54401 (2020)
27. V.K. Prasad, M.D. Bhavsar, Monitoring IaaS cloud for healthcare systems: healthcare information management and cloud resources utilization. Int. J. E-Health Med. Commun. (IJEHMC) **11**(3), 54–70 (2020)
28. P. O'Donovan, C. Gallagher, K. Leahy, D.T.J. O'Sullivan, A comparison of fog and cloud computing cyber-physical interfaces for Industry 4.0 real-time embedded machine learning engineering applications. Comput. Ind. **110**, 12–35 (2019)
29. J. Du, L. Zhao, J. Feng, X. Chu, Computation offloading and resource allocation in mixed fog/cloud computing systems with min-max fairness guarantee. IEEE Trans. Commun. **66**(4), 1594–1608 (2018)
30. Z. Ning, J. Huang, X. Wang, Vehicular fog computing: enabling real-time traffic management for smart cities. IEEE Wirel. Commun. **26**(1), 87–93 (2019)
31. H.A. Khattak, H. Arshad, S. ul Islam, G. Ahmed, S. Jabbar, A.M. Sharif, S. Khalid, Utilization and load balancing in fog servers for health applications. EURASIP J. Wirel. Commun. Netw. **2019**(1), 91 (2019)
32. N. Pontois, M. Kaneko, T.H.L. Dinh, L. Boukhatem, User pre-scheduling and beamforming with outdated CSI in 5G fog radio access networks, in *2018 IEEE Global Communications Conference (GLOBECOM)* (IEEE, Piscataway, 2018), pp. 1–6
33. R. Moreno-Vozmediano, R.S. Montero, E. Huedo, I.M. Llorente, Cross-site virtual network in cloud and fog computing. IEEE Cloud Comput. **4**(2), 46–53 (2017)
34. G. Lee, W. Saad, M. Bennis, An online optimization framework for distributed fog network formation with minimal latency. IEEE Trans. Wirel. Commun. **18**(4), 2244–2258 (2019)
35. M. Ali, N. Riaz, M.I. Ashraf, S. Qaisar, M. Naeem, Joint cloudlet selection and latency minimization in fog networks. IEEE Trans. Ind. Inf. **14**(9), 4055–4063 (2018)
36. S.Z. Tajalli, S.A.M. Tajalli, A. Kavousi-Fard, T. Niknam, M. Dabbaghjamanesh, S. Mehraeen, A secure distributed cloud-fog based framework for economic operation of microgrids, in *2019 IEEE Texas Power and Energy Conference (TPEC)* (IEEE, Piscataway, 2019), pp. 1–6
37. E.B.C. Barros, B.G. Batista, B.T. Kuehne, M.L.M. Peixoto, Fog computing model to orchestrate the consumption and production of energy in microgrids. Sensors **19**(11), 2642 (2019)
38. T. Wang, Y. Liang, W. Jia, M. Arif, A. Liu, M. Xie, Coupling resource management based on fog computing in smart city systems. J. Netw. Comput. Appl. **135**, 11–19 (2019)
39. Y. Dong, S. Guo, J. Liu, Y. Yang, Energy-efficient fair cooperation fog computing in mobile edge networks for smart city. IEEE Internet Things J. **6**(5), 7543–7554 (2019)
40. Y. Zhou, Q. Shen, M. Dong, K. Ota, J. Wu, Chaos-based delay-constrained green security communications for fog-enabled information-centric multimedia network, in *2019 IEEE 89th Vehicular Technology Conference (VTC2019-Spring), April* (IEEE, Piscataway, 2019), pp. 1–6
41. J. Pereira, L. Ricardo, M. Luís, C. Senna, S. Sargento, Assessing the reliability of fog computing for smart mobility applications in VANETs. Future Gener. Comput. Syst. **94**, 317–332 (2019)
42. Y.-S. Chen, Y.-T. Tsai, A mobility management using follow-me cloud-cloudlet in fog-computing-based RANs for smart cities. Sensors **18**(2), 489 (2018)
43. N. Moustafa, A systemic IoT-fog-cloud architecture for big-data analytics and cyber security systems: a review of fog computing (2019). Preprint. arXiv:1906.01055
44. A. Yassine, S. Singh, M.S. Hossain, G. Muhammad, IoT big data analytics for smart homes with fog and cloud computing. Future Gener. Comput. Syst. **91**, 563–573 (2019)
45. M. Nasir, K. Muhammad, J. Lloret, A.K. Sangaiah, M. Sajjad, Fog computing enabled cost-effective distributed summarization of surveillance videos for smart cities. J. Parallel Distrib. Comput. **126**, 161–170 (2019)

46. SmartcitiesWorld, https://www.smartcitiesworld.net/news/news/how-5g-powered-robots-are-helping-china-fight-coronavirus-5154. Last access 22 Oct 2020
47. J. Vora, P. Italiya, S. Tanwar, S. Tyagi, N. Kumar, M.S. Obaidat, K.-F. Hsiao, Ensuring privacy and security in E-health records, in *2018 International Conference on Computer, Information and Telecommunication Systems (CITS)* (IEEE, Piscataway, 2018), pp. 1–5
48. R. Jaiswal, A. Agarwal, R. Negi, Smart solution for reducing the COVID-19 risk using smart city technology. IET Smart Cities **2**(2), 82–88 (2020)
49. S.K. Rao, R. Prasad, Impact of 5G technologies on industry 4.0. Wirel. Pers. Commun. **100**(1), 145–159 (2018)
50. F. Al-Turjman, Intelligence and security in big 5G-oriented IoNT: an overview. Future Gener. Comput. Syst. **102**, 357–368 (2020)
51. The Ericsson Mobility Report, https://www.ericsson.com/en/mobility-report. Last access: 22 Oct 2020
52. J. Cao, M. Ma, H. Li, R. Ma, Y. Sun, P. Yu, L. Xiong, A survey on security aspects for 3GPP 5G networks. IEEE Commun. Surv. Tutorials **22**(1), 170–195 (2019)
53. P. Ameigeiras, J.J. Ramos-Munoz, L. Schumacher, J. Prados-Garzon, J. Navarro-Ortiz, J.M. Lopez-Soler, Link-level access cloud architecture design based on SDN for 5G networks. IEEE Netw. **29**(2), 24–31 (2015)
54. Z. Lü, Y. Lü, M. Yuan, Z. Wang, A heterogeneous large-scale parallel SCADA/DCS architecture in 5G OGCE, in *2017 10th International Congress on Image and Signal Processing, BioMedical Engineering and Informatics (CISP-BMEI)* (IEEE, Piscataway, 2017), pp. 1–7
55. I. Ahmad, T. Kumar, M. Liyanage, J. Okwuibe, M. Ylianttila, A. Gurtov, 5G security: analysis of threats and solutions, in *2017 IEEE Conference on Standards for Communications and Networking (CSCN)* (IEEE, Piscataway, 2017), pp. 193–199
56. H. Zhang, N. Liu, X. Chu, K. Long, A.-H. Aghvami, V.C.M. Leung, Network slicing based 5G and future mobile networks: mobility, resource management, and challenges. IEEE Commun. Mag. **55**(8), 138–145 (2017)
57. O. Aydin, E.A. Jorswieck, D. Aziz, A. Zappone, Energy-spectral efficiency tradeoffs in 5G multi-operator networks with heterogeneous constraints. IEEE Trans. Wirel. Commun. **16**(9), 5869–5881 (2017)
58. I.S. Udoh, G. Kotonya, Developing IoT applications: challenges and frameworks. IET Cyber-Phys. Syst. Theory Appl. **3**(2), 65–72 (2018)
59. M. Alzenad, M.Z. Shakir, H. Yanikomeroglu, M.-S. Alouini, FSO-based vertical backhaul/fronthaul framework for 5G+ wireless networks. IEEE Commun. Mag. **56**(1), 218–224 (2018)
60. I. Ahmad, T. Kumar, M. Liyanage, J. Okwuibe, M. Ylianttila, A. Gurtov, 5G security: analysis of threats and solutions, in *2017 IEEE Conference on Standards for Communications and Networking (CSCN)* (IEEE, Piscataway, 2017), pp. 193–199
61. A. Ahad, M. Tahir, K.-L. Alvin Yau, 5G-based smart healthcare network: architecture, taxonomy, challenges and future research directions. IEEE Access **7**, 100747–100762 (2019)
62. I. Ahmad, T. Kumar, M. Liyanage, J. Okwuibe, M. Ylianttila, A. Gurtov, Overview of 5G security challenges and solutions. IEEE Commun. Stand. Mag. **2**(1), 36–43 (2018)
63. M.R. Palattella, M. Dohler, A. Grieco, G. Rizzo, J. Torsner, T. Engel, L. Ladid, Internet of things in the 5G era: enablers, architecture, and business models. IEEE J. Sel. Areas Commun. **34**(3), 510–527 (2016)
64. S. Li, Q. Ni, Y. Sun, G. Min, S. Al-Rubaye, Energy-efficient resource allocation for industrial cyber-physical IoT systems in 5G era. IEEE Trans. Ind. Inf. **14**(6), 2618–2628 (2018)

Chapter 8
Existing Technologies and Solutions in 5G-Enabled IoT for Industrial Automation

Khalimjon Khujamatov, Doston Khasanov, Ernazar Reypnazarov, and Nurshod Akhmedov

1 Introduction

The current manufacturing industry in the last decade has undergone unprecedented dramatic changes and has attracted a great deal of attention to IoT-based industry automation with the advent and application of modern technologies such as digital twin, cloud computing, big data, edge and service-oriented technology, and advanced sensing technology [2]. At the same time, the world has advanced industrial production methods such as Industry 4.0, Health 4.0, Education 4.0, cloud manufacturing, and Internet + manufacturing, and national strategic manufacturing ideas in many countries, such as Made in China 2025, the US manufacturing strategic plan, and the transition to a digital economy in Uzbekistan are being promoted. The overall goal is to automate the industry through advanced digital production methods.

1.1 The Role of Mobile Network Technology in the Industry

The rationalization of the production process through IoT systems, the configuration of production factors, optimization of the order of shop-floor, production planning and inventory management, data modeling, integration, to implement intelligent management in production in a physical environment of industrial automation, on the other hand, analysis and decision-making in the production of cyberspace, as well as the implementation of high-level, and real-time synchronization in the

K. Khujamatov · D. Khasanov (✉) · E. Reypnazarov · N. Akhmedov
Tashkent University of Information Technologies named after Muhammad al-Khwarizmi, Tashkent, Uzbekistan
e-mail: kh.khujamatov@tuit.uz; dhasanov@tuit.uz; eernazar@umail.uz; Axmedov.N.M@tuit.uz

S. Tanwar (ed.), *Blockchain for 5G-Enabled IoT*,
https://doi.org/10.1007/978-3-030-67490-8_8

production process inwardly cyber environments and physical environments based on modern technologies of data perception and their transmission [2, 21]. Of course, such processes require the use of IoT systems in industrial automation and a network capable of transmitting large amounts of data quickly. Different IoT systems use many wireless technologies such as 2/3/4G mobile network, Wi-Fi, WiMAX, etc., and thousands of devices are connected to each other via those smart wireless technologies [16].

2G mobile networks (covering 90% of society overworld) are intended to voice communication, 3G mobile networks (currently covering 65% of society overworld) to data and voice, and 4G mobile networks to high-speed rate Internet communication. 3G and 4G mobile network have significantly increased capabilities that ensure Internet access to the IoT device. The available 3G and 4G mobile networks are the most effective communication technologies of IoT, offering IoT application designed for industrial automation with a comprehensive, high level of security, low cost, efficient access to the allocated spectrum, and simplicity of management. Also, 4G "long-term evolution" (LTE) has become the quickest and most tenable type of network technology relatively to popular technologies such as Wi-Fi, BLE, WiMAX, LoRa, Sigfox, ZigBee, and others [1, 28, 47]. However, existing mobile networks do not support MTC (Machine-Type Communication), which are the basis of IoT, and are not fully optimized for such systems [16, 17]. Emerging 5G networks provide a potential solution in this case. 5G provides fast mobile network data transfer in the shortest possible time and offers improved coverage for MTC connectivity without negating the current 4G (LTE) network based on IoT requirements. As next-generation communication network, 5G is supposed to address issues that arise in 3G and 4G networks, such as in smart management application (smart city, smart home, smart transportation) or more complex communication, intelligence, and large-scale data processing, which is important for IoT systems designed for industrial automation [10]. The development of 5G is created via 4G LTE-based system, which provides voice and data transmission to users and provides them with a fast connection to the Internet. The 5G network provides reliable and fast connectivity and significantly increases data transfer speeds for IoT applications designed for industrial automation. The current 4G LTE network can provide speeds up to 1 GB/s, but the 4G signal can be easily extinguished due to buildings, Wi-Fi signals, microwave signals, and other wave interference [16]. 5G networks provide users with higher data speeds up to 10 Gb/s than 4G networks; moreover, the network itself provides reliable connectivity to thousands of devices [16, 17].

1.2 Integration of 5G Mobile Network and IoT Technologies

Recent developments in telecommunications and computing have given rise to two different, highly promising, and exciting perspectives: the 5G wireless mobile network and IoT. Taking a closer look at these technologies, together they play an important role in the design of industrial automation and intelligent systems. Many of the arising and rapidly developing concepts such as E-transportation, E-banking, E-agriculture, E-healthcare, E-security, E-industry, E-manufacturing, etc. introduced in these systems are increasingly manifested, and they rely on the development of 5G-enabled IoT technologies [10]. It should be noted that since the emergence of the 5G-enabled IoT concept and its prospects, all stakeholders in the telecommunications sector, especially in the field of education and industry, have intensified their interest in it. Simply put, 5G-enabled IoT describes new and very important Internet-connected tools for the coming future – representing the possibility of using effective autonomous services with minimal human intervention or participation, when various devices or objects (buildings, technical means, machines, structures, equipment, etc.) are interconnected simultaneously and continuously via the Internet [13].

Achievements in IoT and the newly emerging 5G have popularized the development of 5G-enabled IoT applications for industrial automation. Such applications place high demands on high capacity, reliable privacy and security, coverage of various applications, very low latency, optimized use of network resources, and efficient energy management. Although the security architectures currently used in mobile networks and general IoT systems meet the requirements, they are largely centralized [18, 22]. The use of such centralized security solutions for 5G and 5G-enabled IoT applications leads to various barriers, such as genetic heterogeneity, complex and static security management procedures, overuse of network resources, cost overruns due to network failure increase, etc. Thus, the continued use of centralized security solutions for applications managed by 5G and IoT not only poses challenges in meeting requirements but also negatively affects the predictable outlook of 5G and IoT [19].

In this context, blockchain technology will become a promising system for 5G-enabled IoT in industrial automation, as it will ensure that all security and privacy issues are integrated and decentralized. Thus, it is worth exploring the relevant opportunities and challenges in using Blockchain as a decentralized security and privacy solution for 5G and IoT [20, 27].

1.3 Prospects for the Use of Blockchain Technology in 5G-Based IoT Applications

Blockchain, a distributed ledger technology, allows users to interact and interoperate (store and retrieve data) with reliability, consistency, and non-rejection of data. The distributed nature of the blockchain allows industrial enterprises and various 5G-enabled IoT devices to exchange data to meet their central demand. By the support of Blockchain, the 5G ecosystems are able to establish accountability, data validation, and non-rejection for each user. A blockchain contains many blocks and one of them is named the genesis block. This block doesn't include any transactions.

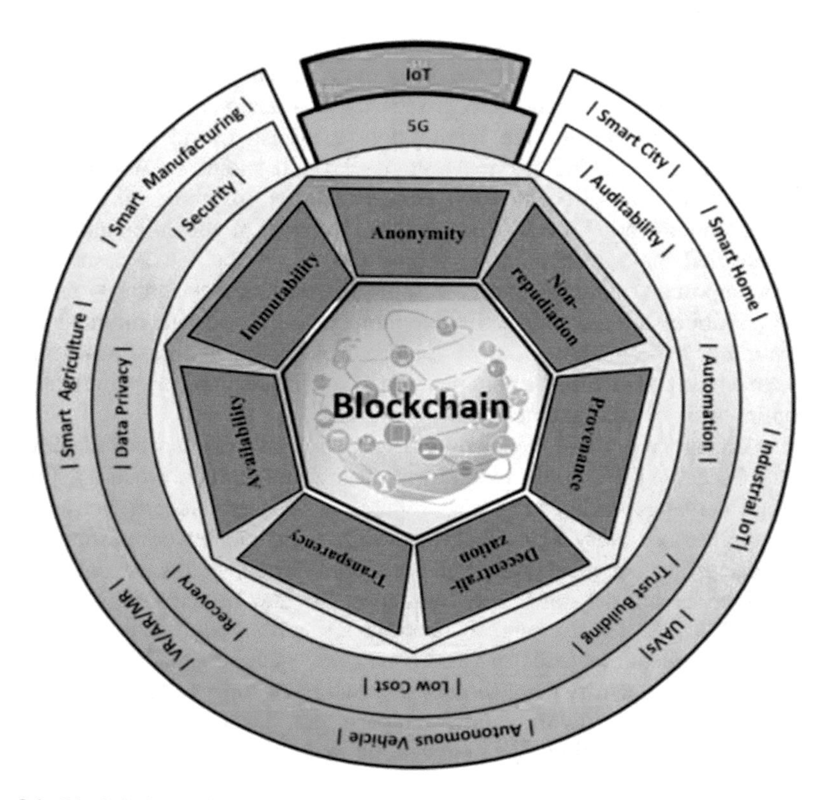

Fig. 8.1 Blockchain and its impact on 5G-based IoT for industry automation

The next blocks include a number of verified transactions and are cryptographically associated with the previous. Figure 8.1 illustrates important points of blockchain with related concepts in 5G-based IoT systems for industrial automation. In this figure, Blockchain technology is the central part of the overall structure, providing important indicators of 5G-based IoT proposed to be used in industrial automation, such as anonymity, non-repudiation, provenance, decentralization, transparency, availability, and immutability [9].

Today's IoT applications are focused on changing daily life quality, including interdependence, such as smart home, smart city, smart environment, or automated industry. In the industrial sector, industrial IoT (IIoT) is still evolving and is facing challenges such as network throughput, resource allocation, network optimization, and data and transaction security as demand for it increases [23]. At the same time, technical problems such as reliability, safety, timing, accuracy, and feedback can be observed in road control and traffic automation, payment system automation and other similar important systems of the industry.

In this case, no matter how promising 5G-enabled IoT is, the lack of resources to manage their operations and the confidentiality of data are potential barriers to their widespread use. Resource allocation (RA), utilization, optimization, and privacy are the most active fields of exploration in 5G-enabled IoT. This article was focused on a comprehensive research of 5G-enabled IoT technologies, their approaches to resource problem solving, their application in industrial automation, and their elimination of security issues through Blockchain technology. Furthermore, in the paper, solutions were categorized, similarities and differences were highlighted in tables and graphs, advantages and disadvantages were identified, and research ways on enhancing resource management solutions for 5G-based IoT networks were discussed. In addition, existing technologies in the field of 5G-based IoT in industrial automation and their solutions have been studied.

2 Opportunities in 5G-Network-Enabled IoT

2.1 Opportunities of 5G Network for IoT Application

Wherewith of mobile technologies, several generational changes have taken place, transforming the mobile communication environment to the global collection of interconnected networks. Currently, the 5G network supports voice and video streaming and highly sophisticated telecommunication services around ten billion users, as well as more than one billion of interconnected devices. Why exactly 5G network? 5G mobile network offers new way to research. This covers a complete network design to establish MTC. In addition, 5G mobile networks offer applications with different operating parameters and efficient support characteristic, which provides high flexibility for setting up services. If compared to previous mobile network generations, the 5G network is a compound of advanced network technologies that developed until nowadays [11, 48] (Table 8.1).

The future of 5G technology will have unlimited possibilities on the delivery of information by any person and everything for the society and businesses, the technological environment, as well as the benefit of individuals at any time and always. 5G technology is basically a set of new technologies that require a large-scale upgrade of equipment/tools or devices compared to the previous generations. This technology's goal is to rely on the achievements of telecommunications systems. Additional technologies (e.g., combination of cloud and core technologies) used in most of the real radio connectivity are applied to provide large amounts of data traffic and other types of devices beneath various operating requirements in diverse 5G network supply conditions. Figure 8.2 illustrates the performance level based on 5G mobile technology required for the above requirements. The main idea of the global agreement is not only to create a new 5G radio technology but also to integrate a number of techniques, devices, and applications based on IoT [11, 12, 14].

Compared to the previous generation technologies, the capabilities of 5G technology in terms of IoT-rated performance are as follows:

Table 8.1 Comparison table of 1G, 2G, 2.5G, 3G, 3.5G, 4G, and 5G mobile network technologies [11]

Network generations	Definitions	Applicable technologies	Bandwidth/speed	Validity period	Features
1G	Analog	AMPS, NMT, TACS	Maximum: 14.4 kb/s	1981–1990	Only designed to transmit audio data over a wireless network
2G	Narrowband digital circuit data	TDMA, CDMA	Average: 9.6 or 14.4 kb/s	1991–2000	Communication channels are compacted by a multiplexer, and data can be transmitted through a single channel with the voice of many users
2.5G	Packet data transmission	GPRS	Maximum: 171.2 kb/s Average: 20–40 kb/s	2001–2004	Supports internet connection. Multimedia services are the first network to provide data flow, and web browsing via mobile phones is available
3G	Digital broadband packet data	CDMA 2000 (1 × EVDO, RTT) EDGE, UMTS	Maximal: 3.1 Mb/s Average: 500–700 kb/s	2004–2005	Provides multimedia services through data flow. Portability and universal connectivity
3.5G	Fully packet data	HSPA	Maximal: 14.4 Mb/s O'rtacha: 1–3 Mb/s	2006–2010	Provides high throughput and speed to support large amounts of data
4G	Digital, broadband, packet data communication	WiMAX LTE Wi-Fi	Maximum: 100–300 Mb/s Average: 3–5 Mbps 100 Mbps (Wi-Fi)	Currently active	High speed and accurate data flow. Supports real-time HD video and audio streaming. It has more improved portability
5G	Fully IP-based network	MC-UWB, CDMA, LAS-CDMA, OFDM, Network-LMDS	With gigabits	It is currently in use	Currently, began to be used in some countries. It provides more efficient bandwidth and higher speed

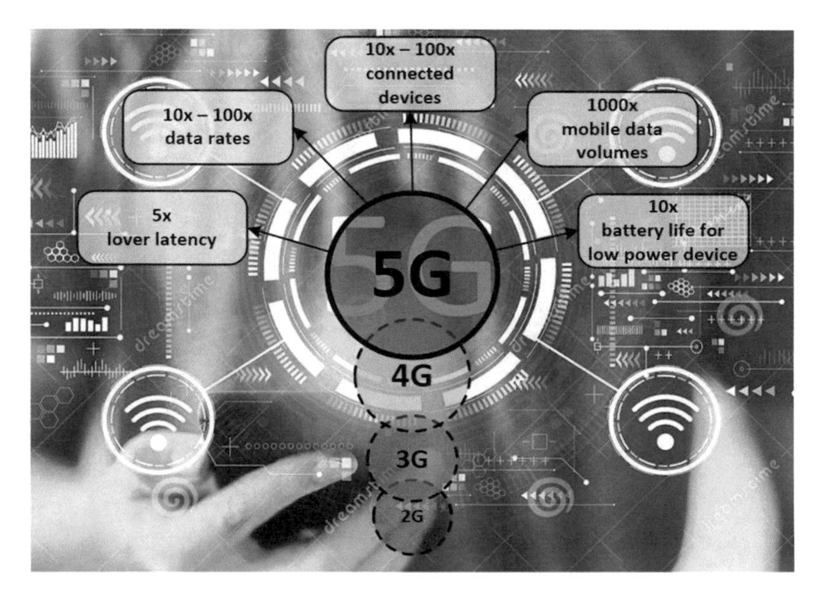

Fig. 8.2 5G demand-side capabilities

- Connected devices number – 10–100× more
- User data transfer rate – 10–100× higher
- Mobile data volume through each area – 1000× more
- Battery life with low-power devices – 10× longer
- End-to-end latency – 5× reduced

One of the key challenges is how to take full advantage of such opportunities. New-generation network technology (5G) mainly allows for a large number of connections. The evolving next-generation network has an important position in shaping different factors such as cost and security and long-term stability and should present connectivity to billions of users and IoT devices. Although it is necessary to establish a comprehensive environment for the 5G network, the use of thousands of applications and IoT devices is one of the key capabilities of the 5G network. The parameters developed by 5G-enabled IoT network technology include:

- Low latency
- Data integrity
- Data transmission
- Traffic capacity
- Smart communication
- Technology convergence
- Energy consumption

The network model of 5G matches to interconnected (ISO/OSI) levels of the open system, which simplifies the methods of connecting any IoT devices to this

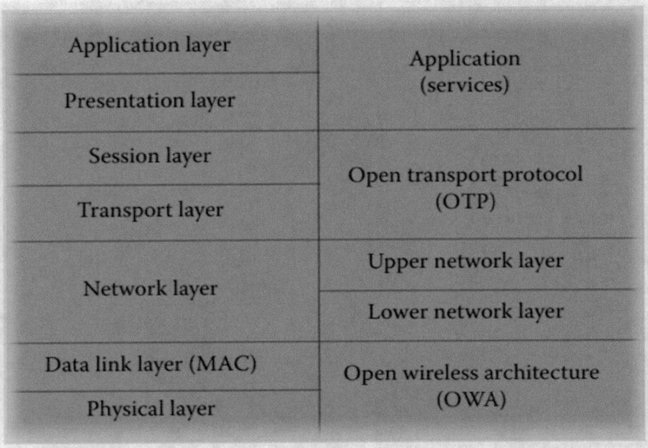

Application layer	Application (services)
Presentation layer	
Session layer	Open transport protocol (OTP)
Transport layer	
Network layer	Upper network layer
	Lower network layer
Data link layer (MAC)	Open wireless architecture (OWA)
Physical layer	

Fig. 8.3 The relationship of 5G open wireless architecture with OSI model [25]

network. Four main layers are used in the 5G network. Figure 8.3 shows the relative levels of the OSI and 5G network layers. The open wireless architecture by 5G mobile network matches to OSI's physical layer, known as Layer I and Layer II, and the data connection layer or intermediate connection management (MAC) layer, respectively [23].

The second network layer of 5G is divided into two parts, the upper and lower network layer, which corresponds to the third layer of the OSI model. This layer is based on IP. Today, two versions of IP are used in the 5G network: IPv4 and IPv6. IPv4 is widespread around the world, so it faces various challenges such as the limited number of addresses and the inability to fully provide quality of service (QoS) for each data stream. These problems are solved through IPv6, but its disadvantage is that the packet header is much larger. Also in this version, the mobility characteristic of the device remains problematic as always. The current mobile IP standard has multiple mobility solutions. Mobile IP is used in all 5G mobile networks. This feature allows mobile phones and IoT apps to connect to several mobiles or wireless networks at the same time [28].

2.2 *Opportunities of Blockchain in 5G-Based IoT*

With the evolution of intelligent systems to change people's lifestyles, the IoT system is playing a crucial role in the industrial automation and services digitization. Many access points are emerging for accessing and exchanging information on the Internet with the fast growth of IoT. Centralized data storage systems (e.g., cloud computing) have added greatly to the evolution of IoT. They are formed in the

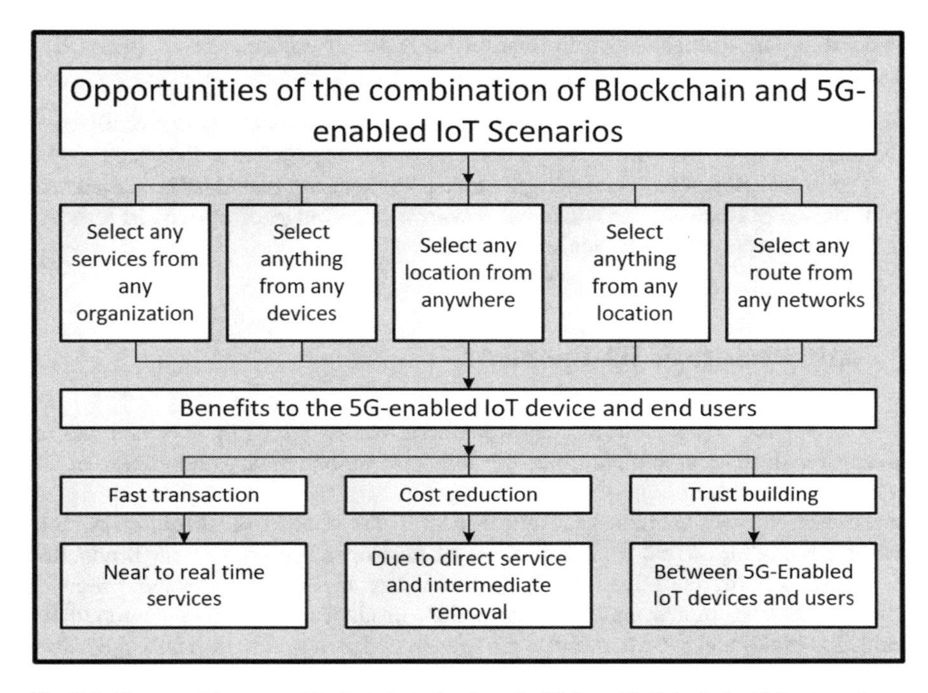

Fig. 8.4 Opportunities to use Blockchain technology in 5G-based IoT for industrial automation

human mind in the form of separate block boxes, but users are unaware of the process of data exchange in this network. Such a centralized construction may not ensure transparency of data [26]. In this case, one of the key solutions to improve security and privacy is to use blockchain technology (Fig. 8.4).

Blockchain is an open, trusted, and verifiable information exchange platform with the ability to make large-scale changes in IoT, where any data exchange can be trusted and controlled. Some of the possibilities of this integration are [15, 20]:

Identity: The use of a general blockchain allows each device to be correctly identified. It can constantly monitor the origin of any required information.

Scalability and Decentralization: The transition of data from a centralized to a decentralized state can reduce some of the shortcomings in improving error tolerance. In addition, several powerful corporations prevent resource oligarchy that can manage the processing and collection of large numbers of user's data.

Autonomy: Devices may interact with each other autonomously without any intermediary involvement by using Blockchain. This can lead the way for the development of devices suitable for IoT applications used in the industry.

Reliability: Reliability integration provides accountability and also enables users to confidently verify any transaction authenticity.

Security: The exchange of information using smart contracts is explained as a transaction that provides a secure connection between devices.

Secure Code Deployment: As a fixed account, the manufacturer can easily track the update history. In addition, it allows them to reliably update IoT devices.

The use of Blockchain technology for IoT applications of industrial automation built on the 5G network provides users with a wide range of options in terms of privacy and security of relevant information.

3 Architecture of 5G-Enabled IoT

This subsection explores the 5G-based IoT architecture, taking into account various areas of industrial automation and the three communication technologies of 5G wireless networks. Initially based on this architecture, the methods of implementing seven related areas of industrial automation in the context of 5G wireless technology, including a real-time collection of heterogeneity pointers in production on the shop-floor, identification of production factors and their location, branched joint production, human-machine interaction, product design, AGV (autonomous guided vehicle) collaboration, the convergence of the digital twin-based shop-floor between the cyber and physical environment, and virtual and augmented reality-based development production and maintenance processes, are studied. It also analyzes the introduction of Blockchain services for 5G-enabled IoT technology in the industry and the expected efficiency from it. Figure 8.5 shows the 5G-enabled IoT architecture for industrial automation.

As can be seen from the architecture, IoT-based industry automation uses three main 5G network technologies: mMTC (massive Machine-Type Communication), URLLC (Ultra-Reliable Low-Latency Communications), and eMBB (enhanced Mobile Broadband), and they operate on the lower physical shop-floor of the architecture [2, 6]. Each of them is discussed below.

A. ***mMTC-based industrial automation application processes*** For industrial automation, the intelligent interconnection of heterogeneous production factors can lead to the creation of CPMS (Cyber-Physical Manufacturing Systems), i.e., production equipment (e.g., robotic arms, machines, supply machines), manufacture of ancillary equipment (e.g., gas or water supply equipment), monitoring source (e.g., product work, material, manpower), and the environment on the shop-floor (e.g., moisture, temperature, toxic gases, dust, etc.), which is the basis for real-time monitoring [2]. The intelligent interdependence of different pointers of production mainly involves two types of production, i.e., the identification and placement of production pointers and the real-time collection of different pointers of production on the shop-floor. There are a high number of nodes that are necessary to be tracked, and it is very difficult for the current 3G/4G mobile networks to support industrial processes that have such high-scale nodes. Or, due to the presence of heterogeneous data structure and various production pointers on the shop-floor,

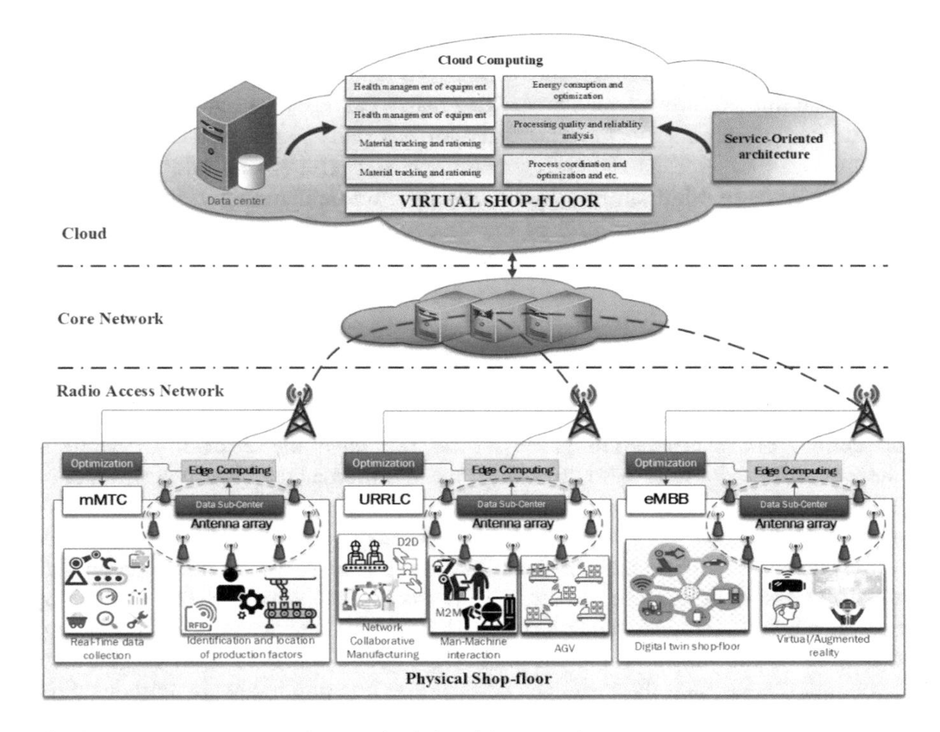

Fig. 8.5 5G-enabled IoT architecture for industrial automation

4G mobile and other available network technologies are not proper for IoT-based industry automation processes with extensive communication networks [15].

Using millimeter wave propagation for the above industrial automation process, 5G technology offers many small-sized antennas, i.e., an antenna matrix. With multi-access (SDMA) technology through efficient spatial division, the antenna matrix method can improve the reuse process from the frequency of the transmitter and receiver. Its multibeamforming technology reduces the noise of the combined channels, improves the quality of the connection, and optimizes the integration in a separate direction by changing the directional transmission to a flexible high-directional radiation pattern. A network architecture that combines remote central base resources has statistical multiplexing benefits, saving energy, and bandwidth. Therefore, a number of connection nodes and their communication coverage have been improved, which allows 5G network technology to meet the demands of the application of giant machine node communication processes for industrial automation.

B. ***URLLC-based industrial automation application processes*** The usual process of applying the CPMS production mode is to co-produce different production pointers on the shop-floor. It is known that automation technology is widely used in shop-floor production, so it is necessary to build large-scale automatic production lines in this. On the one hand, the production process of a component or product includes several sub-processes, and various production types of equipment on the production line perform various processing tasks. On the other hand, a single robot with limited machining functions and structure is usually unable to perform complex machining tasks such as double mounting or performing multiple functions at the same time. In addition, in a high-intensity environment, some manufacturing processes require remote control of the functions of a large number of robots to perform a specific production task. This manufacturing process requires a very precise transmission and a very low latency for two-direction data transmission and instruction control between the control center and the shop-floor equipment. However, it is very difficult for 3G/4G networks to meet low-latency and high-reliability requirements of industrial automation [29].

Another common way to use the CPMS mode is through a human-machine interaction. Some production processes do not have the flexibility feature in some machines, so it is hard to satisfy the requirements of various production operations. In other production processes, due to the need for manual handling of non-ergonomic tasks, workers have to load heavy workloads. With the machine performing specific tasks and the flexibility of the person, the machine can quickly adjust the operating mode according to the instructions given by the workers by pressing a button, body movement, or sound. Besides, when workers enter the range of motion of the machine, the machine (e.g., robot, machine tool) should slow down, and if necessary, the machines have to stop tracking to ensure worker safety. The production model produced by human-machine cooperation not only increases productivity but also makes the production system more strong and adjustable. However, low-latency and very-high-reliability communication technologies are required to ensure safe and efficient cooperation between human and machine.

A variety of AGVs are used on smart shop-floor and warehouse. The AGV not only works jointly with the manipulator and different things for loading and unloading goods but also cooperates with other AGVs and can automatically find a power source for self-charging. In this case, the AGV determines the location of objects in the environment, the condition of other AGVs, the value of battery level, etc., and makes decisions by performing computing quickly. While AGV performs such tasks intelligently, it requires a communication network to transmit the data collected from them to the data center or to receive control information. Therefore, high-reliability and low-latency communication technologies are required for AGV. To implement the above three advanced manufacturing processes, a large number of small antenna matrices operating on the basis of SDMA can realize flexible strong directional electro signal transmission and optimize efficiency in a particular direction. To improve communication quality and reduce channel interference, the antenna matrix uses multi-signal technology, thereby increasing the reliability of transmission. In addition, 5G mobile network architecture is slowly shifting from a centralized base station form to a centralized user mode, and the base network resources are concentrated in centers located at a certain distance [2].

C. ***eMBB-based industrial automation application processes*** The shop-floor of digital twin is causing to attention and widely learning by the scientific and industrial community as a potentially effective method to implement intelligent intercommunication and interaction between the cyber-environment and the physical production environment on the shop-floor. The shop-floor of digital twin with a new generation of manufacturing and information technology realizes the data integration of all factors, real-time interaction between the physical and virtual shop-floor, and the whole business and the whole process through two-way real-world mapping. In addition, the digital twin shop-floor can perform tasks related to mathematical or computational processes between physical store-layer, virtual store-layer, and store-layer service systems for managing production factors, planning production activities, and controlling the production process. Activities such as identifying and uploading production data from the physical shop-floor to the digital twin shop-floor in real time, managing the instructions provided from virtual to physical shop-floor, and generating massive volumes in real time must transmit production data. This requires new communication technologies that provide higher speeds and lower latency compared to 3G and 4G networks [15].

In industrial automation, there are concepts of virtual/augmented reality, and more attention is usually paid to the latter. For achieving better production results, it is necessary to simulate, analyze, and study the product model before completing the project. Therefore, firstly, virtual/augmented reality is used in product design. Secondly, in production processes, operational steps and processes to increase efficiency and reduce errors can be provided to an employee who assembles parts virtually with virtual/augmented reality technologies. Besides, with complexity and the high degree of integration of production equipment, its repair and maintenance are becoming increasingly complex. With the support of virtual/augmented reality, workers can receive real-time maintenance suggestions. But, in production processes based on virtual/augmented reality, the data transmitted is in the form of a very large amount of media data flow. Therefore, in order to reach a seamless connection within virtual/augmented reality, manufacturing, and industrial automation process, a new mobile network technology with lower latency and faster transmission speed features is required than 3G/4G technology [29]. The eMBB characteristic with high frequency of 5G network provides higher transmission speed and throughput than 4G mobile networks to implement the automation of the above two advanced industrial production processes. The 28 Ghz and 60 Ghz frequencies in every frequency ranges of the mm-wave are the most efficient frequencies of 5G network. The maximum bandwidth in a wireless connection is about 5% of the carrier frequency, so it is possible to achieve a transfer rate ten times faster than a 4G network connection. Furthermore, a 5G network connection based on software-configured network (SDN) architecture and C-RAN provides a freely connected management mode between C-Plan (communication plan) and D-Plan (data plan), which increases the efficiency of intact data transmission [2].

D. ***Introduction of Blockchain technology for 5G services*** By adding security features and simplifying service management, blockchains are expected to make 5G network services easier. Blockchain is especially effective in creating a secure data-sharing or spectrum-transmitting environment on 5G networks. Blockchain is seen as a middle layer for performing spectrum trading, checking transaction exchanges, and reliably leasing the spectrum provided by spectrum source suppliers (license holders). To perform access authentication, unlike traditional database management systems that use a centralized server, blockchain can do decentralized user access verification applying the computing power of all legitimate participants in the network with smart contracts [30]. This causes the exchange system very robust to data changes. Numerous researches on blockchain show that it is effective for spectrum management in terms of improved usability with a higher level of security, better scalability for blockchain adoption, and an improved system of defense against threats and DoS attacks [31]. In addition, with a high level of security on 5G networks, blockchain can simplify network virtualization [32]. Blockchain technology provides the necessary features of non-rejection and permanency to overcome the shortcomings of prior centralized architecture in virtual networks. Specifically, the blockchain has the ability to create secure VWNs (virtual wireless networks), so that resource holders on the wireless network can hand over wireless resources (e.g., radiofrequency, infrastructure, spectrum layer) to mobile virtual network operators (MVNOs). In such decentralized virtual networks, smart contracts may be most beneficial to ensure transparency and automation in a distributed manner instead of leaning on a special node or authorization process, which increases the reliability of resource management services. Creating a reliable and fair economic scheme with blockchain may be an important solution for managing network interference, particularly in small mobile blocks [8]. Blockchain platforms provide strong data protection by entering user data into the ledger. In the ledger, data is signed by hash functions and added to blocks. Blockchain has the ability to provide complete control over personal data in a network exchange, which is unusual from all traditional approaches that prevent users from observing data [8]. In addition, the blockchain may be offering a wide range of security capabilities, such as smart contracts, decentralized ledger-based data integrity, and access control through the permission process and smart contract authentication.

4 5G-Enabled IoT in the Field of Industry Digitalization and Automatization

3GPP-based global mobile networks provide things-to-person and things-to-things connectivity. It is effectively taking advantage of mobile network-based IoT in industries such as household electronics, automobile construction, mining, railways, utilities, agriculture, healthcare, transportation, and manufacturing. Today, there are more than one billion mobile network-based IoT connections worldwide, and Ericsson estimates that number will be close to 5 billion by 2025 [33]. Almost

Table 8.2 One 5G network with four multiple-purpose IoT connectivity segments

IoT connectivity segment	Massive IoT (1)	Broadband IoT (2)	Critical IoT (3)	IoT-based industrial automation (4)
Futures	A small volume of data A low-cost device Extremely high coverage	A large volume of data Higher data rates Low latency	Ultrareliable data delivery Limited latencies Ultralow latency	Time-sensitive network Integration of Ethernet protocol Service of clock synchronization
	Battery life of the device, positioning of the device, network exposure, network slicing, network data analytics			

Table 8.3 Industrial digitalization using IoT based on 5G mobile communication

Industrial areas	IoT connectivity segments				Industrial areas	IoT connectivity segments			
	(1)	(2)	(3)	(4)		(1)	(2)	(3)	(4)
Utilities	✓	✓	✓	✗	Entertainment	✓	✓	✓	✓
Transportation	✓	✓	✓	✗	Construction	✓	✓	✓	✗
Automotive	✓	✓	✓	✗	Forestry	✓	✓	✓	✗
Smart city	✓	✓	✓	✗	Oil and Gas	✓	✓	✓	✓
Railways	✓	✓	✓	✗	Agriculture	✓	✓	✓	✗
Ports	✓	✓	✓	✓	Warehousing	✓	✓	✓	✗
Manufacturing	✓	✓	✓	✓	Public safety	✓	✓	✓	✗
Education	✓	✓	✓	✗	Airline	✓	✓	✓	✗
Mining	✓	✓	✓	✓	Media production	✗	✓	✓	✓
Healthcare	✓	✓	✓	✗	Maritime	✓	✓	✓	✗

In the table: ✓ applied; ✗ non-applied

every industry in the 5G market is exploring mobile connectivity options to radically change the business.

In several countries, governments are supporting the acceptance of IoT through indirect and direct means to promote innovation, sustainability, and growth. Based on these requirements, Ericsson proposed the following four segments of IoT connectivity, as shown in Table 8.2 [34]: massive IoT (1), broadband IoT (2), critical IoT (3), and industrial digitalization with cellular IoT (4). Each IoT connection segment is designed for multiple uses in multiple areas. Table 8.3 shows which of the above IoT connectivity segments is applied in the digitalization of various industries with cellular IoT.

Currently, 4G networks support LTE-based broadband IoT and massive IoT based on Cat-M/NB-IoT technology. Massive IoT proceeds to evolve by using Cat-M/NB-IoT connectivity in 5G mobile networks, while broadband IoT is further enhancing through the use of 5G wireless and core networks. At full capacity, 5G networks with ultrahigh-reliability and ultralow-latency capability provide the use of critical IoT for critical time communication. 3GPP standardizes the additional

Table 8.4 Massive IoT modems' comparison

Characteristics	Dual-mode CAT-M1/NB-IoT modem	Single-mode NB-IoT modem
Peak data rates	1.1 Mb/s (UL), 588 kb/s (DL)	158 kb/s (UL), 127 kb/s (DL)
Voice	Supported	Not supported
Connected mode mobility	Supported	Not supported
Coverage	Both modem types are on par	
Battery life	Both modem types are on par	
Guardband carrier	Guradband NB-IoT carrier can be used for both modem types both modem types	
Device positioning	Cat-M achieves better accuracy than NB-IoT due to wider bandwidth	

capabilities offered by IoT connectivity in industrial automation for integrating Ethernet-based industrial wired communication networks with 5G mobile networks. In industrial automation, these four IoT connectivity segments can be used simultaneously in a single 5G network and provide social (people) or nonsocial (devices) access. Some devices may require multiple IoT connection segments to perform one or more tasks, such as an AGV with multiple requirements. The device can use several IoT connection segments used in the mobile network at the same time. Let's take a closer look at the IoT connectivity segments below.

Massive IoT Massive IoT connectivity is designed for narrowband devices that can receive or send small amounts of data quickly at a low price. These devices can be located in complex radio communication conditions that require very extensive coverage and can only rely on a battery source. LTE-M and NB-IoT have been available in the 4G network with LTE since 2017 and meeting all 5G network requirements of 3GPP and ITU for mMTC [35]. LTE-M technology is expanding its LTE coverage to support MTC, which provides access to a simpler category of devices called the Cat-M. NB-IoT is an independent LTE-based wireless connection technology. By early 2020, more than 120 commercial networks, along with millions of commercial users globally, supported access to NB-IoT and Cat-M. It is projected that this connection will exceed 2.5 billion by 2025. Commercial instruments can be used in a variety of industries, including utilities, mechanical engineering, transportation, logistics, agriculture, manufacturing, healthcare, warehousing, and mining, as well as a variety of meters, sensors, motion devices, and plug-in devices.

There are currently two well-known Cat-M/NB-IoT modems: single-mode NB-IoT modem is compatible for very low-cost devices, and two-mode Cat-M1/NB-IoT is compatible for different use cases with low-cost devices. The two-mode modems combine the best features of the two technologies, such as bandwidth, mobility, coverage, device positioning, and voice support, as shown in Table 8.4.

When combined with dynamic spectrum sharing as well as dual-mode 5G Cloud Core, Cat-M1 and NB-IoT will have future-proof evolution and efficient in 5G networks [33].

Broadband IoT Broadband IoT connectivity includes MBB capabilities for IoT systems while providing higher speeds and lower latency than massive IoT. Moreover, broadband IoT connectivity also includes MBB capabilities for IoT systems in providing higher speeds and lower latencies than massive IoTs, as well as provides extra capabilities for IoT, such as increased battery life of device, data transfer rate, increased coverage, improved data transfer speed, and improved device location detection. Thus, broadband IoT is relevant to all industries.

Broadband IoT is actually for all industry application. This year, mostly with LTE connectivity, broadband IoT users have been more than 500 million [36]. Today, private cars, trains, commercial vehicles, accessories, equipment, actuators, sensors, cameras, and portable devices are widely used for commercial purposes. These devices can use MBB connection capabilities, but their traffic content and requirements are sometimes sharply different from ordinary MBB connection capabilities. For example, traffic content may be heavier or longer continuous during loading, and strict requirements such as device location, signal coverage and battery life may cause complications relative to the MBB. LTE has a number of categories of devices (LTE Cat-1 and higher) with a large range of bandwidth, which is suitable for a variety and wide use cases. LTE provides 1Gbit/s transmission speed and up to 10 ms latency. Broadband IoT is capable of connecting at speeds of several tens of Gbit/s with the introduction of 5G NR technology in the old and new spectrum.

If the data transfer speed and latency are adjusted, it is possible to increase the signal coverage relative to each base station: for instance, an LTE device can dynamically change LTE and LTE-M connections depending on the signal coverage. Using the amount of traffic designed for the user, the battery life of the device can be significantly improved. The location detection process of a network-based device can be improved with NR, as location detection usually depends on signal transmission capability and NR can operate in much higher bandwidths than LTE.

Critical IoT The critical IoT connection is designed for critical time communication. This allows information to be delivered within the desired waiting range. It has the strongest capabilities for very-low-latency communication and extremely high reliability at different data rates. Here, reliability is described as the possibility of effective delivery of information over a period of time. Unlike broadband IoT, which provides low latency based on reduced network congestion, critical IoT can deliver data continuously, even in a network with high loading, with the required warranty limits.

Typical situations that require a combination of reliability, standby time, and data speed include AGVs, AR/VR, movable robots, cloud robotics, real-time human-machine interface, haptic ideas, prevention of malfunctions in real-time, machine guidance and control, and other more similar processes. Such processes are important in almost every industry. All elements need to be enhanced in terms of reliability and latency to ensure that significant IoT requirements are met. End-to-end latency from a network perspective is the sum of the individual delay times of core, transport, and radio networks, and the overall reliability cannot be higher than the reliability of the weakest network. Edge computing is necessary to minimize transport delays that require critical IoT use cases. Group management, real-time mobility, and network monitoring can activate a decentralized management unit to optimize latency. It is recommended that the central management unit and the main protection are usually located in one place to prevent thrombosis of the signal flow. The distribution of the central user part reduces the waiting time by keeping user traffic as local as possible and is therefore typically distributed in more local areas than the network functions of the control part. Regional full-core expansion achieves ultralow latency and extreme reliability of allocated resources and provides autonomous operation with local users' data.

A LAN can also interact with a shared network, allowing mobile IoT devices to migrate between networks. Critical IoT may be required in terms of network bandwidth, as any high reliability and low latency typically requires large spectrum sources. NR operates at wider frequencies with greater bandwidth and greater capacity than LTE, which creates NR selection technology. Due to many factors such as the duration of commercial use, the constant expansion of NR URLLC in 3GPP, the acceleration of NR, and the ability to update existing LTE aspects with NR, LTE may not improve at all for critical IoT. MNOs with a flexible spectrum are in the best position to provide significant IoT coverage not only in critical areas but also in local industrial areas. NR supports URLLC in all 5G mobile network frequency bands. Figure 8.6 shows examples of combinations of spectral ranges, along with key features of different ranges in terms of URLLC size and coverage to appeal to a wide range of local users. Given that there is a very limited bandwidth in the 1 GHz band, this value should be used for high-end users. When using TDD, the RAN standby time depends on the TDD signal patterns. The configuration of the TDD to connect to or disconnect from the network has a particularly negative effect on latency at a frequency of 6 Ghz [34].

SA 5G is the ideal solution to meet very complex IoT requirements. Compared to 5GC EPC, 5GC is better in terms of advanced service differentiation, ultra-reliability mechanisms, flexible edge computing, Ethernet connectivity, advanced QoS, end-to-end network distribution features, and network data analysis, and these are important for critical use cases. While providing these cases, LTE has not been adapted or improved for critical IoT, and the NSA 5G also does not provide full access to the URLLC in terms of radio connectivity and core networks. The deployment of NSA 5G will not be effective as previously mentioned for use in local coverage cases such as local industrial facilities. Over time, the transition from NSA-5G to SA-5G application will be achieved, and thus the full potential of critical IoT in a wide range of areas will be achieved.

Industrial Automation with 5G-Enabled IoT IoT-enabled industrial automation allows seamless integration of the wireless mobile network in the wired industry infrastructures used for real-time automation. It involves the ability to integrate 5G systems with time-sensitive network (TSN) and real-time Ethernet used in industrial automation networks. Wireless cellular networks are highly efficient in terms of mobility, flexibility, cost reduction, and digitization when compared to wired communications. However, in some industries, wireless networks cannot be

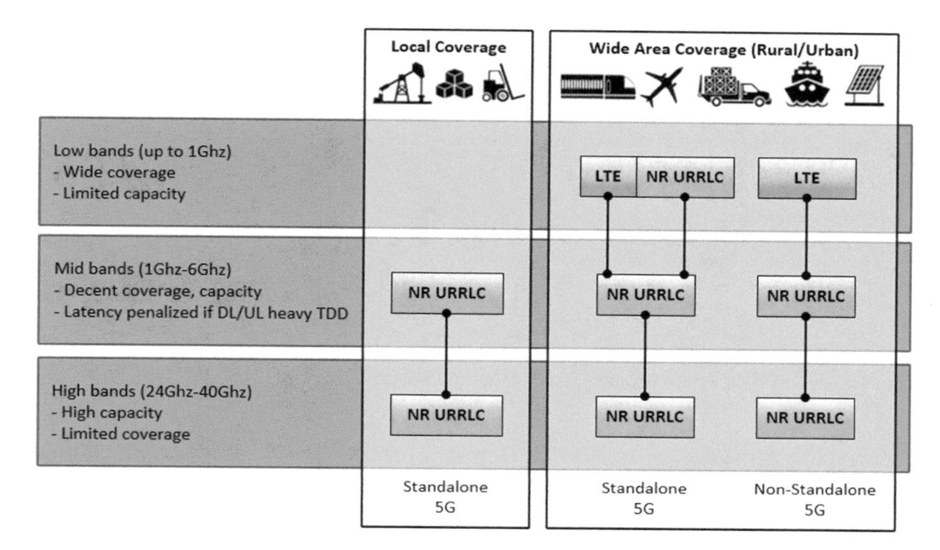

Fig. 8.6 Best positioning of MNOs to activate critical IoT with flexible spectrum quantities

used all at once, and it is possible to gradually implement wireless networks in different systems of these industries over time. Although such industries are in the 5G coverage area, some parts of the system limit wireless connectivity due to various factors, for example, not having the great lack for wireless solutions, possession of a long periodic cycle, and having higher performance needs than 5G's existing capabilities (e.g., deterministic latency at the microsecond level). But the key is to support the seamless integration of current and evolving wired infrastructure of 5G [36].

A number of industries, such as mining, utilities, construction, ports, oil, and gas, use wired connections to transition to modern automation systems. There are several industrial solutions such as EtherCAT, PROFINET, EtherNet/IP, Sercos, Modbus, and Powerlink that support deterministic communication for real-time automation. Rel-15 has standardized support for 3GPP Ethernet sessions. Rel-16, on the other hand, introduces Ethernet header compression for spectral efficiency. Reliable data delivery via 5G URLLC (intended to critical IoT) can be achieved within strict latency limits. With Rel-16, you can set up a 5G virtual network on a 5G network that provides 5G LAN service; in this case, multicast and broadcast personal communication between the same 5G virtual network users is supported on request from UE to UE (user equipment). A common open standard for Ethernet with TSN is being developed to address the challenges facing the fragmented industrial Ethernet market [37]. TSN is designed to meet a variety of QoS requirements, including to provide deterministic and most efficient latency. TSN has been standardized by IEEE, and its industrial automation technologies are being created together by the IEEE and the IEC. To ensure seamless 5G integration

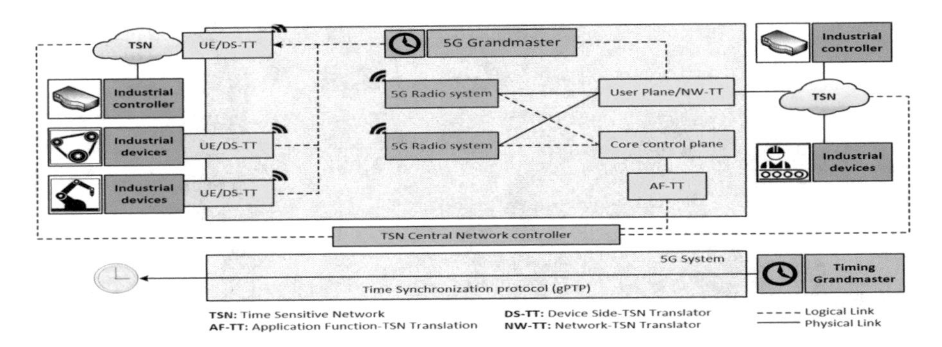

Fig. 8.7 5G-TSN integration and time synchronization service

with TSN, 3GPP has standardized the built-in function in Rel-16 as part of industrial IoT.

The 5G integration with TSN is shown in Fig. 8.7. TSN translator (TT) is introduced in the 5G system for the control plane and the user plane. User-plane translator is deployed to UE (DS-TT, i.e., Device Side-TSN translator) and UPF (NW-TT, i.e., Network-TSN translator). The TSN is centralized and controlled by the TSN Central Network Controller (CNC). The application function of TSN (AF-TT) has been placed in the 5G management plane to demonstrate the capabilities of the 5G system (such as a list of NW-TT and DS-TT ports and waiting for a data exchange between them) to the TSN CNC for adjusting and scheduling TSN data flows across the connection portion of the 5G system. In NW-TT and DS-TT, the process of holding and redirecting buffering is used to deliver traffic flows with a deterministic latency based on planning data taking into account the time taken from the CNC [37].

The TSN nodes are synchronized by the control clock with the generalized Precision Time Protocol (gPTP, defined in IEEE 802.1AS standards) [37]. The wire (brigade) in the 5G system supports the transmission of gPTP synchronization data or can use its own internal clock to provide timely information to the TSN nodes. 5G network can also transfer synchronization signals of a clock as a service to synchronously running industrial applications (e.g., synchronization coordination between multiple controllers in the system).

5 5G-Enabled IoT Requirements for Industrial Automation

IoT systems have radically changed our daily lives by introducing new applications that are widely used in the ecosystems of smart and highly heterogeneous devices. There has been a lot of research on 5G-based IoT over the last few years on a complex and interesting topic. Much of this research has been devoted to the research works of 5G-enabled IoT requirements. The requirements for the 5G network are standardized in the 3GPP recommendation.

The high demands of market participants in the 5G network are not as clear as the 3GPP recommendation. Even though high-level demands were collected from a variety of sources, one of the previous standards created for these requirements was the ITU-R M.2083-0. In describing the key features of the various applications of 5G-based IoT in industrial automation, this standard first identified three major communication technologies: eMBB, URLLC, and mMTC. There are a number of technical requirements for a 5G network, which are distributed across these technologies. In this case, the basic requirements for eMBB include all the requirements for the 5G network and include connection density, latency, mobility, spectrum efficiency, user-experienced data rate, peak data rate, area traffic capacity, and network energy efficiency. The high requirements for URLLC include only two of them, mobility and delay. mMTC, like URLLC, meets only two basic requirements for a 5G network: network energy efficiency and connection capacity [4]. Of course, these technologies must meet other requirements in the network, but the demand for them is relatively low.

Among these requirements, we will identify the most important requirements for IoT systems. They include:

- High data rate: Requires a speed higher than 25 Mbit/s to ensure the performance of future applications of IoT, such as VR/AR, high-definition video streaming, etc. [7].
- High coverage: 5G-based IoT requires high regional coverage over NFV for increase network expansion.
- Very low delay time: 5G-based IoT applications, such as AR, sensitive Internet, video games, etc., require less than 1 ms latency.
- Sustainability of reliability: 5G-based IoT systems require enhanced coverage and delivery efficiency to users of IoT applications and devices.
- Security: IoT technology in the near future will differ from the general security strategy designed for mobile payments and digital wallet applications, connection, and user privacy protection (e.g., blockchain technology), and 5G-enabled IoT requires an enhanced security strategy across the network.
- Connection density: In 5G-IoT, a large number of devices are connected to each other, for which 5G must support successful message delivery in a specific time and region.
- Mobility: 5G-enabled IoT should support a large number of devices while providing high mobility connectivity.
- Long lifetime of battery capacity: Low-power energy solutions are required by this scenario to support billions of low-cost and low-power devices used into 5G-enabled IoT networks.

Modern IoT technologies involve uploading and storing all the resources created with IoT in the cloud, which are processed via cloud servers to retrieve beneficial data using analysis methods [21, 24]. The characteristics of the 5G network and IoT

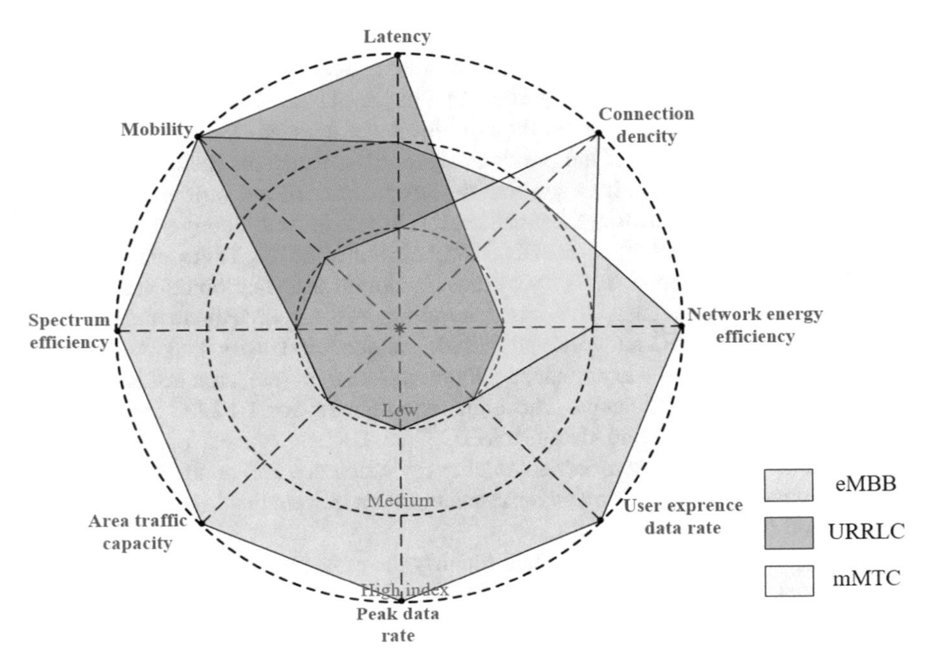

Fig. 8.8 5G-enabled IoT requirements and capabilities for industrial automation

system requirements for industry automation mentioned above are summarized in Fig. 8.8.

Compared with eMBB, massive IoT or mMTC does not have high throughput, but can provide connection of more than one million end devices per 1 km^2 on connection density requirement. As can be seen from the examples of eMBB connection types in Table 8.5, this is close to the broadband connection requirements in a user-dense area (500,000 users per 1 km^2).

Requirements of IoT applications in industrial automation are emerging for a high level of security, real-time operation (low latency), system integrity and interoperability, engineering and efficient deployment, and maintenance.

TS 22.104 recommendation of 3GPP described in detail the requirements for cyber-physical control applications in vertical areas. The detailed level of this quasi-standard should be grounded, i.e., the standardized values for high reliability (between 99.999% and 99.999999%), latency (0.5–500 ms at different connections), and long-term reliability (average failure time between 1 day to 10 years) should ensure the widespread use of 5G. This recommendation involves different requirements for both periodic and nonperiodic deterministic (or nondeterministic) communication, time synchronization, mixed traffic, network operations, and positioning. The requirements are also developed on the basis of the types of connections and cover the various uses of future industry in enterprises, central power generation, power distribution, as well as medical and hospital institutions (in particular

Table 8.5 Data speed and density requirements for different broadband connection types [4]

Scenario	UE speed	Capacity area traffic	Experienced data rate	Overall user density
High-speed vehicle	Passengers in vehicles (max. 250 km/h)	50 Gbps/vehicle 100 Gbps/vehicle	25 Mbps50 Mbps	4000/vehicle
Airplanes connectivity	Passengers in airplanes (max. 1000 km/h)	600 Mbps/plane 1.2 Gbps/plane	7.5 Mbps15 Mbps	400/plane
High-speed train	Passengers in trains (max. 500 km/h)	7.5 Gbps/train 15 Gbps/train	25 Mbps50 Mbps	1000/train
Indoor hotspot	Pedestrians	2 Tbps/km^2 15 Tbps/km^2	500 Mpps1 Gbps	250,000/km^2
Broadband access in a crowd	Pedestrians	7.5 Tbps/km^2 3.75 Tbps/km^2	50 Mbps25 Mbps	500,000/km^2
Urban macro	Passenger in vehicles and pedestrians (max. 120 km/h)	50 Gbps/km^2 100 Gbps/km^2	25 Mbps50 Mbps	10,000/km^2

clear requirements for robotic surgery). In addition, "vehicle-to-everything (V2X)" technology in the industry sector is also expected to require a large amount of 5G network resources, and the relevant requirements for this technology are detailed in TS 22.186 recommendation of 3GPP. Likewise, although railway communications have critical requirements developed in TS 22.289 recommendation of 3GPP, there is little interest by researchers in the specific problems of this transport sector [38]. The requirements and technical capabilities set for 5G technology determine that the industry has the full potential for automation. The latest recommendation named "Service Requirements for 5G" is illuminated in 3GPP TS 22.261 recommendation. It outlines 32 types of performance requirements, key capabilities, as well as supply and security aspects.

Important IoT-related opportunities in the process of industrial automation include:

- Disassembly of the network
- User environment efficiency
- Resource efficiency (IoT mass operations capabilities, IoT management capabilities)
- Use of network capabilities
- QoS, priority, and policy management
- Monitoring of QoS (especially for vertical automation communication, URLLC, and eV2X services)
- Energy efficiency

- Services of location
- Nongovernmental networks
- Cyber-physical management applications by vertical areas

These include low latency, high traffic density, high reliability, high-capacity IoT traffic, high location detection, and key performance indicators for network-connected user equipment (UE) in a 5G system according to TS 22.261 [38]. The current objective measurements and key performance indicators are specified in the various tables placed in Recommendation TS 22.261 [3].

For 5G mobile network, high levels of reliability are critical, as this is a key distinctive characteristic when compared to architectures that use an unlicensed radio spectrum or preexisting evolved heterogeneous networks. An extensive research network on the high reliability of 5G solves the efficiency problems of the existing communication network such as reliability, bandwidth, and standby time. Providing high reliability for new 5G services will actually be higher when the geographically distributed system emerges. The complexity is that all known problems related to networks with a wide area occur, and they affect the performance of applications. In this context, the use of blockchain technology in 5G-based IoT systems in industrial automation on reliability and data privacy requirements can overcome such problems.

6 Existing Technologies Used in 5G-Based IoT for Industrial Automation

This section summarizes the available technologies to help describe, identify, isolate, and generalize 5G and IoT networks designed for industrial automation. It is important to study and implement such technologies, as they help to conduct research in this area, implement it, and then direct it to a new area.

6.1 Existing Technologies for 5G, IoT, and 5G-Enabled IoT for Industrial Automation

For 5G, its core application management technologies are not entirely new. Although the 5G network is considered a generational change, it is being promoted as an idea for the development of previous-generation wireless communications. Therefore, many aspects of 5G have their technological and technical characteristics, which are the basis of previous and current technologies [5, 39]. However, the new generation and currently evolving technologies may not be supported for older communication standards, but they are the main driving force for 5G.

Among the emerging technologies, 5G, IoT, and some of the most suitable for 5G-based IoT in industrial automation are categorized and discussed in Table 8.6. However, in Table 8.6, the key technologies available in 5G networks, IoT systems, and 5G-based IoT applications for industrial automation are classified and identified in a number of ways.

It should be noted that for 5G, IoT, and 5G-based IoT, each of the above-mentioned available technologies has a broad concept, and a large amount of research is being conducted on them. A very important job in them is to actually develop projects that combine their strengths with each other to achieve the goal of industrial automation and the tasks set. The following explores the possibilities of integrating Blockchain technology with existing technologies used in 5G-based IoT for industrial automation.

6.2 Blockchain for Key Technologies of 5G-Based IoT

By analyzing the research work, the blockchain is mainly found to be able to cooperate with the core technologies of 5G-based IoT in industrial automation, which includes the abovementioned cloud computing, edge computing, SDN, NFV, network separation, and M2M/D2D communications. Therefore, in addition to this section, we will briefly focus on the integration of existing technologies of blockchain and 5G-based IoT in industrial automation [8].

A. ***Blockchain for cloud computing (or Cloud RAN)*** In the last decade, cloud computing has received a lot of attention due to its unlimited storage and computing power, which provides efficient and powerful services with minimal control power. Currently, the blockchain has been explored and introduced into the cloud computing system to effectively address security issues in 5G networks based on cloud. In doing so, the blockchain plays an important role, such as ensuring the reliability and security of cloud data in the 5G network. It is used to create a verification platform among blockchain, BBU (baseband unit) system, IoT devices, and industry network, where user login data is permanently saved in the chain; however, user authentication is done automatically through smart contracts. There was a double increase in the use of blockchain in 5G cloud computing networks. It also uses smart contracts and blockchains to test data security in cloud-centric IoT systems. Blockchain enters the cloud computing system to create a complete security network, there IoT metadata is saved in the blockchain, and real data is stored in cloud storage, which provides extensive coverage for densely packed IoT devices.

B. ***Blockchain for MEC*** MEC has emerged as a promising technology to improve cloud computing capabilities and expand 5G network services. Edge computing is also known as mobile cloud, fog computing, or cloud. Edge computing, like cloud paradigm, can offer a range of computing services with capabilities for data storage, task processing, QoS improvements, and heterogeneity support. The distributed

Table 8.6 Existing technology categories addressed to industrial automation

Categories	Technologies	
	Types	Definitions
5G	Heterogeneous network	In order to expand the capabilities of mobile communication using the heterogeneous network, 5G mobile network will be able to get and provide simultaneously two/more network configurations, network architectures, radio connection technologies, base stations, standards, user requirements, data transmission solutions, and more
	Cooperative diversity and relaying	Cooperative diversity offers ways to implement improved wireless channels for 5G. This provides the "virtual" to a significant level by multiplying the number of users working together, named relays or nodes, to form a multi-input, multi-output (MIMO) order, thus increasing the likelihood of diversity among distributed users
	Massive MIMO and beamforming	Numerous additional antennas have been included in the network to help direct energy to small areas of the environment using massive MIMO and improved irradiance capabilities, providing a large amount of improvements in throughput and radiant energy efficiency through this method
	Cognitive radio networks (CRN)	As there is a growing interest in CRN in 5G and next-generation networks, it is expected that the spectrum problem will be addressed by providing access and use of dynamic spectrum for such networks. Using CRN in 5G significantly solves the spectrum problem
	Cloud computing	Large amounts of data that need to be generated and transmitted in a short period of time must first be stored in the cloud and processed in it. For example, many technologies such as 5G-based AGV, AR/VR, and video streaming rely on cloud computing
	Non-orthogonal multiple access (NOMA)	Recent work has rated NOMA as the most efficient technology to provide access to 5G. Because it is a solution to the problem of allocating individual channels, it allows to provide communication sources to several users at the same time through disproportionate use methods
IoT	Embedded system technology	Embedded system technology is the basis of IoT and is present in the structural structure of all devices located in its facilities. It includes sensor technology, computer hardware and software, electronic technology applications, etc. in the production of smart IoT objects
	Network communication technology	Sensors must be able to transmit signals about an object, communicate with each other, or communicate over a network. Wi-Fi, ZigBee, Bluetooth, ultra-broadband, and other common network communication technologies are essential tools for efficient operation of sensors in IoT applications
	Radio-frequency identification	RFID radio waves are used to detect and track elements in real time to ensure the identification of various IoT objects

(continued)

Table 8.6 (continued)

Categories	Technologies	
	Types	Definitions
	Sensor technology	Sensors play an important role in bridging the gap between the physical and the information environment. Sensors are typically located in IoT facilities and receive and collect data from the environment, generate and process data, or keep the situation informed
5G-enabled IoT for industrial automatiza-tion	Device-to-device	D2D notes to a method of establishing a direct connection between two peer nodes. In a D2D communication network, each node is automatically routed and creates a more flexible connection, so that data exchange takes place without the use of additional resources of the base station, which significantly improves the use of network resources
	Machine-to-machine/man	In a general sense, M2M stands for machine-to-machine, man-to-machine, machine-to-man, and machine-to-mobile network connectivity. M2M is not limited to data exchange, but aims to provide a continuous intelligent connection between people, machine, and system
	Big data	Making smart decisions based on big data production is an important indicator of CPMS intellectualization. Using this technology, it is possible to obtain data at the stage of product design, production, assembly, and logistics with 5G communication technology and modern sensor systems
	Service-oriented manufacturing	Based on real-time data on heterogeneous factors in shop-floor and service encapsulation samples, various manufacturing sources in the industry include hardware sources (e.g., machine tools, forging equipment, computing hardware) and software sources (e.g., model, data, software)
	Mobile edge computing (MEC)	Compared to cloud computing, which performs a large amount of data analysis and data management process in an embedded centralized computing resource plane, mobile edge computing is a physical processing method close to the industry in which production data is generated
	Cloud computing technology	Big data of industry including production equipment status parameters, production process data, order data, product quality data, production process data, and thousands of other processes are loaded into the cloud via 5G wireless communication network and computing in it. With the help of cloud technology, the reliability of such important data is strengthened
	Network functions virtualization (NFV)	It can be used to support smart manufacturing in a variety of ways. With this technology, it is possible not only to use a fully virtualized smart industry but also to reduce the interconnection and configuration time of machines and increase the efficiency of large-volume reservations. However, NFV technology can simplify the collection and analysis of machine data

(continued)

Table 8.6 (continued)

Categories	Technologies	
	Types	Definitions
	Network slicing	This technology has been validated in three different use cases, which clearly demonstrate the advantages of the following solution: remote control of industrial production, remote equipment maintenance, and dynamic industrial production. Network slicing technology can also be used in AGVs
	Software-defined networking (SDN)	This technology has aroused great interest in recent years and has been adopted as a mainstay of 5G networks. At the same time, SDN is being touted as a smart network architecture designed to increase the programming and flexibility of networks

structure of edge computing also offers advantages ranging from a variety of computing services to a reduction in network management complexity resulting from a sharp increase in IoT devices and rapid growth in demand for 5G services. However, its safety is an important issue. The data exchange of 5G services can be weak to cyber attacks in a dynamic peak computing environment (e.g., attacks such as network congestion, information theft, denial of services, etc.). In addition, the configuration and setting data provided by edge service providers (ESPs) must be reliable and protected, but in reality, the MEC system cannot fully meet this requirement due to its high dynamics and transparency. Another challenging issue is to ensure the confidentiality and integrity of 5G heterogeneous information from external sources. The existing modern blockchain architecture emerged as a solution to problems such as the confidentiality and reliability of data in a network that could be encountered [20, 40].

The same decentralized characteristics of MEC and blockchain technologies built on networking, data storage systems, computing, and communications ensure that their combination is in a natural state. Recent research has shown that the use of blockchain technology in edge computing systems can support a number of security and management services. In fact, blockchain can be used to optimize the network capability of edge networks. Overall, as shown in Fig. 8.9, blockchains can support 5G-based IoT services for industrial applications in three main areas: network, storage system, and computing.

C. *Blockchain for SDN* The basic concept of SDN is to separate the control level outside the network and provide external data management through a logical software control tool that provides common access between various components of HetNet. The construction of this project not just offers new construction, usage options, and management, although it also enables efficient delivery of user services while making more efficient use of network resources. According to the 5G concept, SDNs are designed to program connection services produced by 5G networks, in which traffic can be dynamically routed and managed, which allows for maximum

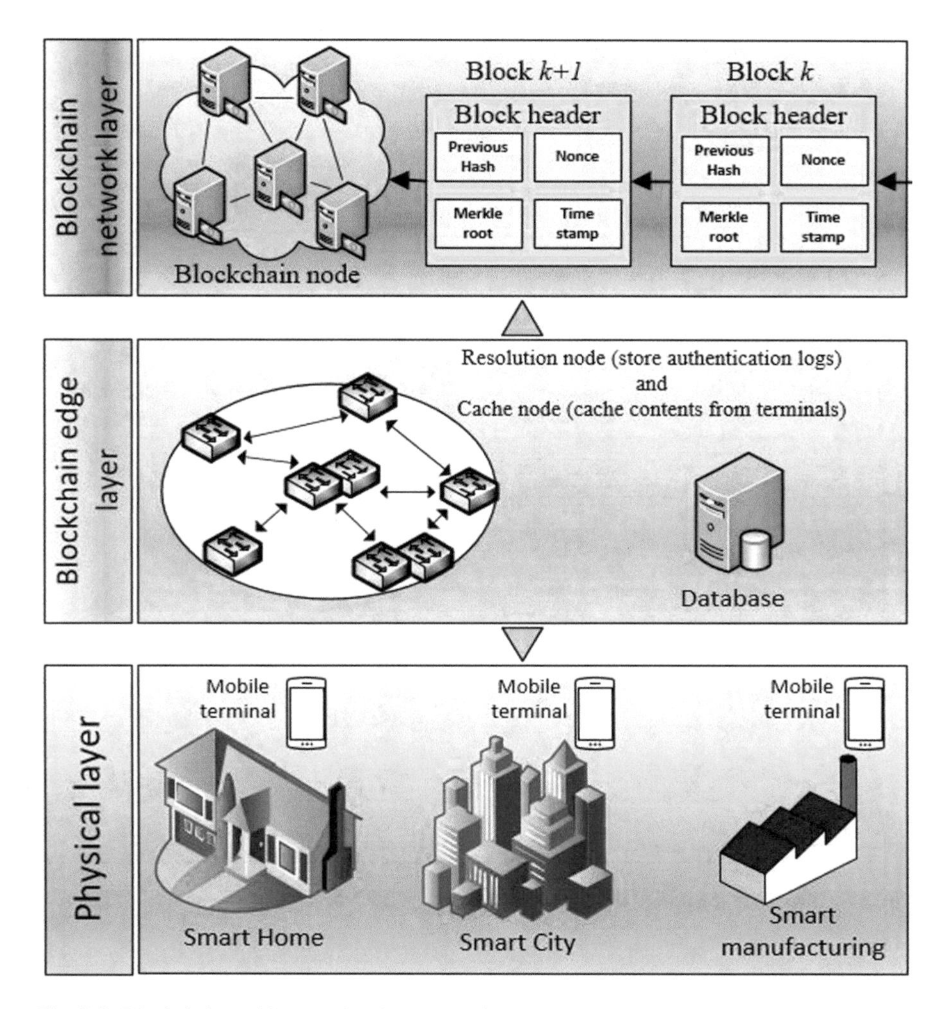

Fig. 8.9 Blockchain architecture in edge computing technology for 5G-based IoT services

efficiency. However, despite the advantages of this network technology, there are some challenges with outdated solutions, particularly in terms of security, flexibility, and coverage [8] (Fig. 8.10).

To overcome these shortcomings in the SDN architecture, many studies suggest blockchain as a decentralized security solution for SDN. Message and transactions obtained from the blockchain can be allocated with the administrator via special keys. Every SDN management device has a dedicated transfer key obtained from the blockchain and is used to send and receive data. Importantly, a hierarchical structure based on the blockchain can effectively solve coverage. The integration of the

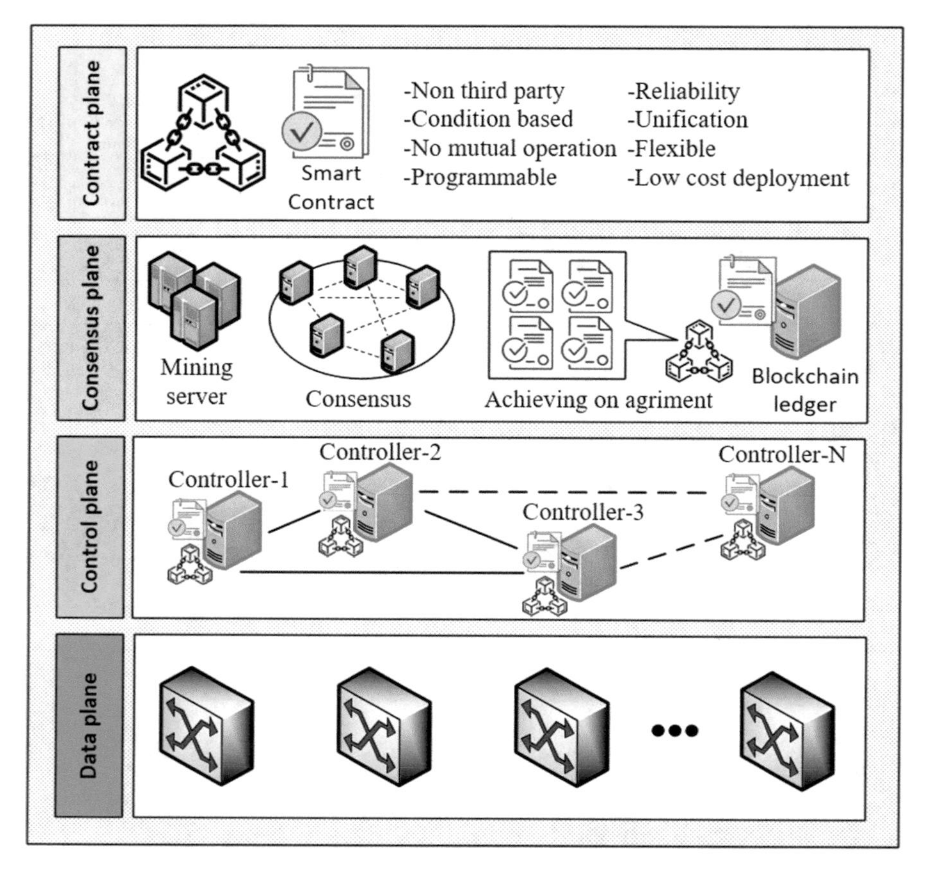

Fig. 8.10 Blockchain-based SDN architecture for 5G-enabled IoT

blockchain into SDN technology is expected to eliminate agents for authentication, reduce the cost of the transaction, and create global access to each user.

D. ***Blockchain for NFV*** NFV is a network architecture concept standardized by the ETSI that uses standard equipment to deploy different standalone and software components of network. NFV mainly covers three principal architectural parts: management and network orchestration (MANO), the network functions virtualization infrastructure (NFVI), and the virtual network functions (VNFs). MANO includes the life cycle management and regulatory process of the software and physical resources, NFVI supports the implementation of VNFs, and VNFs are running in NFVI. NFVs perform network functions (NF) virtually by allocating virtualized gateways, virtualized firewalls, as well as virtualized components of the network that provide flexible networking functions, as well as hardware (e.g.,, gateways, firewalls), to support the functions that run on them. At the same time,

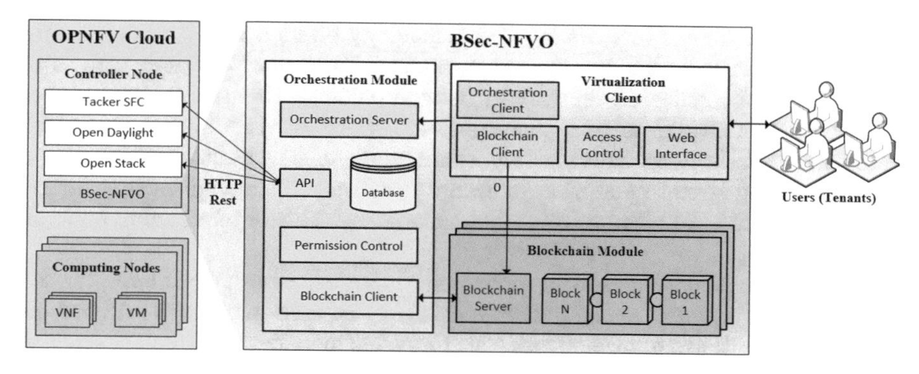

Fig. 8.11 NFV and blockchain integration architecture

the virtualization of network functionality and the creation of a chain of service functions cause new security challenges [41].

In NFVs, virtualization servers can perform separate system operations, such as migrating virtual machines (VMs) or allocating resources using regulated protocols, and VMs can also offer special functions. Furthermore, currently, communication security among the regulator and the physical mechanisms is a major issue. In fact, this architecture is very sensitive to attacks that occur for a variety of reasons. VMs can be created by hackers to run on a server and can also be used to perform external attacks to reject service. In addition, internal attacks by VMs are another problem that negatively affects data integrity and privacy.

In such problematic issues, blockchain has developed as an effective tool to overcome these challenges. By its originality, non-rejection, and integrity feature, the blockchain may simplify NFV networks in three main ways. Firstly, the blockchain allows for better regulation of VNF services and the implementation of reliable, easy, and flexible management for network management. Secondly, the blockchain protects the delivery of network functions and the integrity of the system from internal attacks and external threats, i.e., detrimental DoS attacks and VM modifications. Finally, blockchain can accomplish verification of data and monitoring of system status while networking connection. In the figure below, we can see the architecture of using blockchain to solve the above problems of NFV for 5G-based IoT systems [8].

The BSec-NFVO architecture shown in Fig. 8.11 consists of three main modules: a visualization module (VM), an orchestration module (OM), and a blockchain module (BM). VM provides an interface between users, NFV, and service function chaining (SFC) services. OM executes instructions delivered by users through a visual module. BM checks and validates operations before they are performed by OM. By continuously recording all the instructions that govern the service chains activated by the proposed architecture blocking network, the validity, integrity, and non-rejection of instructions that ensure data validation and tracking in a multiuser and multi-domain NFV environment are guaranteed.

E. ***Blockchain for network slicing*** 5G technology offers a whole new shape for networks to integrate IoT network management. To support different models of IoT systems, 5G relies on the idea of network slicing. This allows operators to allocate their networks to particular applications and services such as industrial automation applications, smart factory, smart home, or transportation. In addition, network slicing leads to a number of security problems, including security threats in network slices and resource allocation among segments of intersection between domains [46].

In such problems, blockchain technology can create big possibilities for 5G cross-slice management security. Blockchain may be used to create a safe end-to-end network slice and allow network slice providers to manage their resources. The blockchain is used to create a mediation mechanism across the network slices to ensure secure and personal agreements between isolated network layer supplier and a resource provider for 5G services. When a network slice supplier sends or receives a request to set up a slice from end to end, it sends it to the blockchain to track and distribute the process. Smart contracts have been developed to support the placement of small slice components, called slice smart contracts (SSCs), where each SSC identifies the resources required for a small slice. In this way, resource suppliers can trade resources under contracts with small slice components. All information related to the placement of additional small slice is permanently stored in an authorized blockchain managed by the network slicing provider. The use of technology based on blockchain not only attaches security capabilities but also supports accountability and privacy in network slicing.

F. ***Blockchain for M2M/D2D communication*** The sharp increase in 5G mobile data traffic has increased the requirement for high-level connectivity services. M2M/D2D communication technology is one such service in the 5G network. Conceptually, M2M/D2D connections represent the type of technology which allows mobile devices/machines (e.g., smartphones, tablets, etc.) to communicate directly with each other without an access point or a core network of mobile infrastructure. M2M/D2D uses device connection compatibility to make efficient use of available resources, which improves overall system throughput, reduces communication interruptions, and reduces power consumption and traffic load [45]. Thus, the M2M/D2D connection can promote new "peer-to-peer" and "location-based" services and applications, making it compatible with next-generation 5G mobile network services.

Furthermore, direct communication among mobile devices makes new difficulties for M2M/D2D-based 5G in terms of network management, security, and loss of performance. Indeed, data exchange between devices can be at risk of data transmission due to various threats in the unreliable environment of M2M/D2D. In addition, M2M/D2D devices have a high level of unreliability, and if the network does not have an authentication mechanism, illegal access to resources on the servers will be possible. In addition, existing M2M/D2D architectures rely on external management to allow data access and require authentication during D2D/M2M

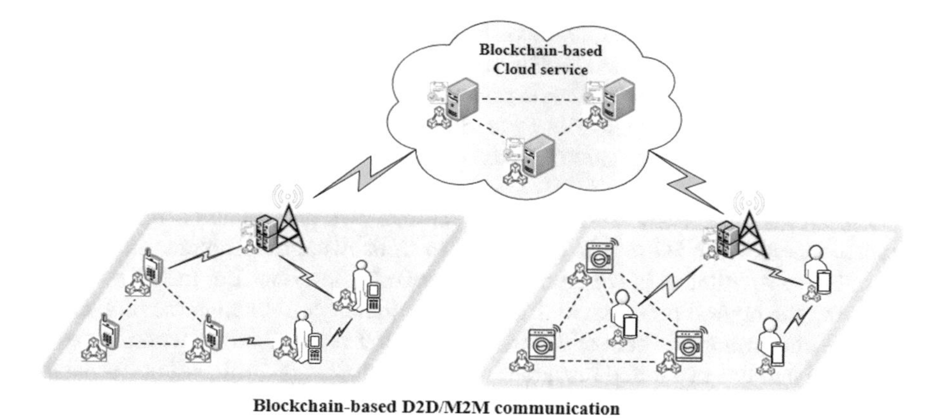

Fig. 8.12 Blockchain integration architecture for D2D/M2M communication technology

communication, which can lead to unnecessary communication interruptions and degrade the overall performance of the network [20] (Fig. 8.12).

In particular, the blockchain demonstrates its effectiveness in providing the required solution, that is, allowing users with caching capability to store and share content with other mobile devices via M2M/D2D to ensure better content sharing between mobile devices. Besides, blockchain technology is used to develop a security mechanism using the consensus protocol to ensure the authenticity of channel status information (CSI) of mobile users under the M2M/D2D mobile network [20, 44]. The M2M/D2D network, based on the blockchain agreement, consists of mobile users and two blockchains: an integrity chain (I-chain) and a fraud chain (F-chain). Mobile users can verify and confirm received CSI messages for share and storage through a mutual agreement mechanism before signing and adding to a decentralized ledger. Research on this technology also concludes that a blockchain-based approach has the potential to significantly increase spectral efficiency while providing effective CSI authentication services for M2M/D2D networks [8].

7 Solutions for Using 5G-Based IoT in Industrial Automation

As mentioned above, 5G and IoT together are a new generation technology that is already known for managing emerging communications and connectivity networks. This means that the expected speed, reliability, and access capabilities for both 5G and IoT are very high. Effective maintenance of these technologies requires a large number of resources (e.g., transmission power, spectrum, bandwidth, time interval, data transfer rate, memory, processor, energy power, etc.) and readiness to make them work and achieve the expected results. However, the network may not always meet this requirement, because the resources to manage real-world wireless technology are usually limited and/or very limited. Many studies have pointed out that limited resource use is a major obstacle to the implementation of modern and promising telecommunications paradigms. The fact that there are resources available to achieve the 5G and IoT goals is a major problem with these technologies [5].

Research is underway to develop appropriate RA and resource management models/schemes for 5G and IoT in order to limit resource problems in the network. The RA solution in communication networks describes the mechanisms for achieving the highest productivity for networks, thereby eliminating the problem of resource constraint. The goal of RA in 5G-based IoT is to effectively coordinate the allocation and use of limited resources to achieve the overall efficient performance of these networks. The availability of resources in 5G-based IoT applications and approaches to addressing existing usage limitations are very diverse, and there may not be an approach that is appropriate for all applications to address resource management issues. This section analyzes the various methods and/or approaches developed and used to solve RA problems in 5G-based IoT applications. Identified resource management solutions are classified and categorized. The advantages and disadvantages of different approaches are highlighted, and ideas for practical solutions are put forward. At the same time, Blockchain technology, one of the new solutions for resource management, has been studied, and the effectiveness of the use of its solutions in the implementation of resource allocation has been clearly demonstrated.

7.1 Classification of RA Optimization Solutions in 5G and IoT for Industrial Automatization

There are many optimization solutions that are well organized and well documented to solve resource allocation problems. However, the specific features and scope of 5G-based IoT do not allow direct application of traditional optimization techniques, so RA problems in 5G-based IoT applications require the development of clear and effective optimization solutions. It should be noted that optimization methods by making the necessary changes or adaptations can also be used to solve RA problems in 5G-based IoTs. The research in this section analyzes various optimization solutions designed to solve RA problems in 5G-based IoT applications. The following describes the various solutions and optimization methods used directly or by any adaptation method in solving RA problems in 5G-based IoT applications [5].

A. *Classical Optimization* Another method of optimization that is widely used in the development and solution of RA problems in 5G-based IoT applications is classical optimization. In classical optimization, RA problems in 5G-based IoT applications are identified and solved using efficiently organized optimization methods such as linear programming and convex optimization. The identified problems of RA fit perfectly into one or more of these classic optimization solutions, so they can be easily solved by making a few modifications as possible.

There are a number of research studies that use classical optimization in solving RA problems in 5G-based IoTs. In one of these research studies, for 5G networks, multiple non-orthogonal access schemes are recommended for a multiuser access scheme that corresponds to the typical OFDM scheme used in LTE/Advanced-LTE [42]. The RA problem identified in it was solved by Lagrange's classical optimization solution of dual and dynamic programming.

B. ***Self-Optimization*** Many researchers point out that due to the complexity and wide usage of 5G-based IoT networks, the most appropriate optimization method for their RA problems is self-optimization. With self-optimization, each device (or user) is able to optimize its resources to provide the required services and improve the overall interests of the larger network [5]. In general, a self-optimization solution for IoT helps to shift the focus from the design and placement of one or more autonomous elements to a large complex ecosystem of the elements network. Additional functions and system computing are often used in distributed autonomous computing systems to achieve self-optimization. Self-optimization in achieving the desired experience quality in a 5G network requires a switch on of these devices. Furthermore, neural networks can be used in 5G applications as an effective method of adaptive assessment of the quality perceived by users or user devices and self-optimization.

Self-optimization is studied in IoT as a solution to the problem of integration and management of various large-scale heterogeneous networks that combine IoT, which is one of the main problems of RA. For example, the control system of a network of self-optimizing sensors will be able to automatically adjust itself with saving energy at the same time. Such an advanced management system needs to be flexible and robustly built to withstand significant structural changes. The development of such management systems is critical to creating an effective IoT network and solving future problems. RA problems using self-optimization methods are an important positive step in realizing the benefits of IoT.

C. ***Heuristics and Meta-heuristics Optimization*** Heuristic and meta-heuristic solutions have been widely used in solving RA problems in 5G-based IoT applications. The heuristic solution does not have to have scientific or analytical conclusions, only the logical basis for solving certain problems in 5G-based IoT. A meta-heuristic solution is used for problems where multiple local "optimal" solutions can be found for a particular RA problem. Although heuristic and meta-heuristic solutions often offer only small optimal solutions, they are usually implemented in a much shorter time interval and with the complexity of computing networks, even in large networks. Therefore, heuristic and meta-heuristic optimization methods have been widely used in solving RA problems.

In particular, objects in IoT need to present their resources in a very flexible way. For example, one of the studies on this topic [43] proposes a distributed optimization protocol based on a consensus algorithm to solve the problem of resource allocation and management in heterogeneous networks of IoT. In the proposed protocol, an IoT technology was created in which nodes containing the same IoT functions had to regulate their buffer capture and frequency functions. The study showed that using the proposed protocol, the network can move to an equally distributed solution of resources between clusters and nodes, thereby significantly improving network performance.

D. ***Multi-objective Optimization*** In multipurpose optimization, the problem usually involves a series of goals that need to be accomplished at the same time. Sometimes these goals conflict with each other, resulting in the RA problem being complicated and difficult to solve. For example, in 5G, the goal of the RA problem may be to simultaneously minimize data transmission time, minimize the power required to transmit that data, and at the same time increase the total amount of data transmitted over a specified time interval. These goals are contradictory, as minimizing power and increasing data transfer at the same time is a challenge. Multipurpose optimization methods are used to solve this type of problem.

Table 8.7 summarizes the optimization solutions discussed above for resource allocation and management in 5G-based IoT applications.

This table shows the available solutions for optimizing resource allocation for 5G-based IoT networks and its application to various applications through a brief systematized review. These brief conclusions can be used to distinguish and apply such solutions to industrial automation applications.

7.2 Blockchain Solution of Resource Allocation in 5G and IoT for Industrial Purposes

Mobile resource (i.e., computing, memory, storage, channel, and bandwidth) is one of the most popular services in 5G networks. The growing diversity of 5G network services is creating unprecedented levels of complexity in mobile resource management [44]. Edge/cloud computing in 5G networks should save resources to meet the growing needs of long-term mobile users and allocate computing capacity to ensure efficient data processing. In virtualized networks, single-layer VNFs can have different sources, namely, CP, bandwidth, memory, and storage system, depending on their user and function requirements.

The resource needs of layers of the same functional type may be different because they serve different mobile users. For example, a provider may use multiple IoT layers dedicated to each specific application. Implementing an optimal allocation to a 5G-based IoT network is a very important task in such situations where different resource capabilities and the necessity for different resources are present. The main thing is that the current resource management architecture relies mainly on a central authority that performs resource allocation and user access control, but such models remain the risk of damage of the one side and safety of the other side. In addition, the tracking capability of existing resource allocation systems is very weak, so this

Table 8.7 Optimization solution summary for RA 5G-enabled IoT applications

№	Categories	Examples	Drawbacks	Features
1	Classical optimization	Linear programming, convex programming, etc.	Proving the convection of problems can be a complex task, and many problems do not fit into any classical model of optimization	It has optimal solutions, and such solutions can serve as a frame for solutions identified on the basis of other approaches
2	Self-optimization	Intelligent controlling traffic lights, optimal mapping between applications and resources in the WSN, smart data flow through network, etc.	Its implementation is difficult due to the complexity of IoT systems operating on the basis of 5G It can be difficult to manage the right balance between rapid system change and the ability of individual users to access resources	An optional object uses auxiliary functions and systematic computing capabilities in optimizing its resources
3	Heuristics and meta-heuristics	Greedy algorithms, genetic algorithms, etc.	In most cases offers lower optimal solutions It may not have solutions or may not be specific to the problem	Offers quick solutions, works effectively with big problems, has the high ability to apply in practice
4	Multi-objective optimization	Cooperative and noncooperative game, Nash bargaining, etc.	Solution models can be complex and analytical modeling of solutions are usually difficult to achieve	Good with problems that have multiple objectives, uses ideas from game theory

leads to the disruption of common resources as a result of various attacks or illegal use by malicious users. All of these issues need to be addressed effectively before 5G services can be implemented in applications.

Blockchains are recognized by scientists as a highly effective solution for solving the above problems and managing resources. The use of blockchain makes distributed resource allocation an effective solution for service providers as well as mobile users/devices. Blockchain simplifies the concept of resource allocation and management while preserving important features of the core network and providing strong security. For example, blockchain can be used in VNFs to implement the allocation of trusted resources that meet user requests in various circumstances, such as user demand, the price [8, 45]. In addition, smart contracts in the blockchain, integrated to create an auction scheme, allow users to implement an optimal resource allocation based on transparency in a dynamic mobile environment.

In 5G-based IoT networks, edge computing plays an important role in improving the QoS of mobile services due to its low-latency and fast computing capabilities. Blockchain is used to develop a decentralized resource allocation architecture, which is an effective solution in terms of latency and service speed of a centralized architecture. When providing IoT data adaptive computing services, resource allocation must be dynamically adjusted without any centralized management to maintain high QoS. Blockchain is very responsive to such scenarios and requires the use of distributed ledgers to update resource information automatically and reliably. In the condition of a lack of resources in the network, a coefficient calculation model may be required to support low-capacity devices [8, 46]. In this context, the blockchain is effective for ensuring reliable resource exchange between edge nodes. Resource requirements can be verified through smart contracts with an access policy without going through strictly centralized control, which reduces resource-sharing time.

8 Conclusion

5G and IoT technologies are a widely used approach in modern industrial automation applications, including promising communication standards and new computing systems in the near future. Both of these network technologies have great potential to help create an interconnected and highly functional digital world. This paper has analyzed the prospects of 5G-based IoT networks used for industrial automation and has explored their technical software capabilities as required by modern applications and technologies. The study has examined the integration of 5G-based IoT networks into the field of industrial automation, their architecture, existing technologies used in them, and blockchain technology, which plays a key role in ensuring the security and reliability of these technologies, and the optimization methods used in the solution of resource allocation problems and the importance of blockchain technology have been discussed.

The key challenge is to use their limited and/or limited resources efficiently to provide autonomous industrial applications, reliable and high-level wireless communications, and computer and Internet services within the maximum demand. Therefore, it is important to study enough models to solve their resource allocation and management problems, and such research is currently underway. In such research, optimization approaches to resource allocation are studied and explored. The approaches have been classified according to their best characteristics. In addition, the strengths and weaknesses of resource allocation optimization solutions have been discussed. Based on this, recommendations have been made on the selection of current RA optimization solutions for 5G-based IoT networks in industrial automation that require in-depth research and based on the high efficiency of blockchain technology for resource management. According to the results of the study, resource management can significantly improve the use of resources in the network and the overall performance of 5G and IoT networks.

References

1. S. Li, L. Da Xu, S. Zhao, 5G internet of things: A survey. J. Ind. Inf. Integr. **10**, 1–9 (2018). https://doi.org/10.1016/j.jii.2018.01.005
2. J. Cheng, W. Chen, F. Tao, C.-L. Lin, Industrial IoT in 5G environment towards smart manufacturing. J. Ind. Inf. Integr. **10**, 10–19 (2018). https://doi.org/10.1016/j.jii.2018.04.001
3. H. Fattah, *5G LTE Narrowband Internet of Things (NB-IoT) (Handbook)* (CRC Press/Taylor & Francis Group, London/New York, 2019), p. 240. ISBN-13: 978-1-138-31760-4
4. P. Varga, J. Peto, A. Franko, D. Balla, D. Haja, F. Janky, G. Soos, D. Ficzere, M. Maliosz, L. Toka, 5G support for industrial IoT applications – Challenges, solutions, and research gaps. Sensors **20**, 828 (2020). https://doi.org/10.3390/s20030828
5. B.S. Awoyemi, A.S. Alfa, B.T. Maharaj, *Resource Optimisation in 5G and Internet-of-Things Networking* (Part of Springer Nature, 2020), pp. 1–32. https://doi.org/10.1007/s11277-019-07010-9
6. S. Ludwig, M. Karrenbauer, A. Fellan, H.D. Schotten, H. Buhr et al., *A 5G Architecture for the Factory of the Future* (IEEE, 2018), pp. 1–8. arXiv:1809.09396v1
7. G.A. Akpakwu, B.J. Silva, G.P. Hancke, A.M. Abu-Mahfouz, A survey on 5G networks for the Internet of Things: Communication technologies and challenges. *IEEE Access*, 1–28 (2017). https://doi.org/10.1109/ACCESS.2017.2779844
8. D.C. Nguyen, P.N. Pathirana, M. Ding, A. Seneviratne, Blockchain for 5G and beyond networks: A State of the Art Survey. IEEE Commun. Surv. Tutor. 2019, pp. 1–45. arXiv:1912.05062v1
9. T. Hewa, A. Kalla, A. Nag, M. Ylianttila, M. Liyanage, Blockchain for 5G and IoT: Opportunities and challenges. *Conference Paper*, 2020, pp. 1–8
10. S. Kumar, G. Gupta, K.R. Singh, 5G: Revolution of future communication technology, in *International Conference on Green Computing and Internet of Things (ICGCIoT)*, (IEEE, 2015), pp. 143–147
11. F. Hu, *Opportunities in 5G Networks. A Research and Development Perspective* (CRC Press/Taylor & Francis Group, London/New York, 2016), p. 538. ISBN-13: 978-1-4987-3955-9
12. E. O'Connell, D. Moore, T. Newe, Challenges associated with implementing 5G in manufacturing. Telecom **1**, 48–67 (2020). https://doi.org/10.3390/telecom1010005
13. B. Mathieu, C. Westphal, P. Truong, *Towards the Usage of CCN for IoT Networks* (Springer International Publishing, Switzerland, 2016), pp. 3–24. https://doi.org/10.1007/978-3-319-30913-2_1
14. F. Tao, Q. Qi, New IT driven service-oriented smart manufacturing: Framework and characteristics. IEEE Trans. Syst. Man Cybern. Syst. **49**(1), 81–91 (2017). https://doi.org/10.1109/TSMC.2017.2723764
15. J. Lee, B. Bagheri, H.A. Kao, A cyber-physical systems architecture for industry 4.0-based manufacturing systems. Manuf. Lett. **3**, 18–23 (2015)
16. G.A. Akpakwu et al., A survey on 5g networks for the internet of things: Communication technologies and challenges. IEEE Access **6**, 3619–3647 (2017)
17. BLE, Smart bluetooth low energy. [Online]: Availability: http://www.bluetooth.com/Pages/Bluetooth-Smart.aspx
18. W. Al-Saqaf, N. Seidler, Blockchain technology for social impact: Opportunities and challenges ahead. J. Cyber Policy **2**(3), 338–354 (2017)
19. M.T. Hammi, B. Hammi, P. Bellot, A. Serhrouchni, Bubbles of trust: A decentralized Blockchain-based authentication system for IoT. Comput. Secur. **78**, 126–142 (2018)
20. I. Mistry, S. Tanwar, S. Tyagi, N. Kumar, Blockchain for 5G enabled IoT for industrial automation: A systematic review, solutions, and challenges. Mech. Syst. Signal Process. **135**, 106382 (2020)
21. K.E. Khujamatov, D.T. Khasanov, E.N. Reypnazarov, Modeling and research of automatic sun tracking system on the bases of IoT and Arduino UNO, in *International Conference on*

Information Science and Communications Technologies ICISCT 2019, Tashkent, Uzbekistan, 2019. https://doi.org/10.1109/ICISCT47635.2019.9011913
22. B. Yu, J. Wright, S. Nepal, L. Zhu, J. Liu, R. Ranjan, IoTChain: Establishing Trust in the Internet of things ecosystem using Blockchain. IEEE Cloud Comput. **5**(4), 12–23 (2018)
23. J. Govil, J. Govil, 5G: Functionalities development and an analysis of mobile wireless grid, in *First International Conference on Emerging Trends in Engineering and Technology*, (IEEE, Nagpur, 2008), pp. 270–275
24. K.E. Khujamatov, D.T. Khasanov, E.N. Reypnazarov, Research and modelling adaptive management of hybrid power supply systems for object telecommunications based on IoT, in *International Conference on Information Science and Communications Technologies ICISCT 2019*, Tashkent, Uzbekistan, 2019. https://doi.org/10.1109/ICISCT47635.2019.9011831
25. A. Gohil, H. Modi, S.K. Patel, 5G technology of mobile communication: A survey, in *International Conference on Intelligent Systems and Signal Processing (ISSP)*, (IEEE, Gujarat, 2013), pp. 289–290
26. A. Reyna et al., On blockchain and its integration with IoT. Challenges and opportunities. Future Gener. Comput. Syst. **88**, 173–190 (2018)
27. H.F. Atlam et al., Blockchain with internet of things: Benefits, challenges, and future directions. Int. J. Intell. Syst. Appl **10**(6), 40–48 (2018)
28. K. Khujamatov, K. Ahmad, E. Reypnazarov, D. Khasanov, Markov chain based modeling bandwidth states of the wireless sensor networks of monitoring system. Int. J. Adv. Sci. Technol. **29**(04), 4889 (2020) Retrieved from http://sersc.org/journals/index.php/IJAST/article/view/24920
29. M.R. Palattella, M. Dohler, A. Grieco, G. Rizzo, J. Torsner, T. Engel, et al., Internet of things in the 5G era: Enablers, architecture, and business models. IEEE J. Sel. Areas Commun **34**(3), 510–527 (2016)
30. S. Bayhan, A. Zubow, P. Gawłowicz, A. Wolisz, Smart contracts for spectrum sensing as a service. IEEE Trans. Cogn. Commun. Netw. **5**(3), 648–660 (2019)
31. K. Kotobi, S.G. Bilen, Blockchain-enabled spectrum access in cognitive radio networks, in *2017 Wireless Telecommunications Symposium (WTS)*, 2017, pp. 1–6
32. I.D. Alvarenga, G.A. Rebello, O.C.M. Duarte, Securing configuration management and migration of virtual network functions using blockchain, in *NOMS 2018–2018 IEEE/IFIP Network Operations and Management Symposium*, 2018, pp. 1–9
33. Ericsson Mobility Report, November 2019. Internet source, available: https://www.ericsson.com/en/mobility-report/reports
34. Cellular IoT Evolution for Industry Digitalization, Ericsson White paper. Internet source, available: https://www.ericsson.com/en/reports-and-papers/white-papers/cellular-iot-evolution-for-industry-digitalization
35. O. Liberg, M. Sundberg, E. Wang, J. Bergman, J. Sachs, G. Wikström, *Cellular Internet of Things – From Massive Deployments to Critical 5G Applications*, 2nd edn. (Academic Press, 2019), p. 774. ISBN-13: 9780081029039
36. Cellular IoT in the 5G era, Ericsson White paper. Internet source, available: https://www.ericsson.com/en/reports-and-papers/white-papers/cellular-iot-in-the-5g-era
37. 5G-TSN integration meets networking requirements for industrial automation, Ericsson Technology Review 2019. Internet source, available: https://www.ericsson.com/en/reports-and-papers/ericsson-technology-review/articles/boosting-smart-manufacturing-with-5g-wireless-connectivity
38. 3GPP, Service requirements for cyberphysical control applications in vertical domains. TS 22.104 v17.2.0, 3rd Generation Partnership Project, 2019. Available online: http://www.3gpp.org/ftp//Specs/archive/22_series/22.104/22104-h20.zip. Accessed 31 Jan 2020
39. X. Meng, J. Li, D. Zhou, D. Yang, 5G technology requirements and related test environments for evaluation. China Commun. **13**(Supplement 2), 42–51 (2016). https://doi.org/10.1109/CC.2016.7833459
40. J. Zhang, B. Chen, Y. Zhao, X. Cheng, F. Hu, Data security and privacy-preserving in edge computing paradigm: Survey and open issues. IEEE Access **6**, 18209–18237 (2018)

41. I. Farris, T. Taleb, Y. Khettab, J. Song, A survey on emerging sdn and nfv security mechanisms for IoT systems. IEEE Commun. Surv. Tutor. **21**(1), 812–837 (2018)
42. L. Lei, D. Yuan, C.K. Ho, S. Sun, Joint optimization of power and channel allocation with non-orthogonal multiple access for 5G cellular systems, in *2015 IEEE Global Communications Conference (GLOBECOM)*, 2015, pp. 1–6. https://doi.org/10.1109/GLOCOM.2015.7417761
43. G. Colistra, V. Pilloni, L. Atzori, Task allocation in group of nodes in the IoT: A consensus approach. in *2014 IEEE international conference on Communications (ICC)*, 2014, pp. 3848–3853. https://doi.org/10.1109/ICC.2014.6883921
44. U. Bodkhe, S. Tanwar, K. Parekh, P. Khanpara, S. Tyagi, N. Kumar, M. Alazab, Blockchain for industry 4.0: A comprehensive review. IEEE Access **4**, 1–37 (2016). https://doi.org/10.1109/ACCESS.2020.2988579
45. M.F. Franco, E.J. Scheid, L.Z. Granville, B. Stiller, Brain: Blockchain-based reverse auction for infrastructure supply in virtual network functions-as-a-service, in *2019 IFIP Networking Conference (IFIP Networking)*, 2019, pp. 1–9
46. C. Xu, K. Wang, G. Xu, P. Li, S. Guo, J. Luo, Making big data open in collaborative edges: A blockchain-based framework with reduced resource requirements, in *2018 IEEE International Conference on Communications (ICC)*, 2018, pp. 1–6
47. I.X. Siddikov, K.E. Khujamatov, D.T. Khasanov, E.N. Reypnazarov, Modeling of monitoring systems of solar power stations for telecommunication facilities based on wireless nets. Int. Sci. Tech. J. Chem. Techn. Control Manag. **2020**(3), 20–28 (2020)
48. A.A. Muradova, Kh.E. Khujamatov, Results of calculations of parameters of reliability of restored devices of the multiservice communication network, in *International Conference on Information Science and Communications Technologies ICISCT 2019*, Tashkent, Uzbekistan, 2019. https://doi.org/10.1109/ICISCT47635.2019.9011932

Chapter 9
Enabling Technologies and Architecture for 5G-Enabled IoT

Parveen Mor and Shalini Bhaskar Bajaj

1 Introduction

The IoT and its utilization in different modern areas is developing at an exponential rate. As indicated by Gartner, the quantity of IoT-enabled gadgets may reach 24 billion by 2020. With the assistance of normalized conventions and layers, engineering can be created to execute the pertinent administrations by IoT devices [1]. These devices have been effectively utilized in the car business to satisfy the needs of the end-clients and to accomplish their moderate business objectives. Lately, there is a requirement for an intensive, high quality item with diminished item cost. The IoT has changed these situations utilizing the 5G framework, in which constant activity between machines, information, and humans is conceivable to address the previously mentioned issues. Currently, most IoT-based frameworks are manufactured utilizing the combined customer, cloud, and solid information databases [2]. IoT is allowed by a range of technologies, including Far Sensor Frameworks, Appropriate Figuring, Big Data Analysis, Embedded Devices, Safety Protocols and Models, Correspondence Shows, Wireless Internet, and Semantic Search Engines. Social networks enable people to use the internet to keep in touch and exchange their data.

P. Mor (✉) · S. B. Bajaj
Amity University, Noida, India
e-mail: parveen.mor@sharda.ac.in; sbbajaj@ggn.amity.edu

S. Tanwar (ed.), *Blockchain for 5G-Enabled IoT*,
https://doi.org/10.1007/978-3-030-67490-8_9

1.1 Scope of 5G-Enabled IoT

The current stage is the Internet of Things (IoT), which provides the means to connect a broad variety of physical articles to the internet. Today, the IoT is increasingly infiltrating the global network room by improving the supply of hardware from tiny consumer devices to massive mechanical machines. Along these lines, IoT can possibly profoundly change the manner in which robotization is seen at the (physical) object level. IoT offers to energize new roads, particularly for businesses, since it inherently bolsters Machine-to-Machine (M2M) communications. Normalization bodies are enthusiastically investigating such M2M correspondences [3]. The development of versatile systems denoted its commencement in 1980 as first age (1G) that simply bolstered voice correspondence. With 2G, computerized frameworks hit the market and in this manner the administrations such as instant message were presented. It was 3G which offered versatile broadband types of assistance with an improved degree of security. 4G, all things considered the current age, upgraded the data rate, security, and QoS while diminishing the inactivity. The versatile correspondence industry is not prepared for 5G. The progressions in IoT and the impending emergence of 5G have advocated the advancement of 5G-enabled IoT applications. Such applications have rigid prerequisites such as high limit, guaranteed protection and security, adaptability of heterogeneous applications, ultra-low inactivity, advanced utilization of system assets, efficient energy management, and low OPEX [4]. Despite the fact that the security architectures that are currently being used form mobile networks and generic IoT systems match the necessary desires, they are concentrated on a basic level [1, 2, 5]. Utilizing such unified security answers for 5G and 5G-enabled IoT applications will prompt different hindrances, such as expanded expense because of innate heterogeneity, mind-boggling and static security of the executive's methods, overuse of system assets, the formation of a bottleneck in the system, specific intention of discontent, and high OPEX. Consequently, proceeding with the utilization of unified security answers for 5G and IoT-driven applications will not only battle to fulfill the needs but will also antagonistically influence the anticipated dreams of 5G and IoT. In this specific situation, blockchain innovation is a promising structure solution because it can provide answers to all security related issues in a unified and decentralized way. To welcome the conceivable business esteem that can be accomplished by supporting blockchain innovation for 5G and IoT, it merits taking a gander at their evaluated singular business esteems. From one viewpoint, IoT's worldwide market is evaluated to reach $457 billion by 2020 [6], and Industry IoT (IIoT) solely will include a $14.2 trillion incentive by 2030 [7]. Then again, 5G will add a business-to-business estimation of $700 billion by 2030 [8]. While, an ongoing Gartner study predicts that $3.1 trillion of business worth will be attributable to blockchain by 2030 [9]. Table 9.1 shows a relative comparison of the state-of-the-art technologies for 5G- enabled IoT.

Table 9.1 Survey of previous work

Authors	Year	Description	Merits	Demerits
Dorri et al.	2016	A secure and lightweight, blockchain-based architecture for IoT was proposed.	Using overlay networks to reduce the computation time for blocks.	No clarity on protection against bugs, such as DoS assaults.
Christidis et al.	2016	Examined the potential of the IoT sector blockchain.	Comprehensive analysis on the role of blockchain and smart IoT contracts.	Challenges relevant to deployment have not been discussed.
Khanet et al.	2018	Based on IoT protection specifications using blockchain.	Discussion of state-of-the-art problems and alternatives to IoT security.	Detailed explanation of the application of blockchain along with IoT is not included.
Florea et al.	2018	Defined the use of blockchain technologies as a data provider for IoT applications.	IOTA network discussed which is labelled as the "IoT backbone".	At present, field instruments are not designed to do PoW.
Miraz et al.	2018	Evaluated the deployment of blockchain for IoT security.	Comprehensive analysis to evaluate the applicability of blockchain for enhanced IoT security.	There was no exploration of blockchain use in industrial automation.
Atlam et al.	2018	A summary of IoT and blockchain integration was presented, thus emphasizing its advantages and challenges.	In depth comparison of developed and blockchain-based IoT networks.	Communication from device-to-device is not considered.
Singh et al.	2018	Focused on the encryption implications of IoT-based blockchain networks.	The strategies to improve IoT security with blockchain is discussed.	The implementation-related challenges are not covered in detail.
Ammar et al.	2018	Fog computing is considered as one of the important research directions for many purposes in healthcare IoT systems.	Inspiration of Cloud to Fog (C2F) computing, which interacts more by serving at the edge of the network.	High latency in service provision to consumers.
Mistry et al.	2019	Comparison of existing proposals in relation to different parameters	Faster data flow to preserve protection.	For high network connectivity, lack of compatibility.

(continued)

Table 9.1 (continued)

Authors	Year	Description	Merits	Demerits
Rathi et al.	2020	An experimental device simulation is undertaken and multiple distributions are used to evaluate the burst and reaction times of the proposed structure.	The solution is designed to boost the system's trustworthiness by using the Hyperledger blockchain to store information on each DEO and MDEO.	The actions of other orchestrators such as ONAP and OSM need to be analyzed.
Mehta et al.	2020	Detailed features, along with security challenges, vulnerabilities, and solutions, on BC technologies and UAV networks.	To fix the above problems, distributed BC technology.	The incorporation of BC into the UAV network has encountered difficulties.
Akhunzada et al.	2020	A systemic layered approach was applied to key IoT supporting technologies that pave the way for the construction of stable future smart cities.	Energy efficiency, scalability, interoperability	More technicalities and methods should be implemented.

1.2 IoT Use-Cases and Applications

For the most part, IoT applications have been downloading, preparing, and consistently accumulating in cloud organizations, and as the amount of "Things" created increases, these organizations disregard helping IoT devices' constant accumulation [10, 11]. It is because these mechanisms work in states of life, across gigantic geographical ranges, therefore requiring low dormancy response times, and having high-density essential/bandwidth intake data [12]. Fog/edge figuring broadens cloud framework limits, by decentralizing asset organization from data centers to edge systems [13]. They are composed as hierarchal systems of fog hubs or cloudlets [14] receiving, processing and handling the executives administrations. Geographic region permits lower reaction latencies and increment's processing transmission capacity by evenly scaling assets, while consuming less energy and empowering asset portability when contrasted with cloud administrations. These qualities allow IoT applications to scale up to both smart size and geographic range, while having constant latencies of reaction, and as such fog/edge figuring could be seen as a potential feature of IoT applications (Fig. 9.1).

The Internet of Energy (IoE) worldview presents the idea of smart matrices and energy management [14], in which dispersed systems energy generators check power utilization and generation, or battery limit, and give coarse-grained insights

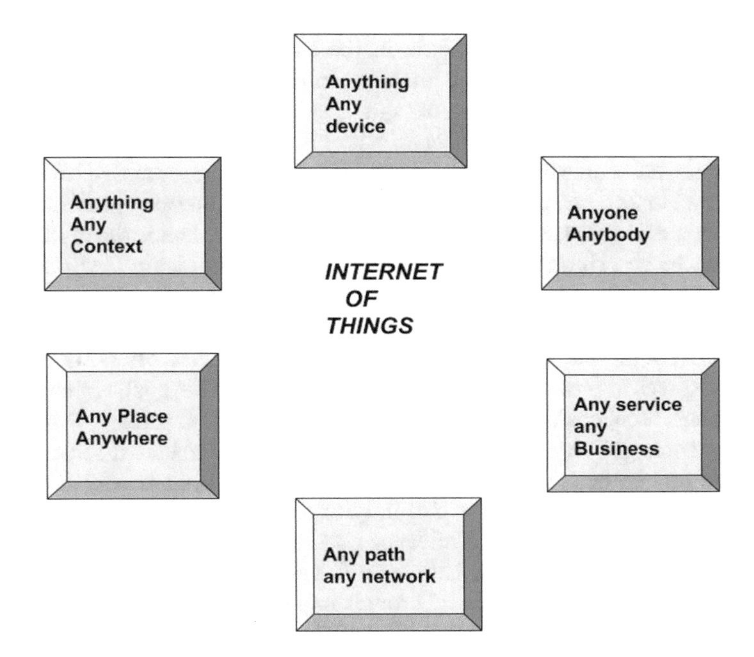

Fig. 9.1 Representing IoT Use-cases and applications

about matrix wellbeing, and "Smart Meters" can screen limit, age, and use at a better granularity and report energy requests to utility suppliers [13]. All things considered, IoT is empowering the innovation of future frameworks, for example, electronic vehicles and smaller-scale lattices. Moreover, such a framework can give more secure, more dependable and vigorous force conveyance, to satisfy changing buyer needs [15]. Currently, more IoT-based frameworks are manufactured utilizing the combined customer, cloud, and solid databases [16] and the internet. The developers considered two major drawbacks of the IoT condensed foundation: (i) a single point of frustration that might conceivably pull down the entire system and (ii) a lack of confidence among the devices connected with the structure [17]. Decentralized platforms may be used for distributed (P2P) interchanges between hubs to resolve the previously stated constraints. In either event, such mechanisms have a range of safety and security issues that can unlock the doorways for gatecrashers to execute various attacks.

Endless apps are available utilizing IoT tools, for example, smart home [18], smart production line [19], smart community, and safe vehicle AdHoc orchestrated [11], to enhance the organization. However, in the future blockchain may be used by 5G-enabled IoT to maintain the IoT protection tools. To be clear of IoT arrangements [20, 21], other blockchain agreements are available for brisk research confided in structures and other research on unpredictable networks. Then IoT devices engaged in 5G can use the same constantly. Security may not be the fundamental task of using blockchain in the same way, as it continues to work faster. However, according to the maker's info, it is difficult to obtain a self-governing numerical confirmation for quick game plans so far. When functional proof is available, an optimum scenario is possible in which stable and agreed center points are connected to the network and can be accessed from 5G by express cloud or fog layers. Another of these systems is an open blockchain-based network. One of the benefits of using blockchain advancement is the ability to store information in an arbitrary way that does not require a centralized database. In addition, in a trustworthy state, it also provides a way to deal with follow-up and execute trades among different individuals. Through utilizing solid cryptography with transparent private key collections, blockchain likewise offers its individuals enhanced degrees of protection. Many decentralized apps (DApps), which were built using IoT and blockchain, are available on the market. Using the IoT framework, app "data exchange" would be feasible with the use of embedded sensors and the appropriate functionality of the network. The ubiquitous system availability can be accomplished using 5G, which is currently exceptionally difficult to accomplish. When contrasted with 4G, these advances decrease the inertia by several times. In addition, incorporating blockchain with IoT enables maintaining a permanent record of exchanges of data trade. Through implementing this in a decentralized P2P manner, the "middle man assault" can be wiped out, which allows clients to collaborate without having to rely on an outsider [14]. Propelled by the above review, this chapter presents a description of the application of the 5G-powered IoT blockchain for computational technology and automation. At this point, we are thinking about some transparent issues and challenges that might hinder blockchain innovation from evolving.

Features of IoT The fundamental characteristics of the IoT are as follows [2, 3]:

Connectivity In regard to the IoT, everything is integrated periodically with the general system for data and communication.

Administrations Relevant to Subjects The IoT is fitted within the parameters to offer thing-based governments, for example, protection monitoring and continuity between real objects and their associated virtual items. In addition, on flexibly thing-related regimes within the parameters, all developments in the physical and knowledge realms can shift.

Complexity The devices within the IoT are heterogeneous, based on the various levels and configurations of the hardware. We may be linked to various devices or to specific phases of administration by different schemes.

Changes in Dynamics The state of gadgets is slowly evolving, e.g., sleeping and waking, connected or also theoretically detached in view of the fact that gadget environment requires area and space. Furthermore, the quantity of devices will vary greatly.

Good Variety The number of gadgets that should be monitored that talk to each other will be at least a significantly larger degree than the gadgets associated with the current internet. Significantly more simple devices are having the ability to be the administration for implementation purposes of the knowledge generated and their translation. This identifies with information semantics, equally as effective information that administers it.

Safety We should not disregard wellbeing as we gain profits from the IoT. We should structure for protection, as both the IoT developers and beneficiaries. It involves the wellbeing of one's own knowledge, and then the protection of one's physical wellbeing. Being sure of the endpoints, the networks, and the data going through all of it means having a philosophy of security that can be scaled.

Network Openness and similarities are balanced by the approving quality. Openness is a program developed although similarity offers the essential ability to spend and generate knowledge.

1.3 Blockchain Technology

Blockchains are updated with transparent and protected computerized documents allowed in a fully centralized way (i.e., though not a focal archive) and without a focal force (i.e., corporation, entity or government) every now and then. In their fundamental point, they require a consumer network to document transactions in an incredibly common database within the structure, seen below the basic activity of the blockchain sorting; no oversight can be modified until written. In 2008, the blockchain plan was combined with various elective progressions and figuring considerations to outline current cryptographic types of cash: electronic money guaranteed through cryptanalytic frameworks rather than a central vault or authority. This advancement ended up being comprehensively known in 2009 with the dispatch of Bitcoin, and is the basics of the various computerized types of cash. In Bitcoin and similar systems, the trading of cutting edge data that addresses electronic money occurs in a decentralized manner. Bitcoin customers will cautiously sign and move their benefits to that data to a substitute customer and thus the Bitcoin blockchain records this trade with no attempt at being subtle, permitting all individuals from the framework to affirm the authenticity of the trades. The Bitcoin blockchain is maintained and managed by a scattered bundle of individuals. This, close to cryptanalytic frameworks, makes attempts to change the record of the blockchain later (altering blocks or advancement trades) extremely difficult (Fig. 9.2).

Blockchain technology allowed the existence of several Bitcoin and Ethereum-like crypto currency systems, and the blockchain infrastructure is mostly seen as secure for Bitcoin or potentially crypto-monetary applications in general. The system is, however, available for a wider range of applications and is being studied for a number of sectors. Alongside its reliance on cryptanalytic primitives and distributed systems, the various components of blockchain technology will make it difficult to grasp. However, each element would merely be interpreted, and it

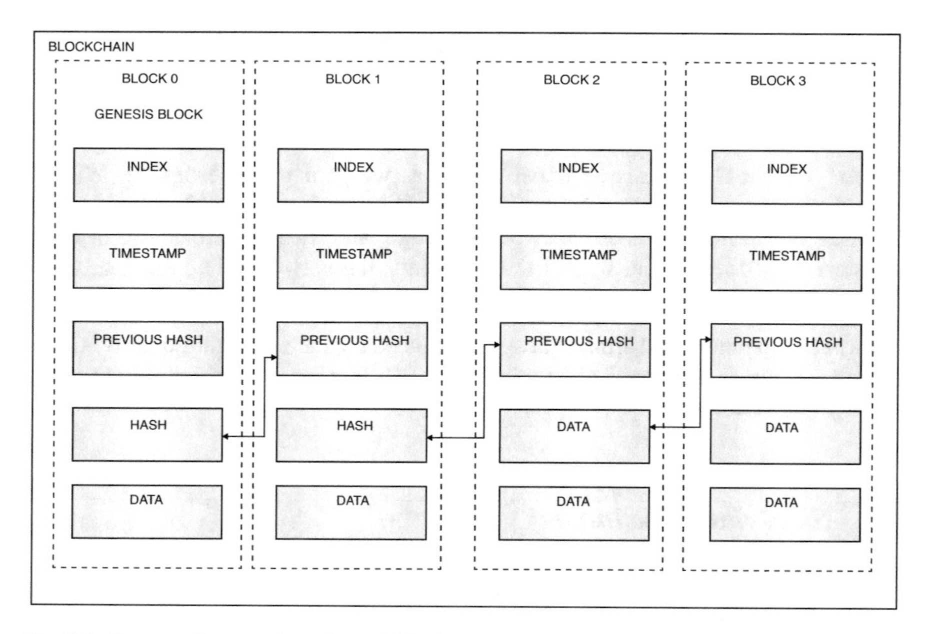

Fig. 9.2 Process of transactions through blocks

was seen as a building block to grasp the larger hierarchical system. Blockchains may be informally described as: blockchain technologies would be spread to public ledgers with cryptographically signed transactions clustered into blocks. A block is cryptographically connected to the previous one (making it clear) until it has been authenticated and sent to an agreement request. Through the introduction of new blocks, more established blocks are more difficult to modify (making protection from obstruction). Blocks are recreated across duplicates of the system record, and any contentions are settled by rules for mechanical exploitation. Trust is a key part of blockchain that is practiced by comparing the hash of the past block to deliver the following hub. So as to accomplish an accord, the "miner" hubs are at risk for approving the subsequent hash and afterward for finding the hash for the following lines. A lot of exchanges are packaged into blocks utilizing the Merkle tree, and just the Merkle root hash is added to the block, as shown in Fig. 9.3. This strategy is viewed as "Proof-of-Work" (POW) for work done on the network [17] and furthermore on mining hubs. These prize models empower mining hubs to connect with mining blocks inside the system to share the processing assets they have. Actually, blockchain is unmistakable from different agreements upheld in incorporated systems with the accompanying properties [13]:

Untrustworthy Inside the program, the elements needed are mysterious to one another. However, they can interact, engage, and operate with each other without knowing each other, which means there is no need for assured developed character to practice any interaction between the hubs.

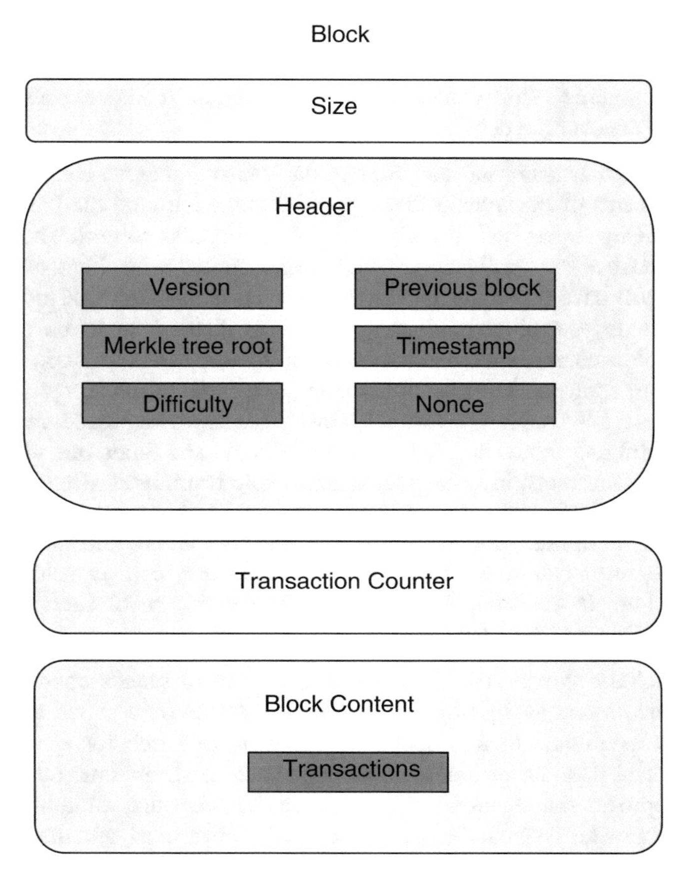

Fig. 9.3 Representation of blocks having all their components used in blockchain

Consent Less Within the program, there is no constraint on who should or cannot operate, i.e., there are no real authorizations.

Free Restriction One may link or operate on the blockchain as a network without regulators. In addition, it is impossible to alter or control any affirmed exchange.

Innovation in blockchain has four basic segments [20] that are analyzed as follows:

Agree The PoW convention is conscious of verifying any operation inside the network that is crucial to avoid one miner center from controlling the whole blockchain framework and therefore to monitor the past of the exchanges [22].

Record It is a growing and shared database comprising almost all exchanges carried out throughout the network. Usually it is unchanging, so that evidence cannot be deleted by any method until it is packed away. It ensures that every exchange is acknowledged as authentic at that point by a larger proportion of the customers required at a specific time [23]. Cryptography: it guarantees that strong cryptographic security preserves all device detail. It enables only authorized customers to uncover the details.

Smart Contract It is normal for program participants to agree and review. There are various forms of blockchain that may be arranged around the basis of edges, knowledge being supervised, benefit potential, and access control. The distinction exists in the ideas of verification, which demonstrates what kind of access the blockchain will have (open versus private) and the permission that indicates what users should do (permitted and non-permitted). If there is to be an event in transparent blockchains, everyone may take an interest in the network, irrespective of some sort of approval. They are either going to want to go out as a plain partaking platform or as a miner center, which helps inside the approval procedure. Miners are paid for by different motivating forces such as Bitcoin and Ethereum, which openly blockchains. Conversely, in secret blockchains, help is minimal, where proprietor's approval is essential to get to the network. Additionally, the assortment of individual blockchains is permitted, which controls the activity that the clients will perform. For example, users can access smart agreements, or they can go inside the device as a miner hub. In addition, there are unregulated private blockchains, such as Hyperledger-Fabric [27] or Ripple [28].

How Blockchain Works Blockchain is a gathering of blocks connected by the cryptographic hash capacity of the resulting hub. At the point when a substitution exchange is mentioned to the blockchain, they make a hub for speaking to the exchange. Then this hub or block is communicated to all or some other hubs and each hub approves the recently made block. In the event that all hubs approval is finished, the new hub is included inside the blockchain record, and the exchange of the block is recorded.

The Blockchain Types Blockchain technologies are organized as approved (private), permission-less (public) as well as consortium [24]. Bitcoin and Ethereum are transparent network experiments that make it easy for anyone to observe. A private blockchain is specifically intended for a single organization, and thus only permitted entities can authorize or challenge and limit the exchanging of subtleties. Transparency, decentralization, and anonymity are the highlights of each blockchain. Block stack, and multi-chain are private blockchain tests. This is clustered and distributed, as seen in Fig. 9.4. Consortium blockchain [25] is semi-decentralized, restricted by the assembly of authorized individuals with approval alongside a single feature, and the negotiation protocol is designed to comply with the gaggle of pre-established hubs within the network. Hyper-record, Corda, and Quorum are part of the system.

Features and Applications The key features of blockchain technology [24, 26] are delivery, decentralization, smart contract, enhanced security, transparency, immutability, confidentiality, and anonymity [22]. Many network transactions come with blockchain technologies due to certain features. The greatest advantage is

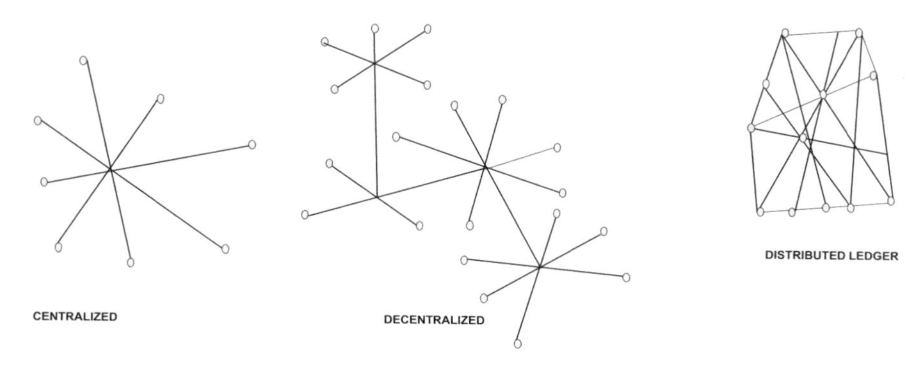

Fig. 9.4 Different categories of blockchain

that in this transaction there is no need for an intermediate and every transaction history is recorded. A smart contract is a computer protocol that is processed in the database, and automatically enforced when any requirement is reached. Blockchain technology has applications in financial as well as non-financial fields.

Exchange Finance Exchange Finance stands to benefit tremendously from the presentation of smart agreements. As of late, Santander Innoventures said that it accepts that blockchain innovation will cause almost $20 billion worth of reserve funds per year by 2022. A gigantic measure of these investment funds will originate from smart agreements computerizing endorsement work processes and clearing computations that now are unimaginably work intensive. This computerization will not exclusively help to decrease work-hours; however, it will additionally significantly lessen errors and therefore the time taken for these counts to require space.

Records Almost every predictable industry on the planet has the chance to utilize smart agreements to help improve the speed and security of its recordkeeping. One industry particularly that stands to benefit immensely is the social insurance industry. As of now, the world's medical services PC frameworks hold numerous patient clinical records. Regardless of the established truth that these social insurance associations have put away gigantic totals of cash on security, current access and capacity techniques are unmistakably more defenseless against digital assaults than their blockchain-based counterparts. Blockchain innovation could permit whole databases of individual medical records to be safely encoded and kept. Another reward is that the innovation likewise encourages the utilization of a private key significance to ensure that only certain people can obtain entrance. An assortment of the other blockchain smart agreements are utilized for giving solutions, storing receipts, general stock administration, storing test outcomes, etc.

Property Ownership Smart agreements have two enormous uses regarding the property showcase. Initially, they will be utilized to record property proprietorship. Because the utilization of smart contracts is faster and more cost-productive, this makes them a better option in contrast to existing frameworks. It likewise implies they will be utilized to record the responsibility for kinds of property from structures, land to telephones and watches. Inside the property market, smart agreements can evacuate the requirement for costly administrations such as those given by attorneys and property intermediaries. This innovation likewise implies, unexpectedly, venders have the facility to deal with the exchange totally without anyone else.

Home Loans The property market likewise stands to exploit less expensive, quicker, and more secure smart agreement-based home loan exchanges. Not only will this permit purchasers to move into the property quicker but it will also help make the whole procedure easier. Smart contract home loans would permit the two players to carefully decide the deal before preparing the installment. When this is frequently regularly done, the agreement would refresh the property possession subtleties to mirror the difference in proprietorship. Since the strategy would require remarkable key code approval in the interest of the essential proprietor, it will make the entire procedure more secure and diminish occurrences of extortion.

Insurance The insurance business expends a great sum every year on preparing claims. Not only that, it also loses money to fraudulent cases. Besides supporting the underlying strategy, smart agreements could likewise help improve the strategy of guarantee preparing here and there. They could permit mistake checks and decide payout sums bolstered by a gaggle of models that considers such an arrangement that was held by the individual or association. Diminished handling times, a decrease in mistakes, and lower expenses are among the first advantages.

Clinical Research The clinical examination industry will appreciate comparative focal points as part of the human services industry. Most importantly, delicate information such as patient records could be moved between divisions/research focuses after having been safely encoded using blockchain innovation. Since a significant number of the patients taking an interest in clinical exploration have delicate ailments that they frequently wish to stay private, keeping these records secure is fundamental.

Casting a Ballot Claims of ballot casting extortion happened after the 2020 U.S presidential race. In spite of utilizing PC frameworks that are sometimes times very expensive, fraudsters find progressively creative approaches to direct them. Smart agreements are a basic and financially savvy answer for this issue. They will be utilized to approve a voter's character and record their vote. This data could then be utilized to start an activity regardless of if democracy had stopped. Since the blocks inside a blockchain are difficult to shift once they have been recorded, manipulation of this record would not be conceivable.

Shared Transactions Smart agreements are regularly utilized for an entire scope of shared exchanges. This thinking is the thing that prompted the making of the Ethereum Project and other such organizations. Clients of every kind can utilize these stages to frame and concur on smart agreements. These agreements stay dynamic until a gaggle of concurred conditions are met. When the smart agreement is sure that every condition is met, it then permits the rest of the consent to be satisfied. Commonly, this is often the exchange of cash yet this is not always the situation.

Advancement Another energizing utilization of smart agreements is to maintain a record with respect to the phases of advancement of an item. Two parties would sign the agreement, which could enact it. Since the settled upon venture was created, the stages and accordingly the other significant data could be recorded to the smart contract. In the event that the parties had consented to such things as part installments, then these achievements would be reached, and the agreement would start their delivery. At the point when it includes property such as thoughts, you simply need to look at the example of the endless patent cases among Apple and Samsung to decide exactly how significant having the ability to demonstrate proprietorship truly is. The rundown of businesses that may appreciate this innovation is huge. For whatever length of time that smart agreements support and secure improvement, such an enterprise included could run from a little startup to an outsized tech organization such as Microsoft or Amazon.

Stocktaking Supply chains are another zone of business that can appreciate blockchain-based smart agreements. IoT gadgets could be utilized all through the chain to record each stage an item takes. Smart agreement supply chains could hypothetically practically dispense with in-house robbery, as supervisors would have the option to follow a missing item back to the exact time and spot that it disappeared. On immense supply chains like those found in huge distribution centers, these smart agreements would empower directors to decide ongoing stock levels and consequently the time it takes items to move through the supply chain. Administrators could utilize this information to oversee stock levels and grow new working practices to fortify conveyance times. For supply chains that work in a few distinct areas or organizations, smart agreements could do the entirety of the abovementioned and even start programmed reorders and installment for orders previously received. Such data as contained in these agreements could even be utilized to help with deciding cutting-edge occupied periods and even which items to stock at various seasons.

1.4 5G Technology

5G innovation [3] refers to the leading edge route that uses CDMA, BDMA, and millimeter gap and is advancing at rates over 100 MP. The key technologies behind the 5G deployment are D2D networks, Large Multiple-Input and Multiple-Output (MIMO), Stronger Applications, Millimeter-Wave, and GFDM [4].

Promote a Broadband Network The first basic portable communications started in the U.S. in the 1950s. After three decades, the first (1G) portable was released.

After the development of the microchip, the second era of portable correspondence (2G) invention was driven by digitized property utilizing 1's SMS efficiency. The GPRS office was connected to the 2G network after a few years, and planned to peruse the internet inside the suitcase. In the twenty-first century, the third age (3G) was implemented in the audio- and informing-equipped for better volume. With the overly fast pace of web correspondence in 4G powered devices, fourth era (4G) was released a few years ago. Today's site use is not only of mobile phones, it is even popular for technological devices, including an icebox, computer, cars, and so on. In order to understand continuous device interaction and access more devices, a rapid site and data transmission functionality is required. This is also the underlying move for creativity in the fifth era (5G) and, as a result, the availability level is up to 25 Mps. The first beneficial circumstances of 5G are fast knowledge delivery, lower inactivity, bolstering the virtual private network, strong data storage efficiency, and so on, the preeminent technology area of 5G is in the healthcare field, the scalable sector, IoT, AI, and so on. 5G will combine every single corner of human existence with innovation through correspondence systems. Data is therefore critical and should be shielded from visual attack on sensitive details. The key health risks of 5G engineering are DoS attack, immersion attack, assault scanning, pacing assault, middle man assault, consumer misrepresentation, directing assault, etc. Therefore, as 5G tech co-ordinates with IoT and blockchain, security engineering is needed [4] (Fig. 9.5).

For almost every sector, a digital revolution, driven by automation, cloud, and connectivity, is occurring, challenging and having us reconsider our ways of working. New usage cases for the platform are growing with the advent of the 5G age when customers and companies continue to work on finding systems and platforms that can improve the quality of their lives and company. At the center of this technical transition is wireless communication. New market prospects are emerging—both to those who have previously invested in the supply chain, such as mobile companies, as well as to entrants from other fields. Attentively, innovative market structures are being built for centralized cloud infrastructure and network programming at the edge. The unprecedented appetite and passion for the sharing of ideas leads to a greater degree of interaction between individuals and different industries of some sort. Solutions are being developed involving a range of fields of specialization, changing existing economic practices, and redefining ecosystems. The cross-industry transition has provided a prerequisite for the advancement of the idea of cellular networking for the fifth generation of mobile technology (5G), providing new approaches to identify performance measurement and assessment as well as services. Compared to previous iterations of wireless technologies, including 4G, the reason for 5G implementation is to broaden the connectivity reach of cell networks to offer new services not only for users but also for separate sectors of the community as a whole, As a consequence, the power of the Internet of Things is released. Ericsson's latest report [27] indicates that a limited number of European and North American operators agree that 5G is being more consumer-driven, while a standardized number in Asia-Pacific and Central and Latin America expect that 5G is more business-driven. The Ericsson Replacement Survey [28] notes the

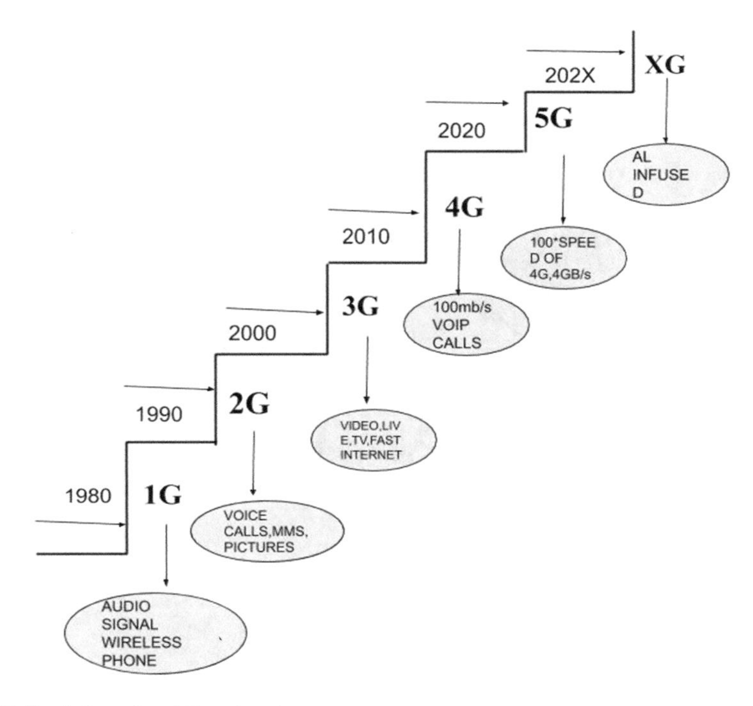

Fig. 9.5 Evolution of mobile network

machine-to-machine (M2M) communications and wireless access. A substantial majority suggested that they will make major improvements to their industries in order to allow a greater benefit of 5G as they launch. Exploring the related market scenarios—in sectors such as agricultural, manufacturing, building, electricity, banking, safety, etc.—the 5G usage cases and their respective specifications are illustrated in this part. The development of the network elements to 5G is then discussed, accompanied by a description of the 5G framework. Finally, the 5G network is exemplified by three specific implementation events.

5G for Humans and Machines Concerning use cases and specifications, many technology groups have established the criteria on what truly constitutes 5G. Here we discuss a range of promising applications that cover a variety of main industries, as seen in Fig. 9.6 below [28]: automobile, building, electricity, safety, engineering, media, retail, and transport.

- Autonomous vehicle control empowers an expansion in self-governing driving, helping people, for instance, and bringing focal points such as an improvement in rush hour gridlock wellbeing, expanded efficiency, and improved personal satisfaction.
- Intelligent transport frameworks (ITSs) promote efficient on-board traffic information, dynamic traffic rerouting, and control of traffic signals.

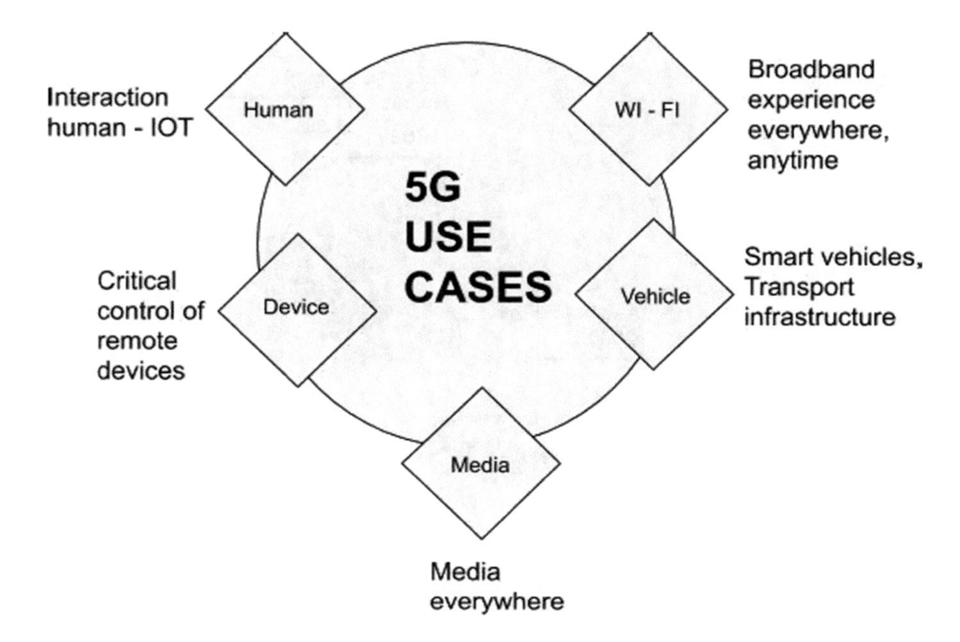

Fig. 9.6 5G use cases and applications

- Emergency communications requires a robust device that can aid in the search and rescue of human beings and, thus, in the detection and resolution of catastrophic situations, particularly the equipment; however, portions of the device are destroyed during the catastrophe.
- Factory cell computerization may also be a tool to recall gadgets for a mechanical manufacturing device communicating to control units with a reasonably high degree of unwavering efficiency and a relatively low inactivity to be able to support specific applications. This could be related to cloud mechanical autonomy.
- Passengers traveling on a rapid train can use the timeframe for recreation or business exercises while getting a charge out of a client experience of a similar quality as though they are either fixed or moving at a slower speed.
- Large open-air occasions held for a restricted period during a drawn out territory could even be attended by a colossal number of individuals. Such occasions incorporate games competitions, presentations, shows, celebrations, firecracker shows, and so on.
- A framework that is ready to convey to and from gigantic quantities of topographically scattered gadgets. Such a framework can detect information, break it down, decide, and control incitation—giving observation, for instance, or actualizing dispersed input control and checking basic segments, etc.
- Media on request underpins an individual client's longing to have the option to appreciate media content (for example, sound and video) whenever and anyplace.

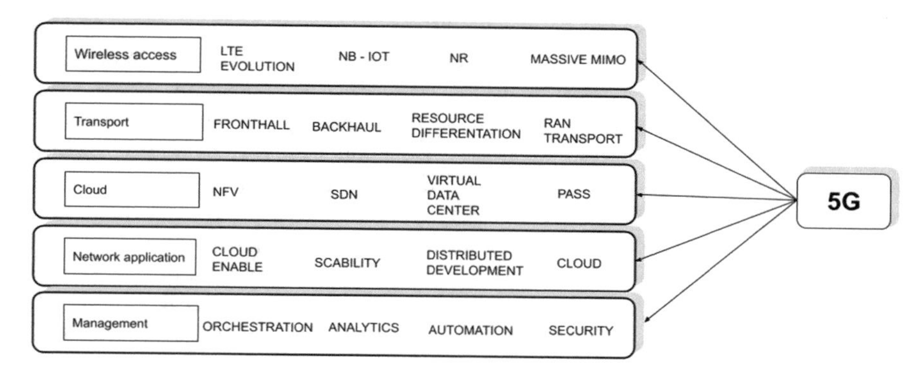

Fig. 9.7 5G application layers and services

- Remote exam and medical procedures empower low dormancy for telehaptic control, and in this manner, the specialist gets material input that is intended to be indistinct from or superior to manual employable methods.
- Shopping centers can permit conveyance of customized shopping encounters.
- Smart city systems incorporate far off checking of city framework, ongoing traffic data, and open security alarms for improved crisis reaction times.

The Services of 5G Deployment of 5G cases is often categorized in terms of criteria for three basic forms of connectivity with dramatically different objectives: large type of connectivity (mMTC), vital MTC, and severe or enhanced mobile broadband (eMBB). The three types that should have been designed as 5G networks are described below (Fig. 9.7).

Huge Machine Type Correspondence In any case, MMTC, referred to as Massive IoT, is intended to be used flexibly in wide region inclusion and profound entrance for a large number of gadgets per square kilometer of inclusion. A further target of MMTC is for graceful universal availability with moderately low programming and equipment unpredictability and low-energy activity. A large number of devices assisted are battery operated or driven by energy supplies, have little payloads, and travel occasionally, all of which together would usually be reasonably delayed for the preeminent portion. Although technologies typically have an all-inclusive life span, administration and programming have been given a chance to scale and exchange relatively rapidly to handle new market openings. Models that fall under this administration classification include the control and robotization of structures and foundations, intensive farming, coordination, follow-up, and armada of executives.

URLLC or Simple MTC The second level of application appeared to be CMTC, which is often called Essential IoT. E2E idleness specifications are small (at the millisecond level) during such an operation, and thus the need for unwavering efficiency is exceptional. The show goals of the CMTC can be extended to work processes such as the robotization of the diffusion of energy during a sound system, in-process monitoring, and sensor management where there are strict criteria for unwavering efficiency and low idleness of the layer of the apparatus. These are in some cases referenced as ultra-dependable low-inactivity interchanges (URLLC) prerequisites. Cautious consideration will get the opportunity to be paid to security inside the instance of both mMTC and cMTC. While the higher system and gadget unpredictability are all the more promptly satisfactory in basic correspondence, mMTC should address digital security confirmation with low-multifaceted nature gadgets. A various leveled way to deal with the system is essential to continuously improve security; thus, start to finish security affirmation are frequently ensured. A terrible lightweight broadband, bringing both an extraordinarily high amount of communication and a small idleness of contact, the enhanced portable broadband (eMBB) [4] often promises enhanced participation—way beyond that of 4G. Network and data delivery are more reliable in the inclusion area, and the implementation of databases is sluggish in the light of the reality that the number of clients is growing.

2 Architectures

2.1 IoT Architecture

The IoT architecture has various layers of IoT-supported advances [7]. It helps, for example, to associate various technologies with each other and to speak about the adaptability, autonomy, and coordination of IoT organizations in specific contexts. Figure 9.4 shows the IoT point-by-point design. The usefulness of each layer is shown below [29]:

Sensor Layer The principal decreased layer is surrounded by quick articles composed of sensors. The sensors allow the physical and propelled universes to be interconnected, allowing constant information to be assembled and prepared. There are different forms of sensors used for a range of purposes. The sensors have the capacity to make temperature, air quality, distance, humidity, strain, wind, shift, and force estimates at this stage. They are going to get a degree of recall every so often, enabling them to log the desired amount of figures. The sensor must check the property and convert it into a sign recorded by the computer. Sensors are built for a range of uses, such as standard sensors, body sensors, family member device sensors, vehicle telematics sensors, etc. Such systems require the target to reach the earth. This can be achieved by a Local Area Network (LAN), for example, Ethernet and Wi-Fi allies, or a Personal Area Network such as ZigBee, Bluetooth, and Ultra-Wideband. With sensors that do not need the use of sensor aggregators, their connectivity to downstream workers/applications is typically allowed through a Wide Area Network such as GSM, GPRS, and LTE. Low-power and low-speed networking sensors usually designed masterminds as wireless sensor networks

(WSNs). WSNs are studying conspicuousness, because they would essentially need more sensor centers while preserving sufficient battery life and covering gigantic areas.

Gateways and Networks The tremendous amount of knowledge to be collected by such tiny sensors demands a robust and superior wired or remote device base as a vehicle medium. Present frameworks, routinely bound to entirely new standards, do not support machine-to-machine (M2M) programs and their implementations. For a policy intended to accommodate a broader variety of IoT administrations and implementations such as rapid value-based administration, prudent network deployment, and so on, various mechanisms for similar technologies and access protocols are expected to communicate with each other during a heterogeneous setup. Such structures are mostly in the state of proprietary, free, or half-breed models and are various inputs (microcontroller, microprocessor, etc.) and door systems (WI-FI, GSM, GPRS), as seen in Fig. 9.8.

The Management Service Layer The company shall express the potential handling of knowledge by evaluation, protection checks, procedure facts and, accordingly, the leading set of devices. One of the many features of the organizational

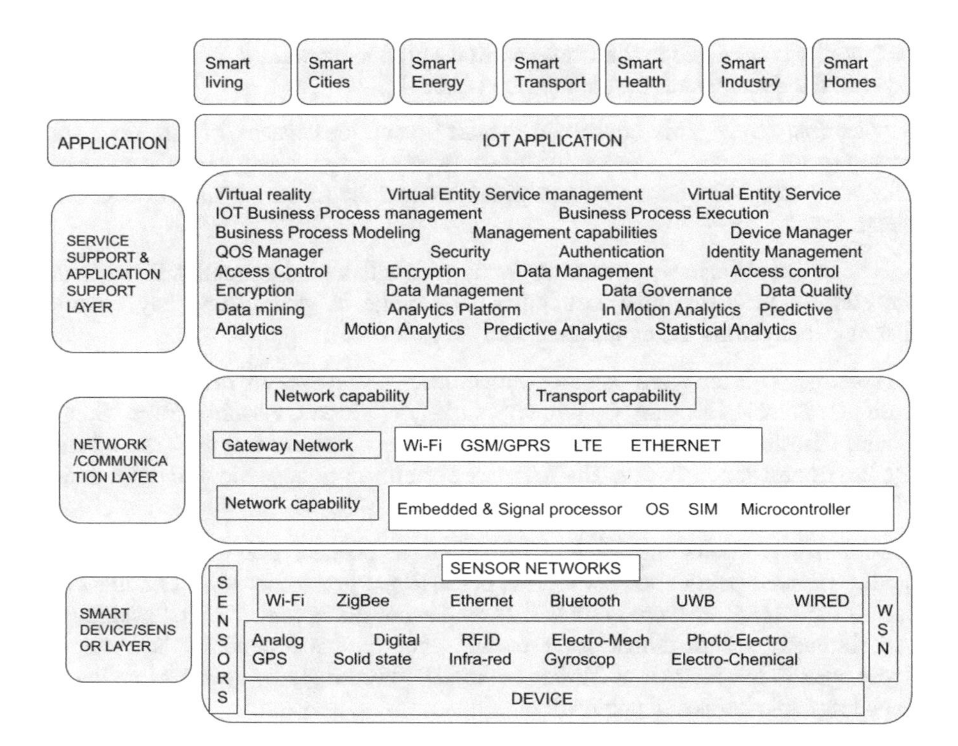

Fig. 9.8 IoT architecture

layer is that the business and methodology rules the engines. IoT ties together the relationship and interaction of objects and structures that include details inside certain activities or legal statistics, such as the weather, the local area, and traffic data. Within the framework of the evaluation, distinctive analysis methods are required to dispose of the appropriate knowledge from a wide proportion of the knowledge and to be handled quickly. In-memory review allows vast volumes of information to be processed in random access memory rather than held in actual circles. In-memory testing decreases the time taken for data and speeds up the collection phase. Streaming evaluation is such a process where the measurement of knowledge, perceived to be moving data, is expected to be continuously managed at the same time that judgments are made as much as possible in only a few seconds. Data on the board is the opportunity to control the flow of knowledge. Information on the board inside the organizational layer is also provided, encouraged, and managed. Higher layer systems are generally protected from the need to process redundant details and reduce the risk of leakage of knowledge source vulnerabilities. Data splitting methods, such as metadata anonymization, data blending, and file matching, are used to hide a wide spectrum of information while merely supplying the details that can be used for a wide variety of purposes. In the convention of the same details, material is always deleted to softly expand the details point of view of the mill to more unmistakable deftness and reuse across spaces. Protection must be lawfully recognized in the whole portion of the IoT program from the sagacious layer of the document to the submission (Fig. 9.8).

Application Layer This concerned "smart" conditions/spaces in the area, such as transport, building, community, lifestyle, shopping, supply chain, emergency services, health care, consumer group, culture and the travel industry, climate, and electricity.

The IoT could also be a luxurious framework with a kind of quality. Its attributes are shifting from one area to another. The range of general and key qualities distinguished during the exploration study is as follows:

Knowledge This includes "smart" situations/environments in fields such as, for example, Transit, Housing, Community, Lifestyle, Market, Farming, Farm, Supply Chain, Disaster, Hospital, Collaborative Consumers, Culture and the Travel Industry, Environment and Power. The IoT may even be a privileged system with a kind of characteristic.

Availability It allows the Internet of Things to connect everyday objects. The quality of such publications is crucial because the grassroots level organizations add to the overall awareness of IoT. It empowers the organization of availability and similarity within things. As a consequence of this availability, new market doors open to the Internet of Things are rendered routinely by systems that handle knowledgeable items and applications.

Dynamic in Nature The core purpose of the Internet of Things is to collect information regarding the world, which is often carried out on a regular basis, with dynamic changes taking place through networks. The condition of these devices is progressively changed, the model is sleeping and waking, associated as well as disconnected because the setting of the gadgets includes temperature, area, and speed. In addition, with regard to the state of the machine, the number of the instrument often varies strongly with privacy, spot, and time. There is enormous variety, and the number of devices that can be monitored communicating to each other is becoming increasingly likely to be substantially greater than the gadgets connected to this website. The administration and translation of the data produced from these devices for application purposes seems to be simpler. Gartner (2015) mentions the significant scale of IoT in the measured study, reporting that 5.5 million additional items will be introduced on a day-to-day basis, and 6.4 billion similar items will be used globally in 2016, which is 30% more than in 2015. In addition, the study predicts that the number of connected devices will exceed 20.8 billion by 2020.

Detecting IoT would not be feasible without sensors that would detect or measure any advancement inside the system to promote intelligence that would include specifics of their position, or even follow the environment. Detecting the developments provides the ability to frame the talents that represent the true nature of the actual universe and, in this sense, the people inside it. Detecting data is simply a basic input from the real world, yet it may provide a rich picture of our stimulating world.

Heterogeneity Heterogeneity of the Internet of Things along with core attributes. Tools of IoT are managed at various levels of the infrastructure and need to be linked to specific devices or to separate levels of administration by different networks. IoT architecture will improve the availability of direct structures between heterogeneous systems. The key technical criteria for heterogeneous artifacts and their systems in IoT are scalability, isolation, extensibility, and interoperability.

Security IoT systems are usually ineffective against security attacks. When we achieve productivity, novel interactions, and various benefits from the IoT, it will be a risk to ignore the protection issues associated with it. IoT provides a great deal of accountability and protection concerns. It is necessary to be optimistic about the endpoints, the apps, and so the intelligence that has progressed to the ordinary little bit of it indicates a protection perspective. IoT advancements have the attributes mentioned above that make human exercises worthwhile and enhanced; they further enhance the IoT's capabilities through shared participation and make it an area of the entire framework.

2.2 Blockchain Architecture

The blockchain architecture is composed of the type of node-user or computer that features a complete record of the blockchain ledger, Block—a knowledge system used to manage transaction aggregation and transaction—the smallest node of the blockchain network (records, documents, etc.) Here are the key components of the blockchain architecture:

Link User or computer within the blockchain network (each one has an internal copy of the blockchain ledger).

Transaction The smallest building block of a blockchain network (records, data, etc.) planned for blockchain.

Block Data system used to handle a series of transactions that is distributed through all or most of the nodes inside the network.

Chain Set of sections in a particular order.

Miners Specific nodes that perform the block verification method before contributing to the blockchain framework.

How Blockchain Works A block in blockchain consists of the details, the hash of the block, the hash of the previous block. The data held within each block depends on the form of blockchain. As an example, inside the Bitcoin blockchain system, the block holds details on the recipient, the sender, and hence the amount of coins. A hash is a kind of special fingerprint that has a long list of a combination of digits and letters. The block hash is generated with the help of a cryptographic hash (SHA 256) calculation. Subsequently, this enables the definition of every block in a blockchain network without any complications. The moment the block is created, it immediately absorbs the hash, while any adjustment made during the block always influences the difference in the hash. As a matter of theory, hashes help to define any changes in blocks (Fig. 9.9).

The last component inside the block is the hash from a past block. This makes a succession of blocks and is the primary component behind blockchain design's security. The absolute first block during a chain might be somewhat unique—all affirmed and approved blocks are obtained from the beginning block. Any degenerate endeavors incite the blocks to shift. All the resulting blocks at that point convey falsehood and render the whole blockchain framework invalid. On the other hand, in principle, it may be conceivable to manage all the blocks with the help of solid PC processors. In any case, an answer eliminates this chance called proof-of-work. This empowers a client to hamper the strategy for the production of most recent blocks. In Bitcoin blockchain engineering, it takes around 10 minutes to work out the necessary proof-of-work and add a substitution block to the chain. This work is finished by miners—exceptional hubs inside the Bitcoin blockchain structure. Miners get the opportunity to keep the exchange charges from the block that they checked as a payment.

A new client (hub) who enters the mutual blockchain network might get a complete replica of the application. When a replacement block is made, it is sent to each hub within the blockchain network. At that point, each hub confirms the block and checks whether the information expressed is correct. In case it is all correct, the block is connected to the blockchain neighborhood of any node. All centers of

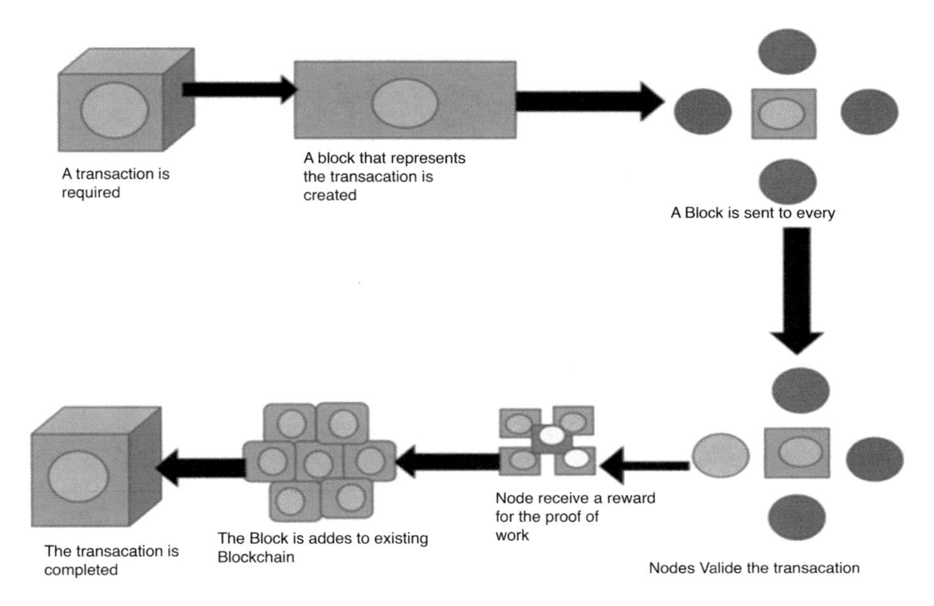

Fig. 9.9 How blockchain works

blockchain systems have an agreement. The arrangement structure might be a ton of protocol guidelines, and thus once anyone submits to them, they are self-authorized within the blockchain. For starters, the Bitcoin blockchain illustrates an arrangement that agrees to declare that the trade volume will be sliced down the center for every 200,000 blocks. This suggests that if a block delivers a check prize of 10 BTC, this value must be divided by 200,000 blocks. Moreover, there will be 4 million BTC remaining to be produced, because there is a cap of 21 million BTC established by the consensus within the Bitcoin network system. When the miners open too many of them, the usability of Bitcoins disappears unless the rule is modified (Fig. 9.10).

To check, this allows blockchain development to be continuous and cryptographically safe by eliminating any outcasts. It is challenging to change the blockchain structure; because it is necessary to play with the sum of its blocks, to recalculate the proof-of-work for each block and, in addition, to monitor increasing 50% of the vast number of center points throughout the diffused process. Blockchain architecture has many advantages for businesses. Here are a variety of built-in features:

Cryptography Transactions are checked and accurate on the basis of sophisticated computational and cryptographic evidence between the parties concerned.

Immutability Any records made during a blockchain process cannot be changed or deleted.

Provenance This applies to the reality that the transaction can be tracked within the blockchain ledger.

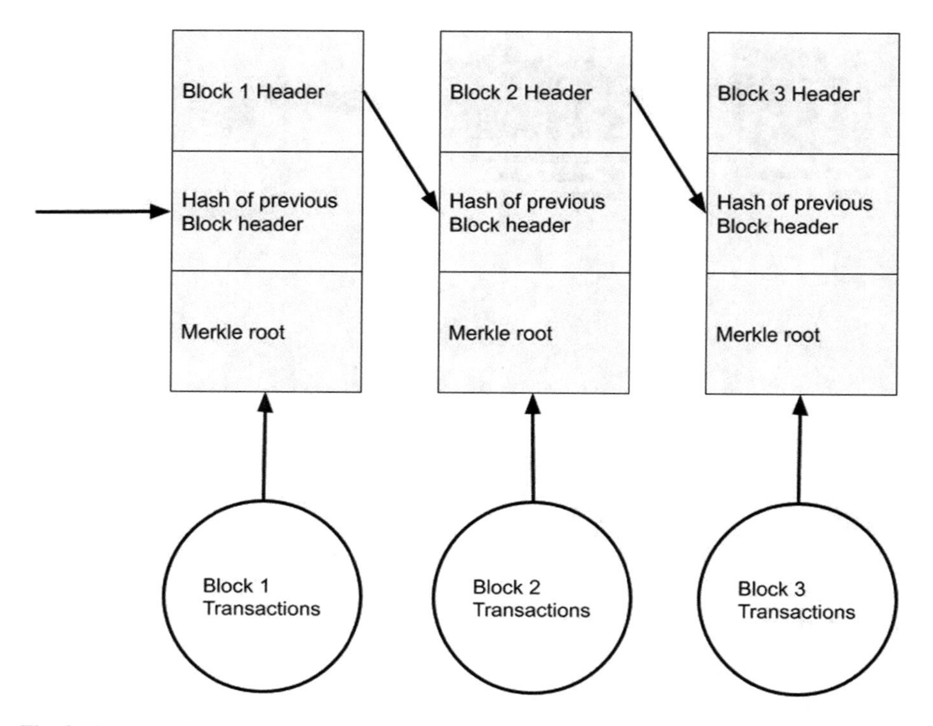

Fig. 9.10 Blockchain architecture

Decentralization Every part of the blockchain system has links to the entire distributed ledger. As for the central-based method, the consensus algorithm enables the network to be managed.

Anonymity Every blockchain network member has an address created, not a user identification. This keeps users safe, particularly during a public blockchain system.

Transparency The blockchain system could not be corrupted. It is often impossible to do so, because it requires a huge amount of computing power to completely overwrite the blockchain network.

2.3 The 5G System Architecture

The 5G architecture, as seen in Fig. 9.11, would be focused on "adaptable" radio connectivity centers, distributed and concentrated server farms making up an adaptable portion of the remaining tasks at hand. Such hubs and server farms are connected by means of programmable vehicle systems. Vehicle networks are connected by spine hubs that relay information from the entrance hubs to data

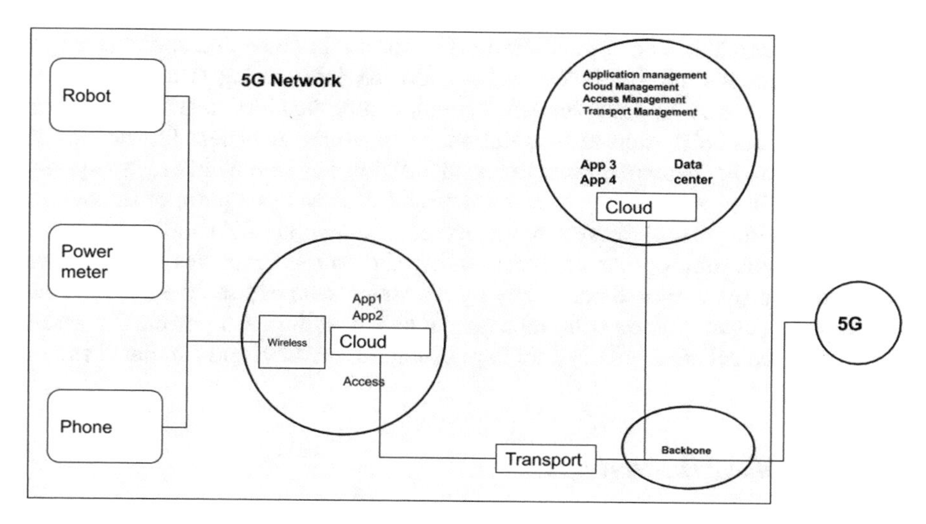

Fig. 9.11 The 5G system architecture

communities where the overwhelming majority of the information is discarded and the network is tracked along these lines (Fig. 9.11). This points out that each program, such as certain device apps, is operating at the edge of the cloud with a specific case with specialized functionality inside the entry hubs. Requirements are either organized on a common framework (Application 3 and Application 4) or distributed (Application 1, Application 2), sometimes on a necessity framework.

In comparison to the existing, server maintenance, storage, transport, and connectivity services are geographically centralized within the data center but may also be flexibly distributed if required.

The 5G framework will fully reinforce the idea of system programmability for a wide range of administrations. Help is deftly allocated to any location within the system, the system hub, the end-client gadget, or the external host on a regular basis. Support obviously will not be held in the framework of the user, and will start from beyond the program region. E2E company is expected to organize external business contributions in order to be successful. For example, to streamline content conveyance in eMBB administration, organizations would put virtualized capacity building on assets that are really close to the supporter's precarious edge.

Network Slicing The method of system cutting considers the meaning of the different coherent systems (or cuts) at the head of the proportional physical framework. Assets are often committed solely to a single cut or to a split between the different cuts. Various types of assets are available, such as registration, stockpiling, equipment, transport, VNFs, etc. The device is cut to control the optimal behavior of the program. Such behavior is commonly associated with protection, disconnection of knowledge sources, essence of administration, unwavering efficiency, and autonomous charging. A network cut may improve one or more administrators;

therefore, it would not be appropriate to plan for a remote administrator to do so, which can provide an altered level of support. System cutting is regularly used for a few purposes: an entire private system, a reproduction of an open system to check the administration of substitution, or an overhead system for the chosen administration. For example, when fixing an individual system within such a system cut, which will be a beginning to end practically confined to a piece of the overall population arrangement, the system uncovered a collection of capabilities in terms of data transmission, dormancy, accessibility, and so on. From this stage on, the cut is frequently monitored personally by the cutter, who can see the machine cut as their own network along with transport centers, handling, and power. The assets assigned to the cut are regularly a mixture of semi-discovered and circulated assets.

2.4 Enabling Technologies

These developments were not initially planned for IoT, but would include IoT implementations such as emerging infrastructure for smart cities and automated vehicles over the next decade. In order to impact these future criteria, certain innovations need to be incorporated into the planned architecture.

Nano-chip In recent years, nano-chip-based devices have provided general applications for the investigation of examples of organic compounds. A tiny chip that places the re-invented cells under the skin and across the electrical field can be a creation inside the shape that we get from the wounded or mature tissue. Nevertheless, the usage of nano-chip devices would not be restricted to healthcare uses; for example, this technique may be seen in military and home mechanization uses that will occupy the enormous area of IoT applications.

Millimeter-Wave (mmWave) Ten years ago, there was a reduction in the transparency of the recurrence spectrum under 6 GHz classes and, along these lines, the desire for a higher rate is growing. Higher frequencies such as the millimeter wave (mmWave) that is occurring in groups of more than 24 GHz are proposed as a possibility for future 5G IoT applications on the grounds that greater transmitting efficiency may be seen to improve the opportunity and enable customers to use high information levels for short-range applications [30].

Heterogeneous (Het-Net) networks It has been established to satisfy the requirements of the 5G-IoT driving system on request. This Model Organizing Tale empowers 5G-IoT to include information on request to the client. In the current year, certain 5G HetNet structures have been established [9], which may convey a gigantic amount of necessary device properties.

Direct Device-to-Device (D2D) Communication This type of statement has been designed as a substitute for the short-term transmission of information that can benefit 5G-IoT with lower power utilization, improved QoS to customer, and payload adjustment. The standard macro-cell base station received a low BS pressure, but the D2D controlled the data.

Fifth Generation Wireless Systems (5G) Such devices are now the pre-eminent behavior for the production of IoT apps. 5G will make major contributions to the subsequent era of IoT by interfacing billions of smart items to call for a real future and a giant IoT. In this respect, the IoT device ID functionality is exceedingly problematic, on the basis that the heterogeneous field of use will satisfy the needs of the program. Compared to the International Data Corporation (IDC) study, the 5G implementation would benefit 70% of companies making up $1.2 billion on the supply of executives. The IoT is rapidly creating innovations, in particular, the new application space. Nowadays, the IoT systems are changing the essence of the way of life that includes connecting smart household devices to smart circumstances. Industry IoT (IIoT) is making strides on various challenges, such as evolving criteria for products and structures, and redesigning action plans [28]. The main known correspondence procedures within the availability of IoT are 3GPP and LTE (4G) systems [29], which flexibly provide IoT frameworks with reliability, long life, strength of association and broad inclusion, low cost of arrangement, high level of security, access to the scope of commitment, and transparent administration [30]. However, predominant cell systems, for example, are not capable of enhancing MTC interchanges, yet 5G-IoT systems could do so. Furthermore, 5G-IoT also provides an extremely low redundancy and improved inclusion rate for MTC correspondence for the quickest cell arrangement rate. In addition, another test took place on 5G-IoT [29]. The CISCO, Microsoft, and Verizon tests 5G, and tailored display output to the requirement of human eye [17]. The images would be manually checked. The 5G-IoT offers persistent, reconfigurable, web-wide, on-demand, and interactive activities for IoT applications. The architecture of the 5G-IoT will be configured and it is hoped that it would be possible to continue to complete supported, smart, and rapid operation at each stage [8].

Coherently Independent Programming Program To replicate network access, organize (RAN) to set up a Cloud-based Network Access Organization (Cloud-RAN) to establish a clear affiliation with the different rules and to upgrade the RAN capability provided by 5G on request. The design of the core system is streamlined to the on-demand form of power tuning.

Machine-Type Communication (MTC) Machine-type communications are computerized information correspondences between the basic framework of information transport and gadgets. Information correspondences grew directly between two MTC gadgets or between a machine-type interchange gadget and a database [31]. It sets out the legitimate scope of use from an oversized organization of independent gadgets to strategic administrations. Cell frameworks (in particular 5G) have been seen as a major contender for the delivery of MTC gadgets to the network. They are constantly transforming into an important part of our way of life.

Remote Software-Defined Networks (WSDN) Virtual Software-Defined Networking (WSDN) is a modern concept that deals with scalable distributed networking that lets management coordinate and empowers the configuration, rather than optimizing network execution or machine management. Such structures need greater adaptability and quick investigation; in order to accomplish this goal, SDN eliminates the lateral absorption of traditional systems and, by way of a single device management, allows freedom to configure the device. SDN is in a position to change the parameters of its flight program while preserving its working conditions [33]. Frequently, 5G applications are run via WSDN worldview to have quicker and more adaptable 5G-IoT implementations.

Advanced Spectrum Sharing & Interference Management (Advanced SSIM) The usable range asset is confined and swarmed. It takes a long time to reclassify a range band for different utilizations, for example, regularization or normalization are difficult. Range effectiveness is one of the significant proficiency measurements in 5G correspondence systems. To reinforce range execution, advanced range sharing strategies are generally utilized. Along these lines, the 5G correspondence systems are required to address the issue with various techniques [34].

Mobile Edge Computing (MEC) System Sensation (fog) processing may also be a shared perception model, which could be a core layer in the middle of the cloud network and IoT gadgets/sensors. Mobile Edge Computing (MEC) has suggested describing the operation of administrations at the edge of the network and seeks to establish a basis for the control and remote communication resources at the edge of the portable device. Model MEC architectures and programs have essential elements to help governments, such as field awareness, radio network information, and program execution.

- Forensic rebellion, MEC, and 5G networks have a chance to be the nucleus of the ensuing IoT;
- In 5G-IoT, MEC can effectively create calculation-related applications involving huge handling, comparable to video game (VR) or augmented reality (AR).

Remote N/w Function Virtualization (WNFV) Remote Network Function Virtualization (WNFV) allows arrangement administrations and capacities to remotely see organization assets, similar to databases, switches, connections, and information, without breaking the overall physical foundation, and to utilize these assets as administration necessities as required. The WNFV splits a physical device into various virtual structures, such that the devices are periodically reconfigured to re-engineer new systems as required. The WNFV as a feature for 5G networks would allow the virtualization of the whole network ability and detangle the organization of 5G-IoT. WNFV offers an adaptable and scalable framework for 5G-IoT applications that will allow the modified device to mold programmable devices for 5G-IoT applications. The WNFV can have 5G-IoT network implementations with the capacity to plan by streamlining the speed, capability, and functionality inside the networks to match the burden of usage. Additionally, the WNFV should boost the reasonableness of the Radio Access Arrangement (RAN) as a whole. Inside the HetNet 5G-IoT infrastructure strain, a number of 5G systems with a multi-RAT network will be used to fulfill the needs of the application administrations [36].

Mobile Cloud Computing (MCC) Program Distributed computing is an exciting theory of computation in the computer and business field that, with the usage of virtualization techniques, applies to a range of network models that patch architectures, from visible fog to non-accessible fog, and from the network as a software device to hardware. In view of the physical separation, cloud administrations cannot transparently access neighborhood conditional data, such as definite client area, nearby system or circumstances of clients versatility conduct. For delay-sensitive applications, such as VR and AR, these necessities (for example, portability bolsters low inactivity and setting mindfulness) are normal. Versatile Cloud Computing (MCC) centers on the marvel of portable assignment because the accessibility of assets in cell phones, the capacity of mass information and thus the execution of computationally escalated undertakings should be appointed to distant elements.

Information Examination and Big Data One of the significant parts of an effective IoT application is Data Analytics. Organizations have an enthusiasm for bits of data, and there is interest in deciphering the gigantic amounts of information collected. The idea of Big Data is a theoretical one, which is typically energetic to the framework's arrangement (for example, RAM and HDD space). Recently, the cost of massive amounts of information has been perceived, and there are various conclusions on defining Big Data. Truth be told, Big Data refers to datasets that are not visibly perceived, collected, monitored, and handled by regular IT and programming devices or equipment gadgets. Advances such as Big Data and Data Analysis are transforming the way we work and generating a range of new possibilities.

Security and Forensics With the combining of 5G and IoT, security issues such as safety, access to power, safe communications, and protected storing of information present challenges in IoT applications. In addition, all devices that have been produced and any information that has been synchronized with the IoT program shall be reviewed. The vast exploitation of IoT hardware and the private existence of the knowledge that has been aggregated and transferred through IoT hubs also renders protection an important issue. It does not take long for individuals to agree to prosecute one another for violating their smart apps, the autonomous automobiles that have mishap, and the crooks that place singular smart sensors in danger. The IoT has built up a stack that contains a variety of valuable legal science antiques, while at the same time identifying facts, range, security, and identification of evidence, equivalent to attack, may be attempted during the process of the inquiry. The IoT will soon be involved in all parts of our life from dealing with our home temperature to autonomous vehicles and smart management of the urban communities and therefore innovations bolster this procedure. The scientists need a normal language to utilize these advances suitably. Presently, these innovations are flawed when used with current IoT engineering. Consequently, modifications are required to style a substitution design that supports these advances. In the following sections, we discuss how these advancements add to the proposed IoT design.

3 5G-Enabled IoT

3.1 Architecture

Here we introduce the structure that is reasonable for the details of potential IoT programming and administrations. The most recent engineering is improved by the advancements that are clarified in the Case Study (Sect. 4) to incorporate a more useful, versatile, and adaptable IoT biological system than current IoT models. The 5G supported equipment, known as the 5G-IoT, has the accompanying features: precise, knowledgeable, organized, adaptable, transparent, and mindful of high demands. This comprises eight interconnected layers of two-way showcase correspondence capacity as found in Fig. 9.12. The following layer and fifth layer contain two and three sub-layers, independently, and in this manner the security layer covers every other layer. These layers are picked to give the primary execution and to keep the structure secluded simultaneously. The most recent advancements actualized in Sect. 3 will be worked inside the setting of this current development to influence the potential troubles suggested in Sect. 1. The developments of

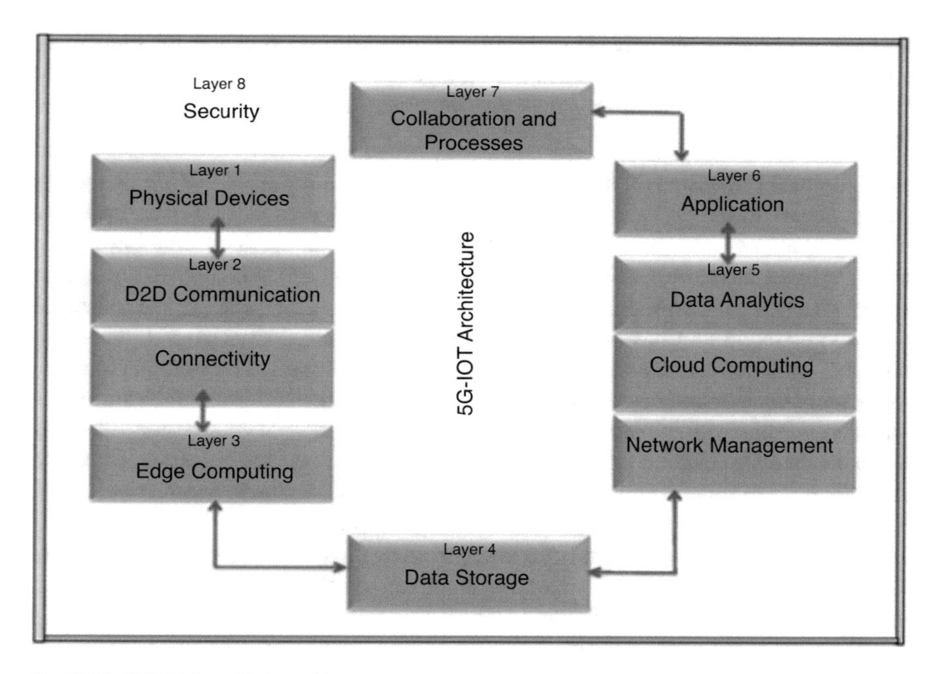

Fig. 9.12 5G–IoT enabled Architecture

altogether various functionalities will be introduced in a couple of layers for simple investigation, versatility, and identification.

Physical Device Layer This layer comprises remote sensors, actuators, and controls, which are essentially "things" of the IoT. Useful gadgets are a characteristic layer out of the frameworks. At this layer, small devices, for example, nano-chips must not be utilized to build the specialized arrangement of assets and to diminish the use of power. Nano-chips can give an elevated level of first-readied data that is proper for wide ranging information on indicative data (layer 7).

Communication Layer This layer has two sub-layers. D2D levels of coordination and synchronization.

Direct Device-to-Device (D2D) Sub-Layer Connectivity As the force and experience of physical devices (center points) expand, they produce their own structure and personality and construct their own insight. To improve the usefulness and limits of the IoT frameworks, such gadgets will shape the HetNet with the end goal that they can address one another. Cut-edge correspondence standards of the far off sensor framework (WSN) are used in this sub-layer. The center points will pack or even pick a pacesetter (bunch head) for the administration of the fitting frameworks. One of the huge progressions that fortify this sub-layer is the mmWave. Moreover, in this sub-layer, 5G is another optional development that is intended to strengthen D2D correspondence during a circumstance. 5G considers the accessibility of MTC devices to be a critical competitor. The solid endorsement levels and other vital features of MTC render it evident that 5G-Plus-HetNet is esteemed to be a steady usage framework inside the arranged 5G-IoT design.

Connectivity Sub-Layer In this sub-layer, systems are related to the accentuation of interchanges such as BSs. Truth be told, they submit and break down their insight into an Intranet association with the Capability Network. For the present, this sub-layer of the IoT has a few specific issues: only a fixed measure of gadget associations are frequently worried about; in frameworks, for example, autonomous vehicles, trade of information for the alternative between information styles is unimportant; high-volume data may hardly be continually arranged as an outcome of colossal correspondence idleness. In the near future, the 5G course of action will brilliantly improve this sub-layer to have unwavering quality, execution, and availability. Another development of this sub-layer is Advanced SSIM, as clarified in a previous section. Through this advancement, IoT devices gain the possibility to pick an ideal set (repeat gatherings) with a moderately low impedance. As a rule, the SSIM conventions are valuable in giving expanded motivations to go trade of checked rational radios.

Edge (Fog) Computing Layer

Here, the information is processed at the edge of the network to settle on choices at the sting stage. With the presentation of 5G innovations and, in comparative lines, the augmentation of portable interchanges (for example, portable applications),

MEC building can possibly be even greater, although still face challenges, and will offer a significant commitment to this system.

The Data Storage This layer includes information storage units throughout which the information collected from the edge nodes is often extracted. This layer includes unique security protection that is related to the enormous amount of information and the traffic of potential applications.

Management Service Layer
It consist of three sub-layers as follows:

Network Management Sub-Layer A system executive includes changed type of correspondence between gadgets and server farms. WNFV is the main major innovation needed in this layer. The WNFV is at the same period in a condition under which the context of the network and the form of communications protocols, such as 5G-IoT or ZigBee, are revamped to reinforce the essence of the IoT framework. WSDN is another important advancement in this sub-layer. WSDN struggles with coordinating IoT and empowers you to coordinate reconfiguration instead of following the normal method for update implementation.

Cloud Computing Sub-Layer At this sub-layer, details and details from the sting registry are (re)processed within the cloud such that the last prepared data is often calculated. By using 5G innovation, cell phones are continuously equipped to perform this kind of calculation, which is referred to as MCC. Consequently, the handling tasks are given the chance to migrate between cell phones helping to make the IoT system more effective, functional, adaptable, and faster.

Data Analytics Sub-Layer Here, modern techniques for the analysis of information are used to include some motivation (manipulable data) to the information. The improvements in Big Data calculations will upgrade the handling of the information in this sub-layer. It is probable that the function of this sub-layer is paramount in the near future as the volume of data generated is enhanced as a consequence of the combination of 5G and the IoT.

Application Layer In this layer, software connects to precedent layers and information, which is very still, so it is not important to see the speed of the system. Applications can disrupt vertical markets and business needs by controlling apps, Vertical and Mobile Applications, and Business Intelligence and Analytics. Truth be told, the Technology layer allows experts to make the best decisions at the right time by getting the right data.

Collaboration and Processes Layer The IoT framework, and along these lines, the information shown in the past layers, is not helpful unless it provides a demonstration. Individuals are engaged in business-based applications. Individuals utilize software and associated knowledge to satisfy their individual requirements. Now and then, different people are using a similar application for different purposes. Truth be told, citizens will have the ability to collaborate to make the IoT useful.

Security Layer Similar to other designs [21, 24], this layer is thought to be a different layer. In fact, this layer covers and ensures every single past layer; however, each area (convergence of this layer with another layer) has its own usefulness. The Engineering Security Layer involves different terms of security efforts, including encoding, client verification, access control, and cloud security [38]. In addition, the security layer also depicts and envisions the dangers and digital attacks, including the crime scene investigation, in order to recognize and deter the type of assault.

3.2 *Characteristics*

IoT is a network of different unmistakable computers or electrical devices capable of communicating with one another through some accessible web, such as the internet. The IoT has expanded the world of omnipresent registration with a range of new technologies operating with various types of sensors. There are, however, several obstacles to the use of the IoT that need to be tackled in order to make it more effective [38, 39]:

Security When the total number of related gadgets increases, the danger of malicious actors abusing weaknesses also increases. This occurs because of the utilization of low standard devices.

Protection The knowledge obtained from IoT hardware is submitted to a centralized focal point for analysis and planning, which involves an observer. Indeed, this sort of dispersal of data without the consent of the customer can trigger data leaks; thus, compromising the security of end clients.

Norms Lack of standards and guidance will contribute to unfortunate outcomes when handling the devices. Notwithstanding the measure of IoT-enabled devices, there is a requirement for advancement that can improve this titanic volume of information movement inside an enormous information development run. Truth be told, the devices themselves will have the chance of addressing such changes over the span of activity, for example, an enormous swapping scale top, expanded information quality, and diminished latencies [40]. The faster-remote arrangement procedure, especially the fifth generation remote (5G) systems, is an impetus for 5G-enabled IoT applications. This additionally permits adapting to a lot of IoT-implemented contraptions [41]. Out of the new 4G development, which uses frequencies under 6 GHz, 5G frameworks utilize astoundingly higher frequencies ranging from 30 GHz to 300 GHz. In addition, it is committed to making explicit Indus starter applications that work past the current versatile broadband range. This pervasive technique is the ping stone battle to look for more prominent transparency, which has been locked in since the beginning of the cell structure [43]. This makes usage of the 5G a focal and energizing impact on the IoT movement. Thus, it supplements IoT by offering higher information levels, lower latencies, lower basic measures, and higher versatility [44]. With the accelerated development of

IoT advances, the capacity to convey considerable preferences to end-shoppers, specifically to clients and business ventures, has increased [31]. Purchasers are given different organizations that depend on their activities. People will drive significantly more efficiently, for instance, by avoiding gridlocks and taking an alternate route when advised by an IoT smart driving device introduced in their vehicle. People can remain healthy by using wearable gadgets that offer their wellbeing characterized data after monitoring their physical activity and body parameters of the day. Associations may use client information to offer various types of help and items. Also, they will utilize field trackers and outer equipment lockouts to maintain their focal points. Government and transparent specialists will limit the cost of social insurance by offering more grounded help for wellbeing remotely, particularly for seniors.

4 Case Study

This case study pertains to how 5G mobile technology will impact the Internet of Things (IoT).

There are currently initial deployments of commercial 5G cellular networks in progress.

A variety of factors are driving the adoption of 5G and IoT, including increased demand from customers and businesses and the availability of more affordable devices. Along with the introduction of global standards, substantial operator investment in 5G technology, spectrum, and infrastructure are also helping to drive growth and increase market interest in the IoT. Today's 5G mobile cellular networks are emerging from current 4G networks, which will continue to support many instances of use. 5G can satisfy current requirements, such as smart energy applications, and foresee use cases that are still some time away, such as self-driving vehicles, expected to last far into the future. Mobile operators would need to ensure that their networks support both existing and future use case requirements as they address the evolution of technology. Prudent operators will manage their investments to ensure customers are supported as networks transition to 5G. Most cases of 5G use can be divided into three major categories: enhanced mobile broadband (eMBB), huge IoT, and essential communications, each with its own specifications for speed, bandwidth, and latency. While 4G will continue to be used for many IoT use cases for customers and businesses, 5G offers a variety of IoT advantages that are not available with 4G or other technologies. This includes the capacity of 5G to accommodate a large number of static and mobile IoT devices that have a wide set of service specifications for speed, bandwidth, and efficiency. The versatility of 5G will become much more important for companies requiring help for the stringent requirements of essential communications as the IoT evolves. 5G ultra reliability and low latency would allow self-driving vehicles, smart energy grids, improved factory automation, and other demanding applications to become reality. Cloud computing, artificial intelligence, and edge computing can all help control

IoT generated data volumes, as 5G improves network bandwidth. Ultimately, more 5G upgrades, such as network slicing, nonpublic networks, and 5G core, would help realize the vision of a global IoT network, supporting a large number of linked devices with diverse demands for versatility and connectivity.

5 Conclusion

5G offers a variety of advantages that other technologies do not provide. These include the simplicity of 5G to accommodate a large range of static and mobile IoT devices that have a variety of specifications for speed, bandwidth, and service quality. The flexibility of 5G will become even more important for enterprises as the IoT evolves. With much stricter efficiency criteria, 5G would enable essential communications. The ultra-reliability and low latency of 5G will help make it a reality for self-driving vehicles, smart energy systems, better factory automation, and other advanced applications. In this chapter, we addressed the 5G-enabled infrastructure, which is considered the future requirements of new applications and their produced data. This also incorporates a new network composed of Nano-chip, Millimeter Wave (mmWave), Heterogeneous Networks (HetNet), Device-to-Device (D2D) connectivity, 5G-IoT, Machine-Type Connectivity (MTC), Wireless Network Virtualization Function, Wireless Software-Defined Networks (WS-DN), Advanced Spectrum Sharing and Interference Monitoring (Advanced SSIM), Mobile Edge Computing (MEC), and Mobile Internet. The architecture was built on the basis of these technologies. The architecture is flexible, efficient, scalable, consistent, convenient, and able to meet high application requirements. It may also assist IoT specialists in the construction of more powerful and flexible IoT systems. Despite the apparent success of IoT and the soon-to-be-released 5G technology, the chapter discusses the problems that will arise with the standardization of the two technologies. In the midst of many possible alternatives, the chapter discusses the use of blockchain to solve all of these problems in the area of 5G-IoT. In the future, the end user will benefit from the capabilities of 5G, followed by the convergence of blockchain with IoT applications. Followed by the respective application architectures and their respective characteristics.

References

1. M.T. Hammi, B. Hammi, P. Bellot, A. Serhrouchni, Bubbles of trust: A decentralized blockchain-based authentication system for IoT. Comput. Secur. **78**, 126–142 (2018)
2. S.T. Mistry, S. Tyagi, N. Kumar, Blockchain for 5G-enabled IoT for industrial automation: A systematic review, solutions, and challenges. Mech. Syst. Signal Process. **135**, 106382 (2020)
3. ETSI. Accessed: 24 Nov 2019. URL: https://www.etsi.org/technologies/internet-of-things
4. M. Liyanage, I. Ahmad, A.B. Abro, A. Gurtov, M. Ylianttila, *A Comprehensive Guide to 5G Security* (John Wiley & Sons, Hoboken, 2018)
5. W. Al-Saqaf, N. Seidler, Blockchain technology for social impact: Opportunities and challenges ahead. J. Cyber Policy **2**(3), 338–354 (2017)

6. Market Pulse Report, Internet of Things (IoT). Accessed: 30 Nov 2019. URL: https://growthenabler.com/flipbook/pdf/IOT%20Report.pdf
7. P. Daugherty, B. Berthon, *Winning with the Industrial Internet of Things: How to Accelerate the Journey to Productivity and Growth* (Accenture, Dublin, 2015)
8. Ericsson 5G report: Industry digitalization could be a USD 700 billion market by 2030. Accessed: 4 Dec 2019. URL: https://www.ericsson.com/en/news/2019/10/ericsson-5g-for-business-a-2030-market-compass
9. Gartner the CIO's guide to blockchain. Accessed: 30 Nov 2019. URL: https://www.gartner.com/smarterwithgartner/the-cios-guide-to-blockchain/
10. F. Bonomi, R. Milito, J. Zhu, S. Addepalli, Fog computing and its role in the internet of things characterization of fog computing, in *Proceedings of the First Edition of the MCC Workshop on Mobile Cloud Computing*, pp. 13–15 (2012)
11. S.N. Shirazi, A. Gouglidis, A. Farshad, D. Hutchison, The extended cloud: Review and analysis of mobile edge computing and fog from a security and resilience perspective. IEEE J. Sel. Areas Commun. **35**(11), 2586–2595 (2017)
12. S.S. Gill, P. Garraghan, R. Buyya, ROUTER: Fog enabled cloud based intelligent resource management approach for smart home IoT devices. J. Syst. Softw. **154**, 125–138 (2019)
13. M. Iorga, L. Feldman, R. Barton, M.J. Martin, N. Goren, C. Mahmoudi, Fog computing conceptual model: Recommendations of the National Institute of Standards and Technology, NIST Spec. Publ., 500-325, 2018. [Online] Available: https://doi.org/10.6028/NIST.SP.500-325
14. M.H. ur Rehman et al., The role of big data analytics in industrial Internet of Things. Future Gener. Comput. Syst **99**, 247–259 (2019)
15. H.-N. Dai et al., Big data analytics for manufacturing internet of things: Opportunities, challenges and enabling technologies. Enterprise Inf. Syst **14**(6), 1–25 (2019)
16. R. Agrawal et al., Continuous security in IoT using blockchain, in *2018 IEEE International Conference on Acoustics, Speech and Signal Processing (ICASSP)*, 2018, pp. 6423–6427
17. W. Ejaz et al., Internet of things (IoT) in 5G wireless communications. IEEE Access **4**, 10310–10314 (2016)
18. J.N. Dewey, R. Hill, R. Plasencia, Blockchain and 5G-enabled internet of things (IOT) will redefine supply chains and trade finance. Secured Lender, 42–45 (2018)
19. A.Y. Ding, M. Janssen, Opportunities for applications using 5G networks: requirements, challenges, and outlook, in *Proceedings of the Seventh International Conference on Telecommunications and Remote Sensing*, ACM, 2018, pp. 27–34
20. M.M. Alsulami, N. Akkari, The role of 5G wireless networks in the internet-of-things (IoT), in *2018 1st International Conference on Computer Applications Information Security (ICCAIS)*, IEEE, pp. 1–8
21. M. Patel, S. Tanwar, R. Gupta, N. Kumar, A deep learning-based cryptocurrency price prediction scheme for financial institutions. J. Inf. Secur. Appl. **55**(102583), 1–13 (2020)
22. A. Kumari, R. Shukla, S. Gupta, S. Tanwar, S. Tyagi, N. Kumar, ET-DeaL: A P2P smart contract-based secure energy trading scheme for smart grid systems, in *IEEE International Conference on Computer Communications (IEEE INFOCOM 2020)*, Beijing, China, April 27–30, 2020, pp. 1–8
23. S. Tanwar et al., An advanced internet of thing based security alert system for smart home, in *2017 International Conference on Computer, Information and Telecommunication Systems (CITS)*, 2017, pp. 25–29
24. R. Gupta, F. Al-Turjman, P. Italiya, A. Nauman, S.W. Kim, Smart contract privacy protection using AI in cyber-physical systems: Tools, techniques and challenges. IEEE Access **8**, 24746–24772 (2020)
25. P. Porambage, J. Okwuibe, M. Liyanage, M. Ylianttila, T. Taleb, Survey on multi-access edge computing for internet of things realization. IEEE Commun. Surv. Tutor. **20**(4), 2961–2991 (2018)
26. R. Gupta, A. Shukla, P. Mehta, P. Bhattacharya, S. Tanwar, S. Tyagi, N. Kumar, VAHAK: A blockchain-based outdoor delivery scheme using UAV for healthcare 4.0 services, in *IEEE International Conference on Computer Communications (IEEE INFOCOM 2020)*, Beijing, China, April 27–30, 2020, pp. 1–8

27. Hyperledger Fabric. https://www.hyperledger.org/projects/fabric (visited on 10/11/2020)
28. Ripple – One Friction-less Experience To Send Money Globally. https://ripple.com/ (visited on 10/11/2020)
29. J. Cheng, W. Chen, F. Tao, C.L. Lin, Industrial IoT in 5G environment towards smart manufacturing. J. Ind. **10**, 10–19 (2018)
30. V. Reja, K. Varghese, Impact of 5G technology on IoT applications in construction Project Management. ISARC. Proc. Int. Symp. Autom. Robot. Constr. **36**, 209–217 (2019)
31. P. Yogita, S. Nancy, S. Yaduvir, Internet of things (IoT): Challenges and future directions. Int. J. Adv. Res. Comput. Commun. Eng. **5**(3), 960–964 (2016)
32. M.K. Jyoti, R. Devakanta, D.S. Parth, A survey on IOT and 5G network, in *International Conference on Smart City and Emerging Technology (ICSCET)*, (2018), pp. 1–3
33. E. Waleed et al., Internet of things (IoT) in 5G wireless communications. IEEE Access **4**, 10310–10314 (2016)
34. J.N. Dewey, R. Hill, P. Rebecca, *Blockchain and 5G-Enabled Internet of Things (IOT) will redefine supply chains and trade finance* (Secured Lender, 2018), pp. 42–45
35. Y.D. Aaron, J. Marijn, Opportunities for applications using 5G networks: Requirements, challenges, and outlook, in *Proceedings of the Seventh International Conference on Telecommunications and Remote Sensing*, (ACM, 2018), pp. 27–34
36. M.A. Mashael, A. Nadine, The role of 5G wireless networks in the internet-of-things (IoT), in *2018 1st International Conference on Computer Applications Information Security*, (IEEE, ICCAIS), pp. 1–8
37. R. Ana et al., On blockchain and its integration with IoT. Challenges and opportunities. Future Gener. Comput. Syst. **88**, 173–190 (2018)
38. B. Aymen et al., Towards better availability and accountability for iot updates by means of a blockchain, in *IEEE European Symposium on Security and Privacy Workshops (EuroSPW)*, (2017), pp. 50–58
39. E.S. Knud, L. Per, Smart home and smart city solutions enabled by 5G, IoT, AAI and CoT services, in *Contemporary Computing and Informatics (IC3I), International Conference on, (2014)*, pp. 874–878
40. R. Rakesh, K.R. Abhay, Challenges and risk to implement IOT in smart homes: An Indian perspective. Int. J. Comput. Appl. **153**, 16–19 (2016)
41. L. Cristian, R. Mariacristina, Smart district through IoT and blockchain, in *IEEE 6th International Conference on Renewable Energy Research and Applications (ICRERA)*, (2017), pp. 454–461
42. H. Donhee, K. Hongjin, J. Juwook, Blockchain based smart door lock system, in *Information and Communication Technology Convergence (ICTC), International Conference*, (2017), pp. 1165–1167
43. D. Ali, S.K. Salil, J. Raja, Towards an optimized blockchain for IoT, in *Proceedings of the Second International Conference on Internet-of-Things Design and Implementation*, (ACM, 2017), pp. 173–178
44. A. Dorri et al., Blockchain for IoT security and privacy: The case study of a smart home, in *Pervasive Computing and Communications Workshops (PerComWorkshops)*, (2017. IEEE International Conference on), pp. 618–623

Chapter 10
Big Data Analytics for 5G-Enabled IoT Healthcare

A. Sivasangari, L. Lakshmanan, P. Ajitha, D. Deepa, and J. Jabez

1 Introduction

According to the HIS market, by 2035, the Fifth Generation (5G) of wireless transmission technology will enable more than $1 trillion worth of products and services for the healthcare sector. The main features focused on in 5G technology, for example, significant increase in speed, coverage, enhanced capacity, network energy and power, and increased bandwidth, impact across many divisions in big data analytics. For people who have diabetes a comprehensive sensing analysis is available. Many mechanisms and personalized building models using 5G smart diabetes testing of smart clothing and smart monitoring using smartphones and using big data clouds are suggested for patients as part of their personalized solution for diabetes monitoring in healthcare. Here, an overall comprehensive process regarding blockchain-based 5G-enabled IoT and also various challenges and integration with blockchain industrial automation with the 5G-enabled IoT are presented. In addition, existing gaps in scalability, interoperability and other challenges in 5G blockchain are discussed. The deficiencies in 5G from all the communication devices and drones and particularly in the field of healthcare are identified and these problems will be overcome with the help of ultra-high reliability.

A. Sivasangari (✉) · L. Lakshmanan · P. Ajitha · D. Deepa · J. Jabez
Sathyabama Institute of Science and Technology, Chennai, Tamil Nadu, India
e-mail: sivasangari.it@sathyabama.ac.in; lakshmanan.cse@sathyabama.ac.in; ajitha.it@sathyabama.ac.in; deepa.cse@sathyabama.ac.in; jabez.it@sathyabama.ac.in

S. Tanwar (ed.), *Blockchain for 5G-Enabled IoT*,
https://doi.org/10.1007/978-3-030-67490-8_10

1.1 Motivation and Scope

The scope of 5G definition relates to potential uses and how those likely affect healthcare. The Internet of Medical Things (IoMT) focuses on the impact of 5G on providers, hospital systems, medical device companies, pharmacy companies and telehealth. Several key companies plan to launch 5G systems as well as 5G wireless networks in healthcare products. In the long term, in the healthcare sector, 5G will help to profoundly transform remote diagnostics and consultations. 5G enables the IoT to have a more powerful bandwidth combined with lower latency; 5G will be the main focus of technologies such as Augmented Reality and Virtual Reality in the healthcare sector. Furthermore, 5G will allow widespread deployment of Artificial Intelligence (AI) which will transform the healthcare sector from manual to smart automation.

Blockchain and 5G are the most hyped technologies emerging in the common marketplace. As mentioned, several features are available in blockchain with 5G, including decentralized approach, immutability, allows localized availability, cost efficiency and security. Also, challenges regarding blockchain with 5G integration focus mostly on scalability of blockchain which needs improvement to deal with the high number of devices and each device must have a unique address. Furthermore, after the 5G technology is deployed worldwide, it is expected that the technology will allow medical professionals to be able to exchange data with patients instantaneously from anywhere. It is an easy way for hospital-like monitoring in patient's homes similar to how intensive care units are monitored nowadays. Blockchain technology is perhaps the silver bullet needed for industry. The blockchain functions as a distributed transaction ledger for various IoT transactions. The blockchain platform supports and uses simple key management systems. The Ethereum platform is capable of managing a more fine-grained way used in many IoT devices with successful smart Turing complete code.

1.2 Research Contribution

In this chapter, we proposed a 5G-enabled blockchain e-healthcare framework. The focus of this framework is a patient e-health management system. E-Health introduced the fog/edge used for easy access of medical data, as well as patient safety and privacy concern. Blockchain is deployed in the e-health system. This consist of three interfaces (1) Near Processing Layer, (2) Far Processing Layer and (3) Data Sensing Layer, and an agent Migration Handler (MH) used to monitor and transfer tasks which will help to locate the client. The current healthcare system is not patient friendly because patients must continually spend time monitoring their illness instead of resting, which is inconvenient for the patients. Wasted patient time has been reduced in our proposed system.

1.3 Organization

In Sect. 2 the literature survey and comparison of existing ideas with 5G technology in healthcare and blockchain is presented, in Sect. 3 our idea is proposed in a detailed manner, in Sect. 4 the performance analysis of the proposed model with graphs is discussed, and conclusions and future work in the healthcare sector are presented in Sect. 5.

2 Related Work

Some studies discussed by Hossain et al. proposed an emotion detection methodology in the healthcare system. They used IoT devices to capture emotion images and speech recognition processes separately, and calculated the value of the detected emotion and validated it [1].

Latif et al. discussed the 5G wireless technology along with emerging technologies that will transform the healthcare system, specifically 5G with cloud computing and 5G with artificial intelligence, and in terms of economist and high potential pitfalls in development of the 5G health revolution [2].

Similarly, Nasri et al. proposed a smart mobile IoT healthcare system using 5G and smart phones to monitor patient's health risk factors. WBSN data was used to monitor and track patient pulse, temperature and oxygen in blood as well as other vital parameters of the patients [3].

Ahad et al. discussed diabetes diagnosis with the solution of comprehensive sensing analysis. They suggested patients could have personalized solutions for diabetes monitoring in healthcare, including many mechanisms and personalized building models using 5G smart diabetes testing on smart clothing and smart monitoring using smartphones and big data clouds [4].

Further, Mistry et al. [5] presented a comprehensive review on blockchain-based 5G-enabled IoT and various challenges stemming from integration with blockchain industrial automation and the 5G-enabled IoT. A comparison of existing gaps between the scalability, interoperability and other challenges in 5G blockchain was also presented.

Ullah et al. discussed the driving with wireless industry and developing the next generation of technology so that mobile technology generation has improved facilities to be efficient in wireless fields. Vehicle-to-everything (V2X) will impact in 5G with all the communication and drones and particularly in the field of healthcare, and they identified deficiencies and overcame those problem with the help of ultra-high reliability [6].

Furthermore, Li discussed how the next generation of wireless remote technology will be useful for healthcare in existing models with respect to the expenses of healthcare services and the imbalance of medical resources and inefficient healthcare system administration. To overcome this, the IoT, big data analysis, artificial intelligence technology and 5G wireless are used to improve patient quality of healthcare service, and the cost inferable method is focused on [7].

Sigwele et al. proposed an information and communication technology utilizing IoT to limit medical errors and cost of healthcare. They discussed the IEE5GG with smartphone gateway connection to save energy, which is executed with the help of MEC while considering QoS and battery level CPU load, and resulting with an energy efficient framework [8].

Chen et al. proposed a 5G-C-sys for healthcare that aims at ultra-low latency in cognitive application and high reliability. They also developed a prototype platform for 5G-C-sys incorporated with speech recognition and emotion detection for the effectiveness in healthcare-based 5G C-sys technology [9].

Similarly, Boban et al. analysed the requirement in 5G technology and identified the gaps with the existing technologies. They overcame the challenges with drone and communication technologies [10].

Latif et al. discussed how 5G technologies, AI, IoT and Big data will revolutionize healthcare, and they provided an overview of how machine learning algorithms are integrated and able to detect the anomalies in the healthcare system. The authors also investigated remote consultation in e-health [11].

Lakshmanan et al. proposed a hybrid approach in combining PSO and ACO & BCO on routing protocol and applying the K-Means algorithm for clustering the nodes [12].

Furthermore, Manoj et al. [13] discussed a congestion adaptive navigation for emergency situations. They also used sensors for locating using GPS and then server takes an action using PIR in emergency areas.

Logeswari discusses the analysis of packets having the fuzzy logic based on the greedy routing protocol. Two characteristics input metrics and fuzzy decision-making system in VANETs were used [14].

Gomathi et al. [15] proposed an energy efficient routing protocol using wireless sensors with dynamic clustering UWSN routing technique. This will be helpful for researchers due to reduced power consumption, response time, avoids overload and improves throughput of the network.

Vignesh et al. discussed fewer deployments in the cloud storage with low cost replication, higher availability and better performance in geo-replicated systems by data centres with these benefits [16].

Ishwarya proposed a project to reduce congestion in traffic and calculated current traffic with normal. If anything unusual is detected, then emergency vehicles pass through the signals; thus, solving the traffic problems [17].

Sivasangari et al. [18] compared security and privacy using fog computing. They also used the fog computing principle to use a smart gateway for an improved big data health monitoring system.

Suganthi et al. discussed security improvement for web based banking and the authentication using fingerprints to avoid hacking or for fault detection [19].

Deepa et al. proposed an idea of detecting road damage by image processing in smart phones and sending the co-ordinate point to the cloud and from the cloud a user can see the road where the damages are because it will show on a map. From this, they can avoid accidents and so on [20].

Keerthi et al. [21] used convolutional neural networks (CNN) to identify dangerous lung disease tumours. The CNN technique has many features and can provide standard representation pneumonic radiological complexity, fluctuation and classification of lung nodule.

Sivasangari et al. [22] proposed major concerns about WBAN regarding the security and privacy of the healthcare sector. The patient health data should reach the physician at the right time. Security has the greatest impact on the lives of humans, and an effective model SEKBAN that ensures security data based on ECG signal was implemented.

Indira et al. [23] implemented an efficient hybrid detection using a wireless sensor network. Wireless devices are spatially distributed over sensors and physical changes. The device network includes multiple detection over sensors with lightweight transport.

Tao et al. [24] discussed V2G technology for enabling renewable energy sources providing power in a smart grid. They proposed a fog and cloud hybrid model.

Vilalta et al. compares the existing approach with the new proposed technologies and distributed field. This paper discussed TelcoFog's benefits and dynamic deployment with low latency, and managing orchestration architecture for TelcoFog service infrastructure [25]. Furthermore, Chaudhary et al. focused on challenges for future demands in integrated fog computing and cloud computing in the 5G environment, in collaboration with SDN, NFV and NSC. They also performed data analytics with device mobility, as well as discussed challenges and potential attacks on the data shared in 5G [26].

Ku et al. [27] discussed advances in the fog radio access network and fog–cloud based in hybrid system issues. GPP is used for communication and computational processing, and it is also used as a simulator for experimental tests.

Furthermore, Yang et al. [28] proposed an SDN-enabled approach for cloud–fog interoperation in 5G, and aimed at quality of optimized network usage.

Crosby et al. [29] shared, in terms of blockchain technology, all criteria that satisfy specific application in both sectors regarding finance. There are many opportunities for revolution in disruptive technology. The digital currency Bitcoin is highly controversial, but blockchain has proved to be useful and has found many applications.

Risius et al. discussed a framework in blockchain which they divided into three group activities and four level of analysis. This paper addresses research predominantly focused on new technologies in blockchain [30].

Dinh et al. proposed a survey about untangling blockchain data processing with its challenges, and they analysed four dimensions in production as well as research systems—distributed ledger, smart contract, cryptography and consensus protocol. They also conducted comprehensive evaluation for major systems such as BlockBench, Parity and Ethereum, and found blockchain performance closer to the database [31].

Huh et al. [32] discussed how to manage IoT devices using blockchain to easily control and configure IoT devices. They also used the RSA algorithm which is capable of managing devices with secret keys. They used Ethereum blockchain for coding in an efficient way.

3 Proposed Work

Fog/Edge computing is tackled by eHealth systems for an easy way of accessing and processing medical data, and ensuring the privacy and safety of patients. All Fog and Cloud have their administers interested in handling medical data that violates the privacy of patients. Blockchain deployed over Fog and Cloud will allow patient data processing and storage. The eHealth architecture consists of three interface levels: sensing, near processing, and far processing. Multiple instances of a Patient Manager include 3-level structures. The Agent Migration Handler (MH) uses Profile Monitoring to transfer a task or execute the task internally, which collects profile information from remote agents.

Sensing layer: Smart devices, implantable sensors, smart watches, mobile devices and other devices monitor patients' body parameters. These applications use Bluetooth or ZigBee to convey to the mobile the physiological sign of a patient.

Near processing layer: A hop away from the sensing devices of the data is where near processing level devices are generally located. Conventional switches, routers and low-profile devices are involved in the near networking layer. In a broad range of formats, healthcare facilities generate vast quantities of data, such as records, financial statements, laboratory findings, imaging tests, such as X-rays and CAD scans, and measurements of vital signs. Blockchain provides the ability to boost the data's authentication and integrity. This also helps to disseminate data inside the network or facilities. Such features have an effect on the cost, quality of data and reliability of providing health care across the system. Blockchain is an open, decentralized, intermediary-removing network. The blockchain healthcare solution does not require multiple authentication levels and provides everyone who is part of the blockchain architecture with access to the data. Data is made open and transparent for customers. Such apps will help to tackle the various issues facing the healthcare industry today. In the healthcare sector, blockchain's role is split into four stages. The proposed architecture is explained in Fig. 10.1.

The inspiration behind blockchain and 5G integration largely stems from the many benefits of blockchain in addressing security, protection, networking and service management issues in 5G networks. The proposed advanced Pos Consensus algorithm is described below:

Data: Performance Transaction (PT), Reputation Transaction (RT), Stake Transaction, Agent Number (Ni) in a cluster
Result: Every fog agent generates PTi, STi and produce RTi from the service provider

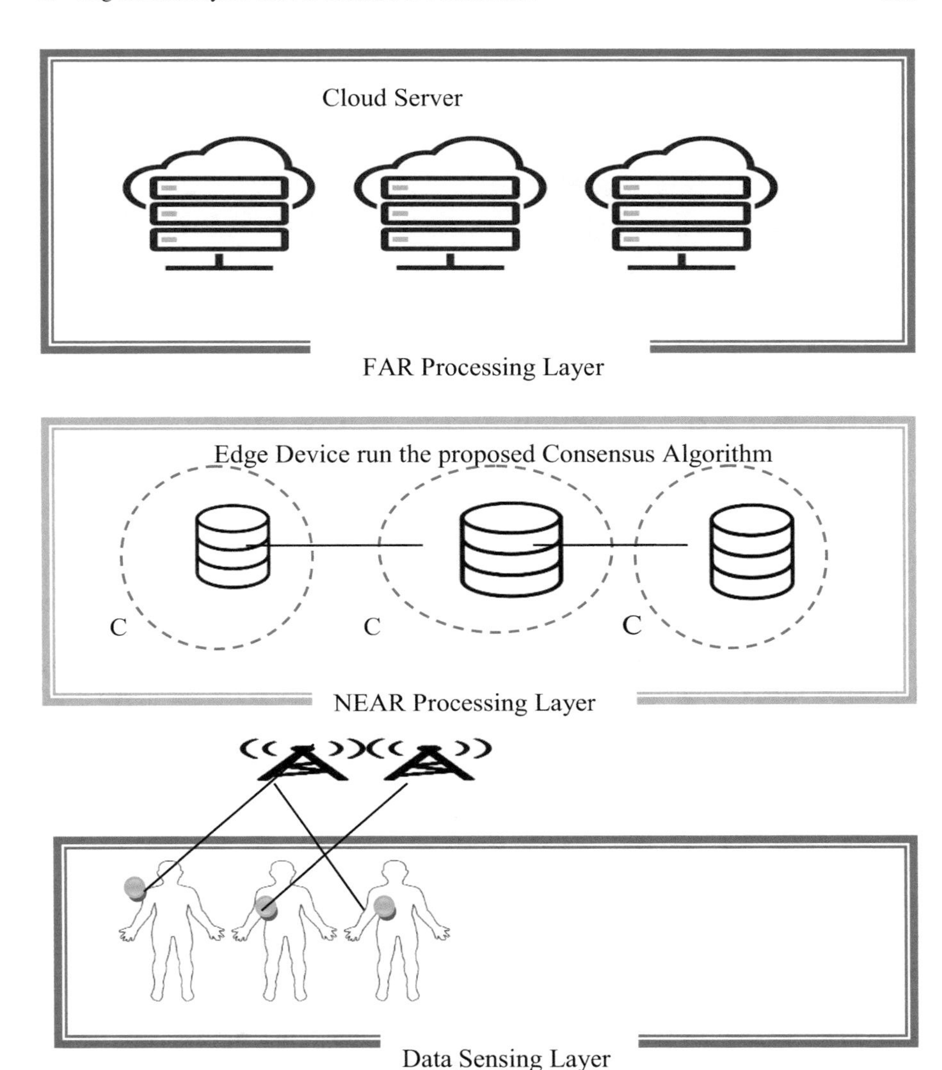

Fig. 10.1 Proposed architecture

Form clusters with fog nodes within a threshold range (R)
For each cluster $k = 1$ to l
Do while
Head election = true do
For each member agent $i = 1$ to n_k of a cluster do
Extract parameters from PT_i, RT_i to produce P_i, R_i and S_i

$P_i = \frac{1}{1+e^{-\frac{\ddot{Y}}{r \times \alpha_i}}}$
$R_i = \frac{1}{1+e^{-r}}$
$S_i = \frac{1}{1+e^{-c}}$
$f_i < -$ *Decision Tree*(P_i, S_i, R_i)
$T_i = \Delta T \times \left(1 - \frac{f_i}{\sum_n^1 f_i}\right)$
Every Member node in the cluster sets their timer (T_i)
If (T_i) is expired
Then broadcast node id to the cluster for approval
End
If approval count [node id] >= 2/3 × n_k
$leader_j < -nodeid$
End
End
End

A cluster within the Near Processing Layer of a certain geographic range (R). A cluster is formed by a fog/edge agent with a different patient member value, where a representative is selected to be the head of the cluster. The cluster head (also called the leader) is involved in running the blockchain consensus protocol by locking a certain amount of stake in the network. From each cluster, a cluster head (CH) is chosen, taking into account the member nodes' multi-criteria. The selection process includes the performance characteristics of a node, its reputation and the stake amount. Criteria are combined to measure a fitness value using a decision tree. The blockchain records the information of each node regarding the parameters listed, and can be retrieved from the blockchain.

The performance parameters include device processing speed, storage capabilities, accessibility, variation distance coefficient and delay in transmission of an Agent. Here, MIPS processing capacities, memory space and availability are symbolized respectively as p1, p2 and p3.

ΔT is the time interval for the selection of cluster head, and where T represents a limited random time period used to distinguish waiting time for the same fitness of the Agent. The Agent broadcasts its identifier across the cluster since its waiting time expires. The other cluster members verify the estimated fitness of the Agent and accept their approval for this Agent. In turn, every node in a cluster will participate in the PoS proposed. This consensus mechanism would be less vulnerable to an attack of 51% from each cluster than DPoS as a leader. The rich node, such as PoS, is less likely to become a cluster leader, as the cluster head is not only selected based on the locked coin.

Decision Tree is a supervised learning method which can be used for problems with classification and regression, but is preferred to solve problems with classification. It is a tree-structured classification where even the internal nodes represent the characteristics of a dataset, the branches represent the rules of decision and each leaf node represents the result.

4 Performance Analysis

The assessment of the proposed programme with simulation settings and evaluation metrics is defined in this section. It also addresses the effects of various parameters such as energy usage and time for block generation. Various scenarios with different configurations are visualized through graph plot.

For the following parameters, the performance of the updated mechanism and the existing mechanism will be investigated.

Energy consumption: energy consumption refers to the energy needed for transmission, receipt of the transaction and simulated validation of the network of a number of blocks.

Block generation time: This refers to the time required for a certain number of simulated network blocks to be uploaded, constructed and validated.

In the simulated network, the updated process is executed five times and the output graphs are shown with average values generated from 10 execution runs. The regular one runs on a horizontal network and is supposed to function on a hierarchical network with the modified one. For both forms of consensus structures, nodes that lock digital coins into the network engage in mining. The energy consumption and execution time necessary for the development of 100 blocks are shown in Fig. 10.2, provided that the variable number of nodes and clusters is taken into account.

For unique clusters and nodes, the block generation time is shown in Fig. 10.3. The illustrated graph in Fig. 10.3 indicates that a pattern which is consistently lower or higher does not follow the period of block generation with a larger number of clusters. With a higher number of clusters, cluster heads collect transactions and construct blocks, with a higher number of blocks per second being generated. On the other hand, because of the delay in testing blocks, higher block generation time was also noticed for some higher cluster numbers.

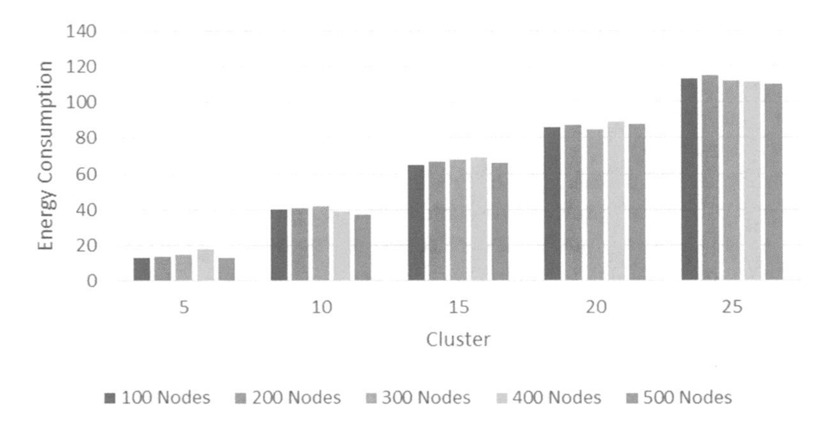

Fig. 10.2 Energy consumption vs cluster

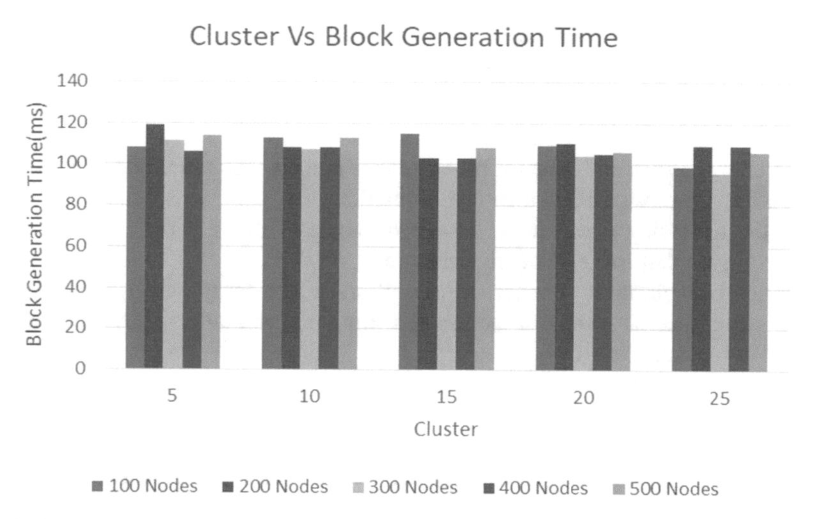

Fig. 10.3 Cluster vs throughput

In terms of power consumption and block generation time, the performance of the modified algorithm is compared to the standard one around. The revised algorithm shows a significant decrease in energy consumption compared to the regular one. A block is validated by a few selected safe miners in the updated one, but the standard one requires more than 50% node participation in the block validation process, resulting in higher energy consumption. Updated energy consumption remains almost constant for a comparable number of clusters with a greater number of nodes, while energy consumption tends to increase as the number of nodes within the network increases (Fig. 10.4).

Figure 10.5 displays the updated and standard block generation period. The graph in Fig. 10.5 illustrates that the time for the standard generation of blocks is greater than for the updated one. As normal, different nodes send their transactions to one leader node for validation, and to broadcast across the network, a validated block is required. The approach thus consumes higher energy and makes it possible to take a longer time for the block's network-wide casting. In addition, some good miners are selected based on reputation, results and stake in the revised one, but a miner based on investment or stake alone is regularly nominated.

We have painted our architecture with an already proven architecture in terms of reliability and overhead touch. The protection protocol is correlated with these two performance metrics. The graph in Fig. 10.6 shows that our eHealth is more robust than the current system due to our decentralized Key Management and several three-layer Patient Agent instances.

On the other hand, the diagram shown in Fig. 10.7 showed that our eHealth security mechanism provided greater overhead communication than the current one. An Agent needs a certain number of data encryption segments to be obtained from

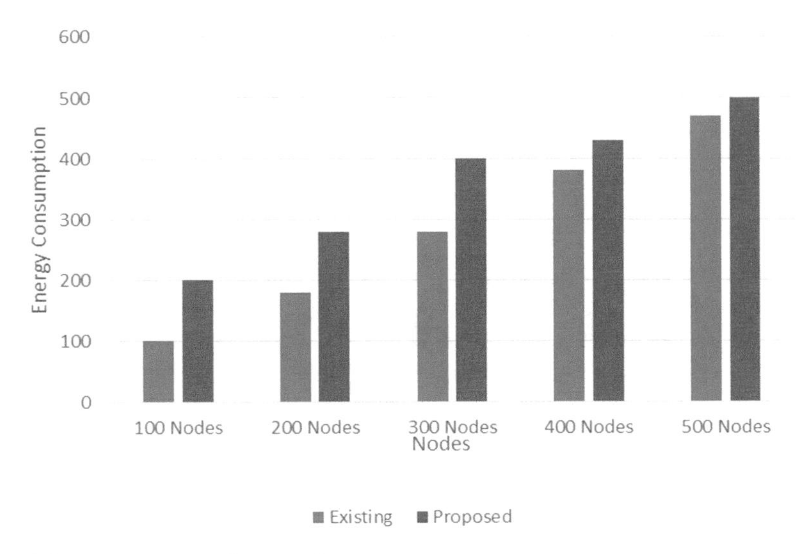

Fig. 10.4 Energy consumption vs nodes

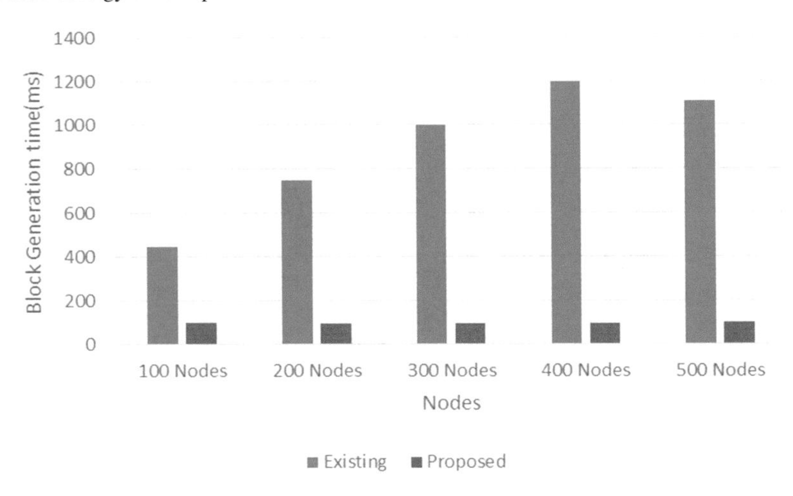

Fig. 10.5 Node vs block generation time

other neighbouring Agents to form the entire secret key. This technique activates overhead communication when exchanging hidden keys and authenticating.

The relation between the different features and the current system [22] is shown in Table 10.1. Similar to the cloud, with different protection strategies or without protection, different stakeholders deploy heterogeneous fog devices. Fog networks, through the identification and analysis of health information, are vulnerable to malicious attacks. In our architecture, the same patient agent replicated in the Mobile, Fog scheme and Cloud will protect wellbeing. To keep a Record of

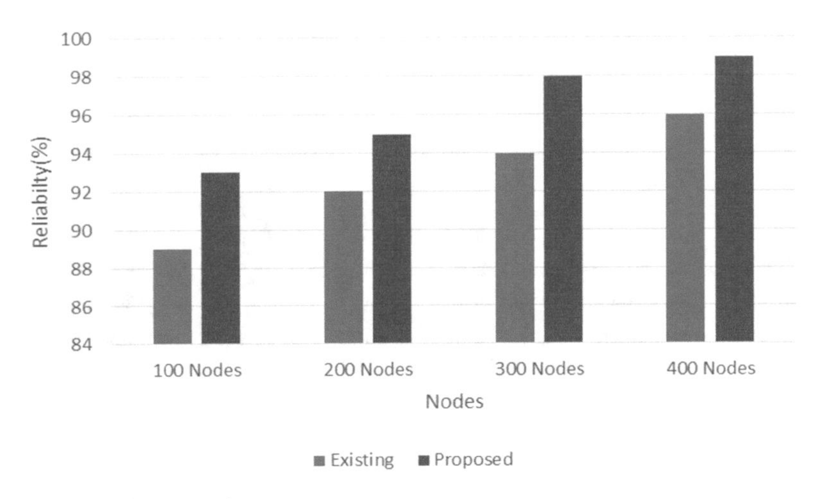

Fig. 10.6 Reliability vs nodes

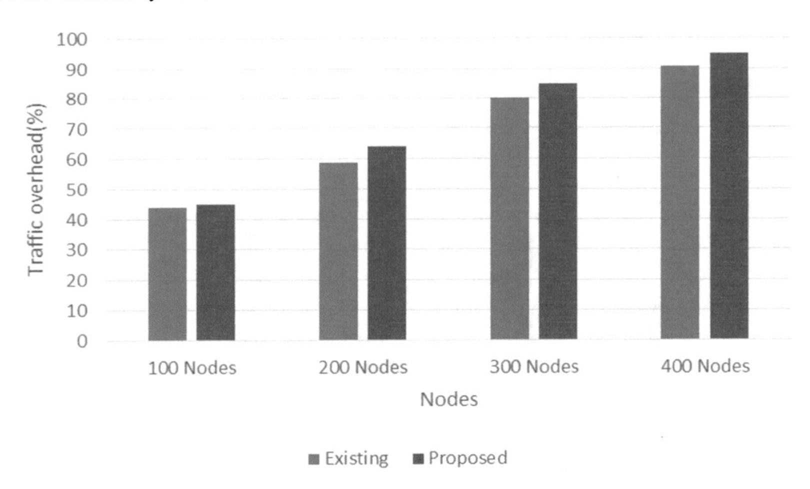

Fig. 10.7 Nodes vs traffic overhead

Malicious Attacks, sensitive medical information is analysed in the homogeneous replicated Patient Agent in order to protect the privacy or confidentiality of the patient.

A dropping assault occurs if a cluster head reduces the transactions. This is unlikely to happen because the cluster head will lose its reputation to share when it is detected as malicious. The consensus mechanism should select the malicious and cluster members who do not collect transactions for verification while the head of the cluster is down.

Table 10.1 Comparison with existing system

Criteria	Proposed	Existing system
Confidentiality, availability and integrity	CAI high, PA homogeneous Responsive medical data task Edge nodes, using the ring signature, blockchain maintenance for metadata ensures several PAs Availability of service	Privacy is average, integrity is high, usability is high, Low because of centralized operations of blockchain Controller
Secure and energy efficient migration	High	Low
Communication overhead	High	Medium
Consensus mechanism	Light weight	Medium

5 Conclusion

In this chapter, we built an eHealth program that deployed several instances of a three-layer Patient Agent software: sensing, near processing and far processing layer, which make the eHealth software more stable and fault-sensitive. We also defined how to implement the Patient Agent on a 5G unit. The dedicated Patient Agent application is able to handle the resources of 5G network slices. A performance analysis has shown that the emerging eHealth program will use blockchain technology to process health data in near-real time. The implementation of blockchain healthcare technology is difficult, with vast volumes of health data constantly being transmitted from wearable sensors.

References

1. M.S. Hossain, G. Muhammad, Emotion-aware connected healthcare big data towards 5G. IEEE Internet Things J. **5**(4), 2399–2406 (2017)
2. S. Latif, J. Qadir, S. Farooq, M.A. Imran, How 5g wireless (and concomitant technologies) will revolutionize healthcare? Future Internet **9**(4), 93 (2017)
3. F. Nasri, A. Mtibaa, Smart mobile healthcare system based on WBSN and 5G. Int. J. Adv. Comput. Sci. Appl. (IJACSA) **8**(10), 147–156 (2017)
4. A. Ahad, M. Tahir, K.-L.A. Yau, 5G-based smart healthcare network: Architecture, taxonomy, challenges and future research directions. IEEE Access **7**, 100747–100762 (2019)
5. I. Mistry, S. Tanwar, S. Tyagi, N. Kumar, Blockchain for 5G-enabled IoT for industrial automation: A systematic review, solutions, and challenges. Mech. Syst. Signal Process. **135**, 106382 (2020)
6. H. Ullah, N.G. Nair, A. Moore, C. Nugent, P. Muschamp, M. Cuevas, 5G communication: An overview of vehicle-to-everything, drones, and healthcare use-cases. IEEE Access **7**, 37251–37268 (2019)
7. D. Li, 5G and intelligence medicine—How the next generation of wireless technology will reconstruct healthcare? Precis. Clin. Med. **2**(4), 205–208 (2019)

8. T. Sigwele, H. Yim Fun, M. Ali, J. Hou, M. Susanto, H. Fitriawan, Intelligent and energy efficient mobile smartphone gateway for healthcare smart devices based on 5G, in *2018 IEEE Global Communications Conference (GLOBECOM)*, (IEEE, 2018), pp. 1–7
9. M. Chen, J. Yang, Y. Hao, S. Mao, K. Hwang, A 5G cognitive system for healthcare. Big Data Cogn. Comput. **1**(1), 2 (2017)
10. M. Boban, A. Kousaridas, K. Manolakis, J. Eichinger, W. Xu, Connected roads of the future: Use cases, requirements, and design considerations for vehicle-to-everything communications. IEEE Veh. Technol. Mag. **13**(3), 110–123 (2018)
11. S. Latif, J. Qadir, S. Farooq, M.A. Imran, How 5G (and concomitant technologies) will revolutionize healthcare. Future Internet **9**(4), 1–10 (2017)
12. L. Lakshmanan, A. Jesudoss, V. Ulagamuthalvi, Cluster based routing scheme for heterogeneous nodes in WSN–A genetic approach, in *International Conference on Intelligent Data Communication Technologies and Internet of Things*, (Springer, Cham, 2018), pp. 1013–1022
13. S.M. Kumar, L. Lakshmanan, A situation emergency building navigation disaster system using wireless sensor networks, in *2018 International Conference on Communication and Signal Processing (ICCSP)*, (IEEE, 2018), pp. 378–382
14. K. Logeshwari, L. Lakshmanan, Authenticated anonymous secure on demand routing protocol in VANET (Vehicular adhoc network), in *2017 International Conference on Information Communication and Embedded Systems (ICICES)*, (IEEE, 2017), pp. 1–7
15. R.M. Gomathi, J.M.L. Manickam, A. Sivasangari, P. Ajitha, Energy efficient dynamic clustering routing protocol in underwater wireless sensor networks. Int. J. Netw. Virtual Organ. **22**(4), 415–432 (2020)
16. R. Vignesh, D. Deepa, P. Anitha, S. Divya, S. Roobini, Dynamic enforcement of causal consistency for a geo-replicated cloud storage system. Int. J. Electr. Eng. Technol. **11**(3), 181–185 (2020)
17. M.V. Ishwarya, D. Deepa, S. Hemalatha, A. Venkata Sai Nynesh, A. PrudhviTej, Gridlock surveillance and management system. J. Comput. Theor. Nanosci. **16**(8), 3281–3284 (2019)
18. A. Sivasangari, P. Ajitha, E. Brumancia, L. Sujihelen, G. Rajesh, Data security and privacy functions in fog computing for healthcare 4.0, in *Fog Computing for Healthcare 4.0 Environments*, ed. by S. Tanwar, (Springer, Cham, 2021), pp. 337–354
19. D.S. Sharmila, L. Lakshmanan, Security improvement for web based banking authentication by utilizing fingerprint. Glob. J. Pure Appl. Math. **13**(9), 4397–4404 (2017)
20. D. Deepa, R. Vignesh, A. Sivasangari, S.C. Mana, B. Keerthi Samhitha, J. Jose, Visualizing road damage by monitoring system in cloud. Int. J. Electr. Eng. Technol. **11**(4), 191–203 (2020)
21. B. Keerthi Samhitha, S.C. Mana, J. Jose, R. Vignesh, D. Deepa, Prediction of lung cancer using convolutional neural network (CNN). Int. J. Adv. Trends Comput. Sci. Eng. **9**(3), 3361–3365 (2020)
22. A. Sivasangari, P. Ajitha, R.M. Gomathi, Light weight security scheme in wireless body area sensor network using logistic chaotic scheme. Int. J. Netw. Virtual Organ. **22**(4), 433–444 (2020)
23. K. Indira, D.U. Nandini, A. Sivasangari, An efficient hybrid intrusion detection system for wireless sensor networks. Int. J. Pure Appl. Math. **119**(7), 539–556 (2018)
24. M. Tao, K. Ota, M. Dong, Foud: Integrating fog and cloud for 5G-enabled V2G networks. IEEE Netw. **31**(2), 8–13 (2017)
25. R. Vilalta, V. López, A. Giorgetti, S. Peng, V. Orsini, L. Velasco, R. Serral-Gracia, et al., TelcoFog: A unified flexible fog and cloud computing architecture for 5G networks. IEEE Commun. Mag. **55**(8), 36–43 (2017)
26. R. Chaudhary, N. Kumar, S. Zeadally, Network service chaining in fog and cloud computing for the 5G environment: Data management and security challenges. IEEE Commun. Mag. **55**(11), 114–122 (2017)
27. Y.-J. Ku, D.-Y. Lin, C.-F. Lee, P.-J. Hsieh, H.-Y. Wei, C.-T. Chou, A.-C. Pang, 5G radio access network design with the fog paradigm: Confluence of communications and computing. IEEE Commun. Mag. **55**(4), 46–52 (2017)

28. P. Yang, N. Zhang, Y. Bi, L. Yu, X.S. Shen, Catalyzing cloud-fog interoperation in 5G wireless networks: An SDN approach. IEEE Netw. **31**(5), 14–20 (2017)
29. M. Crosby, P. Pattanayak, S. Verma, V. Kalyanaraman, Blockchain technology: Beyond bitcoin. Appl. Innov. **2**(6–10), 71 (2016)
30. M. Risius, K. Spohrer, A blockchain research framework. Bus. Inform. Syst. Eng. **59**(6), 385–409 (2017)
31. T.T.A. Dinh, R. Liu, M. Zhang, G. Chen, B.C. Ooi, J. Wang, Untangling blockchain: A data processing view of blockchain systems. IEEE Trans. Knowl. Data Eng. **30**(7), 1366–1385 (2018)
32. S. Huh, S. Cho, S. Kim, Managing IoT devices using blockchain platform, in *2017 19th International Conference on Advanced Communication Technology (ICACT)*, (IEEE, 2017), pp. 464–467

Part III
AI-Assisted Secure 5G-Enabled IoT

Chapter 11
Data Security and Privacy in 5G-Enabled IoT

Darpan Anand and Vineeta Khemchandani

1 Introduction

Today is the era of Internet of things (IoT) in which all the devices are required to be connected through the Internet. Therefore, the connectivity is the requirement of this system [1]. The mobile communication is one of the important communication media for this application. Internet of Things (IoT) is changing the way we live and work. Their success and real value come from the establishment of services on top of the connected IoT devices. According to the Ericsson Mobility Report, there will be over 30 billion connected devices worldwide by 2023, of which around 20 billion will be IoT related devices. Between 2017 and 2023, IoT devices are expected to increase at a CAGR (Compound Annual Growth Rate) of 19%, driven by promising IoT use cases, like smart wearables, smart display, smart metering, robotic control/production automation, robotic surgery, autonomous driving car, and drone surveillance. These applications are usually integrated with wireless mobile communications [2]. Currently, a number of smart IoT devices exploit cellular networks like 3G and 4G LTE to maintain their connectivity and their connection with the cloud data centers. As the skyrocketing of data produced by increasingly large number of IoT devices, there are several burning issues to be solved in the application environments. For example, the transmission latency and reliability of current cellular networks cannot be guaranteed, which in turn limits the effectiveness and feasibility of many emerging IoT applications, such as the autonomous driving car and robotic surgery, which ask for ultra-low latency

D. Anand (✉)
Chandigarh University, India

V. Khemchandani
J.S.S. Academy of Technical Education, Noida, India
e-mail: vkhemchandani@jssaten.ac.in

S. Tanwar (ed.), *Blockchain for 5G-Enabled IoT*,
https://doi.org/10.1007/978-3-030-67490-8_11

and ultra-high reliability. 5G has been introduced with the capability of high throughput, low latency, high reliability, and increased scalability to enable massive number of devices with best QoS and QoE provision of ubiquitous connectivity solution to fulfill diversified IoT application requirements. This brings potentials to deploy more connected devices without worrying about an overcrowded network exacerbating existing issues. The high speed and reliable connectivity underpinned by 5G will create new possibilities for IoT services far beyond those available today. In addition, the enabling technologies of 5G, including functions virtualization and softwarization, software defined networking, massive MIMO (Multiple Input and Multiple Output), mobile/edge computing, and ultra-dense networks, have great potentials to usher in a new era of IoT, aiming to smoothly and flexibly support heterogeneous IoT services with distinct business characteristics under massive smart devices [3]. Furthermore, IoT will bring a rich source of big data. The powerful role of big data analytics in 5G can undoubtedly benefit IoT advancement. This chapter is important to understand the concept of the Cloud Computing and Edge Standards, Security Fundamentals for 5G network, and other security measures. Then, we discussed Architecture of 5G-enabled IoT, Security Threats in 5G-Enabled IoT, Security Analysis, Privacy Threats in 5G-Enabled IoT, Security and Privacy Threats in Specific Domain, and Challenges and Opportunities [4].

1.1 Cloud Computing and Edge Standards

In the past decade, cloud computing [5] is an essential part of IoT. It uses some standard mechanisms to access multiple configurable resources such as application, storage, networks, servers, and services. The first time cloud computing mentioned to the world is in a Compaq internal document as early as 1996. Several years later, cloud computing has come into actuality. In 2006, Amazon introduced the created Elastic Compute Cloud (EC2) [6], which can support Amazon Web Services [7]. Two years later, Google delivered the Google App Engine. In the 2010s, there are various Smarter Computing foundations appeared after IBM released IBM SmartCloud. In these smarter computing frameworks, cloud computing is a critical component. There are many service models for cloud computing; three models are accessible that are Infrastructure as a Service (IaaS) [8], Platform as a Service (PaaS) [9], and Software as a Service (SaaS) [10]. Cloud computing has many advantages such as it enables to reduce the cost, it can be minimal effort to manage the infrastructure, and the users only need to pay when they need process and store data in a private or public cloud. It can adapt to fluctuations.

1.2 Security Fundamentals for 5G Network

Security and privacy are two big issues in 5G-enabled IoT [11]. The cost of security and privacy threats is prohibitive. It is not only considered as potentially damaging to

Table 11.1 Factors affecting coronary heart disease

	Security	Privacy
Target	Against unauthorized access	Protect personal identifiable information
Data type	All data	Private data
Program focus	All information that an organization collected	The personal information such as name, address, SSN, etc.
Relationship	Security can be achieved without privacy	Privacy cannot be achieved without security

the monetary penalty, but also causing other more pressing issues such as consumer confidence, social trust, and personal safety. In this part, we mainly describe the difference between security and privacy. Security indicates against the unauthorized access of data. Privacy refers to protect the personal identification information. For example, a credit report agency enables to protect the customers' personal information with security strategy, but the data scientist may still access this private information. Table 11.1 displays the primary difference between security and privacy.

1.3 Various Security Measures

5G-enabled IoT brings abundant benefits to the end users. However, it also carries some security and privacy challenges. Despite the common traditional security and privacy issues, there are also some unprecedented security and privacy challenges in 5G-enabled IoT.

1.4 Architecture of 5G-Enabled IoT

According to the existing researches, we construct the new architecture of 5G-enabled IoT as Fig. 11.1. The five layers from bottom to top are recognition layer [12], connectivity/edge computing layer [13], support layer, application layer, and business layer.

1.4.1 Recognition Layer

Recognition layer is also known as things layer, which is the foundation layer in 5G-enabled IoT. There are two main parts in this layer: IoT physical objects and hardware. The physical objects can be all the IoT devices such as smartphones, smart wearables, self-driving vehicles, etc. The hardware in this layer includes sensors, controllers, or any electronics hardware that obtain data from IoT devices. For example, RFID readers, barcode, micro-controller units, etc. The main task

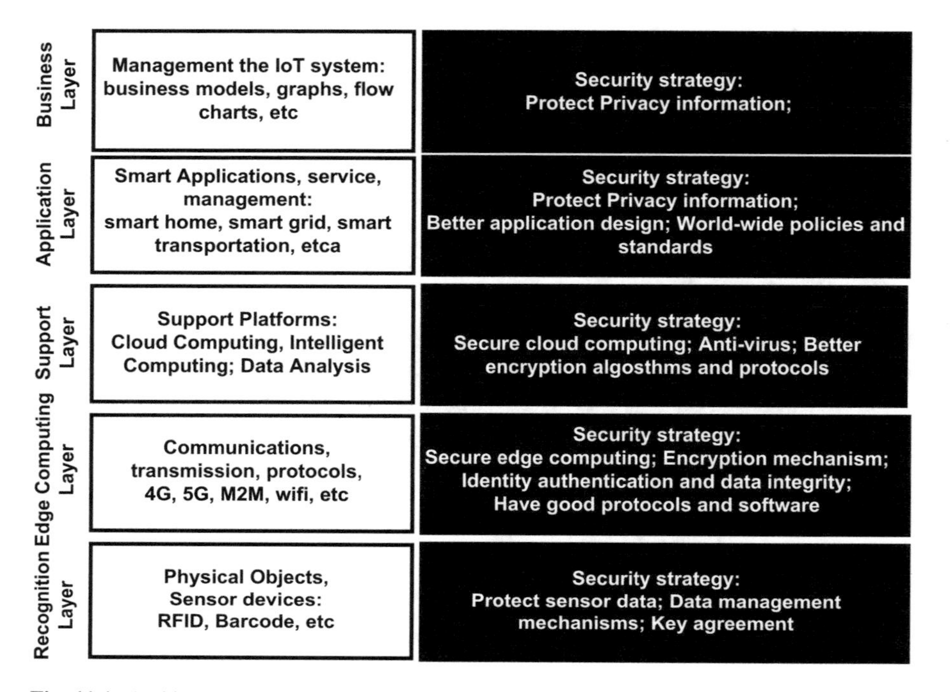

Fig. 11.1 Architecture of 5G-enabled IoT

of this layer is to collect IoT information from physical devices and sensors and recognize and identify the IoT environment.

1.4.2 Connectivity/Edge Computing Layer

This layer task is to help connectivity and edge computing to define the communication protocols and networks. Use networks to transmit collected information from the first layer to the edge and cloud. In 5G-enabled IoT, the underlying networks can be 4G [14], 5G[15], Wifi [16], etc. Another crucial research interest in this layer is how to decide the resource offload and resource allocation between edge computing layer and cloud layer [17].

1.4.3 Support Layer

This layer will offer the support platform for the upward and downward layers. It contains cloud computing powers, data analysis ability, and other intelligent computing abilities [18, 19]. The cloud center has responsibility for the service management and processes the information that edge computing layer cannot handle. In this layer, data analysis will be another vital part; it will support the AI embedding IoT applications more intelligently. For example, storing the users' historical data and implementing AI algorithms to predict users behavior will help 5G-enabled IoT devices to make efficiency and accurate decision in the application layer.

1.4.4 Application Layer

It is a terminal layer that offers the various applications and services based on the information process from the previous layers. The IoT devices can make the personalized decision according to the AI learning results. There are various applications in 5G-IoT, such as smart home, smart grid, smart healthcare, smart transportation, smart wearables, etc.

1.4.5 Business Layer

This layer manages all the IoT systems. It requires to create better business models that improve the IoT service quality. There are many researchers who focus on how to design a better business model. Such as Real-Time Instrumentation [20], return of investment (ROI) [21], zero-capex IoT business models [22, 23], IoT data privacy and trust models, etc.

1.5 Security and Privacy

Any system should be acceptable and adaptable when it is reliable and secure. Same principle will be applied for the health-care system. As far as secure system is concern the security and privacy mechanisms has to be analyzed. The evaluation of the security and privacy will be ensured through discussion about the various related threats as discusses in upcoming sections.

2 Security Threats in 5G-Enabled IoT

In this section, we analyze the security threats in each layer of 5G-enabled IoT and point out the security requirements and threats in 5G-enabled IoT.

2.1 Security Threats as per IoT Reference Model

2.1.1 Recognition Layer

Hardware security: The first essential threat in the recognition layer is the hardware security. This layer consists of multiple sensors, controllers, and different IoT devices. Unauthorized access and cloning the tag can occur in sensors, and the adversary can reprogram the data. The malicious attacker can also control a user-accessible IoT equipment and provide the fake information. The wireless sensors such as RFID [24] are easy for the attacker to get the confidential information such as the password to eavesdrop the information.

Software security: It is important to understand that hardware security alone is not enough as most attacks are via software. Therefore, clarifying the security threats of 5G-enabled IoT is necessary. Any IoT device's integrity, authentication, and availability make software insecure. If the embedding system gets a malicious attack, IoT devices' information can be stolen, monitored, and damaged the software behavior. For the most popular operating systems, such as IOS and Android, the most substantial security threat is hacker attacked. The malicious attackers on smart mobile devices can gain access to enterprise and personal data. The most likely malicious attacks include cryptographic attack and code injection attack.

2.1.2 Connectivity/Edge Computing Layer

Network security: The main security threat in 5G-enabled IoT is Denial of Service (DoS) [25]. This is a common attack through networks. In edge computing scenario, the distributed denial-of-service (DDoS) [26] attacks and wireless jamming are dangerous. Man-in-the-Middle Attack and counterfeit attack still exist in this layer. Malicious attackers can control a section of the network and attack the networks. The unauthorized attacker can even fake the identification and communicate as normal and obtain more IoT users' information. Also, IP theft is dangerous for network security. IPv4 and IPv6 are different versions of Internet Protocol in the network [27]. IPv6 as the upgraded version of IPv4 has some advantages such as better multicast routing, more simplistic header format, no more network address translation, etc. Even though they can provide seamless protection to applications, there still have many security threats, such as DDoS attack, Man-in-the-Middle attack, Packet sniffing [28], etc.

Rogue node: In edge computing [29], after the end user pretends to connect to the edge node, such as fog node, private cloudlets, and edge devices, it gives a chance that malicious attackers can deploy fake gateway devices. The malicious attackers enable to manipulate user requests, collect or tamper user data, and launch further attacks as same as Man-in-the-Middle attack.

Edge data centers: Protect the edge data center to avoid physical damage, and privilege escalation is necessary. For the edge paradigms data center, they are managed by some business organizations. Preventing the attackers from accessing the data centers and damaging the devices is needed. Besides, to block the external attackers, well-training the security management individuals and maintaining the data center professional are necessary.

2.1.3 Support Layer

Cloud Computing: The main security threats in the cloud are Dos attack, shared cloud computing services [30], system vulnerabilities, and malicious or negligence insider. DDoS attacks against cloud platforms are incredible. It will shut down the cloud system and deny the services. Even though there are some security strategies that can detect the DoS attack, with the increasing number of 5G-enabled IoT devices and the high 5G wireless speed, DoS attacks have more chance since the previous method cannot handle the high volume data traffic. For some shared cloud services, they do not provide enough security solution between users and applications. It will lead to another security threat when the users are sharing the resources. The system vulnerability can still exist in some complex cloud computing infrastructures. If the attackers know the weakness, they can easily damage the cloud computing system. Same as edge data center, the cloud computing management team also needs to prevent the internal security human-made security threats.

Data Analysis: The key here is to prevent the data leakage threats that cause more severe privacy issues.

2.1.4 Application Layer

Heterogeneous network: Since in this application layer, there are various 5G-enabled IoT applications [31]. Different application domains have distinctive security threats. However, these applications enable to share data through the heterogeneous network that is easy to cause several security issues: DoS attack and malicious code injection. It is easy to attack and shut down the service if there is no secure authentication and key agreement. Also, it will cause the data breach and privacy problem, which we will be explained in the next section.

Service Management: The people who take the responsibility to maintain and manage the application infrastructures should be professional and educated. Otherwise, it will be another security issue.

2.1.5 Business Layer

Data Leakage: Owing to the requirements in the business model design, the designers will implement some APIs to help their modeling. In the meantime, the data have a great chance to leakage if the APIs lack security protection. The data can include many private pieces of information that will bring a severe cost.

3 Security Analysis

The communication architecture, network, and frame are important to exchange the information among various connected nodes. Because this network is open, security is important for it. There are various security processes that are required to incorporate with the existing mechanism. The analysis of these security features is important to audit, and this audit is important to measure the risk and vulnerability for the known attacks and threats.

3.1 Security Threats

5G vows to offer more noteworthy limit, diminished inertness and quicker speeds—think gigabits per second rather than megabytes. Basic frameworks, transport frameworks, and IoT empowered keen homes will undoubtedly be run on 5G systems. Be that as it may, simultaneously, IoT makers are occupied with a visually impaired pursue for pieces of the overall industry. At the point when time-to-market and minimal effort gadgets are the battleground, security is unavoidably ignored. This profoundly changes the potential assault surface for programmers. By 2025, hackers know the 74 billion "weak connections" in a security chain, bringing down or upsetting basic foundation, vitality networks, or home IoT gadgets. Thus, gadget security is basic and cannot be ignored without gambling extreme results on society.

The great news is that there is help around the corner. Back in February 2019, ETSI [32] Technical Committee on Cybersecurity (TC CYBER) announced a global standard, ETSI TS 103 645 (https://www.etsi.org/newsroom/press-releases/1549-2019-02-etsi-releases-first-globally-applicablestandard-for-consumer-iot-security), cybersecurity for IoT. The standard establishes a security baseline for consumer IoT products and provides basis for future IoT certification schemes. Finland was the first European country to take this next step with the Cybersecurity [33] label, announced in November 2019.

In an eventual fate of IoT in a 5G world, IoT makers need to guarantee that their items meet the necessities of the security guidelines. Also, it is essential that the producers are carefully capable what is more, at risk for assembling secure gadgets.

5G is turning into a necessary piece of IoT administrations. There can be different kinds of administrations, such as M2M correspondence, Vehicular Communication,

Mobile Communication, accessible from a 5G empowered IoT organization. A User Equipment (Vehicle or Machine or Mobile) could interface with 5G network through some access organization for benefiting different administrations from different IoT workers. Normally, a client gadget demands a support of Core Access and Mobility Management Function (AMF), which is answerable for administration designation after the verification of Network Slice Selection Association Information (NSSAI) [34]. UMD is liable for verification qualification age and membership of the executives. NSSF [35] chooses the arrangement of system cut cases serving clients and decides the NSSAI comparing to material system cut occasions. In an open situation of the entrance organization, an interloper may listen to the chosen administration and client accreditations. The security of the user will be in question. Further, an aggressor may utilize the removed information for pantomime and DoS assaults. Henceforth, there is a significant necessity of security convention to protect alongside classification and respectability of the information traded.

Threat on Authentication There is a chance of unauthenticated assailant Trudy, who has the capacity of checking, catching, and imitating as a verified client hardware to send information through the 5G network. The potential assaults by the aggressor are Sybil, pantomime, character based assaults, and so forth.

Threat to Confidentiality A foe is an unapproved individual who could get to and comprehend the unapproved administration question or administration designation in the way to the 5G arrangement. Different conceivable uninvolved assaults incorporate bundle sniffing, phishing, and so forth.

Threat to Integrity An aggressor may have an assault on integrity; he could screen the administration demand/client accreditations in 5G and client devices. The assailant may endeavor to get to and modify the administration before it comes to AMF of the 5G network. Various potential assaults remember man-for-the-center assault, meeting capturing assault, and so forth. Broad Sensitive Information: The extremely high speed of 5G wireless let more and more physical objects to be the IoT devices, such as microwave, robot vacuums, a cloth, even a small ring. These smart IoT equipment significantly improve human's life quality. Meantime, the widespread sensitivity personal information will also be collected by these devices. Sometimes, the connected devices ask end users to input personal sensitivity information, such as name, gender, age, zip code, email address, etc. Even though sometimes, the data are just transmitted by a given endpoint, there still have privacy risks. If the attacker integrates, collects, and uses advanced data mining algorithms to analysis the fragmented data from multiple endpoints, the sensitive personal information can still be achieved. For example, in a smart home, the refrigerator can embed a camera that monitors the grocery types and if they spoil or not. It seems only to collect the grocery data; however, it can definitely use the historical data to predict users' eating habits. The smart robot vacuum, when it works, will collect data to identify the locations of house walls and furniture. This helps them avoid crashing into the furniture, but it can also measure the square of home closely and create a map of the house and share to the cloud.

Location Privacy Location-based services (LBS) are developed fundamentally for many 5G-enabled IoT devices. Every smart phone and vehicle have built-in global positioning systems (GPS) [36]. There are many phone Apps also based on LBS services such as Yelp, Lyft, Uber, etc. The primary purpose of these services is to improve users' life quality and help them to live conveniently, such as recommending the adjoining restaurant, looking up the reviews, etc. However, at the same time, the user's location data is at risk if there is privacy attack. Using the location data, attackers can predict the user's activity routine, interest place, and life habits. There are many researchers who focus on to protect location privacy information. One popular strategy is to employ well-known privacy metrics such as k-anonymity and differential privacy [37]. Nevertheless, most of these strategies could only provide privacy protection with a certain probability. When the attacker integrates the information from different devices and analyzes the hidden correlation of the data, the location privacy [38] can be easily destroyed.

Correlation Privacy With the more personal and identity information collected by 5G-enabled devices, the attackers will achieve more private information. After integrating and analyzing the correlation between fragment data collected from many devices, the attacker will enable to identify the individual information in the specific time, location, etc. For example, if an attacker unauthorized accesses a hospital patient's database. Even the information they get cannot identify the particular individual due to the encryption of the database. After integrating and analyzing multiple sources, the patients' information can still be determined.

3.2 Security Requirements and Security Goals

The security requirements of the capabilities and functionalities are required to make the process strengthen against the securities, while the desired functionalities against the security vulnerabilities are the goal to make the process secure and safe. These security requirements are illustrated in Fig. 11.2, and the security goals are illustrated in Fig. 11.3.

3.3 Security Assumptions

Ideal working suppositions for 5G empowered IoT situation are as follows: Supposition 1: The encoded administration demand/client certification must be unscrambled by a common symmetric key [39], which was utilized for encryption. Supposition 2: AMF is full confided in party and secured. Presumption 3: User hardware is also an authorized device; it is liberated from malware and not genuinely altered.

Denial of Service (DoS) Attack	This attack is due to number of requests generally generated by automatic bots to exhaust the server.
Forgery attack (Impersonation Attack)	In this attack, the malicious attacker, can access the stored information on a smart-card and arbitrary forge a new smart card
Parallel session attack	In this attack, an attacker can masquerade the session as a legal user without any knowledge about authentication parameters like user's PIN, etc.
Password guessing attack:	This attack is based on the vulnerability of password. An attacker can get access the system by guessing the password as a legitimate user
Replay attack:	In this attack, the malicious attacker, can resend the captured message and recieve the crucial and important information.
Smart-card loss attack:	If the smart-card is lost or stolen. Then, this card can easily change the password of the smart-card, or can guess the password of the user ; can easily prove authenticity of the user
Stolen-verifier attack	There is a vulnerability in due to stored hashed password. If anyone can steal the password verifier like hashed password then, the user masquerades as a legitimate user.
Reflection attack:	In this attack same protocol is used in both the directions to authenticate each side. Because in the proposed protocol, there are three layers involved in this full authentication process
Insider attack:	In this attack, malicious user, who knows some legitimate information try to access the user's information and compromise the system.

Fig. 11.2 Security requirements for 5G-enabled IoT authentication process

No verification table	The system should free from a dictionary or verification table which stores the plain text password or hashed passwords of the legitimate user for authentication
Freely chosen password	This security goal is achieved by providing the freedom to the user about selection of the password. A user should have freedom to choose the password.
No password reveal	The insider like an administrator of the server can try to use these login credentials to impersonate user, if the authentication credentials should not be revealed
Password dependent	The authentication is password dependent because user with any random password may access the server, which is provided at the time of registration.
Mutual authentication	Mutually authenticated messages are exchanged at every level and finally, the session key is generated on the basis of these exchange encrypted messages.
Session key agreement:	The objective of the session key agreement is to provide secrecy of messages between various involved entities.
Forward secrecy:	The goal of forwarding secrecy is to prevent the attacker who tries to get passwords or other login information from the stolen smart card.
User anonymity	The prevention of the leakage of private user identities and server identities to malicious attackers.
Smart-card revocation	The protocol gives provision in the system for invalidating the further use of lost smart-card, otherwise an adversary can impersonate a valid registered user
Efficiency for wrong password login	if the user inputs the wrong password by mistake in login phase, without any delay the proposed protocol notifies the user with an error message, instead of sending the user's login request unconditionally to the server.

Fig. 11.3 Security goals for 5G-enabled IoT authentication process

4 Privacy Threats in 5G-Enabled IoT

With the deployment of 5G wireless and the rapidly growing number of IoT devices, the 5G-enabled IoT has more server privacy issues. Edge computing allows that each physical object in the environment has the ability to communicate autonomously over 5G wireless or Internet. In this section, we conclude some main privacy threats in 5G-enabled IoT environment.

4.1 Broad Sensitive Information

With the deployment of 5G wireless and the rapidly growing number of IoT devices, the 5G-enabled IoT has more server privacy issues. Edge computing allows that each physical object in the environment has the ability to communicate autonomously over 5G wireless or Internet. In this section, we conclude some main privacy threats in 5G-enabled IoT environment advanced data mining algorithms to analysis the fragmented data from multiple endpoints, and the sensitive personal information can still be achieved. For example, in a smart home, the refrigerator can embed a camera that monitors the grocery types and if they spoil or not. It seems only to collect the grocery data; however, it can definitely use the historical data to predict users' eating habits. The smart robot vacuum, when it works, will collect data to identify the locations of house walls and furniture. This helps them avoid crashing into the furniture, but it can also measure the square of home closely and create a map of the house and share to the cloud.

4.2 Location Privacy

Location-based services (LBS) are created majorly for numerous 5G empowered IoT gadgets. Each cell phone and vehicle have worked in worldwide situating frameworks. There are many telephone Apps additionally dependent on LBS administrations, for example, Yelp, Lyft, Uber, and so forth. The basic role of these administrations is to improve clients' life quality and assist them with living accommodation. For example, suggesting the abutting eatery, look into the audits, and so forth. Be that as it may, simultaneously, the clients' area information is in danger if there is security assaul. Utilizing the area information, assailants can foresee clients' action schedules, interests place, and life propensities. There are numerous analysts who center around securing area protection data. One famous technique is to utilize notable security measurements, for example, k-namelessness and different protection. In any case, a large portion of these systems could just furnish security insurance with a specific likelihood. When the attacker coordinates the data from different gadgets and investigates the shrouded relationship of the information, the area security can be handily wrecked.

4.3 *Correlation Privacy*

With the more personal and identity information collected by 5G-enabled devices, the attackers will achieve more private information. After integrating and analyzing the correlation between fragment data collected from many devices, the attacker will enable to identify the individual information in the specific time, location, etc. For example, if an attacker unauthorized accesses a hospital patient's database. Even the information they get cannot identify the particular individual due to the encryption of the database. After integrating and analyzing multiple sources, the patients' information can still be determined.

5 Security and Privacy Threats in Specific Domain

To understand the importance and application of the ICT based applications and further the requirements of the security and privacy threats, the upcoming sections describe the concept very clearly.

5.1 *Healthcare*

Healthcare information is very sensitive results in healthcare information security be significant attention for both providers and governments all over the world. In the digital era, most of the patients' information is stored in the database. This carries the patients' records at security and privacy risk. The first security threat is software security. Nowadays, there are more and more health and wellness programs that can be finished on the device side. The patients can download the Apps in the mobile devices, then make an appointment, review the reports, etc. They must be aware of the risk of privacy data breaches and malicious attackers hacking the health information. The second issue is human. Whether it is malicious or ignorant, the information leakage by human all has serious repercussions. It may be caused by the employees steal data or IoT devices, or breach the data by accident. DDoS attacks is the third security threats. Healthcare is a regular target. As the smart healthcare industry continues to deliver new services, the security management leaders in healthcare must always see the big picture. It is also necessary to pay attention to the threats to the supply chain of hospitals. Each vendor a hospital cooperates can present a potential data breach incident. Make sure that the security and privacy in each step in the supply chain transaction are required.

5.2 Smart Home

Confidentiality threats and authentication threats are general security and privacy threats in smart home environment. The undesired personal sensitive information breach is known as the confidentiality threats. For example, the smart home control systems such as Nest can track the home temperature and electric usage and then determine when the house is occupied. Also, losing the password can lead to the unauthorized access threats. When the unauthorized attacker accesses to the system, they enable to reprogram the system and send the fake request to the things, such as releasing the smart lock will cause a more dangerous outcome. Also, in the smart home environment, most of the IoT devices collect the private data. It will cause a series of privacy issues. For example, the AI embedded IoT can predict the users' behavior, and this kind of private information also has a risk to breach.

5.3 Smart Grid

The first important thing in smart grid is privacy. The smart meters enable to transmit the private customers' information to the utility company and service provider. This information can be used to infer customers' behavior and house occupied. Since in smart grid, several devices can manage the electricity supply and network demand at the same time, some attacks can enter the devices through the network. Most of the smart grid devices focus on a specific functionality only; it results in the devices lack enough memory space or capability to deal with security threats. M2M in control systems also can occur in the security issues. When one device is attacked, it may send the fake state to make another device in an unwanted way. Many devices in smart grid need the remote control. It requests the equipment remotely that has a long life cycle and effective security software update ability. The various equipments involved in Smart Grid will stop working when the DoS attack is trying to access the system.

5.4 Smart Logistics

Smart Logistics: In the supply chain IoT, there are three security and privacy issues. The first is data corruption that the attacker accesses to the system, sends the fake request, and causes the device to make the wrong decision. The second threat is the maintenance of the equipment. The DoS attack can damage the equipment or facility if it lacks security protection update or maintenance. Data privacy is the common threat in smart logistics. Supply chain always contains the users' private information that should be carefully protected.

6 Challenges and Opportunities

In the previous sections, we reviewed the security threats in each 5G-enabled IoT layer and illustrated the common privacy problems. In this section, we will provide the security and privacy challenges with the existing researches and a discussion on the potential research areas. As Fig. 11.3 shows, each layer has different security challenges and security strategies. The first recognition layer consists of all the IoT devices, sensors, and controllers. To ensure the security with M2M, how to complete the authentication and identify is a great challenge. Create efficiency and safety key agreement, and data encryption is another challenge to protect the sensor data. Trust management is also needed if the IoT devices are transferred from one user to another. Network security problems are the most critical security challenges in the second layer: connectivity/edge computing layer. All the edge paradigm networks have a security challenge: protocols and mechanisms that have data confidentiality and integrality requirements. When two authenticated entities communicate with each other, make sure that the data confidentiality and integrality are necessary. In addition, how to prevent the unauthorized access to the edge networks and how to avoid the DoS attack will be other challenges. In the third layer, multiple support architectures are important parts. Strong encryption algorithms and protocols are needed in this layer. Also for another support infrastructure, such as cloud infrastructures, secure virtualization mechanisms are necessary. In the next application layer, the first challenge will be how to design secure and efficiency applications. In the meantime, protecting the users' private information is necessary since the application platforms always contain sensitive information. To avoid the authorized access to the heterogeneous network, better authentication and encryption algorithms are required in this layer. In addition, since the smart applications are mainly managed and maintained by the third party, well-training of the management team is essential. Last layer is the business layer, how to create a better business model, and make sure that the APIs are as much secure as protecting the private information are the main challenges. Overall, we summary the important security and privacy challenges and opportunities in the 5G-enabled IoT environment.

6.1 Identity and Authentication

Efficiency and privacy authentication scheme are needed in the 5G-enabled IoT environment. Previously, the authentication protocols are designed based on the single server environment which are not suitable for the new architecture of 5G-enabled IoT environment distributed services environment. Back in 2009, cloud computing as a new technology came to our life. With the numerous growth of users and the request to share the service, many researchers began to pay more attention to design the trust and security authentications between cloud users and the service. Also, they found for most of the users, the past authentications such as SSL Authentication Protocol (SAP) [40] are very complicated to use. Li et al. proposed a new identity-based authentication for cloud computing (IBACC) [41], and their authentication is more efficient and lightweight than SAP.

Around 2015, more and more people found that there are many sensitive data that become digitalized, such as healthcare data. Therefore, many researchers construct more efficiency authentication schemes to protect the E-healthcare database. For example, Wu et al. and Jiang et al. all focus on the proposed three-factor authentication protocols to defense the different types of attacks. After some researchers found the disadvantage of the single server authentication schemes, some pursue the creation of inter-cloud identity management systems that came out, such as OpenID and SAML to provide Single-Sign-On (SSO) authentication. However, these authentication schemes rely on the third party that may bring some new security threats. Therefore, how to design efficiency and privacy authentication schemes will be a challenge in the distributed service environment.

There are some researchers who focus on this field. Lo et al. designed an efficient privacy aware authentication scheme for MCC service. Their scheme allows mobile users to access multiple computing services from multiple providers using a single private key. Recently, He et al. and Jiang et al. found some drawbacks of this Lo's scheme and improved the scheme, such as use identity-based signature scheme. As we mentioned before, the IoT devices may transfer from one user to another. Since most IoT devices need to create a personal account and input the sensitive information, how to provide user anonymity with an efficiency authentication should be another challenge. Li et al., Yang et al., and Wu et al. all focus on how to create an efficiency authentication with user anonymity.

Even though there are many types of research in this domain, there still have some opportunities. The first opportunity in identity and authentication is synergies. We found that most efficiency and privacy aware authentication researches are in the MCC environment. Finding the possible corporations between MCC and other edge paradigms is necessary. The second potential research director is how to control the trade-off between security and privacy when designing an authentication protocol. For example, when using lightweight authentication, how to make sure the user anonymity still works well. For some battery-based IoT devices, how to save the energy and protect the security at the same time is an interesting topic either.

6.2 Trust Management

In 5G-enabled IoT, especially for the different IoT devices, service providers and remote servers exist in the trust issues. How to develop the trust relationship among device–device, device–user, and user–server is a challenge. How to efficiently create the trust management framework is the most attractive research field for many researchers. Previously, the trust management is developed for cloud computing environment to solve the trust issues between the IoT users and the cloud server.

At the beginning of the trust management in cloud computing, Service Level Agreements (SLAs) are the foundation technique but not consistent among cloud provides. This will lead the user unable to identify the trustworthy cloud provider.

Therefore, many trust management frameworks are proposed. Petri et al. proposed a trust model that manages the information and identifies the trust distribution. Hammam et al. employed a trust management system to calculate the trust value for the node in the mobile ad-hoc clouds environment. The trust management system combines various security features like availability, neighborhood's evaluation, response quality and task completeness together to ensure the efficiency. In 2016, Zhu et al. implemented SLAs trust as the base and proposed an innovative trust model to integrate the objective and subjective trust in the MCC manner.

It is evident that most of the trust management researches focused on the centralized service, and from 2016, few trust management pieces of research started to pay more attention to the distributed computing service. Trust management in distributed-based computing in 5G-enabled IoT will be a potential research direction in the future.

6.3 Encryption/Cryptographic Method

Encryption has been the research interest for many researchers in many years. However, the traditional encryption methods such as Triple DSE (3DSE) [42] and Triple Data Encryption Algorithm (TDEA) [43] have drawbacks that the devices must need to know the identities of the information recipients and share credentials with them in advance. In the 5G-enabled IoT environment, in many scenarios, there are many recipients unknown, which means the traditional encryption algorithms cannot handle the complicated IoT system. Therefore, some encryption methods developed for IoT environment contributed many solutions. Attribute Based Encryption (ABE) [44] in one of the encryption algorithms consists of a key authority between data sender and recipient.

How to make the encryption scheme more secure and efficient becomes a challenge when designing encryption algorithms. Wu et al. combined two systems, the hierarchical identity-based encryption (HIBE) [45] and the cipher text attributed based encryption (CP-ABE) [44], and proposed an efficiency encryption scheme for users to share the confidential data in the centralized cloud environment. After 3 years, Li et al. introduced a novel encryption mechanism to protect the healthcare data in cloud servers. They leveraged the ABE techniques to encrypt each patient's PHR file. With the development of edge paradigms, there are many extensive concerns of the edge paradigms encryption method. Alrawais et al. offered an efficient key exchange protocol based on CP-ABE and digital signature techniques in fog computing environment and achieved more efficient performance on confidentiality, authentication, and access control. Jiang et al. designed an encryption scheme based on CP-ABE for fog computing IoT either. Besides KP-

ABE, there is another scheme in ABE: key-policy ABE (KP-ABE) that can also contribute to the big data security.

Unfortunately, from now, most encryption mechanisms focus on cloud computing and fog computing environment, for MEC and MCC, and there still have opportunities to design more efficient encryption algorithms. In addition, most of the researches pay more attention to the CP-ABE algorithms; there may have other encryptions methods that can incorporate with CP-ABE, such as fully homomorphic encryption (FHE), ciphertext policy attribute based proxy re-encryption (CP-ABPRE) [46], etc.

6.4 Access Control

The access control is used to ensure the unauthorized entities cannot access the IoT devices and collect data. Previously, most researches focus on design of the access control system for cloud computing. Yu et al. created the access control system by exploiting techniques of many encryption schemes and built an efficient fine-grained data access control. Also, they proposed another novel framework for access control to the healthcare domain within the cloud computing environment.

However, the researches of the access control mechanisms in edge computing are very few. Implementing effective access control algorithms will be one challenge. Because of the enormous numbers of the edge devices in 5G-enabled IoT, there is a new challenge to access control on how to efficiently find the access model and how to optimize the limited resources especially for some battery-based devices.

6.5 Privacy

Because in 5G-enabled IoT, the IoT architecture is complicated and distributed, the information of the IoT devices collected is sensitive and private. The more complex privacy mechanisms are needed now. In the past years, there are many privacy protection mechanisms such as k-anonymity, differential privacy, quasi-identifier, pseudonymization, etc.

At the beginning of IoT privacy research, there are many privacy protection mechanisms that focus on cloud computing. Wang et al. utilized and uniquely combined the public key based homomorphic authenticator with random masking and then achieved the privacy-preserving public audit requirement. Itani et al. presented the PasS (Privacy as a Service) security protocols to protect the privacy threats in the cloud computing architecture either. After edge paradigms emerge, Lu et al. presented the Lightweight Privacy-preserving Data

6.6 Privacy in the IoT

On account of structure trade-offs as far as cost, multifaceted nature, and vitality required for satisfying their activity, numerous gadgets in the IoT are typically resource constrained. To adapt to unapproved access to, stealing of information, gadgets ought to be furnished with authentication, authorization process, and information protection abilities that guaranteed newness, legitimacy, classification, and uprightness of data. Protection (for example, unlink ability, information mystery, and obscurity) must be precisely saved since individual and delicate data could be taken and manhandled by an attacker. Encryption is central to give sensitive information a fundamental degree of security. Without a doubt, it forestalls that communicated information can be captured and perused by aloof foes. Encoding the data may utilize computationally costly cryptographic techniques (for example, pairing based cryptography), which could not be executed by IoT gadget. So as to distinguish appropriate cryptographic ways to deal with the IoT, Malina et al. [47] estimated the exhibition of the most utilized approach, (for example, RSA, secure hashing calculations, and AES) on probably the most widely recognized micro-controller preparing IoT gadgets. They found that while hashing and symmetric encoding activities take barely milliseconds and can run on extremely restricted micro-controllers, more grounded approaches, for example, RSA (by a 2048-bit private key), can cause delays into several milliseconds, which are unfortunate progressively to IoT applications. Handling of complex activities could be left to the cloud or to communication passages, bringing the decrease of the gadgets' vitality utilization and calculation delays. Be that as it may, this technique requires trustful doors and secure correspondence among parties. A reasonable method to ensure identities (the two gadgets and clients) from being followed is concealing their genuine character by methods for pseudonyms. Anyway, as additionally recommended, when assailants listen to information bundle inside an adequately wide time window of perception, they may unveil genuine casualties identifier. As portrayed later in this paper, at the point when associated with a LTE network, IoT gadgets can translate messages communicated by the network to find a particular endorser. Such messages contain just impermanent identifiers, yet a latent foe might misuse decoded data to recover relationship among brief extraordinary identifiers. Here we contend that despite the fact that the use of security approaches in structure and execution stages, security goals in the IoT could not be accomplished in light of its multifaceted nature. At that point, a far reaching comprehension of inspirations driving protection shortcomings and coming about recognizable proof of suitable relief activities in light of them requires a natural procedure fit for investigating the wide and various IoTs from particular viewpoints.

6.7 UDN and IoT Privacy

UDNs can adequately adapt to the future systems' information prerequisites and likewise give vitality and spectrum effectiveness. Made out of heterogeneous hubs with various radio access advances (for example, LTE, WiMax, IEEE 02.15.x), impart power, and inclusion zone, UDNs are described by a multi-level architecture. In detail, high-power hubs and low-power hubs, with enormous and little radio inclusion, are set individually in large scale cell tiers and in small cell tiers. Cellular correspondence infrastructure, if from one hand, makes it conceivable to offer pervasive network to the most gadgets and, from the other hand, is wasteful for sending little, small information as required by M2M. In addition, cellular correspondence infrastructure could make it conceivable to follow substances engaged with data trade forms, in this manner influencing their location protection. Spatial dissemination of low-power hubs may impact the entire system security. Explicitly, the likelihood of positive mystery rate, that is the limit deviation of the working channel from the spy channel, increments as the thickness of low-power hubs develops (until a basic point, after which is not watched any upgrade in terms of mystery execution). Also, the higher the density of gadgets engaged with correspondence, the higher is the danger of data listening. Evidently, while moving inside a UDN, elements are probably going to be dependent upon more handover forms than in the current systems, making it workable for untrusted subjects to partake in the just referenced procedures.

7 Summary

The chapter has discussed the important facts about Cloud Computing and Edge Standards, Security Fundamentals for 5G Network, and other security measures. Then, we discussed Architecture of 5G-enabled IoT, Security Threats in 5G-Enabled IoT, Security Analysis, Privacy Threats in 5G-enabled IoT, Security and Privacy Threats in Specific Domain, and Challenges and Opportunities. With the development of 5G wireless and IoT device, distributed-based computing is becoming an efficient and possible technology solution to handle the billions of the 5G-IoT devices. The new techniques bring us various convenient and high life qualities. However, they also produce new security threats and numerous users' private information at risk. We construct the 5G-enabled IoT as five layers: recognition layer, connectivity/edge computing layer, support layer, application layer, and business layer. In each layer, we conclude the security and privacy threats. We review the challenges in 5G-enabled IoT. For the future study, we suggest implementing the appropriate existing security and privacy strategies to the edge paradigms domain. Incorporated with AI algorithms to improve the application service quality and defense the unauthorized access, etc.

References

1. J. Gubbi, R. Buyya, S. Marusic, M. Palaniswami, Internet of things (IoT): a vision, architectural elements, and future directions. Future Gener. Comput. Syst. **29**(7), 1645–1660 (2013)
2. X.-Q. Pham, E.-N. Huh, Towards task scheduling in a cloud-fog computing system, in *2016 18th Asia-Pacific Network Operations and Management Symposium (APNOMS)* (IEEE, Piscataway, 2016), pp. 1–4
3. B. Al-Otaibi, N. Al-Nabhan, Y. Tian, Privacy-preserving vehicular rogue node detection scheme for fog computing. Sensors **19**(4), 965 (2019)
4. S. Li, L. Da Xu, S. Zhao, 5G internet of things: a survey. J. Ind. Inf. Integr. **10**, 1–9 (2018)
5. T. Velte, A. Velte, R. Elsenpeter, *Cloud Computing, A Practical Approach* (McGraw-Hill, New York, 2009)
6. S. Yi, D. Kondo, A. Andrzejak, Reducing costs of spot instances via checkpointing in the Amazon elastic compute cloud, in *2010 IEEE 3rd International Conference on Cloud Computing* (IEEE, Piscataway, 2010), pp. 236–243
7. E. Amazon, Amazon, See https://aws.amazon.com/ec2/ (15 June 2018) (2006)
8. S. Bhardwaj, L. Jain, S. Jain, Cloud computing: a study of infrastructure as a service (IaaS). Int. J. Eng. Inf. Technol. **2**(1), 60–63 (2010)
9. M. Boniface, B. Nasser, J. Papay, S.C. Phillips, A. Servin, X. Yang, Z. Zlatev, S.V. Gogouvitis, G. Katsaros, K. Konstanteli et al., Platform-as-a-service architecture for real-time quality of service management in clouds, in *2010 Fifth International Conference on Internet and Web Applications and Services* (IEEE, Piscataway, 2010), pp. 155–160
10. A. Dubey, D. Wagle, Delivering software as a service. McKinsey Q. **6**(2007), 2007 (2007)
11. J. Ni, X. Lin, X.S. Shen, Efficient and secure service-oriented authentication supporting network slicing for 5G-enabled IoT. IEEE J. Sel. Areas Commun. **36**(3), 644–657 (2018)
12. A. Afzal, N. Iqbal, A. Mujahid, R. Schirhagl, Advanced vapor recognition materials for selective and fast responsive surface acoustic wave sensors: a review. Anal. Chim. Acta **787**, 36–49 (2013)
13. W. Shi, J. Cao, Q. Zhang, Y. Li, L. Xu, Edge computing: vision and challenges. IEEE Internet Things J. **3**(5), 637–646 (2016)
14. E. Dahlman, S. Parkvall, J. Skold, *4G: LTE/LTE-advanced for Mobile Broadband* (Academic Press, New York, 2013)
15. M. Shafi, A.F. Molisch, P.J. Smith, T. Haustein, P. Zhu, P. De Silva, F. Tufvesson, A. Benjebbour, G. Wunder, 5G: a tutorial overview of standards, trials, challenges, deployment, and practice. IEEE J. Sel. Areas Commun. **35**(6), 1201–1221 (2017)
16. S. Yun and L. Qiu, Supporting WiFi and LTE co-existence, in *2015 IEEE Conference on Computer Communications (INFOCOM)* (IEEE, Piscataway, 2015), pp. 810–818
17. M.T. Chahine, Remote sounding of cloudy atmospheres. I. the single cloud layer. J. Atmos. Sci. **31**(1), 233–243 (1974)
18. J. Zhang, S.O. Williams, H. Wang, Intelligent computing system based on pattern recognition and data mining algorithms. Sustain. Comput. Inf. Syst. **20**, 192–202 (2018)
19. M. Alvarado, L. Sheremetov, R. Bañares-Alcántara, F. Cantú-Ortiz, Current challenges and trends in intelligent computing and knowledge management in industry (2007)
20. J.H. Kim, D.V. Gunn, E. Schuh, B. Phillips, R.J. Pagulayan, D. Wixon, Tracking real-time user experience (true) a comprehensive instrumentation solution for complex systems, in *Proceedings of the SIGCHI Conference on Human Factors in Computing Systems* (2008), pp. 443–452
21. J.J. Phillips, *Return on Investment in Training and Performance Improvement Programs* (Routledge, New York, 2012)

22. M.R. Palattella, M. Dohler, A. Grieco, G. Rizzo, J. Torsner, T. Engel, L. Ladid, Internet of things in the 5G era: Enablers, architecture, and business models. IEEE J. Sel. Areas Commun. **34**(3), 510–527 (2016)
23. T. Lennvall, M. Gidlund, J. Åkerberg, Challenges when bringing IoT into industrial automation, in *2017 IEEE AFRICON* (IEEE, Piscataway, 2017), pp. 905–910
24. K. Finkenzeller, *RFID Handbook: Fundamentals and Applications in Contactless Smart Cards, Radio Frequency Identification and Near-field Communication* (Wiley, New York, 2010)
25. C.L. Schuba, I.V. Krsul, M.G. Kuhn, E.H. Spafford, A. Sundaram, D. Zamboni, Analysis of a denial of service attack on TCP, in *Proceedings. 1997 IEEE Symposium on Security and Privacy (Cat. No. 97CB36097)* (IEEE, Piscataway, 1997), pp. 208–223
26. S.T. Zargar, J. Joshi, D. Tipper, A survey of defense mechanisms against distributed denial of service (DDoS) flooding attacks. IEEE Commun. Surv. Tutorials **15**(4), 2046–2069 (2013)
27. J. Postel, Rfc0791: Internet protocol (1981)
28. S. Ansari, S. Rajeev, H. Chandrashekar, Packet sniffing: a brief introduction. IEEE Potentials **21**(5), 17–19 (2003)
29. X. Xu, Q. Liu, Y. Luo, K. Peng, X. Zhang, S. Meng, L. Qi, A computation offloading method over big data for IoT-enabled cloud-edge computing. Future Gener. Comput. Syst. **95**, 522–533 (2019)
30. A. Iosup, S. Ostermann, M.N. Yigitbasi, R. Prodan, T. Fahringer, D. Epema, Performance analysis of cloud computing services for many-tasks scientific computing. IEEE Trans. Parallel Distrib. Syst. **22**(6), 931–945 (2011)
31. F. Al-Turjman, 5G-enabled devices and smart-spaces in social-IoT: an overview. Future Gener. Comput. Syst. **92**, 732–744 (2019)
32. M. Mueck, A. Piipponen, K. Kalliojärvi, G. Dimitrakopoulos, K. Tsagkaris, P. Demestichas, F. Casadevall, J. Pérez-Romero, O. Sallent, G. Baldini et al., ETSI reconfigurable radio systems: status and future directions on software defined radio and cognitive radio standards. IEEE Commun. Mag. **48**(9), 78–86 (2010)
33. P.W. Singer, A. Friedman, *Cybersecurity: What Everyone Needs to Know* (Oxford University Press USA, New York, 2014)
34. G. Zhao, S. Qin, G. Feng, Y. Sun, Network slice selection in softwarization-based mobile networks. Trans. Emerg. Telecommun. Technol. **31**(1), e3617 (2020)
35. C.N. Nyororo, Strategic change management and performance of National Social Security Fund (NSSF). PhD thesis, University of Nairobi, 2006
36. M. Ananda, H. Bernstein, K. Cunningham, W. Feess, E. Stroud, Global positioning system (GPS) autonomous navigation, in *IEEE Symposium on Position Location and Navigation. A Decade of Excellence in the Navigation Sciences* (IEEE, Piscataway, 1990), pp. 497–508
37. C. Dwork, Differential privacy: a survey of results, in *International Conference on Theory and Applications of Models of Computation* (Springer, Berlin, 2008), pp. 1–19
38. J. Krumm, A survey of computational location privacy. Pers. Ubiquit. Comput. **13**(6), 391–399 (2009)
39. J. Thakur, N. Kumar, DES, AES and blowfish: Symmetric key cryptography algorithms simulation based performance analysis. Int. J. Emerg. Technol. Adv. Eng. **1**(2), 6–12 (2011)
40. G. Kambourakis, A. Rouskas, G. Kormentzas, S. Gritzalis, Advanced SSL/TLS-based authentication for secure WLAN-3G interworking. IEE Proc. Commun. **151**(5), 501–506 (2004)
41. H. Li, Y. Dai, L. Tian, H. Yang, Identity-based authentication for cloud computing, in *IEEE International Conference on Cloud Computing* (Springer, Berlin, 2009), pp. 157–166
42. P. Patil, P. Narayankar, D. Narayan, S.M. Meena, A comprehensive evaluation of cryptographic algorithms: DES, 3DES, AES, RSA and blowfish. Procedia Comput. Sci. **78**(1), 617–624 (2016)
43. W.C. Barker, E.B. Barker, Sp 800-67 rev. 1. recommendation for the triple data encryption algorithm (TDEA) block cipher (2012)

44. J. Bethencourt, A. Sahai, B. Waters, Ciphertext-policy attribute-based encryption, in *2007 IEEE Symposium on Security and Privacy (SP'07)* (IEEE, Piscataway, 2007), pp. 321–334
45. J. Horwitz, B. Lynn, Toward hierarchical identity-based encryption, in *International Conference on the Theory and Applications of Cryptographic Techniques* (Springer, Berlin, 2002), pp. 466–481
46. K. Liang, L. Fang, W. Susilo, D.S. Wong, A ciphertext-policy attribute-based proxy re-encryption with chosen-ciphertext security, in *2013 5th International Conference on Intelligent Networking and Collaborative Systems* (IEEE, Piscataway, 2013), pp. 552–559
47. R. M. Malina, S. P. Cumming, and M. J. C. e Silva, Physical activity and movement proficiency: The need for a biocultural approach. Pediatr. Exerc. Sci. **28**(2), 233–239 (2016).

Chapter 12
Security and Privacy in 5G-Enabled Internet of Things: A Data Analysis Perspective

S. R. Mani Sekhar, G. Nidhi Bhat, S. Vaishnavi, and G. M. Siddesh

1 Introduction

There is a need to write this chapter to bring awareness about the consequences of the data being transferred every time and how it affects our privacy and security. This works on the measures which are taken and the new technologies which are blooming in the industries in order to protect the data from malicious activities. After that, the author tells well the innovative ideas that could be implemented along with their advantages and disadvantages. In this chapter, the authors have tried to acknowledge the severities in data transfer in 5G-enabled IoT and also have included a few innovations in this field and a comparative study which has helped in securing data.

In this era of advanced technology, connectivity is the utmost requirement for all activities. There has been a continuous exponential development in the field of networking as shown in Fig. 12.1. 2G, 3G, 4G, and 4G LTE have been able to help us connect with people all over the world and now have stepped into the next generation of networking that is 5G. It is a wireless standard where everything including objects and machines is connected. It reduces a lot of drawbacks imposed by the previous networks by providing peak data speeds, low latency, high network capacity, and availability; it explores user experiences and is highly effective. It is estimated that 5G technology is ten times faster than 4G LTE. Its ability is immense. Fields where 5G is most important include autonomous vehicles, improved broadband, healthcare, remote device controller, public protection and infrastructure, and IoT.

In the upcoming section of the chapter, the authors discuss the various security-related issues and their possible solutions followed by privacy section which

S. R. Mani Sekhar (✉) · G. Nidhi Bhat · S. Vaishnavi · G. M. Siddesh
Department of Information Science & Engineering, Ramaiah Institute of Technology, Bangalore, India
e-mail: manisekharsr@msrit.edu

S. Tanwar (ed.), *Blockchain for 5G-Enabled IoT*,
https://doi.org/10.1007/978-3-030-67490-8_12

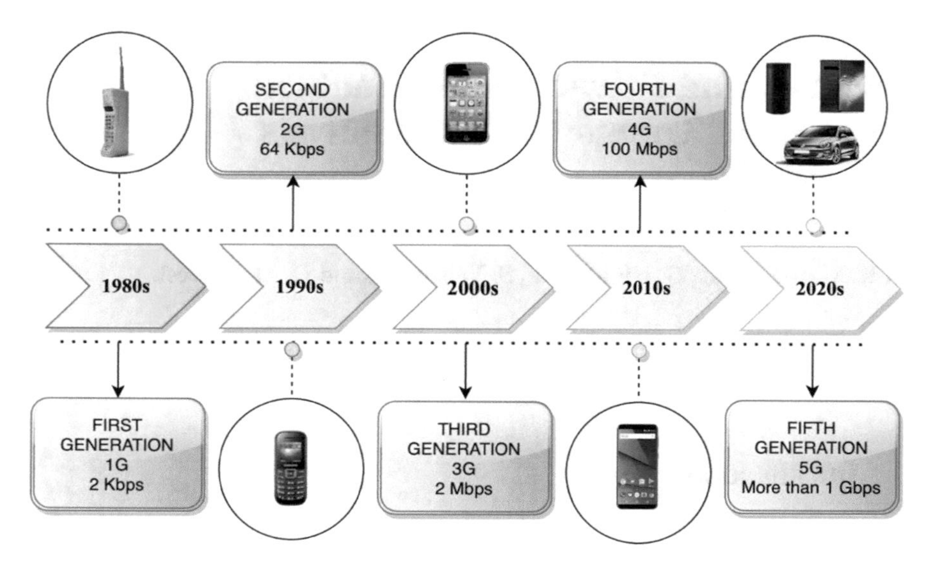

Fig. 12.1 Timeline for evolution from 1G to 5G

discusses the privacy issues and their possible mitigations. Further, the author discusses the various case studies which help in understanding the different approaches to security and privacy issues.

1.1 IoT Devices Working with 5G Networks

IoT devices use a variety of wireless technologies which include Wi-Fi, Bluetooth ZigBee, Z-Wave, GSM, 4G-LTE, and even 5G. Other alternatives like cellular technologies are also used by IoT devices. This offers global connectivity, security, and performance. Connectivity is the heart of IoT devices. With IoT devices having this level of connectivity as in 5G networks, there would be no need for any manual attempt to switch on lights. Although 5G provides a lot of benefits, it still does come with its costs. The case studies help us learn about how privacy and security issues are being tackled in vast areas like transport, communication aids, etc. A lot of data which could be used to do great harms is too easy to reach. A few challenges regarding privacy and security threats are explained with some solutions which would secure our data.

In 2011, the US military lost its national secret data due to a mission over Afghanistan which was contained in American sentinel unmanned aerial vehicle (UAV) to Iranian forces. The theft of such national secret data could cause severe catastrophes which are unimaginable. In such cases, erasure of data is one solution as designed in the paper [1], but it might not be enough at all times.

A wide range of research has been happening to fill in the gap between IoT practical usage and privacy concerns. Various protocols and architectures are being studied to maintain our data in secure hands.

2 Security in 5G-Enabled IoT

Security concerns must be dealt with utmost attention, and this has been a challenge in 5G-enabled IoT. With a number of IoT devices gaining popularity, the significance of dealing with security issues cannot be overlooked. These issues are not only considered to be damaging to the cost but are also causing other pressing issues such as loss of consumer confidence, social trust, and personal safety.

IoT integrates and implements various tools, advancements, and infrastructure. The threats imposed by these network technologies which are utilized in IoT expose the layers of IoT architecture to security challenges [2, 31]. IoT systems raise security concerns, unlike the conventional networks where the security issues can be handled with much ease. The physical layer of the IoT system contains less executing power and less data storage due to which conventional security solutions like encryption and spread-spectrum methods cannot be implemented [3] at all ends. Subsequently, IoT systems are heterogeneous, and hence they have different defense mechanism capabilities; hence, the most vulnerable layer determines the security level of the system.

2.1 Security Issues and Threats of a Layered IoT Architecture

The below section discusses the various security issues of a 5G IoT architecture [4, 5].

2.1.1 The Architecture of 5G-Enabled IoT

Since there is no proper organized layered structure for IoT devices despite the emerging number of IoT applications, let's consider the most common approach using mainly three layers [6]:

(i) Physical layer: It usually includes sensors and actuators to gather information from the environment.
(ii) Network layer: It helps in connecting various smart things, network devices, and servers. Subsequently it is used for sending and processing sensor data.
(iii) Application layer: It is used to deliver application-specific facilities to the end users.

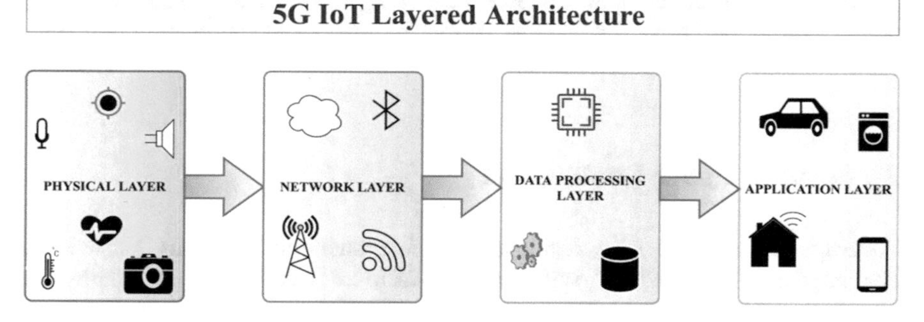

Fig. 12.2 5G IoT layered architecture [32]

Some approaches could consist of a data processing layer which classifies all cloud-based service-related issues [7]. Figure 12.2 represents these layers of the 5G IoT architecture.

2.1.2 Security Issues

(i) Physical Layer

The sensors and actuators of the physical layer are considered to be prone to security breach since they operate where devices can be accessed physically. This means that the devices can be physically tampered and firmware can be replaced or can even be destroyed [8]. Unauthorized access and cloning of the tag can also occur in sensors which may allow the malicious attacker to reprogram the data or provide fake information.

Tampering can also be the initial stage of implementing a denial-of-service (DoS) attack alongside extra nodes in the grid. Closed systems are also subjected to attacks by jamming in the physical layer.

(ii) Network Layer

IoT systems are heterogeneous, and hence in the network layer, it is observed that the security threats differ remarkably. Edge devices and the endpoints in the cloud are more influential devices and hence can adapt to traditional security measures, whereas the nodes lack features like public-key cryptographic methods [9], thus often limiting the defending capabilities of the whole network.

A common attack through networks is a DoS attack. In an edge computing scenario, distributed DoS attack and wireless congestion are dangerous. The other most prevalent form of attack is a man-in-the-middle attack as it violates the discretion, honesty, and privacy of the delimited data. Such attacks are executed in various ways such as eavesdropping on the messages or misusing them; privacy and quality of service are deteriorated by altering unencrypted routing data [10].

The other forms of attacks could be IP theft, counterfeit attack, packet sniffing, and so on.

(iii) Data Processing Layer

The main security threats in the cloud are DoS attacks that can shut down the cloud system and deny the services. With the rise in the 5G-enabled IoT application and 5G wireless speed, DoS attacks are higher-risk factors since the previous detection methods cannot handle the high volume data traffic. System vulnerabilities may also allow attackers to easily damage the cloud computing system [5].

(iv) Application Layer

The application layer clubs different interfaces like web applications, service organization tools, and middleware [11]; in this layer, the major security concern is the traditional software attacks involving a lot of risks. DoS attack and malicious code injection are common security concerns. It is easy to attack and shut down a service if there is no secure authentication and key agreement. At situations where data is not validated enough, attackers would find it a loophole to inject malicious data which as a consequence they get easy access to perform various actions, such as stealing records, negotiating database integrity, or bypassing validation [12].

2.2 *Possible Mitigations for Security Issues*

Some possible mitigations for security issues concerning each layer are mentioned below [7].

(i) Physical Layer

Tamper-resistant packaging can be used to provide maximum security against tampering of the IoT devices. Although there is very little cure for DoS attacks [7], spread-spectrum procedures can be a structured defense mechanism, but this cannot be put into action due to their constraints on computational volume and power consumption [3].

(ii) Network Layer

Passive monitoring (probing), active firewalls, bidirectional link authentication, traffic admission control, encryption, and authorization can be used to prevent eavesdropping and DoS attacks. The edge data centers are to be protected to prevent the attackers from accessing the data.

(iii) Data Processing Layer

Malware detection, traffic monitoring, and appropriately organized firewalls on all system entry points could help prevent various attacks on the cloud system.

(iv) Application Layer

Proper authentication, anti-virus filtering, integrity verification testing, authorization, validation of the inputs, traceability of the data, and process planning and design are some of the possible mitigations to reduce security risks in the application layer.

2.3 *Security in IoT Considering 5G*

The security issues of software-defined network (SDN)/virtual network functions (VNF)-built network fundamentals on the control plane outweigh the benefits they provide which include low latency and high speeds, especially when the authors relate to the nature of communication. A number of proposed solutions for these security issues are emerging, and hence 5G will be a preferred communication platform for IoT. Another major benefit of 5G-enabled IoT is the elasticity of its devices which leads to well-organized, use-case definite solutions which can be useful to the less standardized issues such as low-powered nodes and sensors. Regarding the less-energy nodes which contribute to IoT, an effective technique to advance security is using physical layer security (PLS) solutions and lightweight cryptography systems. Security threats and possible solutions in layered 5G IoT architecture [7] are summarized in Table 12.1.

3 Privacy in 5G-Enabled IoT

5G technology has helped in gaining momentum for IoT directly or indirectly. IoT has gained a lot of popularity through network connectivity and scope provided by 5G technology. 5G assures a faster lifestyle by reducing the downloading and uploading time durations, lowering the latency, and providing more connection density.

Before going ahead with what changes have been made in 5G IoT devices, it is important to note that the basic underlying physical infrastructure of the Internet is kept intact. In other words, the fiber-optic cables run by the service providers are connected with other ISPs and the broader Internet [13]. And hence, in 5G IoT devices, the risk of losing our privacy is still intact. This includes the exploitation of our communication surveillance, data retention, information sharing, and a lot more.

The amplified usage of IoT and the advancement in technology have led to zero private space. For instance, Equifax, a major credit reporting agency, had leaked 143 million American citizens' data such as name, social security number, birth date, address, and a lot more. Similarly, Facebook also was sued for harvesting over 5 million Facebook users' data without permission. This was a major setback in the

Table 12.1 Security threats and possible solutions in layered 5G IoT architecture

Layers/solution	Physical layer	Network layer	Data processing layer	Application layer
Function	Senses and gathers information from physical devices and sensors	Transmits the sensor data	Processes the collected data	Delivers applications based on the processed information to end users
Security threats	Tampering, DoS attacks	DDoS attacks, man-in-the-middle attack, eavesdropping, IP theft, counterfeit attack, packet sniffing	DoS attacks	DoS attack, malicious code injection
Possible solutions	Spread-spectrum procedures, tamper-resistant packaging	Probing, firewalls, encryption, traffic admissions control, authorization	Malware detection, traffic monitoring	Anti-virus filtering, testing, authentication, authorization, traceability

stock prices of Facebook as well as other social media platforms. The 4G-enabled IoT is cloud-based which could impose a lot of security and privacy threats. It comes to the fact that in 5G technology, more data is generated and more network traffic emerges because of features of the edge paradigms and therefore it is prone to more security and privacy concerns [11].

3.1 Privacy Challenges in 5G Networks

The below section illustrates the various privacy challenges in 5G networks [30].

3.1.1 Location Privacy

Semantic information attacks are very common. They can be defined as the usage of incorrect data to cause harm. Each time a user joins a 5G antenna, mobile networks can track the location data of the user, and this can also be done by access point selection systems in 5G.

The international mobile subscriber identity (IMSI) keeps an account of the identities of the mobile device subscribers. By seizing the IMSI of a subscriber's mobile device, an attacker may interrupt his activity like ongoing calls and texts and monitor them. The purpose of these measures is only to improve the quality of life, but the harms seem to be much bigger [14].

3.1.2 Correlation Privacy

At places where our private and sensitive information is being collected like in hospitals, database encryption might still not serve as a solution. Attackers would be able to integrate and analyze the encrypted information from different devices by specifying time and location where the data is fragmented. Hence, attackers would be able to access unauthorized records of the organization [3].

3.1.3 Broad Sensitive Information

As a part of this huge network, users knowingly or unknowingly share a lot of information. This compromises our broad data (like the information we leak every day) privacy because the attackers are given easy access to the information regarding an individual. Various devices which are connected to the Internet collect a lot of information which could be personal and sensitive.

3.1.4 Identity and Authentication

In previous technologies like 3G and 4G, the security and privacy protocols for preserving identity were based on the single server environment which does not suit the trends in 5G technology. The challenge is to design protocols to build trust and privacy between cloud users and the service. Although there were a few protocols designed like SSL Authentication Protocol (SAP) and Identity-Based Authentication for Cloud Computing (IBACC), there were complaints of their data being digitized.

3.1.5 Radio Communications in High-Frequency Bands

With 5G technology being used, the authors see an integration of radio access technologies which are easily able to tap into licensed, licensed shared, and unlicensed spectrum beyond frequency ranges which are in use. A massive antenna array to optimize the frequency ranges with new technology will be used. The more the number of antennas used, the more the signal paths created, and hence, our data is more vulnerable in the hands of the attackers [15].

3.2 Solutions to Privacy Threats

Transparency and consistency have to be maintained in IoT devices. Transparency is letting the user know about what is happening to the data being collected and how it is being used, and consistency is about how consistent the device serves its purpose. The two criteria of any IoT device help in maintaining our privacy [5].

A few common steps which could be taken to protect from the common attacks are:

(i) Selecting manufacturers who provide software support and updates for devices.
(ii) Using software which is updated so that it could show the vulnerabilities so they could manage them appropriately.
(iii) The behavior of the devices has to be monitored; any deviations from the norms have to be taken seriously [17].

End-to-end encryptions in 5G IoT devices must take care of radio transport, IoT and devices Telco cloud, security operations, and slicing security. Reducing the time a hacker stays undetected at cyber-attack approaches can help to secure privacy to some extent.

4 Case Studies

This section illustrates various case studies on IoT-based 5G networks.

4.1 Secure Network Architecture for Smart Grids in 5G Era [18]

Telecommunications and electric energy are two important factors to fuel the future smart cities and economy. With 5G network, the smart grid era presents a multitude of promising applications, particularly demand response, advanced metering infrastructure (AMI), smart house, and so on. The smart grid brought about immense changes on how electricity is manufactured, communicated, or consumed. It associates a two-way information and power flow between the end users and the grid. Smart grid requires the expansion of communication infrastructures and legacy control, supporting the transmission and generation systems, the network boundaries, and the distribution networks to join the complete supply network of the industry. AMI is distributed globally to bridge the gap between end consumers and utilities, establishing two-way communication links that will allow automatic and efficient load management and the adaptation of auspicious smart grid solutions, particularly DER, V2G, DR programs, and so on.

However, the introduction of AMI widens the attack surface and causes the grid to be susceptible to cyber-attacks from malicious attacks. For example, an attacker may initiate a DoS attack against the Supervisory Control and Data Acquisition (SCADA) system or parts of the AMI, like the neighborhood area networks (NAN), wide area network (WAN), and home area networks (HAN). The smart grid lays the foundation for a reasonable electricity marketplace. Hence, value integrity attacks can severely damage the product demand and supply balance, causing grid uncertainty, and can lead to an economy down. Subsequently this could damage the physical components of the grid or cause blackouts over a large geographic region.

To address these security concerns and in order to safeguard the grid against load alteration or price integrity attacks, a secure network framework was proposed. The major part of this network architecture is the intrusion detection system (IDS) scattered across the various places of the information network.

Depending on resource availability, an intruder may target only specific parts of the AMI or a restricted number of links, that is, the attacker may direct the attack at the access links of NAN, HAN, or backhaul links. These contact associations could cause the pricing information to be transmitted from utility companies to HANs and carry back the amount of electricity spent by these households.

WAN, NAN, or HAN could be provided with an IDS that would enhance application security by recognizing intrusions. An IDS is responsible for detecting any security breach. The signature-based systems depend on identified or familiar patterns in the communication packets based on an initial couple of bytes and are mainly used by proprietary-based solutions. Here, an updated record of signatures is maintained for any previously known security attack, and the signature of any given packet is always compared against these maintained signatures. Moreover, these methods experience trouble from two main disadvantages:

(i) Due to checking packet payloads, these techniques may violate the user's privacy.
(ii) As signature lacks in the database, they may be at the risk of new attacks.

ML and statistical learning techniques are merged by IDSs to reduce these problems. These methods can also be utilized for studying the voltage of AC statistics at different circulation buses to notice any peculiarity in the system. Therefore, IDS screens and checks the demand and supply stability in real time and inspects if there is any unexpected rise in the request. This, in turn, can lead to load or price attacks in the AMI. Different IDS organizations could be developed for different components of an AMI; later these mechanisms can be integrated for additional performance enhancement. For example, when an abnormality is noticed in a certain place of the network, the corresponding IDS can communicate it to the IDSs to take suitable actions at the various layers of AMI, that is, WAN, NAN, and HAN.

4.2 *Secure D2D Communication [19]*

D2D technique is a point-to-point communication system without a middle node between devices. In mobile networks, D2D communication has numerous advantages:

(i) A communication bridge could be established for data transmission in a cellular network to expand the coverage of each cell and transmit data to a node inside cell coverage.
(ii) By transmitting data directly between devices, D2D communication reduces the base station energy consumption.
(iii) There is an increase in the efficiency of the same radio frequency being reused.

In D2D communication, the distance between devices is shorter than the distance between a base station and a device which implies that in D2D communication scenario, the interference of radio frequency decreases; therefore, using the same radio frequency allows to transmit multiple data. Furthermore, D2D communication is a fundamental method of 5G vehicle-to-everything (V2X), which is a vital method for self-directed driving.

Typically, in a mobile network, D2D communication has a few security issues. The D2D communication involves mainly two measures, namely, device discovery and data communication. During this procedure, there are no verification steps in authenticating the identity of the device. A device sends in a request for a setup link, after which a node responds with an acknowledgement message. Additionally, D2D communication does not authenticate messages nor maintain integrity. This allows the attacker to direct various attacks like privacy sniffing, eavesdropping, location spoofing, impersonation, and free riding.

To address the service demands, IoT technology collaborates with the 5G network, and this corresponds to Ultra-Reliable Low-Latency Communication (URLLC) and Massive Machine-Type Communication (mMTC). IoT applications handle a considerable amount of sensitive data, but the devices have limited properties in terms of memory, power, and performance intake. Security concerns could be difficult to be processed due to these properties of IoT devices, and it makes it critical to find solutions to these issues. Therefore, there is an essential demand for a secure D2D system that consists of a correct validation process among the IoT devices. It has to be made lightly with the resource-constrained environment in mind.

To address the security concerns, a secure D2D communication is designed. For effectiveness in 5G use-cases equivalent to IoT scenarios, URLLC and mMTC, a lightweight authenticated encryption with associated data (AEAD) cipher and an elliptic-curve cryptography (ECC)-based public-key cryptosystem are used as groundwork for this design.

According to this approach, before the D2D communication is carried out, based on the 5G-AKA given by the 5G network, the identity of the user equipment (UE) is verified, and then a token is generated which is used as the key for the Elliptic Curve Digital Signature Algorithm (ECDSA). The produced token could validate the authenticity of the associated UE in the link setup. This can be achieved without linking to the central network. Also in the secure data communication step, the encrypted communication is done through frivolous AEAD cipher; meanwhile, the authentication of the UE can be accomplished, and integrity/confidentiality of the data can be maintained in each transmission process.

This technique can generate greater energy efficiency and performance, compared to the purpose AEAD cipher-based communication application, and can also ensure protection from eavesdropping, impersonation, site spoofing, privacy sniffing, etc.

4.3 UAV IoT Framework [20]

UAVs contribute in various fields including areas related to military, civilian, governmental, and commercial sectors. Some of the applications include environmental monitoring (pollution, industrial accidents), distribution, and surveillance requests intending on seeking or providing data at locations after an attack or a disaster. This information can be used to deliver medicine and other necessities. The commercial applications include distributing goods and products in rural and urban areas. UAVs are reliant on antennas, sensors, and embedded software; hence, they are considered as a part of the IoT. They provide two-way messages for applications linked to monitor and for remote control.

Considering that UAVs are a part of IoT, it mainly consists of similar security issues as that of mobile communication networks, sensor networks, the Internet, and specific privacy-protection problems. They have to deal with enhanced security concepts, like access control, verification, secrecy, cyber-attack anticipation, data protection, and high authorization, to prevent signal jamming, spoofing, RF and mobile application hacking, physical attacks, firmware hacks/sabotage, and protocol abusing.

The proposed framework incorporates cutting-edge comprehensive advances for approaching the present privacy and security level into a robust, highly protected, and principal environment by integrating dissimilar vision-based methods for scene study. Accordingly, a hybrid centralized-distributed system controls UAV flights and handles the operations such as registration, ranking, identification, and organization of moving objects.

A few solutions the framework proposes in order to solve the abovementioned security challenges are (a) vision-based techniques, (b) privacy anticipation and anonymity methods for mobile things, and (c) a lightweight safety toolbox.

The framework supports multi-domain and multilevel defense mechanisms in safeguarding IoT objects. Here privacy is achieved by an active "crowd of things" approach. Vision techniques aim to strengthen the security of IoT by encouraging machine learning and computer vision solutions.

4.4 Intelligent Transportation System

Traffic monitoring has become a major concern to maintain road safety. Intelligent transportation system (ITS) is one such system which works by the fixed transmission of information between vehicles and subsequently with back-end servers. With the exponential usage of IoT devices in vehicles, personal data is captured and processed. A balance between technology and measures to maintain privacy has to be kept to maintain confidence in digital economy services to further enable societal opportunities for innovation [21].

IoT sensors are widely used in the field of transport to support connected and autonomous vehicles (CAV) and ITS. CAV and ITS have technical and legal challenges in protecting the privacy of commuters. A few problems related to privacy are:

(i) Misuse of data
(ii) Malpractice
(iii) Communication overhead information being tracked, for example, IP address of the sender and receiver

The main focus of this system is maintaining the following:

(i) Anonymity: a person must not reveal his/her identity to use resource.

(ii) Unobservability: a third person should not get to know that a resource is being utilized.
(iii) Pseudonymity: the person using a resource should not reveal his/her identity which is still accountable for it.
(iv) Unlinkability: a person must be able to make several uses of resources without others being able to link with it.

Many security and privacy schemes have been structured to prevent privacy problems [22].

Group/Ring Signature-Based Privacy Schemes It is better when only a single person or a group manager has all the users' information rather than anyone who has access to the Internet, and this feature can be made possible by a cluster sign. Asymmetric cryptography is used here where the group members are given private keys and asymmetric keys. This key is used to generate signs and send communications over the network. Later, at the receiving side vehicle, a set of asymmetric keys are used. There are other schemes, but all of them demand high computational overhead and intermediate security; meanwhile, low computational overhead devices provide low security.

Pseudonym-Based Privacy Schemes Several techniques can be used in this scheme which fall under the two categories of symmetric cryptographic schemes and asymmetric cryptographic schemes. In these approaches, to maintain anonymity, fictitious names are assigned. A few of them are public-key cryptographic and secret key cryptographic methods.

A lot of researchers have come up with several schemes under asymmetric and systematic schemes, but the latter proves to be more efficient when it comes to computation.

4.5 End-to-End Network Slicing for 5G Communication

5G network slicing is one concept which has taken a big leap in the future of technology; it depends on the principles of network functions virtualization (NFV) and SDN. In network slicing, one physical network alone can be partitioned from numerous simulated networks, and each slice signifies an autonomous virtualized end-to-end network. The allocate resources is collected from a physical network; this concept is similar to that of the city transportation system where the authors find many modes of transportation by allocating infrastructure resources like roads, rail tracks, etc. This, in turn, reduces latency and high reliability [23].

With the growth of the IoT, network demands increase immensely, at almost 1000 times more data accumulation and more number of devices, lower latency, and higher bit rates. Network slicing provides a cost-effective solution for these demands. Although this proves as a key technology for simplified networking and has a lot of advantages, security and privacy challenges are a major concern [24].

Shielding, a feature of network slicing, refers to the non-interference of each slice from one another though they share the same infrastructure.

Users are not limited to just a single slice rather than a connection with many slices. Customer can expect more security vulnerabilities in end devices. A lot of distributed DoS are prone to happen. Confidentiality, authentication, authorization, availability, and integrity are the most important security principles which have to be followed. Network slice manager (NSM) should track interactions across slices, and it is responsible for interacting and virtual network function.

A few solutions are proposed to protect the data flow among base stations and devices. Here the authors use the concept of cryptography to focus on privacy preservation and protection of intra-slice ciphers with the help of a stream cipher. To secure communications between 5G networks, public cryptosystems like public key infrastructure (PKI) and certificateless cryptography could be used [24]. The solution provided here is more efficient than the encryption method in which a user encrypts the message before sending it.

Another approach is by using the security controls required for core slice addressing. This is a dynamic, future-proof approach which supports the requirements of network slicing on SDN [25] talks about global security policies. It covers a wide range of issues on network slicing and provides few solutions to meet a secure environment.

4.6 Lip Reading-Driven Secure Hearing Aid

A new approach which makes use of new technology and IoT for audiovisual (AV) aids is used to help people with hearing loss. Lip reading serves as a new efficient method as compared to the already existing audio-only hearing aids [29] has proposed a solution to the challenges on privacy-related aspects as well as low latency. It also provides us with a high data rate and low complexity in computation. 5G cloud-radio access network (C-RAN), IoT, and strong privacy techniques have been integrated to counter cybersecurity attacks like eavesdropping and location of privacy. The information in the form of AV sent by the 5G IoT devices is encrypted using a real-time lightweight encryption method based on piecewise linear chaotic and Chebyshev map, secure hash, and an innovative substitution box method.

There are a few encryption methods which comprise advanced technicalities like Advanced Encryption Standard (AES) and RSA (Rivest-Shamir-Adleman) algorithm that are not compatible with low-power sensor networks. Hence, the authors look into lightweight encryption methods. Generally, there are two stages of encryption: confusion and diffusion. The algorithm used in this study uses piecewise linear chaotic map (PWLCM) in the confusing process to make the decrypting and security attacks very complex. In the model proposed by [29], the encrypted audio signal and video signals are utilized in the cloud designed by the lip reading-driven speech-developed application, complement the concepts of deep learning, and use analytical acoustic modeling.

In the scheme proposed by [29], a system cloud which could enhance speech through lip reading is utilized where a collection of the encrypted audio and video signals takes place. This idea explores the capabilities of deep learning and analytical acoustic modeling. The first level is a deep learning regression model, and the second level is a lip reading enhanced visually derived Wiener filter (EVWF) to estimate the clean audio power spectrum [29] has proposed a lightweight chaotic encryption scheme in his paper which provides an easy enabler for modern digital hearing aids which secures privacy.

4.7 5G AKA Protocols

Authentication and key agreement (AKA) was a protocol designed by the 3rd Generation Partnership Project (3GPP) to standardize 3G, 4G, and 5G technologies in order to establish a secure network with serving networks (SN). AKA works on symmetric cryptography and sequence number (SQN). AKA is used in all 3G and 4G USIM (universal subscriber identity module) which is used in almost all the IoT devices. A few instances of security breach have occurred due to the fake base station attacks like the non-protected identity request mechanism to eavesdrop and inject messages. In 5G AKA protocol, asymmetric encryption is introduced to help in the authentication of identifiers [26].

Table 12.2 represents the key differences in the approaches of 4G and 5G authentication [16].

In [27], 5G AKA protocols have made changes to the protocol to include improving the privacy requirements. Instead of sending the messages directly, the protocol encrypts it using randomized asymmetric encryption. Although this prevents the IMSI attacks, it still is not enough to prevent the attacks applicable to 3G and 4G networks. In IMSI, a commonly observed flaw of AKA protocol in 3G and 4G was that it's very easy to track subscribers in a geographical region by just broadcasting an identity request in that region to the UEs. Hence, to protect messages carrying identity requests, developer uses stronger cryptographic mechanisms in 5G [28] has proposed a new model to analyze the AKA model called Tamarin Prover which has a high level of automation and equivalence properties to maintain privacy properties.

5G AKA provides a SUPI for users with a randomized key, SUCI. Tamarin model makes sure SUPI remains confidential, without which the active and passive users are not able to decrypt the message. Table 12.3 illustrates the brief tabular summary of the above case studies.

Table 12.2 Difference between 4G and 5G authentication

4G authentication	5G authentication
4G defines 4G EPS-AKA as its authentication method	5G defines three authentication methods. They are 5G-AKA, EAP-TLS, AND EAP-AKA
The authentication vector is generated by the Home Subscriber Server (HSS)	The authentication vector is generated by Unified Data Management (UDM)/ Authentication Credential Repository and Processing Function (ARPF) under 5G-AKA and EAP-AKA protocols
The authentication of the UE is decided by the mobility management entities (MME) and Evolved Node B (eNodeB)	The authentication of the UE is determined by Security Anchor Function (SEAF) and Authentication Server Function (AUSF)
The UE identity before being shared in the 5G network is encrypted with the public key of the home network in clear text which could be stolen by any attacker	Before it is sent to the 5G network, the UE makes use of the public key of the HN to encrypt the UE permanent identity
The HN (e.g., HSS) is utilized during authentication to generate authentication vectors and does not have a say in the decisions of the authentication results	The final decision on UE authentication is done by HN, and its results are also transmitted to UDM for logging
The anchor key hierarchy is comparatively shorter	5G has two intermediate nodes, and hence the key hierarchy is longer
The UE identity from UE to SN is IMSI/ Globally Unique Temporary Identity (GUTI) and from SN to HN is IMSI	The UE identity from UE to SN is Subscription Concealed Identifier (SUCI)/5G-GUTI and from SN to HN is SUCI/ Subscription Permanent Identifier (SUPI)
The SN consists of radio access equipment such as MMEs	The SN consists of the SEAF

5 Conclusion

In this chapter, the author has discussed how 5G technology would be used for the development of smart grids and understanding the role of IDS in providing a secure grid for the customers. The author illustrated various methods to make the D2D connections more secure using AEAD ciphers and an ECC-based public-key cryptosystem. They enumerate about the UAV IoT framework and the different methods which could be implemented to secure data. Next, they discussed about how communication between vehicles can be done in a secure way to avoid traffic and accidents using pseudonym- and group/ring signature-based privacy schemes. Network slicing is one of the key requirements in 5G technology to meet its standards, and this is possible by incorporating end-to-end encryption. Although this is prone to security and privacy issues, a few solutions were mentioned like the PKI and certificateless encryption. Later, the author has discussed how 5G has helped the hearing-impaired individuals by lip reading methods, which send the

Table 12.3 Security and privacy concerns and their proposed solutions in the presented case studies

Name of the case study	Security or privacy concerns	Proposed solution
Secure Network Architecture for Smart Grids in 5G Era	DoS attacks against SCADA system or various parts of the AMI as well as NAN, WAN, and HAN Price integrity attacks Load alterations	IDSs merged with ML and statistical learning techniques Different IDS organizations are developed for different parts of an AMI and integrated
Secure D2D Communication	Privacy sniffing Eavesdropping Location spoofing Impersonation Free riding	AEAD cipher ECC-based public-key cryptosystem
Unmanned Aerial Vehicles IoT Framework	Signal jamming Spoofing RF and mobile application hacking Physical attacks Firmware hacks Protocol abusing	Vision-based techniques Privacy anticipation and anonymity for mobile things A lightweight safety toolbox
Intelligent Transportation System	Misuse of data Malpractice Communication overhead information being tracked	Group/ring signature-based privacy schemes Pseudonym-based privacy schemes
End-to- End Network Slicing for 5G Communication	DDoS attacks Compromise in confidentiality, availability, integrity, and authorization	Cryptography for privacy preservation and protection of intra-slice ciphers using stream cipher PKI and certificateless cryptography to secure communication between 5G networks Using the security controls required for core slice addressing
Lip Reading-Driven Secure Hearing Aid	Eavesdropping Attacks against location privacy	Lightweight chaotic encryption scheme based on piecewise linear chaotic and Chebyshev map, secure hash, and an innovative substitution box method
5G AKA Protocols	IMSI attacks Eavesdropping	Asymmetric encryption Tamarin Prover

information in the form of audiovisuals that are encrypted by advanced encryption and use algorithms like PWLCM for a secure network. Finally, the chapter focuses on the various improvements in 5G AKA protocols to solve the privacy constraints which were absent in 4G protocols, as it is challenging to maintain data security and privacy in the field of connectivity, computation, science, and 5G-enabled IoT. This chapter has discussed the overall security and privacy issues in 5G-enabled IoT. Each of these issues has been debated with presently available and possible solutions. At the end of this chapter, the authors have discussed the various case studies which cover the different security and privacy domains of 5G IoT applications and thus helped the reader to gain an insight of the current fields of research on security and privacy of 5G IoT.

References

1. K. Andersson, I. You, R. Rahmani, V. Sharma, Secure computation on 4G/5G enabled Internet-of-Things. Wirel. Commun. Mob. Comput. **2019**, 3978193–3978191 (2019)
2. I. Andrea, C. Chrysostomou, G. Hadjichristofi, Internet of Things: Security vulnerabilities and challenges, in *2015 IEEE Symposium on Computers and Communication (ISCC)*, (IEEE, Larnaca, 2015), pp. 180–187
3. F.A. Alaba, M. Othman, I.A.T. Hashem, F. Alotaibi, Internet of Things security: A survey. J. Netw. Comput. Appl. **88**, 10–28 (2017)
4. P. Varga, J. Peto, A. Franko, D. Balla, D. Haja, F. Janky, et al., 5G support for industrial IoT applications–challenges, solutions, and research gaps. Sensors **20**(3), 828 (2020)
5. L. Liu, M. Han, Privacy and security issues in the 5G-enabled Internet of Things. 5G-Enabled Internet of Things, 241 (2019)
6. P. Sethi, S.R. Sarangi, Internet of things: Architectures, protocols, and applications. J. Electr. Comput. Eng. **2017**, 1 (2017)
7. P. Varga, S. Plosz, G. Soos, C. Hegedus, Security threats and issues in automation IoT, in *2017 IEEE 13th International Workshop on Factory Communication Systems (WFCS)*, (IEEE, Trondheim, 2017), pp. 1–6
8. R. Mahmoud, T. Yousuf, F. Aloul, I. Zualkernan, Internet of things (IoT) security: Current status, challenges and prospective measures, in *2015 10th International Conference for Internet Technology and Secured Transactions (ICITST)*, (IEEE, London, 2015), pp. 336–341
9. W. Wei, A.T. Yang, W. Shi, K. Sha, Security in internet of things: Opportunities and challenges, in *2016 International Conference on Identification, Information and Knowledge in the Internet of Things (IIKI)*, (IEEE, Beijing, 2016), pp. 512–518
10. N. Neshenko, E. Bou-Harb, J. Crichigno, G. Kaddoum, N. Ghani, Demystifying IoT security: An exhaustive survey on IoT vulnerabilities and a first empirical look on internet-scale IoT exploitations. IEEE Commun. Surv. Tutorials **21**(3), 2702–2733 (2019)
11. J. Lin, W. Yu, N. Zhang, X. Yang, H. Zhang, W. Zhao, A survey on internet of things: Architecture, enabling technologies, security and privacy, and applications. IEEE Internet Things J. **4**(5), 1125–1142 (2017)
12. A. Mosenia, N.K. Jha, A comprehensive study of security of internet-of-things. IEEE Trans. Emerg. Top. Comput. **5**(4), 586–602 (2016)
13. Welcome to 5G: Privacy and security in a hyperconnected world, https://privacyinternational.org/long-read/3100/welcome-5g-privacy-and-security-hyperconnected-world-or-not
14. M. Agiwal, N. Saxena, A. Roy, Towards connected living: 5G enabled internet of things (IoT). IETE Tech. Rev. **36**(2), 190–202 (2019)
15. D.M. West, How 5G technology enables the health internet of things. Brookings Center Technol. Innovation **3**, 1–20 (2016)
16. A comparative introduction to 4G and 5G authentication, https://www.cablelabs.com/insights/a-comparative-introduction-to-4g-and-5g-authentication#:~:text=EAP%2DAKA'%20is%20another%20authentication,the%20UE%20and%20the%20network

17. Zeljka Zorz: 5G IoT security: Opportunity comes with risks, https://www.helpnetsecurity.com/2019/12/02/5g-iot-security/#:~:text=%E2%80%9CLarge%20numbers%20of%20vulnerable%2C%205G,some%20aspect%20of%20the%20infrastructure.%E2%80%9D
18. F.B. Saghezchi, G. Mantas, J. Ribeiro, M. Al-Rawi, S. Mumtaz, J. Rodriguez, Towards a secure network architecture for smart grids in 5G era, in *2017 13th International Wireless Communications and Mobile Computing Conference (IWCMC)*, (IEEE, Valencia, 2017), pp. 121–126
19. B. Seok, J.C.S. Sicato, T. Erzhena, C. Xuan, Y. Pan, J.H. Park, Secure D2D communication for 5G IoT network based on lightweight cryptography. Appl. Sci. **10**(1), 217 (2020)
20. T. Lagkas, V. Argyriou, S. Bibi, P. Sarigiannidis, UAV IoT framework views and challenges: Towards protecting drones as "things". Sensors **18**(11), 4015 (2018)
21. D.H. Cruickshank, Security and privacy open issues in 5G connected IoT devices. Guildford, Surrey: Institute for Communication Systems (ICS), http://www.charisma5g.eu/wp-content/uploads/2016/07/Security-and-privacy-open-issues-in-the-5G-connected-IoT-devices.pdf
22. Q.E. Ali, N. Ahmad, A.H. Malik, G. Ali, W.U. Rehman, Issues, challenges, and research opportunities in intelligent transport system for security and privacy. Appl. Sci. **8**(10), 1964 (2018)
23. N. Cranford, Understanding end-to-end network slicing for 5G (2018), https://www.rcrwireless.com/20180404/understanding-end-to-end-network-slicing-for-5g-tag27-tag99
24. V.A. Cunha, E. da Silva, M.B. de Carvalho, D. Corujo, J.P. Barraca, D. Gomes, et al., Network slicing security: Challenges and directions. Internet Technol. Lett. **2**(5), e125 (2019)
25. Z. Kotulski, T.W. Nowak, M. Sepczuk, M. Tunia, R. Artych, K. Bocianiak, et al., Towards constructive approach to end-to-end slice isolation in 5G networks. EURASIP J. Inf. Secur. **2018**(1), 2 (2018)
26. R. Borgaonkar, L. Hirschi, S. Park, A. Shaik, New privacy threat on 3G, 4G, and upcoming 5G AKA protocols. Proc. Privacy Enhancing Technol. **2019**(3), 108–127 (2019)
27. A. Koutsos, The 5G-AKA authentication protocol privacy, in *2019 IEEE European Symposium on Security and Privacy (EuroS&P)*, (IEEE, Stockholm, 2019), pp. 464–479
28. D. Basin, J. Dreier, L. Hirschi, S. Radomirovic, R. Sasse, V. Stettler, A formal analysis of 5G authentication, in *Proceedings of the 2018 ACM SIGSAC Conference on Computer and Communications Security*, (ACM, Toronto, 2018), pp. 1383–1396
29. A. Adeel, J. Ahmad, A. Hussain, Real-time lightweight chaotic encryption for 5G IoT enabled lip-reading driven secure hearing-aid. arXiv preprint arXiv:1809.04966 (2018)
30. Privacy international, https://privacyinternational.org/long-read/3100/welcome-5g-privacy-and-security-hyperconnected-world-or-not
31. I. Mistry, S. Tanwar, S. Tyagi, N. Kumar, Blockchain for 5G-enabled IoT for industrial automation: A systematic review, solutions, and challenges. Mech. Syst. Signal Process. **135**, 106382 (2020)
32. A.K. Sikder, G. Petracca, H. Aksu, T. Jaeger, A.S. Uluagac, A survey on sensor-based threats to internet-of-things (IoT) devices and applications. arXiv:1802.02041 (2018)

Chapter 13
Adversarial Artificial Intelligence Assistance for Secure 5G-Enabled IoT

Mohammed Husain Bohara, Khushi Patel, Atufaali Saiyed, and Amit Ganatra

1 Introduction

With the advancement of various technologies and increased capability of communication technologies, many things in terms of computing devices are connected with each other. Today, the Internet of Things (IoT) plays a major role to provide a comfortable and easier lifestyle to human beings by physically connecting sensors and other embedded devices. Moreover, machine learning (ML) and artificial intelligence (AI) add intelligence and self-learning capabilities in machines which significantly attract many users and devices to get connected with the Internet which significantly affects the automation of industry [1]. The globally connected IoT devices generate billions of data on a daily basis which can be used for training the models through machine learning techniques to incorporate intelligence with devices. By applying various AI techniques to the large amount of data generated by various IoT sensors, more automated IoT devices can be designed [1]. The expansion of big data and, recently, the applications of AI technologies in different fields have expanded rapidly. In object detection, image recognition, computer translation, speech control, and more advanced areas such as drug structure analysis, due to high intelligence, high availability, and high performance,

M. H. Bohara (✉) · K. Patel · A. Saiyed · A. Ganatra
Devang Patel Institute of Advance Technology & Research (DEPSTAR), Faculty of Technology and Engineering (FTE), Charotar University of Science & Technology (CHARUSAT), Changa, Gujarat, India
e-mail: mohammedbohara.ce@charusat.ac.in; khushipatel.ce@charusat.ac.in; saiyedatufaali.ce@charusat.ac.in; amitganatra.ce@charusat.ac.in

S. Tanwar (ed.), *Blockchain for 5G-Enabled IoT*,
https://doi.org/10.1007/978-3-030-67490-8_13

artificial intelligence technologies have been applied. Moreover, deep learning (DL) is attracting many industries in the field of computer vision to add momentum to AI-based applications. However, AI and ML with IoT are the most revolutionary technologies today; security concerns become the center of attraction for the major researchers.

Conventionally, the data with identical statistical parameters are used to train and deploy the ML models with a benign environment which is vulnerable to tampering of statistical parameters of the ML model by some capable intruder that leads to incorrect prediction. Some recent studies have indicated the neural networks are open to malicious attacks, current research on adversarial technology of artificial intelligence has slowly become a hotspot, and studies have regularly proposed new methods of adversarial attacks and methods of protection.

The adversarial attacks can be divided into three classes as per the target model's different phases: attacks in the training phase, attacks in the testing phase, and attacks in the model's implementation phase. Presently, artificial intelligence has several different sub-fields, such as intelligent optimization, biomedical devices, and machine learning [2, 3]. Here, the machine learning (ML) sub-field utilizes the learning mechanism which makes it the artificial intelligence's major sub-field. A typical method of machine learning uses certain metrics, which are knowledge and practical adjustment according to a set of target data, if previously obtained or evaluated instantly. It is for that reason the machine learning background is highly associated with statistics, mathematics, and logic [4, 5]. It is important that all the context provides machine learning with both certain benefits and drawbacks. Essential benefits for even modern, complex real-world problems are becoming versatile and simple to use. Alternatively, drawbacks include needing thorough review of the data, often using training phase for the data, and often needing reliable expertise for the true modeling for the problem. As with cyber security threats, today's and future intelligent systems can employ some gaps that make them vulnerable against attacks built with a powerful logical and mathematical context.

The Internet of Things (IoT) connects various things (sensors, actuators, etc.) with each other using communication technologies like the fifth generation (5G) to fulfill the demand of latency-sensitive applications. The data generated by various sensors with some identical parameters are provided to the ML models for training and deployment under some benign settings to make some decisions. The ML requires the reliable, trusted, and secure platform of data storage for working more effectively. The security and privacy of the data collected from the sensors during transmission and building the model using ML techniques plays an important role. Blockchain is a trusted and distributed ledger which provides integrity, resilience, and tamper-proof environment. Integration of such technologies like 5G (networking), Internet of Things (IoT), machine learning (ML), big data analytics (BDA), blockchain technology (BT), etc. plays a significant part in the revolution of automation in industry with a transparent environment.

1.1 Motivation

The motivation comes with the integration of IoT, blockchain, and machine learning. 5G is evolving to enhance the performance, security, and connectivity in IoT device. The 5G network targets to support the application of IoT. Blockchain can be used to solve the challenges of 5G-enabled IoT. Various concepts of blockchain with IoT are explored in the papers; some of the papers give brief ideas about the adversarial effects of machine learning. The data generated with the IoT devices can be trained with the machine learning models.

1.2 Contribution of This Survey

This chapter provides the detailed understanding about role of blockchain and 5G technology in integration of IoT and AI.

1. The evolution of blockchain technology with its architecture, types, and features is covered in this chapter.
   - The 5G offers numerous benefits to the IoT applications in terms of performance, fault tolerance, and decentralization. The overview of 5G-enabled IoT with the use of blockchain technology is covered.
   - The research in the area of adversarial attacks and types of adversarial attacks on machine learning models is in trend. The various examples of adversarial attacks, threat landscape, and their impacts are covered in detail. Various adversarial effects of machine learning techniques while the same set of parameters for the data is used for training and deploying the model are also covered.
   - The way of offering trustworthy machine learning approach to industry has been analyzed.
   - The various scenarios of integration of blockchain with IoT, 5G-enabled IoT, and AI along with their challenges and opportunities are covered.

1.3 Organization

The chapter structure is as follows: In Sect. 2, the basic theory of blockchain, evolution of blockchain, and types of blockchain with blockchain framework are explored. In Sect. 3, 5G-enabled IoT with architecture and use case of 5G-enabled IoT in blockchain is discussed. Section 4 explains various adversarial effects of machine learning techniques while the same set of parameters for the data is used for training and deploying the model. Blockchain technology is one of the trending solutions for providing an immutable decentralized environment for data storage and transactions which can be utilized to provide a secure, trusted, and reliable environment to ML models for solving the adversarial effects. In the next section, the basics of blockchain technology, 5G communication, IoT, and adversarial AI techniques are covered. In Sect. 5, the threat landscape and how a trustworthy machine learning approach can be offered to industry are covered. Section 6 covers the attacks. Section 7 covers the role of blockchain in establishing trust among various parties in the machine learning model for industrial automation with the case study and challenges.

2 Background Theory

In this section, blockchain technology along with 5G-enabled IoT is covered to understand how to provide security against adversarial effects on machine learning and artificial intelligence (AI) models.

2.1 Blockchain Technology

The digital currency Bitcoin [6] which is based on blockchain is the first application of blockchain published in 2009 by a person named Satoshi Nakamoto. Blockchain was initially put forward as an underlying technical framework of Bitcoin. Due to excessive fluctuation and regulatory management, Bitcoin was deprived in many countries. The reason for the acceptance of blockchain by the public is the three security measures of confidentiality, reliability, and integrity. The era of Bitcoin was Blockchain 1.0.

The blockchain is an emerging area. In Fig. 13.1, the evolution of blockchain from 1.0 to 4.0 is depicted. The introduction of smart contracts in blockchain was the evolution of Blockchain 2.0. The smart contract is the set of instructions which can be configured in such a way that it gets executed automatically by miners on the occurrence of some event. Smart contract is used in Ethereum which is another well-known cryptocurrency. Nowadays, many industries get attracted toward developing various platforms for smart contracts which leads to adoption of smart contracts in various areas like ML, AI, IoT, and big data analytics for integration of blockchain.

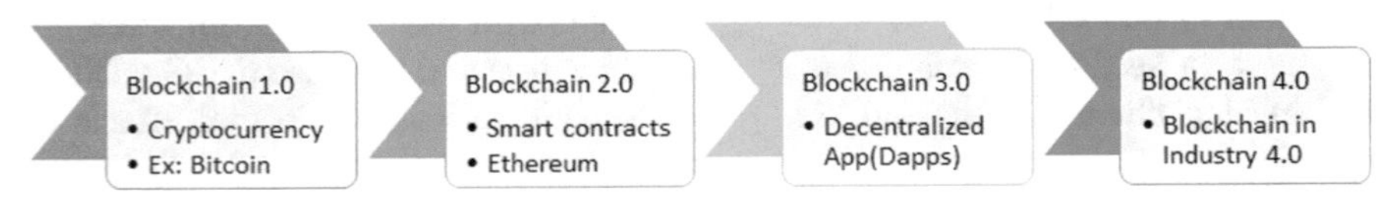

Fig. 13.1 Evolution of blockchain technology from Bitcoin to blockchain in Industry 4.0

Fig. 13.2 Type of blockchain

Blockchain is used to transmit the data and transfer of data is done in a decentralized manner, without the involvement of any third party. From cryptocurrency to the numerous industries, blockchain is growing day by day. The blockchain provides the technical solution for the verification, authentication, and data storage. The blockchain is used in many applications such as financial market, voting system, IoT, medical, supply chain, and agriculture.

Today the blockchain is a trending technology. The data transfer is efficient using blockchain technology. Privacy is maintained using blockchain, and the trust of different parties increased, as it reduces the chance of victimization and provides the record of transaction done. Blockchain reduces the risk when the untrusted parties come into the finance business which generates the reliability. Using the Internet, we can transfer the data, image, movies, etc. in any corner of the world. Similarly, in the transaction, the sender trusts the unknown party and makes the transaction. But in the blockchain security is ensured for the transaction. Block in the blockchain transactions is recorded, and the time of the creation and modification of the block is also recorded.

The blockchain network can be categorized into different categories on the basis of access to the network which is permissionless and permissioned. Permissioned blockchain network is only accessible to authorized users in consortium or cloud-based environments, while permissionless blockchain is open for all the users. Figure 13.2 shows the various types of blockchain network.

Private and public blockchain are the types of blockchain. The classification is based on the access control, data storage, and parameters. In the public blockchain, any node can participate in the process, but in the case of private blockchain, the restriction is provided, and to enter the process, approval is required. Consortium blockchain falls under the category of permissioned blockchain but multiple organizations are involved.

2.2 Blockchain Framework

The blockchain is considered as the link of block data forming a chain-like structure; the representation of blockchain framework is explained in detail as shown in Fig. 13.3. The blockchain framework is divided into data link layer, network layer, and

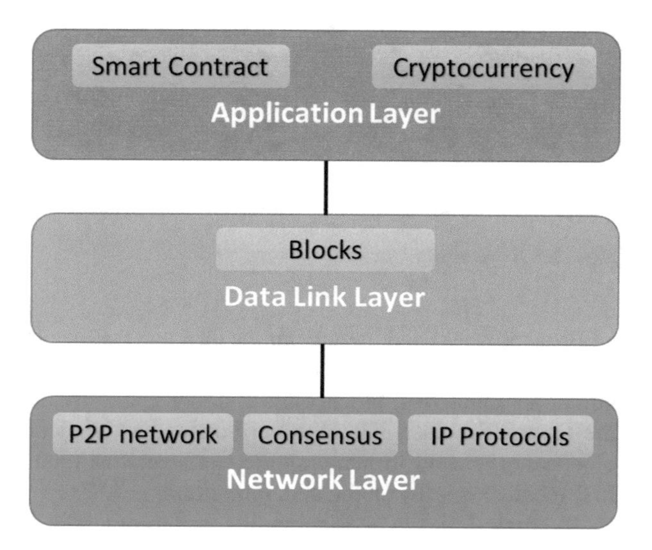

Fig. 13.3 Blockchain framework

application layer. In the data link layer, the block is created which includes the data structure, hash, Merkle root tree, and hash pointer. The network layer is used to interact with the blockchain which includes P-P network protocol and consensus algorithm. The network layer also enables the network distribution among the users. Lastly, the application layer integrates the blockchain with the application that enables the smart contact, Hyperledger, etc.

Blockchain is the technology that combines different algorithms, cryptography, and distributed networks. The blockchain works on the principle of the agreement of transaction without any central authority. In the blockchain, the blocks are linked to one another, and each block is linked with the hash of the block at the time stamp. The hash value is generated to verify the identity of the block.

3 5G-Enabled IoT

The Internet of Things, commonly referred to as IoT, connects the Internet with the electronic devices of different capabilities and structures. This is mainly done by wireless sensor, RFID, machine to machine, and Zigbee. The IoT comprises five components which are sensor, computing node, receiver, actuator, and device. To improve the efficiency of the system, the limitation with the usage of IoT needs to be resolved. Due to the increase in the usage of IoT devices, there should be a large amount of data transmission support with high bandwidth.

IoT is the technology to set the vision of connected living. It is not only to enhance the quality of life but new revenue generation. Capabilities of IoT promise to save people's and organizations' money and time while at the same time contributing toward enhanced outcomes in a wide range of novel application

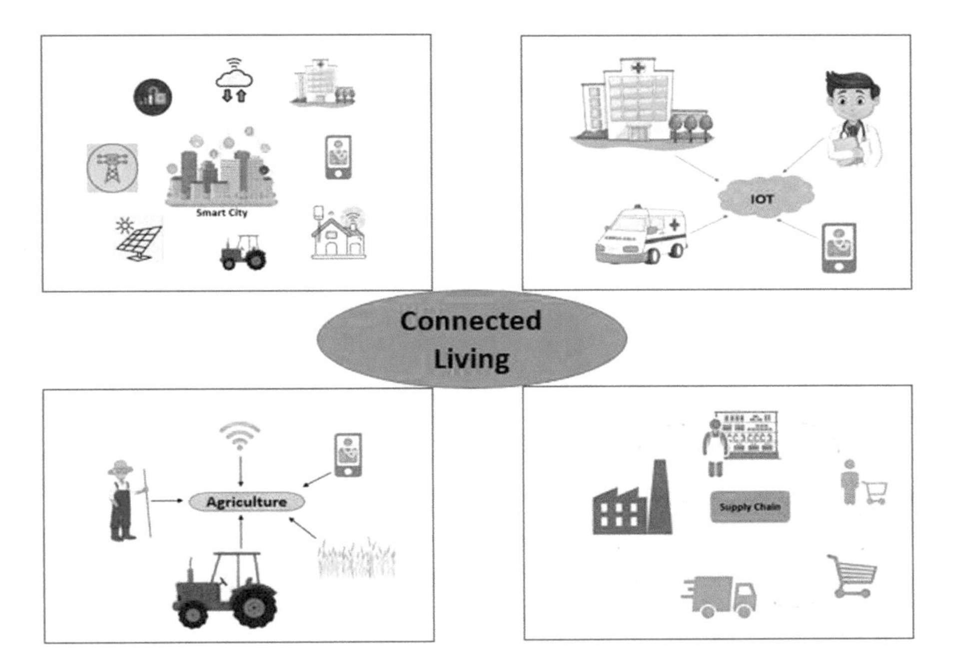

Fig. 13.4 IoT with connected living [8]

areas. The new era of living in the different areas like healthcare, smart city, agriculture, smart homes, etc. connects with IoT as shown in Fig. 13.4. The lifestyle would connect to IoT-enabled devices with the various applications. The advance of wireless technology which is the fifth-generation wireless systems (5G) is a driver for the 5G-enabled IoT applications. Next-generation 5G is rapidly growing as the massive machine-type communication transfers to high-level 5G technology. This would lead to IoT with 5G technology.

5G is evolving to enhance the performance, security, and connectivity in IoT device. The 5G network can provide faster speed than any other generation. In the past years, many researches have been done. The CISCO, Intel, and Verizon have jointly worked upon a research project on 5G and designed a novel set of "neuroscience-based algorithms" that adapt video quality to the demands of the human eye, suggesting that the future wireless networks would have built in human intelligence [7].

3.1 5G IoT Architecture

During each phase, 5G IoT is expected to provide end-to-end data transfer with real-time applications. The architecture provides independent network and use of

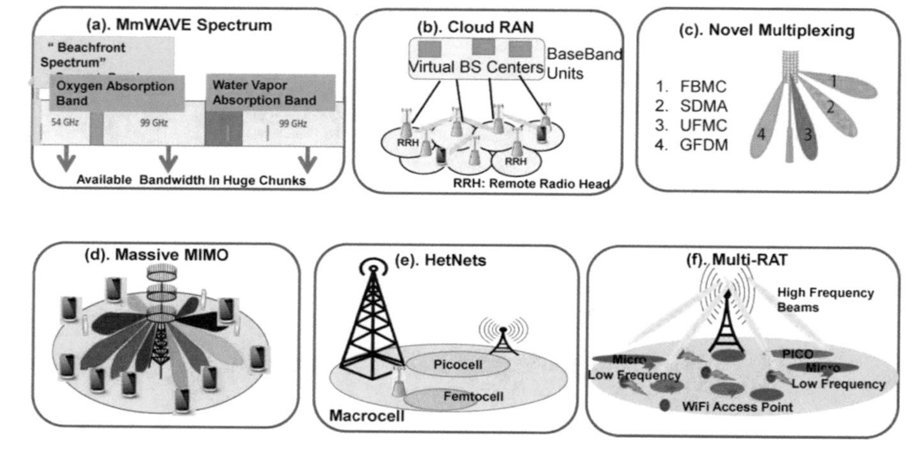

Fig. 13.5 5G-enabled IoT [8]

cloud-based radio access network (CloudRAN) to provide on-demand deployment of the applications. 5G required some key enabling technologies such as high bandwidth, high battery life – fundamental requirement in 5G, flexible and novel time-frequency multiplexing, antenna array technology for narrowband operation, HetNets, and massive MIMO for IoT architecture, network virtualization, self-organizing network (SON), and coexistence of multiple radio access.

3.2 Use of 5G-Enabled IoT in Blockchain

Cloud computing is the centralized data storage system and is used in the development of IoT. But the issue in cloud computing is data privacy. The user is not aware about where the data is stored in the network. Therefore, such systems fail to meet the requirement of the user. To secure the data and maintain the privacy, blockchain is the efficient solution. Blockchain has the ability to revolutionize IoT with an open, trusted, and auditable sharing platform, where any information exchanged is reliable and traceable. Some of the benefits of this integration are as shown in Fig. 13.5 [9].

4 Adversarial Artificial Intelligence

In this section, we present mainly adversarial samples and types of attack, including the causes and features of adverse samples and the capabilities and objectives of the adversarial attacks.

Here we define the common technical terms relating to adversarial machine learning attacks.

- *Adversary* – More commonly, adversary refers to the operator who produces an example of adversity.
- *Adversarial Attack* – Current research incorporates techniques on machine learning and deep learning for adversarial attacks. This deals fundamentally with the concept of fooling train models.
- *Adversarial Example* – It is a changed version of a data cleansing that purposefully adds noise to confuse a model of machine learning.
- *Adversarial Training* – Apart from the clean data, adversarial training uses adversarial data set to build machine learning models.
- *Black-Box Attack* – A type of attack that feeds adversarial examples to a specific model which are created without that model's awareness.
- *Threat Model* – The model of threat refers to different types of potential attacks which a strategy considers, e.g., black-box attack.
- *White-Box Attack* – An attack that presumes the full understanding of the target model, including its values for parameters, design, training process, and in some cases even training data.

4.1 Types of Attack

In this section, we are going to explain a few attacks with examples for better understanding of the topic.

4.1.1 Box-Constrained L-BFGS

The authors in [10] first show the existence of minor disruptions to the photos, so the disrupted photos could trick the classification errors of deep learning models. These contradictory findings have thus provoked a wide interest among researchers in the area of computer vision adversarial assaults using deep learning (Fig. 13.6).

4.1.2 Fast Gradient Sign Method (FGSM)

Szegedy et al. [10] found that adversary training could increase the robustness of deep neural networks against the adversarial instances. Goodfellow et al. [12] introduced a strategy for efficiently calculating an adversarial disturbance for a given image in order to allow effective adversarial training. The Python program in [13] tricks the regression model as the method of target machine learning (full source code can be downloaded from the GitHub folder of the developer:

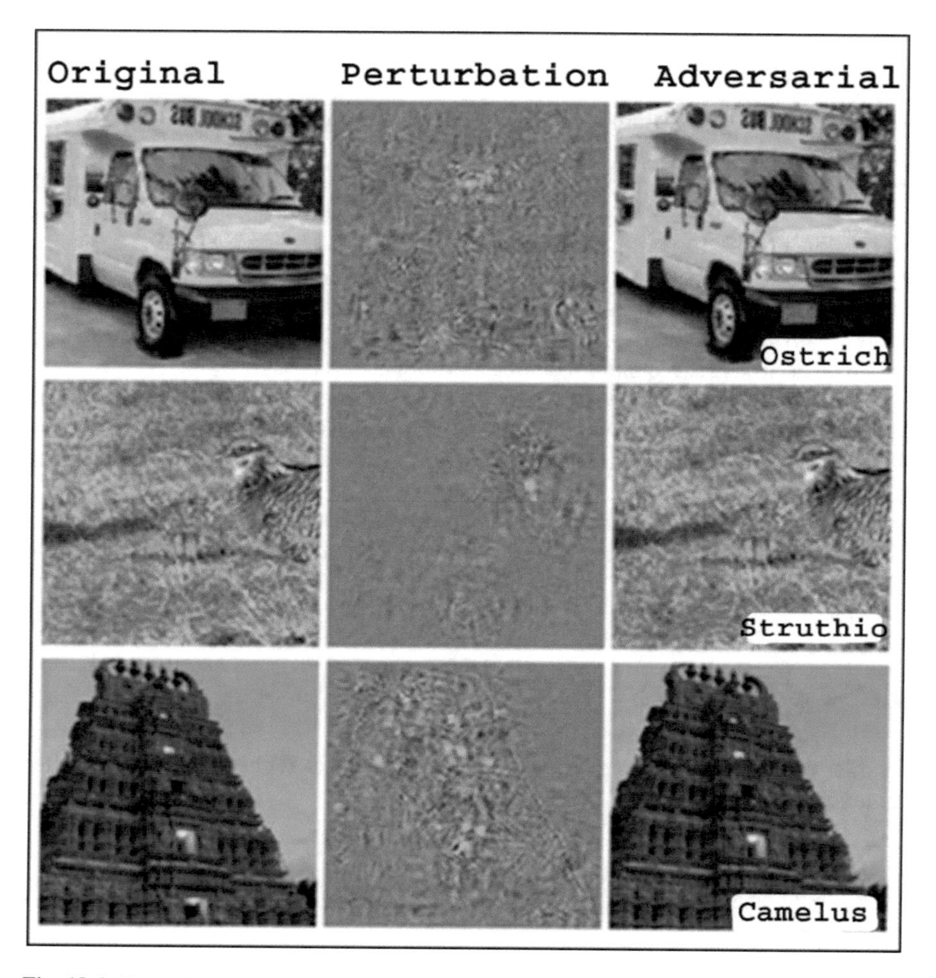

Fig. 13.6 Examples of adversarial created using AlexNet [10, 11]

https:/github.com/kenhktsui/adversarial examples). As can be seen from Fig. 13.7, the true data was categorized by the regression analysis into two groups.

4.1.3 One-Pixel Attack

An intense attack scenario on the opponent is when just a single pixel is modified in the picture to deceive the classifier. Su et al. [14] confirmed that three different types of neural network models were successfully fooled on 70–97 percent of the images tested by only changing a single pixel in each frame. Figure 13.8 shows that every image contains the appropriate label, and the resulting predicted label is shown in brackets.

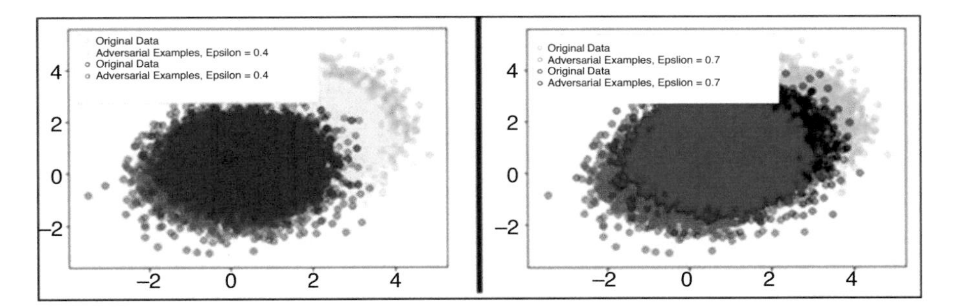

Fig. 13.7 Actual classification outputs and erroneous classifications for two distinct 0.4 and 0.7 epsilon values, respectively (as stated by Tsui in [13] application)

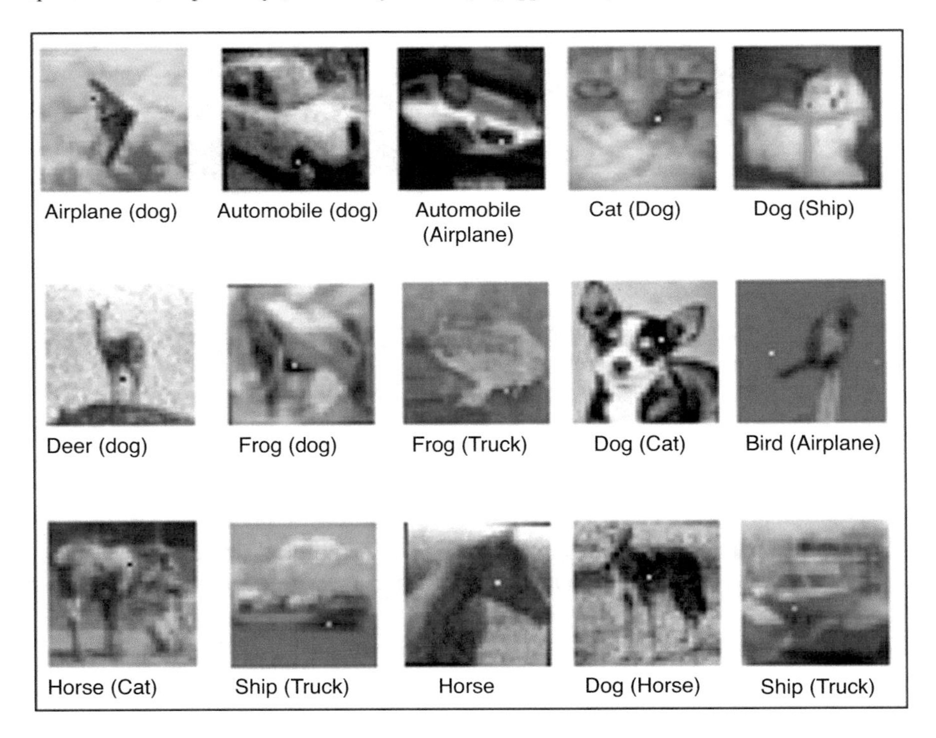

Fig. 13.8 Single-pixel adversarial attacks [14]

4.1.4 Carlini and Wagner Attacks (C&W)

In the wake of defensive distillation against the adversarial perturbations, Carlini et al. [15] introduced a set of three adversarial attacks [16] (Fig. 13.9).

Corresponding code for Nicholas Carlini and David Wagner's paper "Towards Evaluating the Robustness of Neural Networks," presented at

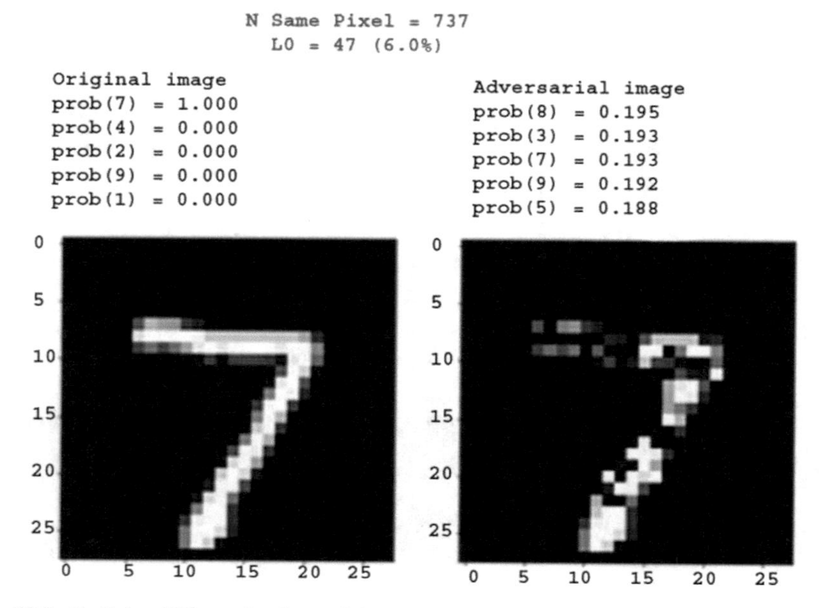

Fig. 13.9 Carlini and Wagner's adversarial attack on MNIST [15]

the IEEE Security and Privacy Symposium, 2017 is available online at https://github.com/carlini/nn_robust_attacks.

4.1.5 DeepFool

Dezfooli et al. [17] suggested iteratively calculating a minimum standard of adversarial disturbance for an input image. Their technique, that is, DeepFool, initializes the clean picture presumed to be located in an area confined to classifier decision limits.

5 Trustworthy Machine Learning for Industry

Breakthroughs in computational capacity have made online quantitative machine learning a useful and practical tool in many systems and networking domains to solve decision-making problems, spam filtering, virus detection, and network intrusion detection. A machine learning algorithms such as a support vector machine (SVM) and Bayesian learner can be useful in these domains [18].

5.1 Machine Learning Technique and Issues

In actual life, people are really going to classes, learning from lessons, and trying to offer some tests to show they've learned something right. Similarly, the machine learning techniques' pre-learning phase is performed along with training data, and few other databases have been used to check whether the model has been properly trained [19]. Because machine learning's logical and mathematical architecture will cause regularization, it means to remember the data pattern. In that case, the method is not effective because it has merely been memorized. It's similar again for the learning cycle of humans as if we remember something, it can be successful in the short term but it doesn't mean we're going to pass the main exam. When a machine learning approach works in the testing process as well, then it can be approved to be implemented in actual applications. Basically, it can be said for all machine learning techniques the process involves the "reading," "testing," and "download" phases (Fig. 13.10a) [20].

Some important issues in machine learning are data for learning and shaping the target problem. It is also essential to choose the most effective approaches for machine learning, as some mechanisms may not be successful on specific problems. Moreover, because machine learning techniques may use their own required features, selecting the most suitable value for these parameters is another problem. Eventually, some other problems that should be addressed are the total iterative process number for improved learning or a data preprocessing step for information collected (Fig. 13.10b) [20].

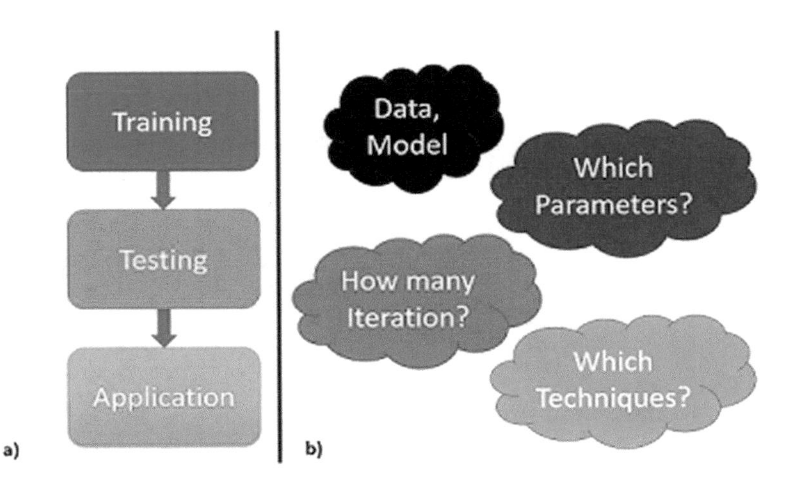

Fig. 13.10 (**a**) Machine learning cycle. (**b**) Machine learning issues [20]

5.2 Learning Paradigms

Data obtained from the different sources may not all be in the similar form, although the ML algorithm uses various learning steps to make them ready for use. At present, there are three different learning paradigms that can be defined as follows.

5.2.1 Supervised Learning

It is a type of learning process, where results are identified for each individual data input. In these, each given set of input cases is labeled with known target output classes for the problem under consideration, which means that a successfully trained model with supervised learning can forecast a result and also classify according to a newly discovered combination of inputs. Figure 13.11 reflects traditional (a) mathematical and visual regression and (b) nonlinear classification [21, 22].

5.2.2 Unsupervised Learning

Unsupervised learning occurs when results are not identified for the target data from which to learn. Basically, it calculates similitudes-distances between gathered data and group them into a certain similar number of clusters. Mathematically, similarity distance between data is calculated by Euclidean distance [23]. Figure 13.12 explains the traditional clustering method applied to a data set [24].

5.2.3 Reinforcement Learning

With potential robotic systems in particular, reinforcement learning is recognized as a primary learning method. Within that learning model, it is the duty of the

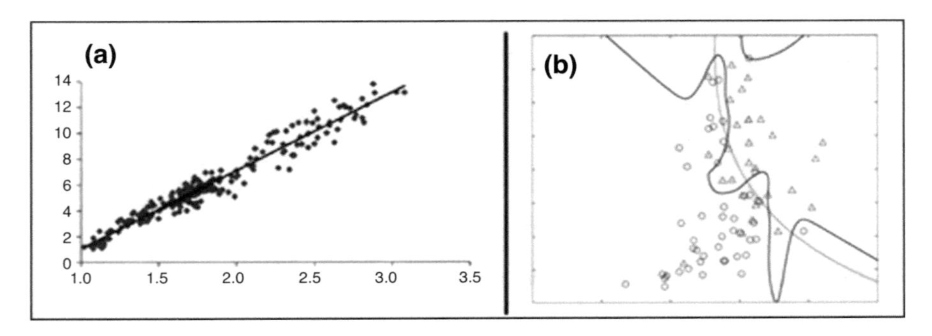

Fig. 13.11 (**a**) Linear regression. (**b**) Nonlinear classification with supervised learning [21, 22]

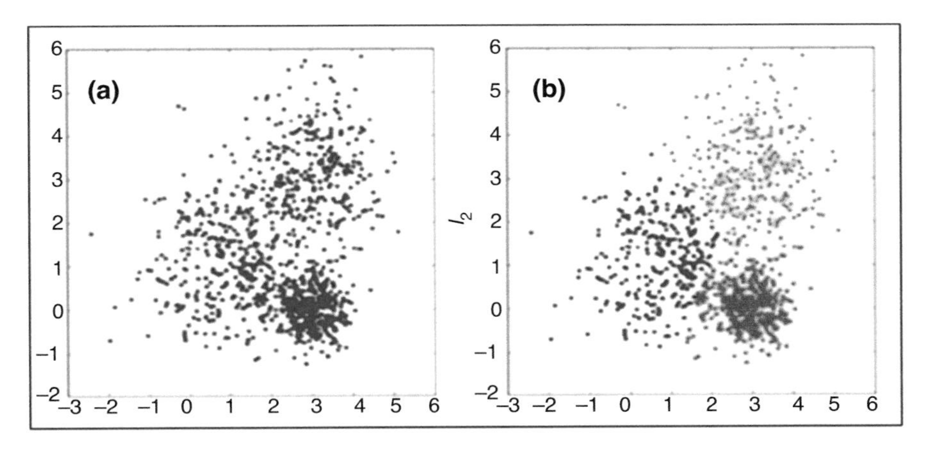

Fig. 13.12 (a) Row data scattered. (b) Cluster data using an unsupervised learning [21]

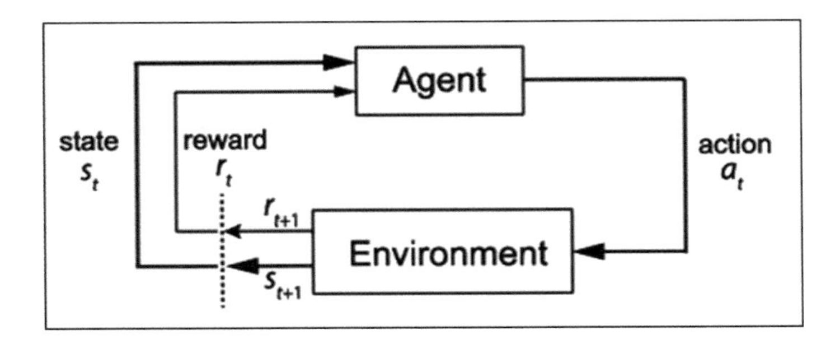

Fig. 13.13 General flow of reinforcement learning [25]

machine learning model to adjust some of its feature space against the effects of its behavior according to the feedback provided. Mathematically, this learning process includes estimating incentive values for such goal acts preparing for future considerations a scheme or data model of behavior. Figure 13.13 reflects a basic flow of reinforcement learning [25].

6 Understanding the Threat Landscape

Adversarial machine learning means designing ML algorithms that can withstand these sophisticated attacks and studying the attackers' limitations and capabilities.

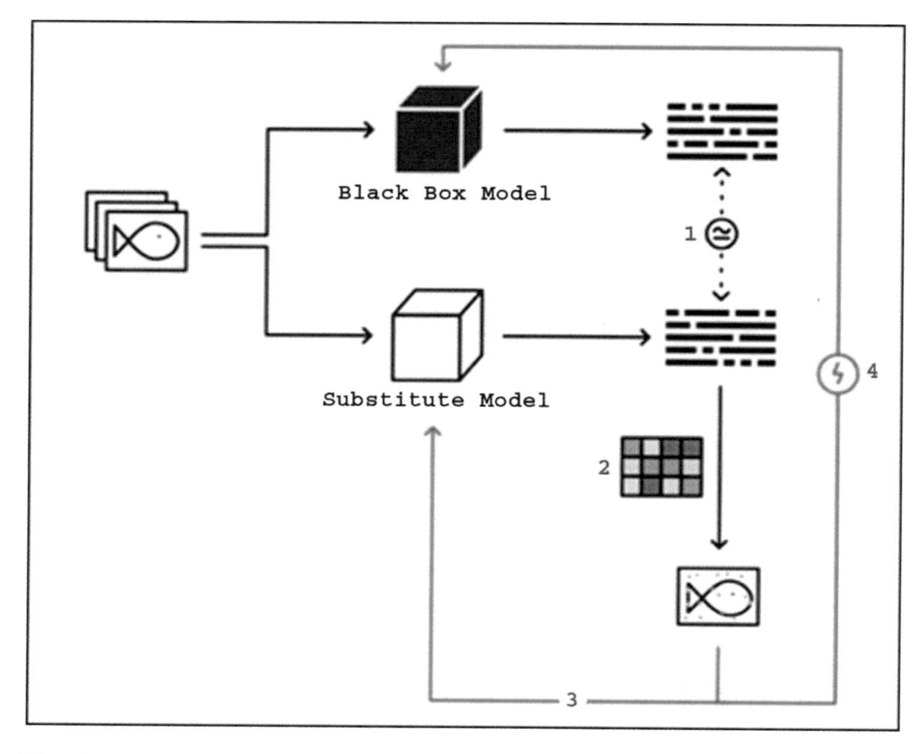

Fig. 13.14 Flow of the substitute attack [27]

6.1 Case Study: Anomalous Traffic Detection

Adversaries may use substitution attacks to interrupt regular user behavior and also to escape by causing the monitor to have a lot of false negatives via an integrity attack. Through doing so, these opponents will limit the risk of detecting their unauthorized activities. When the target machine learning framework is known, some data sets for the artificial training can be generated to allow for decision-making boundaries. As Papernot et al. tested, this technique of black-boxing is considered a substitute attack [26]. Figure 13.14 displays a clear flow of the substitute attack [27].

6.2 Real-World Attacks

Current study said that adversaries feed perturbed data to known models. In addition, the impact of the attacks is assessed using image feature data sets.

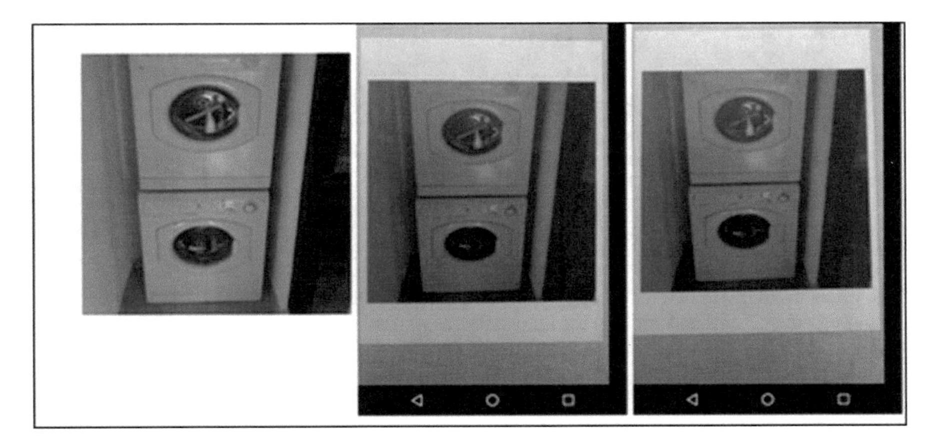

Fig. 13.15 (**a**) Data set image. (**b**) Clean image. (**c**) Adversarial image

Although those environments have proved yourself to be enough to persuade many investigators, this adversarial attack is a major concern for AI. In fact, we find this also from the literature [28, 29], situations where this issue is downplayed and where adversarial instances include real-time issues.

6.2.1 Attack on Cell Phone Camera

A. Kurakin, I. Goodfellow et al. [30] initially revealed that there are risks of threats and attacks even in the physical world as well. They identified adversarial photographs to demonstrate this and took screenshots from a cell phone picture. It has been shown that even when viewed through the camera, a significant fraction of images is misclassified. As Fig. 13.15 explains, the attack on camera pictures is seen here.

6.2.2 Attack on Road Sign

Building on the attacks proposed in [15, 31], Etimov et al. [32] modeled robust perturbations for the physical universe. We also showed the probability of effective attacks on physical environments, such as difference in angles of view, range, and solving. In this work, two attack groups of visible road sign posters were added: (a) poster, at which the attacker designs a distorted sign poster and positions it over the actual sign poster as shown in Fig. 13.16, and (b) sticker disturbance, where the road sign poster is located. The printing takes place on a ledger, and the ledger is stuck over the real label.

Fig. 13.16 A road sign poster attack [32]: 100 percent fooling LISA-CNN [32] classifier. Even indicating the distance and angle to the image. The classifier is educated for road signs using LISA data set [33]

7 Role of Blockchain for Industrial Automation

Blockchain offers a distributed, immutable, and trusted environment for data storage and transaction processing by using cryptographic concepts to sign, validate, and verify each transaction by miners. Blockchain provides the platform to smart contracts to get executed automatically using consensus algorithms. Currently, many researchers are combining blockchain in various fields like machine learning, IoT, cloud computing, big data analytics, etc. in order to secure the system in a decentralized manner.

Smart contracts attract many industries to automate many things and to eliminate third-party verification. Smart contracts build the trust among unknown entities by providing verification and validation of transactions by special entities called miners. Smart contracts with blockchain attract many industries like health insurance, healthcare, finance, governance, fraud detection, supply chain management, logistic management, natural disaster, self-sovereign identity, and real estate in order to automate various processes and improve efficiency of the overall system [34, 35]. Smart contracts are used in health insurance to automate the process of claiming insurance; for digitization of patient records for surgeries, organ transplant, and OPD in healthcare; to provide a trusted, reliable, and transparent environment to automate transactions in an open distributed system; etc. [35]. Figure 13.17 shows us the various use cases of blockchain where smart contracts can be used for automation, security, and transparency [34, 35] (Fig. 13.18). Table 13.1 shows the various approach that combines 5G, blockchain and IoT for Industrial automation.

7.1 Smart Home

Smart home automation is a prominent technology which aims to change the lifestyle of the people. The application of smartphones provides comfort to the user and also provides security by single access from phone. The architecture of smart

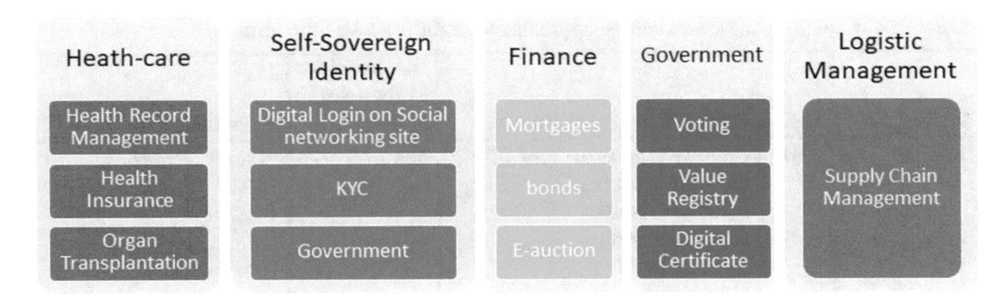

Fig. 13.17 Use cases of blockchain

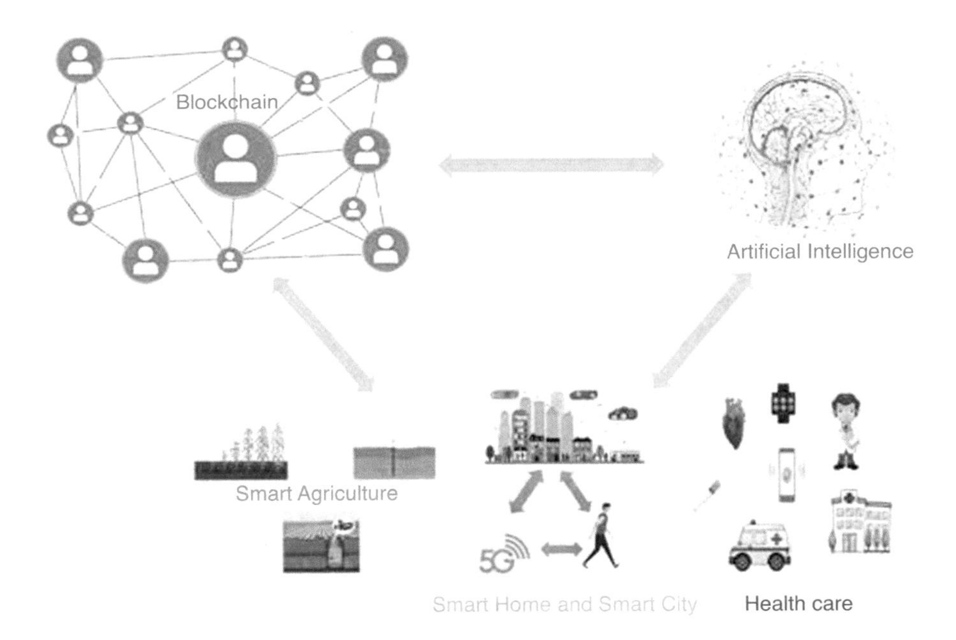

Fig. 13.18 Application of blockchain

homes consists of the network connectivity, sensor, and application. A smart door system plays an important role in the smart home automation. The data is stored in the centralized server so there are chances of the unauthorized person to access the data. To solve this problem, the idea of the blockchain is proposed. Blockchain provides the integrity and authentication to smart door applications. The sensors are used to detect the motion of the person entering into the house through a smart door. Due to blockchain implemented in the smart door, any intruder cannot access the data. Oftentimes IoT devices cannot address the issue; therefore, 5G technology overcomes the limitations.

Table 13.1 Comparison of blockchain approaches for industrial automation

Author	Year	Description	Blockchain technology used	Technique
Jong-ho Noh [37]	2019	Proposed the smart city with blockchain technology. Also listed the issues in the smart city project and solve the security problem using blockchain	Yes	5G and blockchain
Daniel Minoli [38]	2020	The challenges for the smart home with the blockchain for the IoT. The research needs to be done for making use of blockchain in the smart home automation	Yes	IoT and blockchain
Srinivas Jangirala [39]	2019	Limitation in the blockchain-enabled 5G authentication in the supply chain. Proposed an efficient protocol (LBRAPS)	Yes	5G, IoT, and blockchain
Mohamed Amine Ferrag [40]	2020	Proposed the security solution of IoT-based smart agriculture using blockchain	Yes	IoT and blockchain

7.2 Smart City

The migration of the people of villages to the cities increased the number of populations in the urban areas. This leads to the scarcity of the needs to water, food, electricity, etc. To address this problem, smart city projects are implemented which include the use of the IoT and cloud technology. In addition to this, blockchain technology is applied in smart cities. The role of the blockchain is to protect the data of the user from the malicious attack, and blockchain framework allows the entity to communicate in the smart city.

7.3 Supply Chain Management

The activities in the lifestyle of the product involved the organization; resources are the supply chain management. The chain starts from the development of the product to deliver at the user end. The impact of the blockchain and 5G increases the transmission of the product. Blockchain provides the integrity of the product from the development of the product to the end user. The smart contract is the main component of the blockchain which eliminates the use of the third party of having the decentralized network. Using 5G with blockchain, the tracking of products can be easily done.

7.4 Smart Agriculture

IoT is used nowadays in the agriculture field. Hence smart agriculture comes into the picture, the resultant to improve the product quality and quantity. Sensors can detect the temperature of the soil, water level, moisture, etc. Blockchain is used in the field of agriculture, and the main focus is on the food supply chain. The food supply is connected to agriculture as the raw materials from farmers are taken and used as a product in supply chain management.

7.5 Smart Healthcare

Healthcare is the most essential part in the entire world. With the rise in the population in the world, healthcare conditions are a crucial matter for the government. 5G-enabled IoT is the considered solution in the healthcare domain [36]. The remote system with sensor monitoring system of the patient is useful in healthcare.

7.6 AI and Machine Learning

AI and machine learning algorithms work on the data to train, learn, and make decisions, but data must be secured during the transaction, training, and testing. AI and machine learning work on the set of parameters which must be secured from the unauthorized changes; otherwise, it leads to adversarial effect. Many AI models work on some sensitive information like patient's health data, biometric data for authentication, etc. which must be secured from the intruders as well as to train and deploy the model built on these data using some statistical parameters that need to be secured to prevent adversarial effects. The intruder may manipulate the statistical parameters/environment variables, the devices from where data need to be collected, or the input samples. The attack can be performed on input samples by inserting noise by silently moving the decision boundaries during the training process [41].

Many AI-driven applications like recommendation systems, object detection systems, medical diagnosis, robotic surgeries, etc. use large scale of data which needs to be stored either at some central server or at some distributed environment. Storing data at the central server decreases the chances of external attacks on data and statistical parameters but has the risk of a single point of failure which redirects to adopt a distributed environment. Distributed environment offers many benefits in terms of efficiency, data sharing, and fault tolerance, but due to the involvement of multiple unknown parties, it leads to trust and security issues. Blockchain solves this problem by providing a tamper-proof distributed environment to build trust among unknown peers without including third-party authority.

In [42], the authors have proposed collaborative training which works in a distributed environment to detect the adversarial attack on any statistical attributes or input samples for convolutional neural network (CNN). This model provides security at network-level attack. The cryptographic concepts and decentralizing properties of blockchain are employed to the CNN model to provide security and accountability. By implementing a hash chain of blocks of CNN models and by hiding the parameters and network scenario, the prevention of threat to any white-box attack is possible. Due to the implementation of hash chain, tempering in any block effect the hash of the current and subsequent block can be easily detected.

In [41], the authors have proposed the secure and decentralized blockchain-based framework for explainable AI (XAI) which is the new and trending field of AI nowadays. This approach provides facilities to record and govern interactions and also allows consensus for predictions and their explanation through the smart contracts to provide security against some adversarial attacks. Blockchain in this framework offers decentralized, reliable, secure, and immutable storage which is built on the top of decentralized applications (DApps). Blockchain can provide transparency and visibility, immutability, traceability, and nonrepudiation using the smart contracts for an explainable AI system.

In [43], the authors proposed the blockchain-based solution for biometrics recognition systems. They covered the traditional architecture of biometrics recognition systems; vulnerabilities at every level of architecture such as feature extraction, template matcher, etc.; and possible solutions for the same. The proposed solution gives alert when alteration occurs at any component of the system. The time complexity of the proposed solution is higher due to complex cryptographic techniques being included to improve the security level.

The integration of blockchain technology with AI models improves security, efficiency, and auditability. It also improves trust among peers for decision-making and fault tolerance. AI enumerates intelligence into the machines over the decentralized blockchain network (Table 13.2).

7.7 *Case Study*

With the demand of Internet-enabled devices, IoT is also in demand to offer flexible and easy lifestyle to the customers nowadays in the field of healthcare, grids, cities, agriculture, etc. Increasing demand of IoT also increases the requirement of automatic decision-making and intelligent machine to make accurate and efficient decision-making by processing the data collected through various sensors that can be achieved by ML algorithms. IoT network uses resource-constrained devices with smaller memory and smaller computation power that leads to the usage of cloud-based infrastructure for processing of data on cloud-based environment. Various

Table 13.2 AI/ML techniques used in 5G-enabled IoT with key features and advantages

Types of AI/ML methodology	Key techniques	How it's helpful in adversarial attack	Advantages
Classification algorithms-supervised learning	Linear classifiers Logistic regression Support vector machines Naive Bayes classifier Nearest neighbor Neural networks Decision tree Etc.	Monitoring network parameters such as throughput and network fault logs to detect anomalies	Flexible modeling of algorithms with emerging features Agile and self-evolving nature of frameworks for defense
Clustering algorithm-unsupervised learning	Gaussian mixture model K-means Dimensionality reduction Markov decision model	Categorize risks and loopholes of different kinds in network security	Extremely complex data sets for automatic clustering Real-time data discovery
Reinforcement learning	Q learning Gaming algorithm Robot navigation Deep Q learning.	Model learns from the environment Adaptive in nature No training and testing threshold	Highly robust Agent learns from action taken and reward received

machine learning techniques can be used to train the data collected by various sensors to offer intelligence to machines for automatic decision-making. In the area of healthcare, supply chain management, cities, agriculture, etc., many latency-sensitive applications ask for higher bandwidth which can be achieved by 5G network. As seen in the above sections, machine leaning models which use the same set of parameters for training and testing are vulnerable to adversarial attacks which can be attempted by changing the parameter values. Moreover, privacy and lack of control are the major issues in the cloud-based environment. Blockchain is the promising solution which offers security in terms of integrity and transparency. Figure 13.19 shows us the scenario of integration of blockchain with 5G-enabled IoT system which uses cloud infrastructure for processing and storing data.

Here the scenario of embassy which approves the visa of various countries after health examination of the person is taken to show the role of blockchain in integration of 5G-enabled IoT with machine learning through cloud environment.

In traditional visa approval process, medical examination plays an important role. The embassy-certified doctor can examine the person by reviewing immunization and medical history of the person. The patient needs to go through various medical tests like tuberculosis test, urine and blood test, etc. for the fulfillment requirement to get visa of various countries. Nowadays, IoT-based smart healthcare devices can

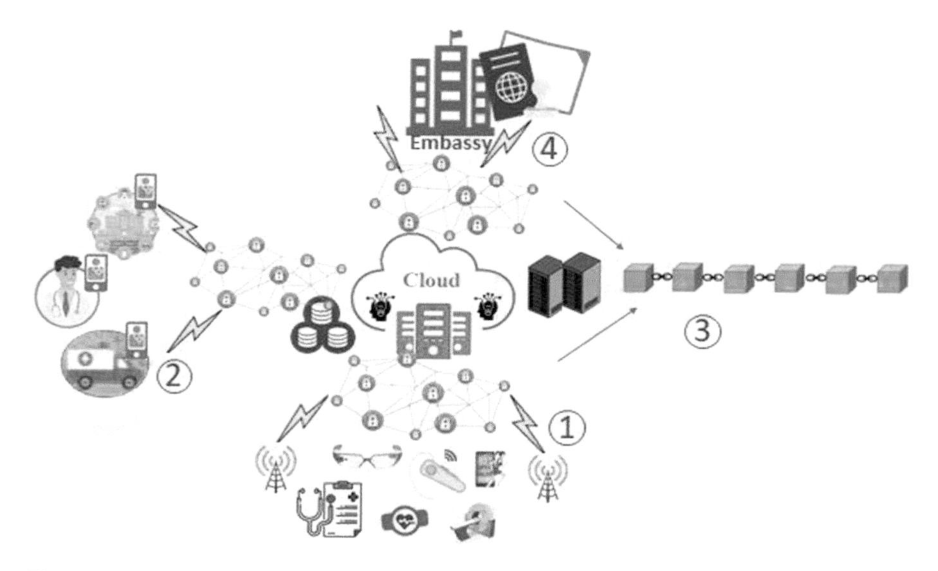

Fig. 13.19 Blockchain-based smart heath examination system for visa approval

examine the person and send the data to cloud for processing. The cloud-based system can share the data received from sensors with doctors and embassy officers for reviewing and visa approval. However, the process of data sharing through cloud saves time and money; the adversarial attack can be easily applied due to lack of control in cloud.

The whole system can be moved on blockchain-based environment with 5G as a communication protocol to achieve integrity, transparency, and higher bandwidth. The scenario is depicted in Fig. 13.19.

8 Challenges of Integrating Adversarial Artificial Intelligence Techniques to Secure 5G-Enabled IoT

The new advances in 5G networks promote the immersive growth of data connectivity through higher data speeds. Such a big rise in data traffic and connected devices suggests further bugs, risks, and attacks that result in financially devastating harm. Artificial intelligence (AI) and machine learning (ML) are intended to play a key role in solving difficult problems of optimization in this context. There is a list of challenges to integrate it with 5G-enabled IoT such as:

- High Bandwidth Demand – The vulnerability and privacy environment from personal devices to the service provider network has also been broadened by increased bandwidth, higher spectrum usage, and fast data speeds in 5G networks. The network should also be clever enough to solve these issues in real time; AI/ML approaches should help model these robust complex algorithms that can help recognize network problems and offer real-time solutions to them which is still challenging.
- Higher-Level Network Heterogeneity – Smart cars, smart houses, smart buildings, and smart cities are supported by 5G-enabled IoT networks. It would require more stable and flexible strategies to deal with essential security problems on both the network and system sides. Intrusion detection using AI/ML by classifying the unauthenticated links would be useful to offer security solution. 5G IoT security and safety can span all levels, such as defense of identity, safety, and machine-to-machine communication. In key authentication, AI techniques found an essential role along with effectively mitigating masquerading attacks.
- Affordable Infrastructure and Lower Cost of Computing – AI/ML consume huge amount of data and train itself. It takes a significant amount of time to process the data collected from IoT devices and store into the cloud. Still optimization is the biggest challenging in the existing algorithm to improve the performance in real time.

9 Conclusion

In this chapter, artificial intelligence has a significant role in automation. Industry 4.0 is the current requirement to develop smart solutions securely. IoT-enabled devices collect huge data and store it over the server. Every transaction between the user to machine and machine to machine can be secured by blockchain technology in terms of integrity. Adversarial machine learning is recent research issues to identify the vulnerability in model and noisy data. Machine learning paradigms provide different learning mechanisms depending on the type of data, continuous data, categorical data, or image data. Types of attack on deep learning algorithms and machine learning algorithms are open issues in computation study. A case study covered in this chapter will be helpful to understand the real-time attack on traffic direction and road signal identification. Some of the researchers have proven the impact of attack in Industry 4.0. To make it 5G-based, IoT-enabled blockchain-based system can be adopted to bypass the various types of attacks on the system. For the enhancement in study, a small survey can be carried out by readers to aware industry to check if their automation systems are preventing attacks and safe for data alteration.

References

1. R.R. Reddy, C. Mamatha, R.G. Reddy, A review on machine learning trends, application and challenges in internet of things, in *2018 International Conference on Advances in Computing, Communications and Informatics (ICACCI)*, (IEEE, Piscataway, 2018)
2. S. Tanwar, S. Tyagi, N. Kumar, *Multimedia Big Data Computing for IoT Applications: Concepts, Paradigms and Solutions, Intelligent Systems Reference Library* (Springer Nature, Singapore, 2019), pp. 1–425
3. K. Thakkar, R. Thakor, P.K. Singh, M-tesla-based security assessment in wireless sensor network, International Conference on Computational Intelligence and Data Science (ICCIDS 2018), NorthCap University, Gururgram, 07–08th April, Procedia, Computer Science, Elsevier, 1154–1162 (2018)
4. S. Tanwar, J. Vora, S. Kaneriya, S. Tyagi, Fog based enhanced safety management system for miners, in *3rd International Conference on Advances in Computing, Communication & Automation, (ICACCA-2017)*, (Tula Institute, Dehradhun, 2017), pp. 1–6
5. I. Bhudiraja, S. Tyagi, S. Tanwar, N. Kumar, et al., Tactile internet for smart communities in 5G: An Insight for NOMA-based Solutions. IEEE Trans. Ind. Inf. **15**(5), 3104–3112 (2019)
6. S. Nakamoto, Bitcoin: A peer-to-peer electronic cash system. bitcoin.org (2008)
7. S. Li, L.D. Xu, S. Zhao, 5G internet of things: A survey. J. Ind. Inf. Integr. **10**, 1–9 (2018)
8. M. Agiwal, N. Saxena, A. Roy, Towards connected living: 5G enabled Internet of Things (IoT). IETE Tech. Rev. **36**(2), 190–202 (2019)
9. I. Mistry, S. Tanwar, S. Tyagi, N. Kumar, Blockchain for 5G-enabled IoT for industrial automation: A systematic review, solutions, and challenges. Mech. Syst. Sig. Process. **135**, 106382 (2020)
10. C. Szegedy, W. Zaremba, I. Sutskever, J. Bruna, D. Erhan, I. Goodfellow, R. Fergus, Intriguing properties of neural networks. arXiv preprint arXiv:1312.6199 (2013)
11. A. Krizhevsky, I. Sutskever, G.E. Hinton, Imagenet classification with deep convolutional neural networks, in *Advances in Neural Information Processing Systems*, (Morgan Kaufmann Publishers, San Mateo, 2012)
12. I.J. Goodfellow, J. Shlens, C. Szegedy, Explaining and harnessing adversarial examples. arXiv preprint arXiv:1412.6572 (2014)
13. K. Tsui, Perhaps the simplest introduction of adversarial examples ever, Medium.com, 21 Aug 2018 [Online]. Available: https://towardsdatascience.com/perhaps-the-simplest-introduction-of-adversarial-examples-ever-c0839a759b8d. [Accessed 2020]
14. J. Su, D.V. Vargas, K. Sakurai, One pixel attack for fooling deep neural networks. IEEE Trans. Evol. Comput. **23**(5), 828–841 (2019)
15. N. Carlini, D. Wagner, Towards evaluating the robustness of neural networks, in *2017 IEEE Symposium on Security and Privacy (SP)*, (IEEE Computer Society, Los Alamitos, 2017)
16. N. Papernot, P. McDaniel, X. Wu, S. Jha, A. Swami, Distillation as a defense to adversarial perturbations against deep neural networks, in *2016 IEEE Symposium on Security and Privacy (SP)*, (IEEE, Piscataway, 2016)
17. M. Dezfooli, S. Mohsen, A. Fawzi, P. Frossard, Deepfool: a simple and accurate method to fool deep neural networks, in *Proceedings of the IEEE Conference on Computer Vision and Pattern Recognition*, (IEEE, Piscataway, 2016)
18. N. Cristianini, J. Shawe-Taylor, *An Introduction to Support Vector Machines* (Cambridge University Press, Cambridge, 2000)
19. S. Lawrence, C.L. Giles, Overfitting and neural networks: Conjugate gradient and backpropagation, in *Proceedings of the IEEE-INNS-ENNS International Joint Conference on Neural Networks. IJCNN 2000. Neural Computing: New Challenges and Perspectives for the New Millennium*, (IEEE Service Center, Piscataway, 2000)
20. U. Kose, Techniques for adversarial examples threatening the safety of artificial intelligence based systems. arXiv preprint arXiv:1910.06907 (2019)

21. R. Gupta, S. Tanwar, S. Tyagi, N. Kumar, Machine learning models for secure data analytics: A taxonomy and threat model. Comput. Commun. **153**, 406–440 (2020)
22. S. Yang, S. Cai, F. Zheng, Y. Wu, K. Liu, M. Wu, Q. Zou, J. Chen, Representation of fluctuation features in pathological knee joint vibroarthrographic signals using kernel density modeling method. Med. Eng. Phys. **36**(10), 1305–1311 (2014)
23. S.K. Halgamuge, L. Wang, *Classification and Clustering for Knowledge Discovery*, vol 4 (Springer Science, Cham, 2005)
24. A. Gattal, F. Abbas, M.R. Laouar, Automatic parameter tuning of K-means algorithm for document binarization, in *Proceedings of the 7th International Conference on Software Engineering and New Technologies*, (ACM, New York, 2018)
25. R.S. Sutton, A.G. Barto, *Introduction to Reinforcement Learning*, vol 135 (MIT press, Cambridge, 1998)
26. R. Gupta, S. Tanwar, S. Tyagi, N. Kumar, Tactile internet and its applications in 5G Era: A comprehensive review. Int J Commun Syst **32**(14), 1–49 (2019)
27. R.R. Wiyatno, Tricking a machine into thinking you're milla jovovich and other types of adversarial attacks in machine learning, Medium, 9 August 2018 [Online]. Available: https://medium.com/element-ai-research-lab/tricking-a-machine-into-thinking-youre-milla-jovovich-b19bf322d55c. [Accessed 2020].
28. J. Lu, H. Sibai, E. Fabry D. Forsyth, No need to worry about adversarial examples in object detection in autonomous vehicles. arXiv preprint arXiv:1707.03501 (2017)
29. A.A.R. Graese, T.E. Boult, Assessing threat of adversarial examples on deep neural networks, in *2016 15th IEEE International Conference on Machine Learning and Applications (ICMLA)*, (IEEE, Piscataway, 2016)
30. A. Kurakin, I. Goodfellow, S. Bengio, Adversarial examples in the physical world. arXiv preprint arXiv:1607.02533 (2016)
31. Y. Liu, X. Chen, C. Liu, D. Song, Delving into transferable adversarial examples and black-box attacks. arXiv preprint arXiv:1611.02770 (2016)
32. K. Eykholt, I. Evtimov, E. Fernandes, B. Li, A. Rahmati, C. Xiao, A. Prakash, T. Kohno, D. Song, Robust physical-world attacks on deep learning visual classification, in *Proceedings of the IEEE Conference on Computer Vision and Pattern Recognition*, (IEEE Service Center, Piscataway, 2018)
33. J.A. Alzubi, B. Bharathikannan, R. Manikandan, A. Khanna, C. Thaventhiran, Boosted neural network ensemble classification for lung cancer disease diagnosis. Appl. Soft Comput. **80**, 579–591 (2019)
34. B.K. Mohanta, S.S. Panda, D. Jena, An overview of smart contract and use cases in Blockchain technology, in *2018 9th International Conference on Computing, Communication and Networking Technologies (ICCCNT)*, (IEEE, Piscataway, 2018)
35. S.R. Mani Sekhar, G.M. Siddesh, S. Kalra, S. Anand, A study of use cases for smart contracts using Blockchain technology. Int. J. Inf. Syst. Soc. Change **10**(2), 15–34 (2019)
36. K. Sheth, K. Patel, H. Shah, S. Tanwar, R. Gupta, N. Kumar, A taxonomy of AI techniques for 6G communication networks. Comput. Commun. **161**, 279–303 (2020)
37. J.-H. Noh, H.-Y. Kwon, A study on smart city security policy based on Blockchain in 5G age, in *2019 International Conference on Platform Technology and Service (PlatCon)*, (IEEE, Piscataway, 2019)
38. D. Minoli, Positioning of Blockchain mechanisms in IoT-powered smart home systems: A gateway-based approach. Internet Things **10**, 100147 (2020)
39. S. Jangirala, A.K. Das, A.V. Vasilakos, Designing secure lightweight blockchain-enabled RFID-based authentication protocol for supply chains in 5G mobile edge computing environment. IEEE Trans. Ind. Inf. **16**(11), 7081–7093 (2020)
40. M.A. Ferrag, L. Shu, X. Yang, A. Derhab, L. Maglaras, Security and privacy for green IoT-based agriculture: Review, blockchain solutions, and challenges. IEEE Access **8**, 32031–32053 (2020)
41. M. Nassar, K. Salah, M.H. Rehman, D. Svetinovic, Blockchain for explainable and trustworthy artificial intelligence. Wiley Interdiscip. Rev.: Data Min. Knowl. Discovery **10**(1), 1340 (2020)

42. A. Goel, A. Agarwal, M. Vatsa, R. Singh, N. Ratha, DeepRing: Protecting deep neural network with blockchain, in *Proceedings of the IEEE Conference on Computer Vision and Pattern Recognition Workshops*, (IEEE, Piscataway, 2019)
43. A. Goel, A. Agarwal, M. Vatsa, R. Singh, N. Ratha, Securing CNN model and biometric template using Blockchain, in *IEEE 10th International Conference on Biometrics Theory, Applications and Systems (BTAS)*, (IEEE, Piscataway, 2019)

Chapter 14
Machine Learning, Data Mining, and Big Data Analytics for 5G-Enabled IoT

Puneet Kumar Aggarwal, Parita Jain, Jaya Mehta, Riya Garg, Kshirja Makar, and Poorvi Chaudhary

1 Introduction

The evolving of fifth era (5G) networks is turning out to be all the more promptly available as a significant driver of the development of Internet of Things (IoT) applications. Agreeing to the International Data Corporation (IDC) report, the worldwide 5G administrations will drive 70% of organizations to burn through $1.2 billion on the network the management arrangements. For the new applications and models in the future, IoT require new execution rules such as huge network, security, dependability, inclusion of remote correspondence, super-low latency, throughput, super-reliability, etc. for a colossal number of IoT gadgets. To meet these necessities, the developing Long-Term Evolution (LTE) and 5G innovations are relied upon to give new availability interfaces to the future IoT applications. The advancement of up and coming age of "5G" is at its beginning phase, which focuses on new radio access innovation (RAT), reception apparatus upgrades, utilization of higher frequencies, and re-architecting of the networks. Nonetheless, principal advances have been made, and the development of LTE should be supplemented with an extreme change in the following years in the essentials of remote networking – a generational move in innovation, structures, and business measures. Wireless technology has seen rapid growth in recent years. 4G/LTE is one of the most successful mobile communication networks. It is created in such a way that it is

P. K. Aggarwal (✉)
ABES Engineering College, Ghaziabad, Uttar Pradesh, India

P. Jain
KIET Group of Institutions, Ghaziabad, Uttar Pradesh, India

J. Mehta · R. Garg · K. Makar · P. Chaudhary
HMRITM, Delhi, India

S. Tanwar (ed.), *Blockchain for 5G-Enabled IoT*,
https://doi.org/10.1007/978-3-030-67490-8_14

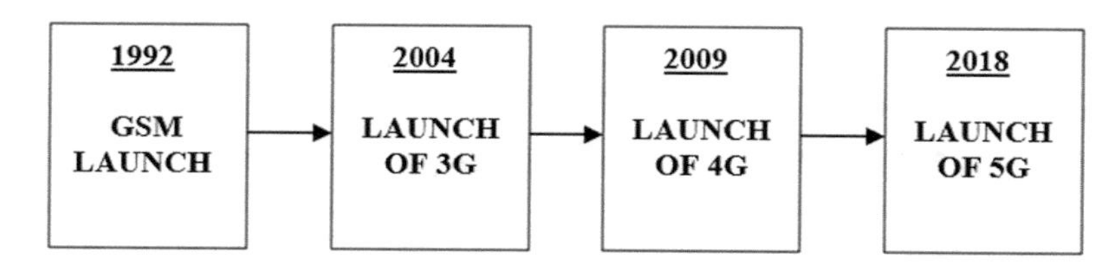

Fig. 14.1 Evolutions of 5G technologies

capable enough to fulfill some of the most critical 5G requirements. The evolution of 5G technology can be seen in Fig. 14.1.

The IoT worldview is both progressive and an empowering influencer of robotized and advantageous ways of life for advanced people. The development of the IoT can be credited to a conversion in progresses that took place over the previous decade in correspondence, computing, and designing applications. The subsequent influence of IoT has extended quickly to cover the entire human race. IoT gadgets are used to encourage human's everyday exercises which incorporates the PDAs, home assistants, smart brilliant vehicles, building computerization frameworks involving smart lifts and temperature control frameworks. Furthermore, automated airborne vehicles, for example, drones for ecological observing and relaxation are also used as IoT gadgets.

The huge scope multiplication of IoT gadgets stretch past these gadgets to the storage centers, for example, backend cloud frameworks which are topographically scattered. Thus, a huge volume of information is created by IoT gadgets and their supporting stages, for move and ensuing storage and preparing at backend distributed storage communities. IoT gadgets produce a steady stream of crude information, which can't be perceived for important information except if the information is prepared through utilization of strategies, for example, information discovery, that is, through data mining and machine intelligence. The heterogeneity of the information produced from different IoT organizations is reliant on the application space, involving smart medical care, online media, e-agribusiness, e-wellbeing, savvy power lattices, and shrewd vehicles. IoT gadgets are planned with custom conventions that consider the asset obliged nature of these gadgets, so as to save power utilization related with gadget activities.

IoT gadgets produce a huge volume of information that is privately prepared, in a restricted way, and also moved to a concentrated processing hub or a cloud storage, where it tends to be additionally prepared or dissected to create information. Machine learning (ML) is characterized as a group of methods for dissecting information, wherein the cycle of model structure on preparing information is mechanized, i.e., expects almost no human mediation. Thusly, the way toward classifying information into different classes is completely mechanized. The function of big data analytics in IoT information handling can't be downplayed. It allows information extraction with the help of ML algorithms. Machine learning algorithms provides an extremely solid support in speedy preparing of enormous volume information rising up out of IoT gadgets, which helps in creating patterns that is important from analytics point of view [24].

1.1 Research Contributions

The existing 4G networks have been broadly utilized in the IoT and are constantly developing to coordinate the necessities of things to come in future IoT applications. The 5G networks are relied upon to monstrous extend the present IoT that can support cell operations, IoT security, and network difficulties and driving the Internet future to the edge. The current IoT arrangements are confronting various difficulties, for example, huge number of connection of hubs, security, managing data, extracting knowledge, performing data analytics, and new standards for the 5G-enabled IoT. The 5G-empowered IoT (5G IoT) will interface monstrous number of IoT gadgets and make commitments to satisfy market need for remote administrations to invigorate new monetary and social improvement. The new necessities of applications later on IoT and the advancing of 5G remote innovation are two huge patterns that are driving the 5G-empowered IoT. The convergence of big data, data management, IoT platforms, and machine intelligence is empowering the next advancement of data analytics wherein venture will acknowledge critical substantial and elusive advantages from IoT information.

This chapter surveys the current research state of the art of 5G IoT; key empowering technologies including ML, data mining, and big data analytics; and primary exploration patterns and difficulties in 5G IoT.

1.2 Motivation for the Chapter

Presently, various IoT gadgets exploit networks, for example, the third era (3G) and 4G LTE, to keep up their network and their association with the cloud server centers. With the exponential development of information delivered by progressively enormous quantities of IoT gadgets, a few of the serious issues remain unsolved in application development. The 5G communication framework has been presented with the abilities of high-throughput, low-latency, and high unwavering quality. These capacities can empower countless gadgets, with best quality of service (QOS) arrangement of omnipresent network solutions for satisfying their various IoT application necessities. 5G can possibly permit the arrangement of more web-associated gadgets without worry that current issues would be exacerbated by a packed network. The rapid and dependable network supported by 5G will make additional opportunities for IoT services far beyond those accessible today.

Moreover, the empowering advancements of 5G can possibly introduce another IoT era, planning to easily and deftly uphold heterogeneous IoT applications with particular business qualities under countless smart gadgets. Also, the 5G-enabled IoT will bring a rich wellspring of enormous information. The amazing part of enormous information examination in 5G will without a doubt profit IoT progression.

However, how the empowering advances in 5G whether coordinated into the entire or as an aspect of a framework can consistently fuel the IoT unrest is a challenge. This raises new contemplations of network plan, asset arrangement, the board, nature of involvement, and guideline of 5G-enabled IoT. It is, hence, a basic significance to devise novel solutions by planning keen 5G-empowered IoT ideal models incorporated with the empowering advances of 5G. The IoT has been set up as another cross order research subject requiring the expectation of the specialized and reasonable difficulties looked by blended researchers considering that cross different disciplines.

1.3 Organization of the Chapter

The rest of the chapter is organized into different sections. Section 2 discusses about the concept of data mining and how it is useful for discovering information and hidden patterns from large data sets. In Sect. 3, the concept of machine learning is explored to understand how it will support IoT devices to operate in a better manner, creating a pleasanter lifestyle for society. Then in Sect. 4, big data analytics is explained to understand how the variety of data collected gets gathered, stored, and processed. In Sect. 5, the convergence of data mining, machine learning, and big data analytics gets done, and how it helps in 5G IoT is explained. In Sect. 6, two case studies based on this convergence are explained to understand actually how all the technologies are an advantage to the society. Then in Sect. 7, conclusions are drawn from the chapter. Finally, references are at the last.

2 Data Mining

Data mining (DM) is a process of taking out useful information from a large raw data set which in turn affects the decision-making process [10]. It is employed for the discovery of new information and patterns hidden inside that large data set by including the use of the intersection of artificial intelligence (AI), ML, and statistical analysis as shown in Fig. 14.2. It is considered as a field which is a combined form of computer science and statistics with the basic aim of extracting information and representing it adequately for further use [11].

It also involves a database, data pre-processing, models, complexity, post-processing structures, and online updating. In a nutshell, data mining is the process of digging through a large number of data sets by using different levels of human rules and algorithms to extract hidden information and obtain unique and helpful future outcomes as well. This process of searching includes discovering of

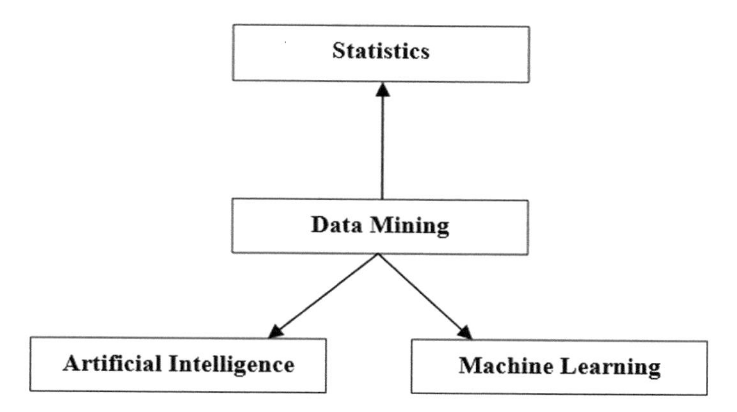

Fig. 14.2 Data mining using AI and ML

aberrations, different kinds of patterns, and relationships among the data which can be further used in different ways. The aim of any data mining procedure is mainly to prepare efficient as well as effective prediction models that even generalize the data. The process of data mining mainly consists of five stages:

Stage 1: Data Collection and Data Loading. This is the first step in the process of data mining where necessary and useful data is collected and loaded in a data warehouse.

Stage 2: Data Storage and Data Administration. After the data collection process, now comes the next important stage, i.e., data storage and its efficient management. The information is mainly stored on the servers or by using the cloud servers.

Stage 3: Information Association. This is the stage where the role of business annotators, administrative teams, and information technology sector comes into play. They have complete access to the data and provide an efficient as well as an effective way to organize a large amount of data.

Stage 4: Data Sorting. Data sorting is performed by the application software based on user results.

Stage 5: Data Representation. In this step, the data is presented in the most user-friendly manner that is easy to read and gather information in less time by the end-user by involving various charts, graphs, or tables. The data mining process further breaks down to mainly three models which are as follows:

(i) Definitive Modeling: It discloses all the similarities and bunches of already existing historical data to understand the antecedents behind success or failure, for example, distinguishing clients by product choices or sentiment.
(ii) Predictive Modeling: It mainly focuses on the classification of the events that will happen shortly, for example, client churn and campaign demand and response.

(iii) Prescriptive Modeling: This modeling deals with internal and external variables and restrictions to study further the course of action, for example, marketing offer to the client.

Information extraction offers the way to deal with a wide measure of machine created and regularly unstructured information. In this manner, huge information innovations and prescient examination permit smoothing out of industrial processes. In this way, the combination of the two sees that there is an upgrade in choice making which is the sole motivation behind both unstructured information and IoT advances. Current studies explore how information mining can take care of information on control and the executive modules.

2.1 Data Mining in 5G-Enabled IoT

Currently, DM is viewed as one of the basic elements for the up and coming age of versatile networks. Through exploration and information investigation, there are desires that multifaceted nature of these networks will be survived and it will be conceivable to do dynamic administration and activity exercises. In order to fully grasp the specifics of 5G network, there are particular sort of data that ought to be assembled by network segments so as to be investigated by an information mining plan. Versatile networks are now essential for our everyday lives; however, what we anticipate from them is ever-evolving. Changing the present networks to 5G is vital to staying up with the requests of a developing network society, where openings range new high-data transmission applications, low inactivity IoT administrations, and beyond. 5G means to fulfill these advancing needs by giving pervasive network for any gadget or application that may profit.

DM is used to extract the hidden information from the data generated. Since IoT promises to provide the caliber in various sectors such as smart cities for the waste management or efficient traffic control or smart homes for monitoring of people or collection of data through the sensors, therefore different data require different analysis techniques [12, 15, 16]. Data mining technique handles not only various manifolds of IoT data and data in bulk but also the speed at which the information is produced. Data mining for the IoT brings certain advantages with some challenges as well which are explained as follows:

(i) *Data admittance and extraction from different locations:* When dealing with the data which is in bulk, having a lot of noise is itself a challenging issue, but detecting a fault and correction is even more difficult. Modification of the data mining algorithm in big data is a tough task.
(ii) *Mining of partial data:* The second challenge is all about mining uncertain data for large data applications. For this, data effectiveness and data security is the most important aspect which needs to be taken into consideration while

transferring the information from various applications such as medical records or transaction records.

In a nutshell, the IoT turns out to rise from the need to automate, investigate, and cope with all the appliances, instruments, and sensors. To frame decisions effectively, IoT and data mining technologies are integrated. This integration supports the optimization of the system.

In the recent years, it has been seen an enormous exertion placed over the span of planning the fifth generation of versatile networks (5G). The development of 5G networks has been pointed toward giving tailor-cut answer for various types of businesses especially the intelligent industries, telecommunication sectors, and wellbeing part and even in smart production lines. On the other hand, the logical network has understood that huge information arrangements can essentially improve the activity and the executives of both current and future portable networks. Normally, DM is utilized throughout finding examples and connections between various factors especially in huge informational indexes. Using factual investigation, AI and ML are utilized in the information set throughout separating essential information from the inspected information.

Data mining is considered to be the most important and valuable aspect for the future generation of networking in mobile. It has proved to be an integral part of 5G technology as it facilitates the process of framing decisions. 5G has helped in evolving many industries mainly the telecommunication sector, transportation industries, and health sector and in smart factories as well [12–14]. Moreover, the big data and predictive analysis have made it even simpler for the execution of these industrial processes. Therefore, the sole purpose of the integration of the two is fulfilled as it provides an effective and enhanced level of decision-making.

It has not only helped in overcoming the complexity issues in networking; rather, it provides a solution to carry out operational activities as well as robust administration. Moreover, the world is alchemizing at a much faster pace and so do the demands and expectations in terms of the mobile network as well. Therefore, the transformation of today's network to 5G is the need of the hour. It adds a great advantage to the existing network such as improving the properties like that of high bandwidth application, less inactive state IoT services, and many more by providing common connectivity. It has been analyzed that nearly 6 years from now, there will be nearly a 34 percent hike in revenues.

Recent research and studies have an entire knowledge and analysis of the fact regarding how data mining handles the control and management techniques. However, along with the success, there come challenges that need to be taken into consideration. The only focus of the research is based on the study of the data mining algorithm to make the best use of it in mobile networking and communication and not on explaining detailed information and examples of the data to be gathered. Also, the number of times of data collection and the various mechanisms of data minimization that need to be exchanged among the network are unknown.

2.2 Applications of Data Mining

Various applications of data mining include [10–12]:

(i) *Descriptive Modeling:* It discloses all the similarities and bunches of already existing historical data to understand the antecedents behind success or failure, for example, distinguishing clients by product choices or sentiment.

(ii) *Banking:* The need for generating, capturing, storing, and transforming data into a useful piece of information has encouraged the industry to use data mining. It acts as a strategy tool for the banks to use it effectively by generating and utilizing the technology for the benefit of the bank. Data mining not only enhances the customer benefit but also used in accessing information. It also helps in finding of the frauds, know about the payment defaults, and market risks as well while doing the different analysis in business. Data mining overall enriches the business processes in the banking industry.

(iii) *Manufacturing:* Data mining is used in the manufacturing field as well. To predict the process of production, error spotting, the quality of the product produced in manufacturing, damage of production assets, and the maintenance of the production, the role of data mining comes into play. There are five major domains in manufacturing where data mining can be used:

   (a) Product design
   (b) Route time evaluation
   (c) Quality
   (d) Delivering system management
   (e) Manufacturing environment

(iv) *Communication:* Data mining is considered one of the most important building blocks of next-generation mobile networks. With the massive increase in telecommunication and multimedia field, a large amount of data is generated such as call details network data or client information. Therefore, data mining can be used to extract and gather a useful piece of information that will benefit the telecommunication industry by predicting the client's experience. It also helps in identifying the frauds, enriching marketing strategies, and gathering prior knowledge of the network faults.

(v) *Beamforming:* It is a traffic-flagging framework for cell base stations that recognizes the most effective information conveyance course to a specific client, and it diminishes impedance for close by clients all the while. Contingent upon the circumstance and the innovation, there are a few different ways for 5G organizations to execute it.

All in all, it is certain that the use of DM in 5G-empowered IoT will see the cell network advancements changed into proactive instead of receptive networks. A portion of the areas that will profit by this reconciliation contain the nature of network load, profile of supporters, accounting data, and design of flaw signs among others.

3 Concept of Machine Learning

Machine learning can be referred to as a subcategory of the AI field, the main aim of which is to focus on examining and recognizing patterns and arrangements in data to facilitate features such as training, thinking, decision-making, learning, and researching without interference from human synergy. Machine learning allows the user to apply a computer algorithm on an enormous sum of data in order to examine and analyze the data and to make recommendations based on that data. If some features require improvement, they are recognized and incorporated for a better design for the future. Machine learning utilizes computational algorithms to do core work for making decisions [1]. Also, the variables, features, and innovations are responsible for making decisions. Core awareness toward the answer is required for the better learning of the embedded system which enables the system to learn.

Actually, the design feeds the machine with data for which the result is already known. The algorithm then runs, and changes are made until the output matches with our result. Increasing volumes of data are fed to boost the system to acquire more eminent decisions. The most important component of the business world is data. Decisions that are derived based on data make the difference in this fast-growing society. Machine learning can be considered as the key to the evolution of data interpretation and formulating decisions for better services.

3.1 Types of Machine Learning

Machine learning can be categorized mainly in three major categories [2]:

(i) *Supervised Machine Learning:* This enables the collected data or composes a knowledgeable output from a former ML progress. It also, managed training as it provides the program with the capacity to attain human-like intelligence. In supervised learning tasks, one provides the program or the machine with a huge set of inputs to attain a fruitful result.

(ii) *Unsupervised Machine Learning:* It aids by getting all sorts of hidden designs in data. In this module, the algorithms attempt to acquire any crucial method of the data with unlabeled samples only. The job is to combine and dimensionally reduce the data. The process of combining is known as clustering, i.e., it classifies data intents into meaningful groups so that components inside a given cluster are similar to each other but dissimilar to those from other batches. Clustering is beneficial for marketing segmentation for businesses. The dimension reduction patterns decrease the number of variables in a set of data by clubbing similar or related properties for better understanding.

(iii) *Reinforcement Machine Learning:* It refers to a machine application that communicates with a progressive atmosphere or dynamic environment to perform a specific objective (like to order food, to book a cab, or to play a game). As it is applied on different problem areas in order to train and test the data which may or may not be similar to the required output to achieve maximum efficiency through machine learning. Distinct methods have been acquired which may not fit into these categories, and from time to time, there are instances where one uses a combination of more than one category to reach the required aim.

This machine learning technology when gets integrated with IoT can make a huge impact on the economical aspect of society. The real benefit can be experienced when there is an addition of 5G networks in the same field. Now, next we will be looking at the aspects which make this collaboration of machine learning, Internet of Things, and 5G networking possible and how they proved to have the potential to create all forms of tasks easy, feasible, affordable, and efficient.

Machine learning is a support to many businesses as it allows understanding from the data, automate trading methods, boost productivity, and gradually benefit too [3]. And while corporations and firms are ardent on utilizing machine learning algorithms, they sometimes find it to be challenging. All the organizations are diverse and their courses are unique. But importantly, they often face problems in machine learning which involve basic concerns like company purpose arrangement, individual mindset, and more. The biggest challenges in the adoption of machine learning include:

(i) Remote data and data safety
(ii) Infrastructure essentials for trial and experimentation
(iii) Inflexible business models
(iv) Lack of talented individuals' time
(v) Utilizing implementation
(vi) Cost

3.2 Machine Learning and 5G IoT

IoT device utilization is growing day by day. This extensive usage of IoT produces enormous volumes of data. This data can be efficiently treated by the use of machine learning to infer several beneficial insights that can enhance services and influence the lives and technologies deeply. In the modern-day world, the growth is being witnessed in the consumption of IoT devices. As the number of devices increases, more data will be generated as the data is proportional to the consumption of these devices. Data concerning social patterns of trade and sustenance makes the machine learning ready for the interpretation that can benefit our work and life in a positive way. Machine learning will support IoT devices to operate in a better manner, creating a pleasanter lifestyle for society.

The domain of machine learning is developing and undeviating, accompanying the growth of the IoT. IoT elements such as sensors and nano cameras are now omnipresent; installed in mobile phones, laptops, PCs, parking stations, and shopping stores; and used for traffic control, medical industry, and even in-home appliances [4]. There is a huge amount of IoT devices prevailing in the world, and there is no change of its decrement soon. These devices collect large measures of data that is supplied to machines through the Internet, permitting devices to "learn" from the information given and make them more useful. In IoT, it is necessary to record that an individual method/factor can generate gigantic quantities of data at each instant. This information from IoT is dispatched to servers to build better machine learning standards. By the year 2021, the world will witness a huge share of IoT devices which are estimated to be around 20 billion. Data gathered by those devices often concerns the development of automation. Furthermore, machines can absorb more efficiently and can backside their shortcomings [4, 5].

Let us take an example; suppose there are a set of people visiting a doctor and we have scanners, sensors, and nano cameras to get the data about those patients. If the data is fed to the system via the Internet, then the data can be analyzed. Now, if the graphical representation of patients having fever is visible, then it may be considered that the machine or device has not imbibed it fully. Parallel to this, if the system can identify other diseases that a patient possibly has by observing their face and body structures, then it can be considered that the machine is fully developed and has learned the needful and is now intelligent. Collecting, processing, interpreting, and remaining ready to "think out" practicing IoT data need a lot of infinite computational and commercial support to achieve marketing and machine learning benefits. Today, the IoT joins numerous areas such as production industries, healthcare, construction industry, transportations, traffic, purchasing, and so on. Data accumulated from these regions can positively make the foundation of learning meaningful and efficient.

One major trait of 5G is the capacity to foretell action over the network and control them. Machine learning is accommodated to serve in 5G networks as it needs large volumes of data batches to foretell the accuracy of the activity. Machine learning seems to be a necessary component of any networking system as it is more complex than the few initial network generation. It works at higher frequencies, offers more elaborate configurations, and operates more advanced connectivity devices. The MIMO antennas which stand for multiple inputs multiple outputs used by 5G networks can manage multiplied data "communications" simultaneously across the same data signal [6].

This implies that more data transmission can be done without affecting the transfer of other data across the same network. As more machines tend to connect to the 5G networks, it creates a difficulty in managing all the traffic without the cooperation of machine learning. The 5G networks will be ready to examine data patterns and communication, enabling a dynamic transmission of data. A completely working 5G network will not happen without AI and ML as they are capable of making judgments by themselves. It gives supplementary penetrations into the shape of the network by giving other skills and functionalities for the error, administration, and issues related to security as shown in Fig. 14.3.

The factors can be quickly controlled and resolved for the networks and systems are informed about the handling issues before they become difficult. For instance, a weak network plan is found more quickly based on the machine learning analysis

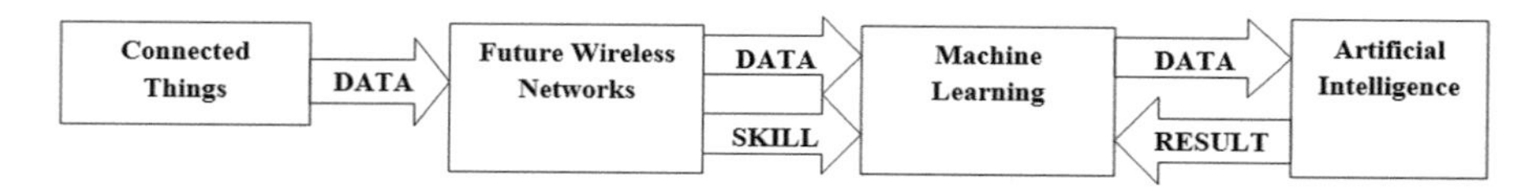

Fig. 14.3 Machine learning and 5G networks

of past traffic models. System failures are speedily rectified as the identification is done even before showing a negative impact across the network and are alleviated efficiently, something tough to do in profoundly complex operations without machine learning.

To check the number of hardware tools required to empower a 5G network by introducing the ideology of virtualization or network slicing. Network slicing refers to the usage of a single shared network with parallel running multiple virtualized networks. For example, firms can run a single network slice to lower the delay in the system that requires higher reliability; on this particular device, the employee traffic is handled by another slice. A 5G network-stimulated car has one network slice for the scenarios like autonomous driving facility and other slices for other functions and a different slice for the car features like radio, navigation system, and AC. Installing network slices is confined on a 5G network as there is a requirement of manual configuration. ML and AI will aid this service and find a way to reduce the complexity of the systems. This will assure that all information exchange is routed based on the need of the device and ensures the configuration settings to be appropriate.

3.3 Applications of Machine Learning

The potential and prevailing applications are [5–7]:

(i) *Agriculture:* Agriculture is undoubtedly one of the most vital human pursuits. More helpful technologies imply higher yield. This holds the human race more satisfied and healthier. According to some surveys, global food generation will rise by 70 percent by the mid-twenty-first century to keep up with global demand. The selection of IoT and ML in the agricultural area is also progressing swiftly where the total number of related device demand seems to grow at a very high rate reaching 75 million in 2020. In smart agriculture, all communications between farmers and agricultural methods are converted to be data-driven. Even analytic tools are presenting the correct information at the appropriate time. Gradually ML will provide the reason to surmount and automate the agricultural sector. It is encouraging to discover patterns and derive information from enormous quantities of data.

(ii) *Healthcare:* Today, smart, supported living circumstances are required for patients with chronic health issues, and to the rescue comes home healthcare.

This connects the patient's medical history and semantic illustration of the individual care process with the powers to watch the living situations with the integration of technologies like machine learning, Internet of Things, and 5G networking. The combined healthcare structure can produce significant additions while enhancing general wellbeing.

IoT devices in this sector are the nano sensors connected to the patient's skin to measure sugar levels, heartbeat, blood pressure, etc. This inexperienced data is forwarded to the database that remains in the highly guarded cloud platform. The doctor can obtain the data of the earlier prescriptions and use advanced ML algorithms to prescribe specific medications to patients at distant places if demanded. Thus, subjects at the house can be protected from sudden health hazards like heart attacks.

(iii) *Maintenance:* In 2017 "Internet of Trains" project was launched. Embedded sensors in tracks and the trains were set up in countries like Russia and Spain. Employing the data input from the sensors, they raised a machine learning design to recognize signs if the track or the rain was about to fall. They later use the overtaken observation to find the area which requires repairs.
(iv) *Surveillance from CCTV:* This is the largest in scale example of IoT with ML. This has enabled facial identification software to be appointed with the CCTV cameras. The technology is nowadays being used in cities, airports, railway stations, and malls to catch hold of criminals. This technology also has its drawbacks as it interferes with personal space or privacy.
(v) *Supply Chain:* As businesses now have a lot of data and want to know about the responses from their customers, they are practicing this technology to adapt their list accordingly. They guide various companies like Uber and Walmart to attain a better position in the service market.

3.4 Various Challenges Encountered While Applying ML

These days companies are overwhelmed with data that arises from IoT projects and are exploring ML and 5G networking to help and maintain these devices. It is difficult to control and secure vital data from these operations [8]. There are features to IoT similar to a data warehouse, connectivity, safety, app advancement, integration of the system, and even methods that are growing in this area. An extra layer of complexity among the IoT takes functionality to the next level.

Significant trials that businesses face amidst IoT and ML are the application, access, and interpretation of IoT data. If one has a set of input from varied origins, then one can run some analytical process. Despite the need for prediction regarding the events, the companies need to learn the process of using this technology. Numerous firms are shifting to the cloud platform providers as these businesses offer a variety of services to store data and serve it for data analytics and machine learning models. They also support in building plans, dashboards, and other designs to imagine the data these models produce. Overall, IoT, ML, and 5G networking are coupled to implement high clarity and power of the wide array of devices attached to the Internet [9].

Futurists assume that ML and the IoT will convert the market deeply showcasing revolutions. Are there distinctive kinds of uncertainties in this technology? This is the new question to which one can only answer that with any developing technology, one ought to admit both the advantage and risks that occur with the main adoption. The technology that is tested against all the odds can be considered safer and more profitable.

5G networking will facilitate a modern age of possibility for everyone. The capacity to manage and transfer data from anyplace in the world to wherever one want that to in the shortest possible time will change the way one used the technology. It will unbar the minds of technology specialists as they consider new, progressive, and innovative ideas to develop the potential and subsistence of our markets. In this age of connection and connectivity, people have many technologies to sustain their daily necessities. In this situation, IoT, ML, and 5G networking are developing as a possible answer for queries faced by the service sector in several sectors. Adding machine learning and IoT to mix with 5G will make matters even more impressive. Intelligent machines will be able to experience learning with methods intended to assist and optimize our experiences. With the precise succession of machine learning and the Internet of Things, one can enjoy unparalleled levels of accomplishment and computerization with the help of 5G networks. It will stimulate discovery in all phases of growth, improving the way one abide and operate for ages to come.

4 Big Data Analytics

As the world is becoming more and more digitalized, it involves the generation of data from different resources which in result has led to the growth of big data. Big data refers to a large and complex set of data that can be present in any format, and the analysis of such data is called big data analytics [17, 18]. Big data and DM are two unique things; while the two of them identify with utilization of huge data sets to deal with the information that will fill our need, they are two distinct terms in the part of activity they are utilized for. Enormous data alludes to an assortment of huge data sets (e.g., data sets in Excel sheets which are too huge to possibly be dealt with without any problem). Data mining then again alludes to the movement of experiencing an enormous lump of information to search for applicable or relevant data. The difference between the two can easily be understood through Table 14.1.

Big data is symbolized by 3Vs which are velocity, volume, and variation. Volume can be referred to as the quantity of data that is generated every day by various sources around us, while velocity means the amount of development and in what speed the data is accumulated for analysis. Variety is responsible for the flow of information or data which is structured, unstructured, semi-structured, etc. The not

Table 14.1 Difference between big data and data mining

Features	Big data	Data mining
Definition	It is a term used to refer to a large amount of data set	It is a technique used to excerpt useful data from a raw data set
Aim	It mainly aims at gathering, storing, and processing a variety of data	It mainly aims at analyzing data and finding relationships between them
Type of data	It consists of structured, unstructured, and semi-structured data	It consists of relational databases and structured data
Applications	It is used for dashboards and prophetic means	It is used for strategic judgment purposes

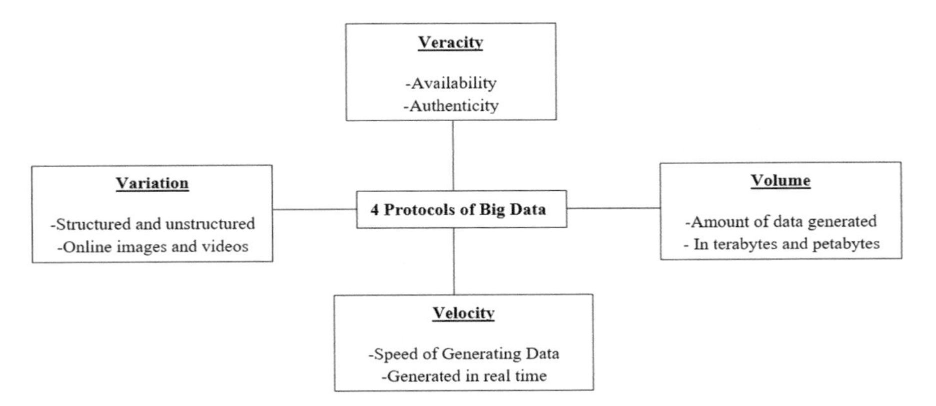

Fig. 14.4 Big data protocols

so popular V, that is, veracity, which comes fourth is actually the possibility and liability. The major aim is to successfully achieve and follow these 4Vs as shown in Fig. 14.4. Big data analytics has become such an important aspect because all the IT and computer industries are producing a huge amount of data that can only be taken in use after its analysis as that all information present in the form of big data is not necessarily required for analysis or decision-making process. Industries are fascinated to invest in customized findings of big data.

4.1 Challenges in Big Data Analytics

Initially, data warehouses were used to process big data, but precision and accuracy were the main concerns that were faced [19, 20]. The major issue in the analysis of big data is the deficiency of coordinated communication between database systems and analytic tools. The first challenge for big data analysis is the unavailability of storage methods which can be due to the increased speed of input/output. In such scenarios, the data approachability should be the main concern for better representation. In earlier times, hard drives were used to store the data, but it became steady and became less reliable due to the random input/output production. To avoid this problem, the idea of solid-state drive (SSD) was introduced.

The second challenge in the implementation of big data analysis is the multiplicity of data, which refers to the growth of data sets in an undefined pattern. As a result, the already existing algorithms might not respond in an expected way and insufficient time whenever dealing with high-dimensional data. The only solution possible in such situation is to shift the attention to designing storage systems and to use appropriate big data analysis tools which can guarantee the result when the data comes from multiple sources.

The next third most important challenge is data scalability and security. All establishments have their proper policies to protect their sensitive and private information, which is the main concern for big data analysts as there is an immense security risk which is related to big data due to which the reliability of data and information is becoming a major problem. The possible ways in which security can be improved are by implementation of the techniques of authorization, authentication, and encryption.

For processing the high-dimensional data and solving the above issue, various tools have been developed. The major three tools are [21, 22]:

(i) *Apache Hadoop and MapReduce:* The commonly known and successful software platform that can be used in big data analysis is Apache Hadoop and MapReduce. MapReduce is a programming representation for large data sets which are based on the famous divide and conquer algorithm. This method is generally executed in two steps which are Map step and Reduce step. Hadoop has two types of nodes which are master node and worker node.
(ii) *Apache Spark:* Apache Spark is an open-source processing structure of big data which is built for speed processing and sophisticated analytics.
(iii) *Apache Drill:* Apache Drill is a specially built system for interactive of big data analysis. It has much more flexibility and the ability to assist different sorts of query languages, data formats, and data sources.

4.2 Big Data Analytics and IoT

Big data analytics has widely proven itself as a major asset in the field of IoT to increase the potential of the decision-making process. The most important feature of IoT that should be taken care of is the examination of information related to "connected things." Big data analytics with the integration of IoT aims at filtering a large set of data on the fly and storing it using diverse storage methods/technologies. It is well known that most of the data collected are unstructured as it is collected directly from web-enabled "things"; thus, big data usage becomes necessary to perform rapid analytics with large data to permit organizations to obtain fast intuitions; they can build speedy decisions that can interrelate with both humans and virtual devices. The incorporation of IoT and big data analysis allows sensing and triggering devices that in turn provide the ability to share information across platforms and develop a successful operating image for enabling unconventional applications [14, 23].

It is required to embrace big data in IoT applications as these technologies have been individually acknowledged already in all the fields related to IT and business. These technologies are correlated, and if they are jointly developed, then it would lead to discoveries and a series of achievements. They act as a way of opportunity for each other. For example, the deployment of IoT has displayed an increase in the amount of raw data generated and, thus, offers chance of development. The advantages of this interaction don't end here; the application of big data technologies in IoT also facilitates speed to the research and business-related models of IoT. The relation between IoT and big data can be classified into three categories to ensure proper administration of data.

The first major step includes the management of IoT data resources, so that various connected devices can use big data applications to interconnect with each other. For example, the huge quantity of data sources is generated due to the interlink age of devices such as CCTV cameras, traffic lights, and smart home devices which produces the data in different formats. The second step involves the generation of "big data" that are based on their characteristics (3Vs) velocity, volume, and variety.

These large amounts of data are then stored in big data files that are by default in shared, dispersed, and fault-forbearing databases. The last step involves the application of various analytics tools such as MapReduce, Spark, Splunk, and Skytree that can inspect the storage of big IoT databases. All four levels define how the raw data generated in IoT is successfully managed using big data analytics as shown in Fig. 14.5.

4.3 IoT Architecture for Big Data Analytics

The architecture of IoT can be defined and established on IoT domain abstraction and recognition. The architecture is like a reference model that specifies the relationships among different IoT levels, such as traffic, smart home, transportation, and smart health. On the other hand, the architecture for big data provides an idea of data abstraction. There are various kinds of architectures depending upon its application. For example, there is an IoT architecture of IoT associated with cloud computing at the center, and it represents a model of end-to-end interconnection among shareholders in a special kind of framework also known as cloud-centric IoT framework which leads to proper comparison with the suggested or default IoT architecture [25]. The architecture can be only accomplished by logical pervasive recognition, data analytics, and proper representation of information with IoT which seems like a unifying architecture. Nonetheless, the present architecture focalizes on IoT concerning communications. The proposed or default architecture is considered the one that integrates IoT as well as big data analytics.

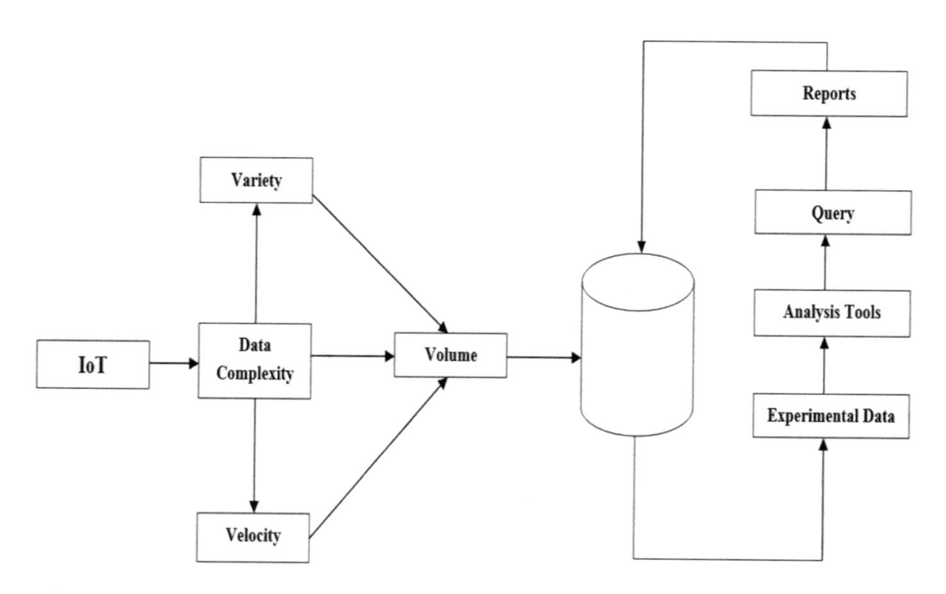

Fig. 14.5 Correlation between big data analytics and IoT

The IoT architecture combined with big data analytics includes the sensor layer that as the name suggests has the sensor devices along with the objects that are connected with each other through a wireless network as shown in Fig. 14.6. This communication and connection is a wireless network that can be RFID, Wi-Fi, ultra-wideband, and Bluetooth. The interaction of the Internet and other web browsers is possible today due to the IoT gateway. Big data analytics come into play in the upper layer in the architecture and a huge amount of data obtained from sensors and are then stored in the cloud which, later on, become accessible through big data analytics-based applications. These applications include API management to aid in communication with the processing engine.

The correlation among IoT and big data provides different utilizations in various fields [25–27]:

(i) *E-Commerce:* Big IoT data display non-uniformity, volume, and real-time data-related processing characteristics. It has numerous applications in almost every industry. The major achievement areas of analytics are in growth of revenue, the increment of customer size, the precision of sale forecast results, product enhancement, risk management, and upgraded customer segmentation.
(ii) *Smart Cities Applications:* A large amount of data is collected from applications used in smart cities which provide new opportunities that can only be gained through suitable analytics to analyze big IoT data. Today, all devices can link to the Internet and create an environment that is smart, and then the detail is passed and processed using big data. Therefore, big data plays a major role in transforming every zone of the economy of a country.

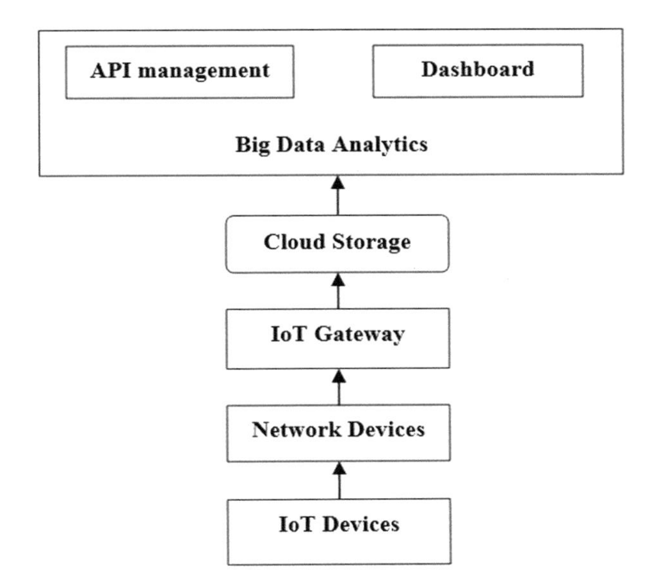

Fig. 14.6 IoT Architecture with big data

(iii) *Market and Logistics:* IoT plays a major part as an emerging technology in the field of retail, market, as well as logistics. In logistics, containers are tracked, platforms and the container is kept using RFID. Moreover, development in IoT technologies can smooth dealers by giving various benefits. IoT devices produce huge amounts of data on a daily basis. Therefore, big data analytics helps to categorize the data and hence allows enterprises to attain insights from the huge sets of raw, undefined data generated through IoT-enabled technologies. Application of data analytics can uplift the shipment experience of users. Besides, retail enterprises can gain large amounts of profit by analyzing user data and by predicting the trends as well as demands of products.

4.4 Big Data Analytics in 5G-Enabled IoT Systems

The interaction of big data techniques with 5G distributed applications shows up an inconvenience, as these methods generally require a huge amount of technical resources. The convergence of 5G cellular, IoT, and data analytics has to lead to great advancement in the information and communications technology (ICT) industry. The combination of such technologies has successfully built a leading path for new business models. Moreover, it also helps in technical innovation and innumerable opportunities for applications of all those industries that are dependent on Telecom and IT services [28, 29]. 5G ensures the creation of intelligent networks, while IoT ensures connectivity of application services to remote sensors, and the huge amounts of data produced by these IoT connections are analyzed further by big data analysis.

Big data analytics is now no longer considered a future aspect or something which requires a second thought; in fact, it plays a very important part in the growth of 5G standards which further enabled intelligence over the applications and business. Today the world traverses the role of analytics in the conditions of 5G, to understand major technology trends or patterns and business leaders that outline the path to 5G services and applications, and, finally, provides a maturity model of 5G analytics and networks. It has to lead to the transformation from being context-aware to becoming cognitive along with intuitive [35].

Data analytics is at the highest spot seeking full advantage from 5G network characteristics. For example, high-bandwidth, low-latency, mobile edge computing (MEC). The big data at rest and the data in motion can come into reality due to the ability of 5G to aid gigantic connectivity over multiple devices which is further supported by the distributed computer architectures. It also produces the potential to convert big data into real-time insights and allows interaction with intelligence. Big data analytics plays a dual role in 5G. While, on the one hand, analytics resume to help diverse business applications of 5G webs, on the other hand, it also plays an important role in the roll-out of 5G network operations. Data analytics is compatible with 5G technology systems.

Big data analytics along with machine learning are used to develop an energetic mechanism by anticipating the approval distribution of the data in 5G networks. It is believed that this method can enable the structured usage of network resources and return a better experience for users. The first task is accumulating the raw data, i.e., the user traffic, and then the big data program plays a role in predicting the customer requests by pulling out the useful data, like Location Area Code, Hypertext Transfer Protocol, Tunnel Endpoint Identifier TEID and (TEID)-DATA to control the data levels. The content popularity can be achieved by the information collected from the collected raw data. This task has been experimentally tested on 16 base stations, as part of an operation of network, and it successfully resulted in complete request results and 98% backhaul offloading [30].

Various applications in which the goals of association among these technologies have been seen are:

(i) *Mobile Cloud Computing:* Today a smart and intelligent world is a gift to us due to mobile cloud sensing and 5G network. Major important applications such as the safety of the customers and the healthcare domain require analytics in actual time. Due to promises made by a new 5G interface such as prioritization-based MEC-supported local analytics, and the reaction time improvements, 5G is now able to lay an establishment for supporting mission-analytical edge analytics as well as tangible Internet applications. 5G allows not only sensing and analyzing at the edge but also activating actuators to activate responses within minimum required time due to which all data proceed from cloud to an excess of endpoints and vice versa.

(ii) *Comprehension/Cognitive Analytics:* Alternatively, conventional illustrative analytics with regular BI reporting and analytics in 5G transform itself into machine/deep learning. Analytics in 5G in the future will evolve to such an extent that it will be able to not only learn from the context but also predict what is going to happen next and advise the following finest step that needs to be taken. It will also help to learn from the past behavioral mistakes for picking the most suitable decision. For completely autonomous applications, it will automate the next step. Data analytics has already-built-in insights that can accelerate decision-making, and due to the convergence of 5G, a huge amount of information will be accumulated and processed, which will be fastest in the history, thus leading to cognitive intelligence applications.

In the case of a sober network with limited capacity, data analytics is like a boon with utmost importance: the network can collect so much data, but before the exposure of big data analytics, there were not that many resources to process it. 5G networks are not anything but limited as well as simple, making it possible for analytics to provide on the 5G assurance while also taking full advantage of 5G resources. Gartner had predicted that by upcoming years, some percentage of 5G networks will be commercially launched by networks-based mobile communications service providers (CSPs). According to the Hype Cycle report, it is mentioned that in upcoming years,

5 Convergence of Machine Learning, Data Mining, and Big Data Analytics for 5G-Enabled IoT

When 4G was proving to be a new era of expansion then the convergence of Machine Learning, Data Mining, and Big Data Analytics for 5G enabled IoT sets as a new bar of extension. With this comprehensive integration of technologies, 5G has increased efficiency, reduced decision time, and also created strong real-time analytics. The whole idea of 5G-enabled Internet of Things has emerged an adequate amount of satellites to send signals from highly populated places to the poorest places of the world. These satellites need data to be processed, and the outputs are generated on the same; in this scenario, big data analytics play a vital role. They harmonize the ground data concerning the satellite, and every microscopic detail is transferred to the satellite as inputs, and then ultimately high speed and efficient data are provided to the users [31, 32].

Researchers, industrialists, and organizations are investing in these upcoming technologies by providing them more aggregate of machines and unstructured data so that further refurbishing of the results takes place. Technology is increasing rapidly, things are getting more complex and user-friendly, and to meet the new demands of the customers, it is very supreme to adapt to the new significant function of these technologies. Communication is the foremost way to connect people; 5G has proved to be viaduct for millions of people. Machine learning, data mining, and big data analytics have been pillars of the stability of 5G networks, and by virtue of contribution of Internet of Things platforms, the whole mechanism has become efficient, and the output time is reduced. Ultimately, this revolutionary convergence is a stepping stone for the upcoming generation of networks.

6 5G-Enabled IoT Case Studies

5G will be most often utilized by organizations to achieve IoT communications. It is believed that 5G will be an essential program that will enable artificial intelligence widely [33, 34]. Some of the case studies through which one can able to understand about how all the technologies are correlated to each other are explained as follows:

(i) *Case Study: Smart City*

Till now it has been discussed how machine learning makes systems adapt to data and create outputs from the same, how data mining provides the amorphous data to the systems to be used as inputs, and how data analytics deals with the manipulation and organization of the data. According to the smart city architecture, the three major constituents which are the pillars of the smart cities are:

(a) Technological Stack – It deals with the backend situation of the cities which includes servers, machines, algorithms, etc. The convergence of machine learning, big data analytics, and data mining focuses on the satisfaction of people's needs by providing them with swift networks that can make their communication fluent.
(b) Human Resources – It is more of like the frontend of the whole system. The IoT services are now ready to cater to the needs of everyone. There are open resources for learning, observing, and creating new projects from these networks. They simulate a crucial role improving the economy of the country.
(c) Organization Factors – Once both backend and frontend are ready, it's time to link them, and hence, the organization factors work as the middleware for both ends. For close analytics, they test the networks at different frequencies to get superior results. The advancement of smart cities has been assisting in time-saving, energy-saving, and money-saving as well. The IoT has proved to be a stable manifesto for the smart cities concept. It has held the back of databases, pieces of equipment, algorithms, etc. Earlier it was an enormous challenge to summarize the whole strategy of providing a stable network system to the people into a single primitive network.

(ii) *Case Study: Mobile Networks*

Mobile Networks is one of the major necessities of today which provides the advantage to millions of users to operate any application, services, etc. Technologies like machine learning, big data analytics, data mining, and the

Internet of Things are responsible for connecting them with the high-tech servers. These servers will supervise the traffic and also analyze them.

Network providers like JIO, Airtel, Vodafone, etc. are one of the biggest companies in terms of users and services. These companies operate on the substantial data principles which redeem and then examine the data generated through the mobile Internet provided to the users. The data which is produced is in millions, and to handle it and store it in a particular order is impractical for a single person; therefore, devices with massive storage space are employed. As compared to normal storage devices, these high-tech devices have ten times more capacity to store any sorts of data.

7 Conclusion

The 5G-enabled IoT paradigm has become an indispensable aspect of our everyday lives. However, IoT gadgets are obliged in communication and computation which are the bottlenecks in the advancement of versatile, smart solutions utilizing AI methods. Also, in making new innovations and stage upgrades for the future includes quick IoT advancement, application development, and solid examination of high volume IoT information through data mining and big data analytics.

Throughout the chapter, it has been discussed how, with the help of machine learning, the system can react and produce output based on data provided and also analyzed the role of data mining which provides unstructured data as the input and how big data analytics simulates a crucial role for monitoring and analyzing data provided by data mining techniques. 5G networks are a new era of technology whose integration with machine learning, data mining, and big data analytics can be transformed into further reactive networks.

The chapter also focused on IoT platforms, convergence of machine learning with IoT platforms, convergence of data mining with IoT platforms, and convergence of big data analytics with IoT platform. The concept includes how machine learning enhanced efficiency of 5G networks, how data mining furnish data for 5G networks, and how big data analytics reduced the time consumption of the 5G networks. With the detailed theories about the 5G networks, it was time to explore some real-time case studies for a better understanding of the concept. In the chapter, two case studies are presented which will bestow a closer look at the mechanism of 5G networks with the help of these revolutionary technologies. The first case study is about smart cities in which the role of 5G networks is highlighted, and the second case study is about mobile networks where the concept of MSNs (mobile social networks) is elaborated.

Lastly, how the convergence of machine learning, data mining, and big data analytics with 5G networks can be done is explained and how they are shaping the world to be more user-friendly. The chapter is a complete package of information that will allow users to explore new things. Technology is improving every second and the sapiens have to adapt the change for their survival in the world.

References

1. M. Mohammed, M.B. Khan, E.B.M. Bashier, *Machine Learning: Algorithms and Applications* (CRC Press, Boca Raton, 2016)
2. M. Kubat, *An Introduction to Machine Learning* (Springer, Cham, 2017)
3. M. Ghadge, D. Pandey, D. Kalbande, Machine learning approach for predicting bumps on road, in *IEEE International Conference on Applied and Theoretical Computing and Communication Technology (iCATccT)*, (IEEE, Piscataway, 2015), pp. 481–485
4. L. Atzori, A. Lera, G. Morabito, The internet of things: A survey. Comput. Netw. **54**, 2787–2805 (2010)
5. T. Koreshoff, T. Robertson, T. Leong, Internet of Things: A review of literature and products, in *the 25th Australian Computer-Human Interaction Conference: Augmentation, Application, Innovation, Collaboration*, (ACM, Adelaide, 2013), pp. 335–344
6. M.A. Alsheikh, S. Lin, D. Niyato, H.P. Tan, Machine learning in wireless sensor networks: algorithms, strategies, and applications. IEEE Commun. Surv. Tutorials **16**(4), 1996–2018 (2014)
7. A. Smola, S. Vishwanathan, *Introduction to Machine Learning*, vol 32 (Cambridge University, Cambridge, 2008)
8. Maruti Techlabs, https://marutitech.com/challenges-machine-learning/
9. M.S. Mahdavinejad, M. Rezvan, M. Barekatain, P. Adibia, P. Barnaghid, A.P. Shethb, Machine learning for internet of things data analysis: A survey. Digital Commun. Networks **4**(3), 161–175 (2018)
10. J. Han, M. Kamber, J. Pei, *Data Mining: Concepts and Techniques*, 3rd edn. (Morgan Kaufmann, Boston, 2012)
11. A.L. Buczak, E. Guven, A survey of data mining and machine learning methods for cyber security intrusion detection. IEEE Commun. Surv. Tutorials **18**(2), 1153–1176 (2016)
12. G.V. Krishna, Data mining processes, applications and challenges for IoT. Int. J. Trend Res. Dev. **4**(3), 157–160 (2017)
13. K.K. Lammatha, Data mining on 5G technology IoT. Int. J. Eng. Comput. Sci. **8**(5), 24655–24660 (2019)
14. A. Kaloxylos, Application of data mining in the 5G network architecture, in *The Thirteenth International Conference on Digital Telecommunications*, (IEEE, Piscataway, 2018), pp. 39–44
15. A. Savaliya, A. Bhatia, J. Bhatia, Application of data mining technique in IoT: A short review. Int. J. Sci. Res. Sci. Eng. Technol. **4**(2), 218–223 (2018)
16. A.M. Njeru, M.S. Omar, S. Yi, S. Paracha, M. Wannous, Using IoT technology to improve online education through data mining, in *International Conference on Applied System Innovation (ICASI)*, (IEEE, Piscataway, 2017), pp. 515–518
17. M. Hilbert, Big data for development: A review of promises and challenges. Dev. Policy Rev. **34**(1), 135–174 (2016)
18. H. Hu, Y. Wen, T.S. Chua, X. Li, Toward scalable systems for big data analytics: A technology tutorial. IEEE Access **2**, 652–687 (2014)
19. S. Madden, From databases to big data. IEEE Internet Comput. **16**, 4–6 (2012)
20. R. Casado, M. Younas, Emerging trends and technologies in big data processing. Concurrent. Comput. Pract. Exp. **27**(8), 2078–2091 (2015)
21. Apache. Hadoop MapReduce, https://hadoop.apache.org/
22. Apache. Apache HBase, https://hbase.apache.org/
23. Y. He, F.R. Yu, N. Zhao, H. Yin, H. Yao, R.C. Qiu, Big data analytics in mobile cellular networks. IEEE Access **4**, 1985–1996 (2016)
24. I.A.T. Hashem, V. Chang, N.B. Anuar, K. Adewole, I. Yaqoob, A. Gani, E. Ahmed, H. Chiroma, The role of big data in smart city. Int. J. Infrastruct. Manage. **36**, 748–758 (2016)
25. D. Puthal, R. Ranjan, S. Nepal, J. Chen, IoT and big data: An architecture with data flow and security issues, in *International Conference on ICT Infrastructures and Services for Smart*

Cities International Conference on Cloud Networking for Internet of Things Systems, IISSC & CN4IOT, (2017)
26. S. Talari, M. Shafie-Khah, P. Siano, V. Loia, A. Tommasetti, J. Catalão, A review of smart cities based on the internet of things concept. Energies **10**, 421 (2017)
27. J. Yang, Y. Han, Y. Wang, B. Jiang, Z. Lv, H. Song, Optimization of real-time traffic network assignment based on IoT data using DBN and clustering model in smart city. Future Gen. Comput. Syst. **108**, 976–986 (2017)
28. R. Chaiken, B. Jenkins, P.K. Larson, B. Ramsey, D. Shakib, S. Weaver, J. Zhou, SCOPE: Easy and efficient parallel processing of massive data sets. Proc. VLDB Endow. **1**, 1265–1276 (2008)
29. A. Imran, A. Zoha, Abu-Dayya. A.: Challenges in 5G: How to empower SON with big data for enabling 5G. IEEE Netw. **28**(6), 27–33 (2014)
30. F. Chang, J. Dean, S. Ghemawat, W.C. Hsieh, D.A. Wallach, M. Burrows, T. Chandra, A. Fikes, R.E. Gruber, Bigtable: A distributed storage system for structured data. ACM Trans. Comput. Syst. **26**(2), 1–26 (2008)
31. B. Ryder, F. Wortmann, Autonomously detecting and classifying traffic accident hotspots, in *ACM International Joint Conference on Pervasive and Ubiquitous Computing and 2017 ACM International Symposium on Wearable Computers, Maui, HI, USA*, (2017), pp. 365–370
32. M.A. Al Mamun, J.A. Puspo, A.K. Das, An intelligent smartphone based approach using IoT for ensuring safe driving, in *IEEE International Conference on Electrical Engineering and Computer Science (ICECOS)*, (IEEE, Piscataway, 2017), pp. 217–223
33. Research and Markets, https://www.researchandmarkets.com/reports/4968358/5g-artificial-intelligence-data-analytics-and
34. Business Wire, https://www.businesswire.com/news/home/20200207005390/en/2020-2025-Worldwide-5G-Artificial-Intelligence-Data-Analytics
35. S. Tanwar, S. Tyagi, N. Kumar, *Multimedia Big Data Computing for IoT Applications: Concepts, Paradigms and Solutions, Intelligent Systems Reference Library* (Springer Nature, Singapore, 2019), pp. 1–425

Chapter 15
Smart Secure Telerehabilitation Apps for Personalized Autism Home Intervention Using Blockchain System

Nurnadiah Zamri, Zarina Mohamad, Wan Nor Shuhadah Nik, and Aznida Hayati Zakaria Mohamad

1 Introduction

In the era of Industrial Revolution 4.0 (I.R. 4.0) escalates the advancement of the healthcare industry. Patient monitoring and sensors with the Internet of Things (IoT) devices have opened up new opportunities and chances in the healthcare industry. Wireless body area network (WBAN) is one of the subsets in the IoT healthcare trend. WBAN offers many assuring new applications in the IoT healthcare industry, including home/healthcare and remote health monitoring, and offers the increase of freedom of movement healthcare. Upon freedom, WBAN can take real-time measurements of vital indicators such as an electrocardiogram (ECG), heart rate, respiration, blood pressure, temperature, and many more. All the collected data from the WBAN will be transmitted using Wi-Fi or Bluetooth to a master device (it can be a personal computer, laptop, or mobile phone) directly toward the healthcare providers. Time-consuming doctor's appointments can be reduced by this remote patient monitoring (RPM). Thus, it permits patients to go about their regular lives more freely.

Since 2016, RPM is readily used as part of 7.1 million patients worldwide for their health management tracker. By 2021, this number is estimated to grow up to 50.2 million patients. Various smart healthcare devices with the RPM monitoring have been produced nowadays such as monitoring chronic diseases [1], high ventricular rates [2], automatic pacemaker [3], glycemic control for diabetes patient [4], vital signs monitoring on general surgery wards [5], implanted cardioverter defibrillators [6], knee arthroplasty [7], deteriorating patients [8], chronic dis-

N. Zamri (✉) · Z. Mohamad · W. N. S. Nik · A. H. Z. Mohamad
Faculty of Informatics and Computing, Universiti Sultan Zainal Abidin, Besut, Terengganu, Malaysia
e-mail: nadiahzamri@unisza.edu.my; zarina@unisza.edu.my; wnshuhadah@unisza.edu.my; aznida@unisza.edu.my

S. Tanwar (ed.), *Blockchain for 5G-Enabled IoT*,
https://doi.org/10.1007/978-3-030-67490-8_15

ease [9], etc. This chapter focuses on autism with RPM for personalized home intervention. These smart secure telerehabilitation apps offer online rehabilitation (known as telerehabilitation) for self-autism intervention. Besides, patients and medical professionals (including therapists, doctors, and psychologists) will receive notifications regarding medical interventions and real-time patient monitoring. All collected data from the RPM must be accumulated, systematized, and managed together in order to instruct combined health management. However, these apps still suffer with the issues of security of the medical transmission data.

Healthcare record is a worthwhile goal for hackers, and there is a definite reason in government law to ensure Protected Health Information (PHI) transmission [10]. Therefore, there is a need to a new way to protect and preserve patient data privacy and their electronic health records (EHRs). This includes protecting the integrity among the patients and therapist and securing all the electronic treatment and services, besides maintaining all the manageable and transferable data timely. Therefore, this chapter highlights these concerns by integrating WBAN systems with Autism Telerehabilitation Apps (ATA) using a blockchain. Blockchain generates an immutable log for every connection between the WBAN devices and the autism therapist for a distributed data processing service. With this proposed system, all the patient monitoring, therapist notification, and immutable ledger are all secured and precise. The smart contract from blockchain reacts an innovator approach that consents healthcare contributors while simply automating the autism notifications from multiple devices using ATA.

Autism is a neurodevelopmental disorder categorized by repetitive behaviors and limited attentions and dissimilarity in societal communication and social interaction. Roughly 1 in 59 children is detected with autism spectrum disorder (ASD) since 2018. Based on the previous research, almost 50% of children with autism can catch up to their colleagues if they can obtain intensive behavioral treatment at an early age. ATA is built for Universiti Sultan Zainal Abidin (UniSZA)'s Psychology and Rehabilitation Centre in line with the UniSZA's aspiration to help the autism kids. This proposed system offers parents with a set of questionnaires detecting early ASD among kids using fuzzy rule-based techniques, besides offering automated and computerized managements based on the capacity taken by the devices. Thus, it would support health interventions and concurrent autistic children (patient) observing by sending notifications to patients and therapists (medical professionals). All the data transactions and notifications are secured with blockchain security.

The idea of blockchain comes from the blocks of cryptocurrencies connected by chains [11]. Satoshi Nakamoto first developed blockchain in 2008, where he used a hashcash-like process to add blocks to the chain without a consigned third party [12]. Since 2008, blockchain technology has gone through three cohorts of development, which are Blocks 1.0, 2.0, and 3.0. Block 1.0 focuses on a currency where its successful project is Bitcoin. Block 2.0 covers loans, futures, bonds, smart contracts, mortgages, and cash transactions. Block 3.0 offers in wide application such as technology [13, 14], smart applications [15], financial [16], vehicles [17], tourism [18], and many more. Thus, it leads to three main layers in blockchain: peer-to-peer (P2P) network, databases, and applications. Besides, there are three categories of blockchain: public block, private block, and consortium block. Public block offers broad permissionless blockchain to initiate the transaction. Private block offers more strict permissioned blockchain and only involves members. Consortium block, quite similar to a public block, offers only for federated blockchain.

Our ATA system proposed an automatic execute code based on predefined conditional triggers using smart contracts. Smart contracts in ATA are used to collect all the autism data based on custom threshold amounts for each patient via WBAN devices. All the patient data are transferred based on set-up measurement and scale. This smart contract can identify if these data are higher or lower than the set-up measurement. Besides, it can give alert for appointments and therapy schedules, battery charging, etc. Moreover, it helps to secure activities among participating autistic over the public network. Any autism transaction (in nodes) among the participating autistics is added to the block only if they are agreed upon by the smart contracts. The smart contract also helps in maintaining privacy through permissioned anonymized accounts and consortium management. Only certified individuals can admittance the blockchain for block verification and inspection. To allow the patients to improve and manage their autism data and promote data transparency, each certified user will also have their personal account that can only be trailed by their discretion.

1.1 Contributions

In this chapter, we highlighted various important keywords such as PHI, EHRs, WBANs, RPM, I.R. 4.0, IoT, ATA, blockchains, smart contracts, Ethereum, and ASD. Based upon the above keywords, the following are the significant contributions of this chapter.

- We propose a systematic and comprehensive ATA system with WBAN devices for ASD patients.
- We collect all the autism data, including patients' health status, diagnosis, and intervention, from ATA systems and WBAN devices.
- We present the blockchain-based smart contract solution to protect data, promote data transparency, and mitigate each patient's security and privacy issues.

1.2 Motivations

The privacy and authenticity of the ASD data are the highest motivation in the proposed of this chapter. Blockchain is the best method nowadays for verifiability and authenticity. It helps to maintain privacy through permissioned consortium management and anonymized accounts. Besides, blockchain provides high security and utilizes the consensus or agreement of nodes to authorize the additions of blocks to the chain. It operates as an all-purpose ledger for all transactions. Blockchain technology has facilitated in the improvement and efficacy of many industries. It can be used to document the procedures for each experience from its origin to the recent condition in an unalterable log. Blockchains can also implement smart contracts, which are pieces of code that can automatically execute based on predefined conditional triggers.

1.3 Organization

The chapter highlights as follows: "Preliminaries and Related Research" is divided into two, (1) related work on autism with technology-supported provision and (2) related work on Healthcare 4.0 environment with blockchain. Section 3 focuses on "System Architecture and Construction" with a detailed explanation of the ATA's conceptual model. Section 4 discusses on "System Development and Implementation." Next, Sect. 5 provides "System Analysis and Contributions." Section 6 discusses "Results and Discussions." Section 7 highlights some of the "Limitations and Challenges" of ATA with blockchain security. Lastly, Sect. 8 concludes.

2 Preliminaries and Related Research

Section 2 starts by explaining on related work on autism with technology-supported provision. Next, it continues with the discussions on the Healthcare 4.0 environment with blockchains. The theories and ideas for each research are discussed briefly with the objective, method, and results explanation.

2.1 Related Work on Autism with Technology-Supported Provision

Many researchers focus on several autism difficulties from different angles. One of the angles is on the diagnosis and treatment of autism health problems. Still, most of the researchers handle face-to-face therapy and intervention sessions. The new IR4.0 era has seen a dramatical change in therapy and intervention sessions toward technology-supported provision. For example, Williams et al. [19] surveyed and produced new research directions on education, neurology, psychology, and critical disability studies using wearable technologies. Pérez-fuster et al. [20] enhanced washing saucers skills and doing laundry skills for adults with ASD using digital technology (DT)-mediated intervention and compared them to a treatment-as-usual (TAU) intervention. So et al. [21] examined whether the robot-based drama intermediation had better narrative capabilities and gestured for children with ASD than the other ASD who did not receive the intervention. Maskey et al. [22] used computer-delivered virtual reality and flat-screen graded exposure with cognitive behavioral therapy in young people with ASD for the intervention of phobias and fears. Law et al. [23] investigated whether intervention fidelity can be increased or not among special preschool educators using the developed initially mobile app for parent training. Koumpouros et al. [24] reviewed the use of mobile and wearable technologies for ASD-related interventions and provided guidelines and insights to researchers to create more closer and useful to market products. Torrado et al. [25] targeted to ease behavioral issues on mental health of individuals with ASD using a smartwatch system that infers outburst patterns from physiological signals, movement, and self-regulation strategies. Amiri et al. [26] designed a modern smartwatch called WearSense assembled with IoT framework to detect stereotypic behaviors in children with ASD. Bittner et al. [27] enlarged physiologic responses to physical activity using the proposed ExerciseBuddy application (EB app), a continuous measurement of heart rate and energy expenditure for children with ASD. Campbell et al. [28] processed the implementation and improvement of a digital screening form in quality of care for children at risk for ASD. Artoni et al. [29] implemented coaching and observing learning for autistic children using an open-source Web application named ABA programs with a discovering analytics tool for rehabilitation therapy for children with autism. This technology combined the ABA programs with a discovering analytics tool, an open-source Web application, and was executed for monitoring and teaching-learning for autistic children. Alzrayer et al. [30] used an iPad loaded with Proloquo2Go to children with ASD and other evolving disabilities to determine the effectiveness of orderly instruction on teaching multistep requesting skills. Ahmed et al. [31] notified the emergency alerts using temperature and pulse rate and detected coordinates of the autistic child using a wearable device. Gomez et al. [32] addressed the self-regulation process of individuals with ASD using the potential of interventor smartwatches. Pain et al. [33] designed, pilot tested, and reported multiple technologies on how educational technologies and therapeutics for ASD populations contributed to the dissemination of high standards of an iPad app design for very young children with ASD. The International Society for the Study of Trauma [34] developed a useful robot-based approach to assess autism risk factors for diagnosing children with ASD.

This subsection successfully reviewed comprehensively on the previous autism with technology-supported provision. The list synopsis of them is listed as in Table 15.1.

Based on Table 15.1, we listed a Table 15.2 where the comparison of each method is highlighted based on the purpose and technology-supported provision. From Table 15.2, we can conclude that four different purposes based on examining, diagnosing, intervention, and emergency alert are created for the ASD. Each purpose is using a different technology-supported provision.

Next, Sect. 2.2 demonstrates the concepts and methods of how blockchain is so critical for Healthcare 4.0 situations. Further explanations are described in detail in each study and research.

Table 15.1 List of autism with technology-supported provision

No.	Author	Year	Description
1.	Williams et al. [19]	2020	Surveyed and produced new research directions on education, neurology, psychology, and critical disability studies using wearable technologies
2.	Pérez-fuster et al. [20]	2019	Enhanced washing saucers skills and doing laundry skills for adults with ASD using DT-mediated intervention and compared them to a TAU intervention
3.	So et al. [21]	2019	Examined whether the robot-based drama intervention had better narrative abilities and gestures for children with ASD compared to the other ASD who did not receive the intervention
4.	Maskey et al. [22]	2019	Used computer-delivered virtual reality and flat-screen graded coverage with cognitive behavioral therapy in young people with ASD for the intervention of phobias and fears
5.	Law et al. [23]	2019	Investigated whether intervention fidelity can be increased or not among special preschool educators using the developed initially mobile app for parent training
6.	Koumpouros et al. [24]	2019	Reviewed on the use of mobile and wearable technologies for ASD-related interventions and provided guidelines and insights to researchers to develop more closer and useful to market products
7.	Torrado et al. [25]	2017	Targeted to ease behavioral issues on the mental health of the individuals with ASD using a smartwatch system that infers outburst patterns from physiological signals, movement, and self-regulation strategies
8.	Amiri et al. [26]	2017	Designed a modern smartwatch called WearSense assembled with the IoT framework to spot stereotypic behaviors in children with ASD
9.	Bittner et al. [27]	2017	Improved physiologic responses to physical activity using the proposed EB app, an unceasing measurement of heart rate and energy expenditure for children with ASD
10.	Campbell et al. [28]	n.d.	Processed implementation and improvement of a digital examining form in quality of care for children at risk for ASD

(continued)

Table 15.1 (continued)

No.	Author	Year	Description
11.	Artoni et al. [29]	2017	Implemented teaching and monitoring learning for autistic children using an open-source Web application named ABA courses with a learning analytics tool for rehabilitation therapy for children with autism. This technology combined the ABA programs with a learning analytics tool, an open-source Web application, and was executed for monitoring and teaching-learning for autistic children
12.	Alzrayer et al. [30]	2017	Used an iPad loaded with Proloquo2Go to children with ASD and other evolving disabilities to determine the effectiveness of organized instruction on teaching multistep requesting skills
13.	Ahmed et al. [31]	2017	Notified the emergency alerts using temperature and pulse rate and able to detect coordinates of the autistic child using a wearable device
14.	Gomez et al. [32]	2016	Addressed the self-regulation process of individuals with ASD using the potential of interventor smartwatches
15.	Pain et al. [33]	2016	Designed, pilot tested, and reported multiple technologies on how educational technologies and therapeutics for ASD populations contributed to the dissemination of high standards of an iPad app design for very young children with ASD
16.	International Society for the Study of Trauma [34]	2011	Developed a useful robot-based approach to assess autism risk factors for diagnosing children with ASD

2.2 *Related Work on Healthcare 4.0 Environment with Blockchain*

Since 1760, our healthcare has been transformed into four phases of Industrial Revolution, starting from Healthcare 1.0, where, during this time, people were focusing more on public health solutions. Vaccines were created to solve primary endemic diseases. The appearance of broad production perceptions and technology brought healthcare to 2.0 [35]. During this time, better hospitals and medical education were built, and more qualified doctors were produced for better treatment quality with better facilities. Next, Healthcare 3.0 came after the size of the computer shrunk. During this time, healthcare became faster and more efficient due to the fast development in information technology and computer. Now, we are

Table 15.2 Comparison of autism with technology-supported provision

No.	Main purpose	Technology-supported provision	Author
1.	Examining	Wearable technologies	Williams et al. [19]
			Amiri et al. [26]
			Gomez et al. [32]
		App	Bittner et al. [27]
			Alzrayer et al. [30]
		Digital examining form	Campbell et al. [28]
2.	Diagnosing	Robot	International Society for the Study of Trauma [34]
3.	Intervention	Digital technology	Pérez-fuster et al. [20]
			Maskey et al. [22]
		App	Pain et al. [33]
			Artoni et al. [29]
			Law et al. [23]
		Robot	So et al. [21]
		Wearable technologies	Koumpouros et al. [24]
			J. C. Torrado et al. [25]
4.	Emergency alert	Wearable technologies	I. U. Ahmed et al. [31]

facing a new era of Healthcare 4.0 where innovations lead to smart medicine. The emergence of artificial intelligence and IoT leads to smart healthcare.

Due to the complexity and urgency in the Healthcare 4.0 environment, the necessity to propose an accurate method of answering healthcare security is vigorous and has expected important courtesy from scholars worldwide. Thus, much research has projected a variety of ways with different methodologies to solve uncertainty that arises during the evaluation process and model development. Zhang et al. [36] increased robust and powerful blockchain technologies using AI-mediated data exchange on blockchains with artificial intelligence in real-world healthcare problems. Abujamra et al. [37] improved the healthcare systems' cost efficiencies that are associated with the information technology using blockchain system. Tanwar et al. [38] developed an access control policy algorithm for enlightening data availability among healthcare providers. Farouk et al. [39] discussed the future and vision opportunities of the blockchain platform for industrial healthcare. These blockchain technologies help increase the information security and manage and analyze the healthcare data while preserving the data's security and privacy. Islam et al. [40] collected health data from users using an uncrewed aerial vehicle and stored on the nearest server-based blockchain technology. Comput et al. [41] proposed a smart healthcare system framework using blockchain technology to provide the system's integrity and security, besides discovering the social and technology barriers by examining the user's view and expert perception of the smart healthcare system. Hasselgren et al. [42] enhanced data availability on healthcare providers using an access control policy algorithm and implemented the Hyperledger-based EHR sharing system using the concept of a chain code.

Pandey et al. [43] improved services and health education processes, health sciences, and healthcare using an assessed, synthesized, peer-reviewed, and systematically reviewed proposed blockchain technology. Mcghin et al. [44] exploited a graph neural network (GNN) to manage users' trust in malicious node detection. It prevents anonymous data sharing and storage without permission using smart contracts and blockchain technology. This smart contract and blockchain security are suitable for the smart healthcare system. Onik et al. [45] provided a comprehensive explanation of blockchain challenges and opportunities in healthcare problems and solutions. Besides, it discussed the availability of its regulations and identified any major principles, plan, and prerequisites needed when developing the blockchain system. Tripathi et al. [46] discovered the social and technological barriers of SHS in proving integrity in security and system. Brunese et al. [47] proposed a magnetic resonance image to protect data and information exchanged within the hospital networks and then developed the blockchain technology to authenticate the transmitting data network, check the perform validation, and model the magnetic resonance images for automata radiometric validation. Islam et al. [48] improved the accuracy in videos supporting the cloud computing and fog-based blockchain to recognize and monitor framework in e-healthcare services. Alam et al. [49] highlighted many of the potential research opportunities in healthcare applications that used the blockchain experiments. Gordon and Catalini [50] explored five mechanisms on how blockchain technology facilitated the transition which are (1) digital access rules, (2) data aggregation, (3) data liquidity, (4) patient, and (5) data immutability and next explored the barriers to blockchain-enabled patient-driven interoperability as an exciting trend in healthcare. Koshechkin et al. [51] reviewed public healthcare among the Russian Federation community using the blockchain method. Zhang et al. [52] focused on recognizing suitable blockchain method to be used in healthcare, offering a real case that accomplishes the blockchain method, assessing appropriate design architecture when operating this blockchain technology in healthcare cases.

Further explanation is summarized as Table 15.3.

3 System Architecture and Construction

The use of IR4.0 technology as an educational aid for autistic children consists of the combination of videos, graphics, text, and games that are suitable for them to play and learn at the same time as their therapy session. These technology-supported interactive therapy modules can foster the skills of these autistic children. Based on the study by Stanford University School of Medicine, most autistic children excel in mathematical skills compared with non-autistic children in the same level of IQ range. Recovery in autistic disorder is rare but possible after a few years of regular therapeutic intervention. It is proved by Albert Einstein and Bill Gates, who were once diagnosed with autism disorder problem. Therefore, there is a need to secure

Table 15.3 Summary of blockchain for Healthcare 4.0

No.	Author	Year	Description
1.	Zhang et al. [36]	2020	Increased robust and powerful blockchain technologies using AI-mediated data exchange on blockchains with artificial intelligence in real-world healthcare problems
2.	Abujamra et al. [37]	n.d	Improved the cost efficiencies of the healthcare systems that are associated with the information technology using blockchain system
3.	Tanwar et al. [38]	2020	Access control policy algorithm for enlightening data accessibility between healthcare providers
4.	Farouk et al. [39]	2020	Improved the accuracy in videos supporting the cloud computing and fog-based blockchain for recognizing and monitoring framework in e-healthcare services
5.	Islam et al. [40]	2020	Collected health data from users using an uncrewed aerial vehicle and deposited on the nearest server-based blockchain technology
6.	Comput et al. [41]	2020	Proposed a smart healthcare system named GuardHealth using blockchain technology in providing integrity and security of the system
7.	Hasselgren et al. [42]	2020	Enhanced data accessibility on healthcare providers using an access control policy algorithm and implemented the Hyperledger-based EHR sharing system using the theory of a chain code
8.	Pandey et al. [43]	2020	Improved services and health education processes, health sciences, and healthcare using an assessed, synthesized, peer-reviewed, and systematically reviewed proposed blockchain technology
9.	Mcghin et al. [44]	2019	Exploited a GNN for managing the trust of users in malicious node detection. It prevents anonymous data sharing and storage without permission using smart contract and blockchain technology
10.	Onik et al. [45]	2019	Provided a detailed explanation of blockchain challenges and opportunities in healthcare problems and solutions, besides discussing on the availability of its regulations and identifying any major principles, plan, and prerequisites needed when developing the blockchain system
11.	Tripathi et al. [46]	2019	Technological and social barriers in the adoption of the SHS

(continued)

Table 15.3 (continued)

No.	Author	Year	Description
12.	Reginelli et al. [47]	2019	Proposed a magnetic resonance image to protect data and information exchanged within the hospital networks and then developed the blockchain technology to authenticate the transmitting data network, check the perform validation, and model the magnetic resonance images for automata radiometric validation
13.	Islam et al. [48]	2019	Improved the accuracy in videos supporting the cloud computing and fog-based blockchain for recognizing and monitoring framework in e-healthcare services
14.	Alam et al. [49]	2019	Highlighted many of the potential research opportunities in healthcare applications that used the blockchain experiments
15.	Gordon et al. [50]	2018	Explored five mechanisms on how blockchain technology facilitated the transition which are (1) digital access rules, (2) data aggregation, (3) data liquidity, (4) patient, and (5) data immutability
16.	Koshechkin et al. [51]	2018	Reviewed on public healthcare among the Russian Federation community using blockchain method
17.	Zhang et al. [52]	2018	Focused on recognizing suitable blockchain method to be used in healthcare, offering a real case that accomplishes the blockchain method, assessing appropriate design architecture when operating this blockchain technology in healthcare cases

an app that covers therapy strategies that can have positive impacts on an autistic's socio-emotional well-being such as self-efficacy, social relationships, and identity.

3.1 Architecture Model of the Autism Telerehabilitation Apps (ATA)

We design the app as follows. An autistic child tenuously observed by a therapist is equipped with several medical devices, such as thermometers, blood pressure monitors, and heartbeat rate sensors. This autistic child follows personalized autism home telerehabilitation using ATA guided by his/her parent and online therapist. The raw data is taken from both medical devices and ATA and sent to a UniSZA's cloud data bank for aggregation and formatting. Once complete, the formatted information and customized threshold values are sent to the relevant smart contract for full analysis. The source of information fed in the Ethereum protocol to the smart contracts using ATA is stored in UniSZA's cloud server. The smart contract will process and calculate the collected data and subject notifies to both the patient and therapist provider, and automated treatment instructions for the actuator nodes if desired.

The proposed ATA is designed with blockchain security support's capability to trace and trail the device transactions performed during the managing of the therapy session. Blockchain security not only provides privacy and security features but also provides accessibility of the features and seamless connectivity of the proposed smart apps. The blockchain security may secure the concept of central storage between medical devices and ATA. Figure 15.1 highlights the proposed personalized autism home intervention between sensors, smart devices, actuators, users, networks, and UniSZA's cloud storage. A copy of blockchain secures all data from the medical devices and ATA act as nodes and each node. If the patients are first timers in ATA, they need to register; each node must verify the transaction. Each transaction is validated and each node is recorded and permitted to occur with a hash code in blockchain. Then, each of these nodes is copied and sent to the blockchain for further use and security. Blockchain-based IoT model helps secure all the storage with the data from medical sensors and ATA devices and deliver them to users upon requests with secure transactions.

Figure 15.2 shows the whole of ATA architecture starting from the device and perception layer, network and connectivity layer, management layer, and application layer. All these layers interrelate with each other to offer secure communication in medical sensors and ATA devices for the seamless working of the proposed ATA.

The device and perception layer contains all the hardware devices such as sensors for temperature, blood pressure, heart rate, ECG, respiration, movement, and ATA devices and interconnects them with each other using Wi-Fi connectivity. Next is network and connectivity layer. This layer provides Bluetooth and Wi-Fi components. All the information including data from medical sensors and ATA between autistic children, parents, and therapists is transmitted using this layer. Besides, it provides access to UniSZA's cloud server. Furthermore, it handles remotely and causes high-level processing. The next layer is the management layer. This layer controls all device management, including the cloud server and the proposed blockchain security.

Blockchain offers security protected channels for transmission of data between hardware devices and users (autistic children and parents). IoT application, apps, and system control are the interfaces provided by the application level. All three layers interact with each other and end to the last layer which is the application layer. Our proposed personalized autism home intervention named ATA is parked in the appilcation layer.

Figure 15.3 presents the architectural design of the physical components of the proposed apps. It contains all the medical sensors, hardware devices, ATA, and the overall personalized autism home intervention system in two different parts: Part A and Part B. There are blood pressure monitors, thermometers, and heartbeat rate sensors connected to a raspberry pi microcontroller, a centralized device from Part

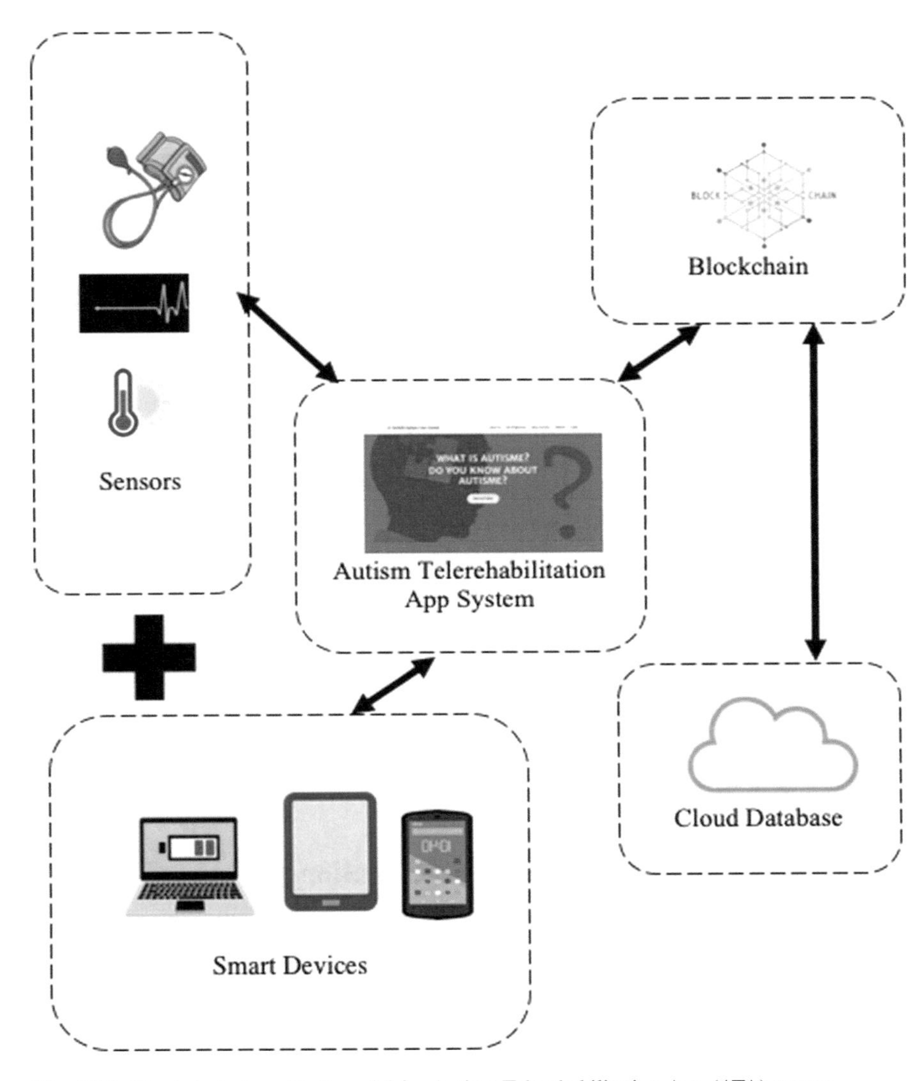

Fig. 15.1 Blockchain-based IoT model for Autism Telerehabilitation App (ATA) system

A. Part B refers to an android application (ATA). All the sensor's data are collected from Part A and transmitted to ATA (Part B) via Bluetooth or Wi-Fi connection. Then both data sensors and therapy data are transformed via Bluetooth or Wi-Fi to UniSZA's cloud database. Next, Sect. 4 elaborates more details on how medical sensors and ATA are linked with blockchain and UniSZA's storage database.

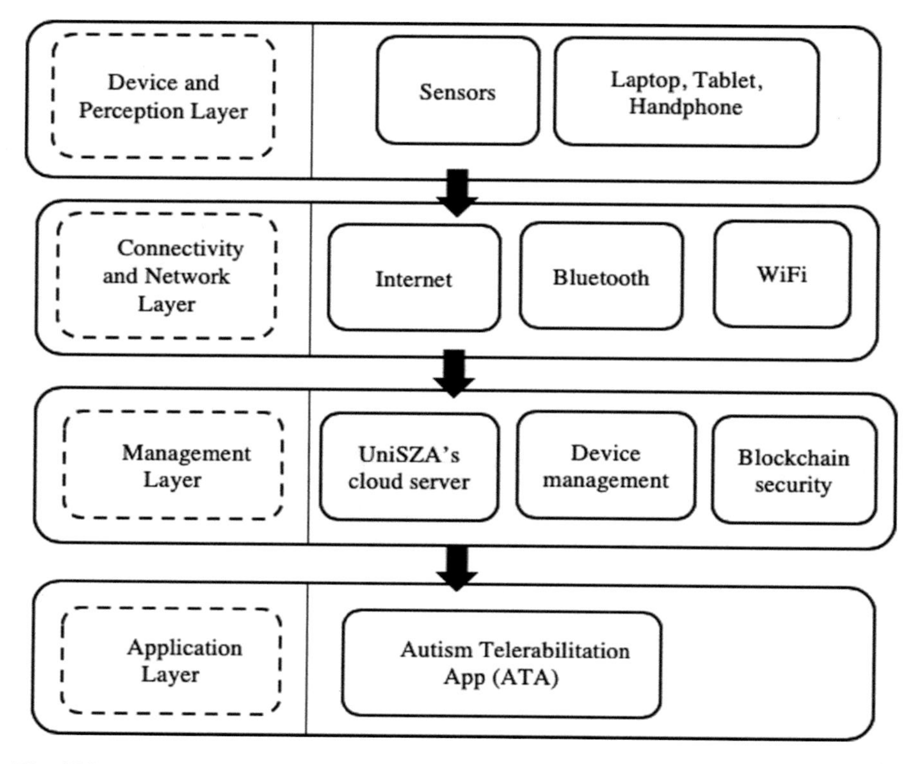

Fig. 15.2 Autism Telerehabilitation App (ATA) architecture

4 System Development and Implementation

We outline the construction and development of the system as one case study as follows:

A new patient came and registered to the ATA; they need to fill up their background details such as their name, age, contact numbers, weight, height, and parent info. Once they submit these details, the admin will provide them with a new id and password, which means that the registration is succeeded. Using the WBAN devices, the blood pressure, temperature, and heartbeat rate will be collected and recorded in a system. Then, the system will suggest this patient take an early diagnosis test for autism. If the results shown are high, they will directly be referred to the therapist. The therapist will monitor every single aspect in each patient account. The therapist will suggest suitable therapy for the intervention. Each diagnosis and intervention data will be recorded and kept well in the system.

Details of the system development and implementation are explained as follows:

A patient remotely observed by a therapist is supplied with medical sensors to monitor their blood pressure, temperature, and heartbeat rate which are connected

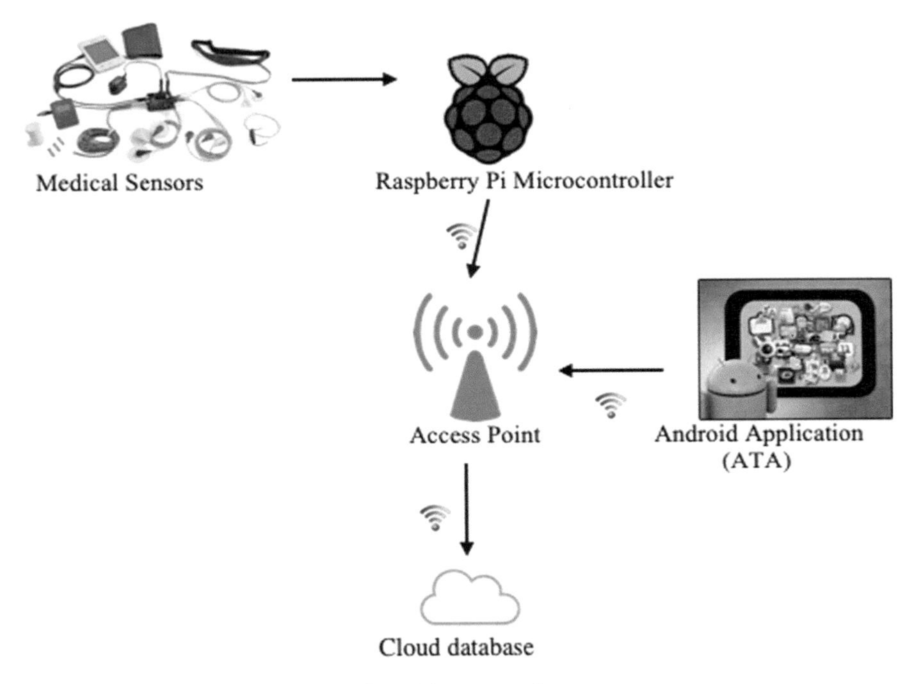

Fig. 15.3 Design of hardware integration of the proposed system

to a raspberry pi microcontroller. The raw data from sensors are combined with the raw data taken from the ATA once all the therapy sessions complete. The raw data from ATA are taken from smart devices, either smartphones or tablets. Both data from sensors and smart devices are aligned and combined, for aggregation and formatting. For full investigation along with customized threshold values, the constructed information is sent to the relevant smart contract once it is completed.

This chapter created a smart representation of all medical records taken from medical sensors and ATA and kept on the network within individual nodes using smart contracts from Ethereum. Contracts are built based on data integrity, record ownership, permissions, and metadata. Each contract carries out policies for state-transitioning functions and the legitimate transaction compelling data alternation. Smart contracts for ATA have been created by using autism medical workflows and then organizing data access authorization between the different individuals in the ATA ecosystem. All the important data need to grant a third party (in this case, an ATA admin) for viewing permission. All data authorization rules are embedded in smart contracts. These smart contracts also help in tracking all the autistic activities (including diagnosis and therapy) and the history of all patients. Besides, we need different permissions to retrieve all of the data. Smart contracts evaluated the retrieved data from the medical sensors and medical devices with Oracle's help. Oracle is the smart device that interconnects straightly to the smart contracts in the

Ethereum protocol. Patients' identification card number is used as a unique id for the tracking activities. All the autism record data is securely stored in UniSZA's database storage to preserve performance and security. If there are new therapy sessions or new news, the admin will add a new record and update in the ATA. All the information receives an automated notification and needs verification of the projected record before the data is rejected or accepted. This will help in creating high trust and better interaction between patients and therapists.

All the activities, therapy sessions, and events are recorded using blockchain technology as a ledger. Next, they are forwarded to an authorized EHR UniSZA's storage database. EHR offers patient-centric privacy based on the readiness of patients. EHR enhances a vital role in minimizing patient operating costs and offers accurate and better healthcare intelligence and patient care quality. Once each patient finishes their therapy session, it automatically transmits therapist's results through the smart contract system. All treatment records and commands are automatically recorded in a blockchain transaction. All the patient's medical data and history are recorded with care using blockchain transactions connected to the EHR to supply data authentication. This authentication helped to detect any amendments to a patient's EHR. The therapist must view and respond to the patient's therapy results and refer the patients to the specialist when appropriate for further care. All the patient's therapy data and information should be reported to the EHR. These data are collected and kept in an UniSZA's local database with the specific rules where only authorized people can access to this record. The smart contract on Ethereum blockchain governs all the authorization. Here, blockchain helps to store, analyze, and gain health data without disturbing the privacy feature on online medical data. Otherwise, the therapist will be suggested directly to the hospital if any emergency occurs.

5 System Contributions

Once the system development and implementation succeed, there is a condition that highlights that this proposed system contributes an excellent effect to the autism home intervention. Thus, we listed five significant improvements from this system compared with the existing system. These differences and comparisons are highlighted into four benefits: (1) speed up the process and reduce the error; (2) reduce the cost, (3) high the security aspects, and (4) privacy and transparency.

1. Speed up the process and reduce the error

Traditional intervention therapy sessions usually took place for 1 to 3 hours per week over 12 consecutive weeks or till the patients show any improvement. All sessions need to be seriously monitored by both patients and therapists and cannot be skipped. Otherwise it will interrupt the consequent sessions. Parents and autistic children usually take half an hour before and after to prepare for the therapy center. Using this ATA, parents and their autistic children will get ready anytime and anywhere for the therapy session without waiting for the therapist to respond. They can choose a suitable time to do this therapy, while the automated smart ATA system will monitor the therapy session. Besides, this ATA can reduce the error made by the human with the help from smart medical sensors and smart medical devices. Again, it can reduce the inaccurate interpretation of manually written prior authorization forms upon requests.

2. Reduce the cost

This proposed automated ATA process would ensure in noteworthy cost saving for patients, which presently consumes extensive amounts on manually sending and waiting process to the therapy center. Meanwhile, the traditional therapy sessions usually offered high fees compared to the online therapy session fees due to many side charges. This side charges for therapist salary, and rental spaces including center fixtures, equipment, inventory and data system, operating expenses and many more. For time issues, patients will be able to schedule their exact time to do the therapy session. Besides, therapists will be able to continue speedily with therapy remedy instead of having to stop their patient's care while waiting in line for another patients' therapy session.

3. High the security aspects

With this ATA, all patients' data, identity, and therapy records are well kept and secured in UniSZA's cloud database. Each patient has the authority to access their data. Nevertheless at the same time, the smart contacts help in the sharing of data by using secure data-sharing features. All the patients' data collected from the sensors and medical devices has been protected against integrity and confidentiality attacks using a scalable and secure storage layer. Each patient has the authority to view their latest and previous data. In contrast, the healthcare professionals (including therapists and doctors) have the authority to edit the thresholds of their patients for the smart contracts. Moreover, each individual patient's data are interlinked and independent with each other and well kept in the management layer. This management layer also processed and evaluated all the incoming and outgoing data using the data access gateway. This data access gateway is also able to manage all the heterogeneous patients' data in the database engine. Although each patient can preview and access their own data, these smart contracts are still required to achieve a majority signatures (to make a block valid) from consortium members to prevent one partaker from controlling the ledger. The full observing privileges of the blockchain itself are inhibited to only permitted parties involved in HDG such as the government, the Ministry of Health, recognized medical officers, researchers, and others.

4. Privacy and transparency

Blockchain nodes and ledger increase the privacy and transparency on data sharing between patients and medical officers. Our ATA system offers immutable and authenticated records of a patient's observing to settle disputes or investigate. Anonymous addresses will protect the identity of patients; hence, no involvements can be made between patients and their data. Moreover, patients can associate remote monitoring actions straight to their medical records while preserving control and privacy.

6 Results and Discussions

Using blockchain technology, our smart contract for ATA system has been developed to manage large-scale autism data (including patient data, diagnosis data, therapy, etc.) and streamline complex autism procedures. We proposed an advanced approach from autism record handling, affording interoperability, offering auditability, and easy accessibility via these smart contracts. Besides, this ATA system allows data sharing in safe mode sharing among permitted parties. Using blockchain technology, availability, security, privacy, and fine-grained control of EHR data access can be certified. The blockchain in this chapter helps to increase the healthcare practices and the patient managements. Our proposed ATA system uses blockchain-based smart contract technology to produce a secure, private, trustworthy, accessible environment for patients and healthcare providers. Patients will feel secure and safe especially in terms of data sharing with the therapist, doctors, researchers, and so on. This ATA system helps to solve concerns on data management, data security, and privacy, high administrative costs, and network security issues.

7 System Limitation and Challenges

Although blockchain helps increase privacy and security, there are some limitations on this ATA system. First is the size of blockchain. The size of blockchain is a very critical part, and data continues to grow for every transaction. The addition of new blocks data into old data records increased day by day upon the transaction, and this size will be accumulated each year. Once the addition of data is occurring based on the increase in the number of nodes, and this data is broadcasting to all nodes, then the cost for a bigger size database will also be increased, besides the increasing number of patients. New data space will be created once the new patients registered. The number of nodes is also growing once these patients continue the therapy transactions. Each of the nodes will be increased once they updated their therapy record. Besides, another challenge is to maintain security at every individual node. This security includes the data transaction and transmission between a patient's medical sensors and patient's smart devices as if used local or public Wi-Fi and relies on standard channel encryption. Moreover, other challenges are the management issues on verification of the next block due to multiple nodes from the smart devices' broadcasting transactions. This could be resolved by producing large numbers of keys using a key management system. However, these limitations can still be overcome with future development and research.

8 Conclusion and Future Work

In this chapter, we described an ATA along with medical sensors in a WBAN for personalized autism home intervention using blockchain system-based smart contracts to perform log transaction metadata and real-time analysis. Blockchain-based smart contracts have highlighted security concerns about the logging and transfer of telerehabilitation of autistic data transactions. Once the patient is registered via the ATA system, this blockchain managed and executes smart contracts to evaluate all the information retrieved by each patient based on customized threshold values. Each transaction of recording details on the blockchain is verified, and the smart contract would spark signals for the patient and therapist for verification of EHRs. Blockchain has the potential to automate the delivery of health-related notifications in a compliant manner and improve security for personalized autism home intervention. It helps collect and manage patient data, clean this data, secure them using blockchain, handle big data, and come out with significant results. It also can trigger alerts for the unusual data taken. For future work, introducing anonymizers to increase patients' privacy by the difficulty of linking transaction together. Lastly, applying new applications in other healthcare services such as mental health services, physiotherapists for stroke, and many more.

References

1. M.K. Hassan, A.I. El Desouky, S.M. Elghamrawy, A.M. Sarhan, A Hybrid Real-time Remote Monitoring Framework with NB-WOA algorithm for patients with chronic diseases. J. Future Gener. Comput. Syst. **93**, 77–95 (2018)
2. A. Isath, V. Vaidya, V. Yogeswaran, A. Deskmukh, S. Asirvatham, D. Hayes, S. Kapa, Long term follow-up of patients with ventricular high rate events detected on remote monitoring of pacemakers. Indian Pacing Electrophysiol. J. **19**(3), 92–97 (2019)
3. K. Curila, J. Smilda, O. Leseticky, D. Herman, P. Stros, P. Osmancik, J. Zdarska, R. Prochazkova, P. Widimsky, Cost effectiveness analysis of out-patient and remote monitoring of patients after pacemaker replacement from the perspective of the health care payer. J. Cor Vasa **60**(4), e387–e392 (2018)
4. T.L. Michaud, M. Siahpush, K.M. King, A.K. Ramos, R.E. Robbins, R.J. Schwab, M.A. Clarke, D. Su, Program completion and glycemic control in a remote patient monitoring program for diabetes management : Does gender matter ? J. Diabetes Res. Clin. Pract. **159**, 107944 (2020)
5. C.L. Downey, J.M. Brown, D.G. Jayne, R. Randell, Patient attitudes towards remote continuous vital signs monitoring on general surgery wards : An interview study. Int. J. Med. Inform. **114**, 52–56 (2018)
6. J. Ng, S.F. Sears, D.V. Exner, L. Reyes, X. Cravetchi, P. Cassidy, J. Morton, C. Lohrenz, A. Low, R.K. Sandhu, R.S. Sheldon, S.R. Raj, Age, sex, and remote monitoring differences in device acceptance for patients with implanted cardioverter defibrillators in Canada. CJC Open **2**(6), 483–489 (2020)
7. P.N. Ramkumar, H.S. Haeberle, D. Ramanathan, W.A. Cantrell, S.M. Navarro, M.A. Mont, M. Bloomfield, B.M. Patterson, Remote patient monitoring using mobile health for total knee arthroplasty: Validation of a wearable and machine learning e based surveillance platform. J. Arthroplast. **34**, 2253–2259 (2019)

8. L.M. Posthuma, C. Downey, M.J. Visscher, D.A. Ghazali, M. Joshi, H. Ashrafian, S. Khan, A. Darzi, J. Gildstone, B. Preckel, Remote wireless vital signs monitoring on the ward for early detection of deteriorating patients - a case series. Int. J. Nurs. Stud. **104**, 103515 (2019)
9. R.C. Walker, A. Tong, K. Howard, S.C. Palmer, Patient expectations and experiences of remote monitoring for chronic diseases: Systematic review and thematic synthesis of qualitative studies. Int. J. Med. Inform. **124**, 78–85 (2019)
10. K.N. Griggs, O. Ossipova, C.P. Kohlios, A.N. Baccarini, E.A. Howson, Healthcare blockchain system using smart contracts for secure automated remote patient monitoring. J. Med. Syst. **42**, 130 (2018)
11. V. Chang, P. Baudier, H. Zhang, Q. Xu, J. Zhang, M. Arami, How Blockchain can impact financial services – The overview, challenges and recommendations from expert interviewees. Technol. Forecast. Soc. Chang. **158**, 120166 (2020)
12. Y. Li, Emerging blockchain-based applications and techniques. SOCA **13**, 279–285 (2019)
13. P. Mehta, R. Gupta, S. Tanwar, Blockchain envisioned UAV networks: Challenges, solutions, and comparisons. Comput. Commun. **151**, 518–538 (2020)
14. A. Kumari, R. Gupta, S. Tanwar, N. Kumar, A taxonomy of blockchain-enabled softwarization for secure UAV network. Comput. Commun. **161**, 304–323 (2020)
15. S. Tanwar, Q. Bhatia, P. Patel, A. Kumari, P.K. Singh, W.C. Hong, Machine learning adoption in blockchain-based smart applications: The challenges, and a way forward. IEEE Access **8**, 474–488 (2020)
16. N. Kabra, P. Bhattacharya, S. Tanwar, S. Tyagi, MudraChain: Blockchain-based framework for automated cheque clearance in financial institutions. Futur. Gener. Comput. Syst. **102**, 574–587 (2020)
17. R. Gupta, A. Kumari, S. Tanwar, A taxonomy of blockchain envisioned edge-as-a-connected autonomous vehicles. Trans Emerging Tel Tech. 2020;e4009. https://doi.org/10.1002/ett.4009
18. U. Bodkhe, P. Bhattacharya, S. Tanwar, S. Tyagi, N. Kumar, M.S. Obaidat, BloHosT: Blockchain enabled smart tourism and hospitality management, in *International Conference on Computer, Information and Telecommunication Systems (IEEE CITS-2019)*, Beijing, China, August 28–31, 2019, pp. 237–241
19. R.M. Williams, J.E. Gilbert, Perseverations of the Academy: a survey of wearable technologies applied to autism intervention. Int. J. Hum. Comput. Stud. **143**, 102485 (2020)
20. P. Pérez-fuster, J. Sevilla, G. Herrera, Research in Autism Spectrum Disorders Enhancing daily living skills in four adults with autism spectrum disorder through an embodied digital technology-mediated intervention. Res. Autism Spectr. Disord. **58**, 54–67 (2019)
21. W. So, C.-H. Cheng, W.-Y. Lam, T. Wong, W.-W. Law, Y. Huang, K.-C. Ng, H.-C. Tung, W. Wong, Robot-based play-drama intervention may improve the narrative abilities of Chinese-speaking preschoolers with autism spectrum disorder. Res. Dev. Disabil. **95**, 103515 (2019)
22. M. Maskey, H. McConachie, J. Rodgers, V. Grahame, J. Maxwell, L. Tavernor, J.R. Parr, An intervention for fears and phobias in young people with autism spectrum disorders using flat screen computer-delivered virtual reality and cognitive behaviour therapy. Res. Autism Spectr. Disord. **59**, 58–67 (2019)
23. G.C. Law, A. Dutt, M. Neihart, Research in Autism Spectrum Disorders Increasing intervention fidelity among special education teachers for autism intervention : A pilot study of utilizing a mobile-app- enabled training program. Res. Autism Spectr. Disord. **67**, 101411 (2019)
24. Y. Koumpouros, T. Kafazis, Research in Autism Spectrum Disorders Wearables and mobile technologies in Autism Spectrum Disorder interventions: A systematic literature review. Res. Autism Spectr. Disord. **66**, 101405 (2019)
25. J.C. Torrado, J. Gomez, Emotional self-regulation of individuals with autism spectrum disorders: Smartwatches for monitoring and interaction. J. Sensors (Basel) **6**, 17 (2017)
26. A.M. Amiri, N. Peltier, C. Goldberg, Y. Sun, A. Nathan, S.V. Hiremath, K. Mankodiya, WearSense: Detecting autism stereotypic behaviors through smartwatches. J. Healthc. (Basel) **5**(1), 1–9 (2017)
27. M.D. Bittner, B.R. Rigby, L. Silliman-french, D.L. Nichols, S.R. Dillon, Use of technology to facilitate physical activity in children with autism spectrum disorders: A pilot study. Physiol.

Behav. **177**, 242–246 (2017)
28. K. Campbell, K.L.H. Carpenter, S. Espinosa, J. Hashemi, Q. Qiu, M. Tepper, R. Calderbank, G. Sapiro, H.L. Egger, J.P. Baker, G. Dawson, Use of a digital modified checklist for autism in toddlers – Revised with follow-up to improve quality of screening for autism. J. Pediatr. **183**, 133–139.e1 (2017)
29. S. Artoni, L. Bastiniani, M.C. Buzzi, O. Curzio, S. Pelagatti, C. Senette, Technology-enhanced ABA intervention in children with autism: A pilot study. Univers. Access Inf. Soc. **17**, 191–210 (2017)
30. N.M. Alzrayer, D.R. Banda, R. Koul, Teaching children with autism spectrum disorder and other developmental disabilities to perform multistep requesting using an iPad. Augment. Altern. Commun. **33**(2), 65–76 (2017)
31. I.U. Ahmed, N. Hassan, H. Rashid, Solar powered smart wearable health monitoring and tracking device based on GPS and GSM technology for children with autism, in *2017 4th International Conference on Advances in Electrical Engineering (ICAEE)*, (IEEE, 2017), pp. 111–116
32. J. Gomez, J.C. Torrado, Using smartwatches for behavioral issues in ASD, 2016, pp. 10–11
33. H. Pain, S. Hammond, A. Humphry, H. Mcconachie, Designing for young children with autism spectrum disorder: A case study of an iPad app. Int. J. Child-Comput. Interact. **7**, 1–14 (2016)
34. International Society for the Study of Trauma, Guidelines for treating dissociative identity disorder in adults, third revision. J. Trauma Dissociation **12**(2), 115–187 (2011)
35. C. Chen, E.-W. Loh, K.N. Kuo, K.-W. Tam, The times they are a-changin' – Healthcare 4.0 s coming! J. Med. Syst. **44**(40), 1–4 (2020)
36. P. Zhang, M.N.K. Boulos, Blockchain solutions for healthcare, in *Precision Medicine for Investigators, Practitioners and Providers*, (Academic Press, London/San Diego, 2020), pp. 519–524
37. R. Abujamra, D. Randall, Blockchain applications in healthcare and the opportunities and the advancements due to the new information technology framework. Adv. Comput. **115**, 141–154 (2019)
38. S. Tanwar, K. Parekh, R. Evans, Blockchain-based electronic healthcare record system for healthcare 4.0 applications. J. Inf. Secur. Appl. **50**, 102407 (2020)
39. A. Farouk, A. Alahmadi, S. Ghose, A. Mashatan, Blockchain platform for industrial healthcare: Vision and future opportunities. Comput. Commun. **154**, 223–235 (2020)
40. A. Islam, S.Y. Shin, A blockchain-based secure healthcare scheme with the assistance of unmanned aerial vehicle in Internet of Things R. Comput. Electr. Eng. **84**, 106627 (2020)
41. J.P.D. Comput, Z. Wang, N. Luo, P. Zhou, GuardHealth: Blockchain empowered secure data management and Graph Convolutional Network enabled anomaly detection in smart healthcare. J. Parallel Distrib. Comput. **142**, 1–12 (2020)
42. A. Hasselgren, K. Kralevska, D. Gligoroski, S.A. Pedersen, A. Faxvaag, Blockchain in healthcare and health sciences — A scoping review. Int. J. Med. Inform. **134**, 104040 (2020)
43. P. Pandey, R. Litoriya, Implementing healthcare services on a large scale: Challenges and remedies based on blockchain technology. J. Health Policy Technol. **9**, 69–78 (2020)
44. T. Mcghin, K.R. Choo, C. Zhechao, D. He, Blockchain in healthcare applications: Research challenges and opportunities. J. Netw. Comput. Appl. **135**, 62–75 (2019)
45. M.H. Onik, S. Aich, J. Yang, C. Kim, H. Kim, Chapter 8 – Blockchain in healthcare: Challenges and solutions, in *Big Data Analytics for Intelligent Healthcare Management*, (Academic Press, London/San Diego, 2019), pp. 197–226
46. G. Tripathi, M. Abdul, S. Paiva, S2HS- A blockchain based approach for smart healthcare system. Healthcare (Amst) **8**(1), 100391 (2019)
47. L. Brunese, F. Mercaldo, A. Reginelli, A. Santone, A blockchain based proposal for protecting healthcare systems through formal methods. Procedia Comput. Sci. **159**, 1787–1794 (2019)
48. N. Islam, Y. Faheem, I. Ud, M. Talha, M. Guizani, A blockchain-based fog computing framework for activity recognition as an application to e-Healthcare services. Futur. Gener. Comput. Syst. **100**, 569–578 (2019)

49. F. Alam, M. Asif, A. Ahmad, M. Alharbi, H. Aljuaid, Blockchain technology, improvement suggestions, security challenges on smart grid and its application in healthcare for sustainable development. Sustain. Cities Soc. **55**, 102018 (2020)
50. W.J. Gordon, C. Catalini, Blockchain technology for healthcare: Facilitating the transition. Comput. Struct. Biotechnol. J **16**, 224–230 (2018)
51. K.A. Koshechkin, G.S. Klimenko, I.V. Ryabkov, P.B. Kozhin, ScienceDirect scope for the application of blockchain in the public healthcare of scope for the application of Blockchain in the Public Healthcare of the Russian Federation the Russian Federation. Procedia Comput. Sci. **126**, 1323–1328 (2018)
52. P. Zhang, D.C. Schmidt, J. White, G. Lenz, Blockchain technology use cases in healthcare, 2018, https://www.dre.vanderbilt.edu/~schmidt/PDF/blockchain-bookchapter-2018.pdf. Accessed 1 Sept 2020

Part IV
5G-Enabled IoT Models, Solutions and Standards

Chapter 16
A Hybrid Blockchain-Secured Elderly Healthcare Environment

Aishwarya Gupta, Pooja Khanna, and Sachin Kumar

1 Introduction

The healthcare industry is that sector of the economy that provides healthcare goods and services to treat people or the patients with rehabilitative and curative care. Healthcare industries are termed as one of the largest and fastest growing industries that consume most part of the economy.

The healthcare services are governed by the private hospitals, government hospitals, and private clinics. Most people prefer private clinics as they are convenient to visit and provide hassle-free care to its patients. Healthcare systems presently functional are managed by various authorities and laws as put down by state and central government; sometimes, services provided are not up to the mark leading to improper care of the patients, citing availability of experts, infrastructure facilities, population density, and remote location as primary reasons. The private sector surpassed the public sector in terms of the healthcare services, and it became a preferred choice of the patients as they gave more assurance to their consumers.

The healthcare industry is growing at a rapid pace during these years due to the increase in the income of the people and the growing population of the elderly people. Elder people play a major role in the expansion of the healthcare sector as this section of the society needs more care and convenience than the people of any other age group. In addition to these, the changing demographics as well as new lifestyle has led to the outburst of diseases which in turn have led to vast research and development in this field. Despite the presence of many healthcare facilities, these all seem to be inadequate due to various reasons including the overburden of the dense population and the mismanagement of the available resources [1–6].

A. Gupta · P. Khanna · S. Kumar (✉)
Amity University, Lucknow Campus, Lucknow, Uttar Pradesh, India
e-mail: aishwarya.gupta7@student.amity.edu; pkhanna@lko.amity.edu; skumar3@lko.amity.edu

S. Tanwar (ed.), *Blockchain for 5G-Enabled IoT*,
https://doi.org/10.1007/978-3-030-67490-8_16

Thus, all these have led to the requirement of the healthcare assistance system to provide convenient healthcare to people especially to the elderly people who find it difficult for themselves to find and have healthcare assistance. During the 1970s onward, the globe witnessed the emergence of modular IT systems in the healthcare industry, a period addressed as Healthcare 1.0. Throughout the next decade and a half, healthcare systems at diversified platforms started getting networked with integrated EHRs and clinical imaging, giving doctors an optimized perspective for diagnosis; this era was addressed as Healthcare 2.0. During 2005 onward, the world witnessed the development of genomic information and emergence of wearables and implantable, integrated with EHRs, termed as Healthcare 3.0. Lately, the advent and coming together of disruptive technologies like AI, augmented reality, blockchain, IoT, robotics, etc., coupled with real-time data collection for efficient healthcare support systems, gave rise to the era of Healthcare 4.0. Focus with Healthcare 4.0 was more on collaboration, coherence, and convergence for concluding to more predictive, personalized, and informed decision, with anytime anywhere access and enhanced analytics. Healthcare 4.0 facilitated accurate differential diagnosis and timely medical responses even to remote locations [7–11].

The healthcare system generally included an app-based service which has a unique feature of search, contact, compare, evaluate, and select accordingly. This system focuses on the convenience for all but aims at providing a little more help to the elders for the medical assistance. The chapter is organized as follows: Section 2 presents the motivation for taking up the work; Section 3 provides the methodology proposed for the work; Section 4 presents the implementation details of the proposed work; Section 5 explores about the benefit of proposed model, such as enhanced features of security and reduced latency; Section 6 discusses about future scope of the healthcare support systems for the elderly and work conducted by the authors; and finally the paper concludes with optimum choices for design in Sect. 7.

2 Motivation

According to literature from the World Bank and UN, 12% of the population in developed region was above 60 years in 1950, and the value reached 23% in 2013 and is expected to reach 28% by 2025 and 32% by 2050. These numbers will decide infrastructural, technological, and financial support challenges. The globe would be confronted too. Changes in population demographic will lead to implications on healthcare systems, workforce, social systems, and economic systems. Healthcare systems for the elderly population would require convenient, flexible, and user-friendly solutions in addition to support from medical professionals.

With the Fourth Industrial Revolution, connectivity and technological awareness level across the globe has risen. Even people in remote locations now have access to smart technologies. Healthcare is a sector which is expected to have a wide reach, even to people in remote areas, and innovations with smart communication devices and technology have made this possible. The market is already having a number of existent medical services, quite a few online portals are there, and these websites and applications assist in providing medical aid to the patients through video chats, messaging support with physicians and other professional doctors, and online pharmacy support and electronically generated prescription services. Whereas some of these services facilitated digitization of hospitals, very few provided home-based check-ups and medicine home delivery. One common trait seen in most of them was the considerably high consultation and medicinal charge mainly because of the services being provided by a small group or organization. The proposed system in a way overcomes and minimizes this issue of unaffordable medical aid: firstly, because the services are not exclusively provided by a group of trained doctors but it rather directs its users to the already practicing doctors who charge a nominal amount and secondly, because an automated option for the online home delivery of medicines at a minimum possible cost from a number of nearby pharmacies would also be provided to the patients [9–15].

Other important factors include the fog layer being employed for interaction with the user interface while fetching the medical tests needed for a specific disease and also displaying the options for the nearest pathologies available. This fog layer reduces the latency in getting the data. Based on the last visit and the patient's illness (along with its stage), his/her recovery time is predicted, and the relevant precautions and tips are provided on the home screen itself in a user-friendly graph format (say, Gantt chart). Moreover, collecting feedback after each doctor visit would further improve the search results of the patient with the doctors being displayed popularity-wise (doctors having the most number of feedbacks displayed first and so on).

The framework proposed offers many benefits as compared to already existing solutions; some differentiating characteristics include electronic prescription and report generation, personal information of the patient is automatically scanned and extracted from the QR code, and no manual entry is needed for it. Entry of symptoms, medications and tests are made easy with the help of Auto-Complete suggestions. Check-box entries are employed for common frequently occurring names and Apps are used for locating directions to nearest medical clinic and pathologies for visit, which is potentially one of the major problems faced by the old people and patients in general. Due to hassles in finding the doctors through their specialization, symptom based medical centers, procedure for automated ordering of medicines and easy to use platform for Online transaction, patients often end up skipping their medicine dose which might ultimately lead to health complications and late recovery of the patients, generally patients do not have the patience, time, and energy to run for their medications, therefore easy to use support system is the need of hour. The framework proposed automates and eases the process of procuring medicines with one push delivery of medicines at home, the framework provides a technique for maintaining a huge amount of patient data at one place, and the technique also incorporates blockchain for the transfer of medical records and patient details to ensure that the process takes place with complete confidentiality and any unauthorized intrusion doesn't take place. Blockchain also maintains synchronous and tamper-proof medical record as case studies which may be utilized

Table 16.1 Existing mobile-aided medical support systems

App/online portals	Merit details	Ref.
SteadyMD.com	Fully available online service. It charges a subscription fee of $99 per patient or $169 per family. Each Doctor has just 200–400 patients total for managing	[16]
PlushCare	Fee-for-service video chat exams with physicians. The per visit fee is $99. PlushCare doctors accept most major insurance plans. PlushCare works with the established specialists. Books appointment 24 hours a day	[17]
DoctorOnDemand	Offers short video consultations 24 hours a day with doctors. Patients have an option of scheduling an appointment for a later time as well as seeing the next available doctor "on demand"	[18]
BetterHelp.com	Focuses exclusively on electronic counseling. Patients who prefer a live session can schedule it via their smartphone, tablet, or desktop. Patients using "BetterHelp.com" have to pay a fee of $65 per week that covers all of their communication with their therapist	[19]
First Opinion	Completely free service. Offers unlimited messaging 24*7 with their Matched Doctor Team (MDT). Following are not permitted: video chat examination, an online prescription, or get a sick note. Only health-related questions are permitted	[20]
Amwell	An online doctor it is. Once we enroll in the service and choose from the available doctors, one enters a virtual waiting room. Online doctor do prescription and these prescriptions are sent electronically to pharmacy for procuring medicines, so there's no hassle	[21]
Online Registration System (ORS)	Framework links hospitals with Aadhaar-based online registration and appointment system. Portal facilitates online appointments with various departments of hospitals using eKYC data registered with UIDAWE	[22]
Practo	Provides consultation services to its patients over the app and allows the customer to leave a feedback. Allows an easy edition of displayed information, such as timing, fees, services, etc., from anywhere allows it to be manageable	[23]

for research and reference purposes. Table 16.1 depicts already available existing mobile-aided medical support systems in the domain:

In most of the systems and online services, usually patient's personal data along with the treatment duration and at most medications are stored at one place. What makes our application stand apart is that the storage of patients' entire medical record is at one place in the same database every time a patient visits a doctor for check-up and consultation either online or in person, the medications could be ordered online or through any local pharmacy, prescriptions are saved online, and the app proposed is extremely user-friendly since the users not only have the ease of searching the nearest doctor and pathology on the go but can also view and contact them in case of an emergency or so. The patient himself can view and share his previous medical history for disease prevention and the precautions to be needed.

3 Proposed Methodology

Evolutionary prototype software development lifecycle model was employed for the development of the proposed healthcare system. The system developed was huge owing to so many dimensions it had. We started with only a minimal set of requirements, and since the visibility of requirements became more clear as the design progressed, therefore evolutionary prototype application development model seemed the most suitable for the scenario. Similar to the spiral model, work took a start from the most important and complex requirements first and then moved on to the less important and simple ones. Figure 16.1 depicts the evolutionary prototype software development flow diagram.

Design started with the major module for the authentication and verification of doctors and then moved on to the patient registration with the help of the Aadhaar UID. Establishing a strong connection between the database tables was a huge task for the later easy retrieval and addition of data. After the creation of the prototype based on the given set of requirements, the prototype was tested at each step of design for the then system functionalities, and feedbacks were taken in order to make further improvements and get a new set of requirements to work upon.

Instead of starting from scratch every time, we worked upon the shortcomings of the system software. This cycle of requirement gathering, prototype creation, verification, and correction continued till the final satisfactory application was obtained.

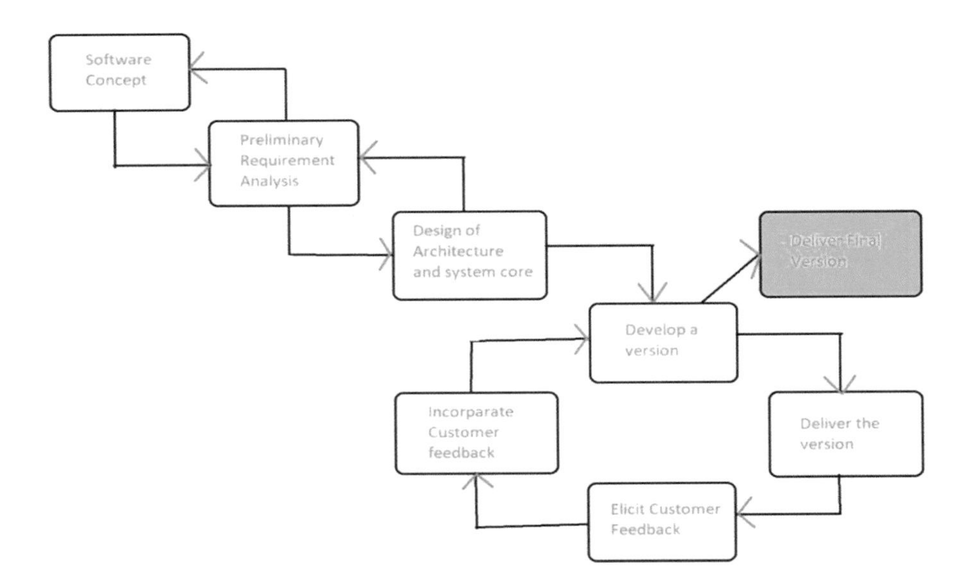

Fig. 16.1 Evolutionary prototype software development flow

Reasons for choosing this approach are as follows:

(i) All of the requirements were not gathered at the very beginning but developed gradually.
(ii) Systems so designed were robust with efficient maintainability because of the iteration cycles that the prototype underwent before the final application software was ready.

3.1 Model Design

Preliminary work started with the identification of necessary functionalities required for the work.

3.1.1 Use Case and Dependencies

In the initial phase of the work, basic functionalities were realized for implementation in the application, as depicted in Fig. 16.2.

In spite of the diverse nature of work and various dimensions, the broad idea is still the same, that is, to provide easy and enhanced medical support, especially for the elderly patients.

The application's target audience can be categorized into two types of users:

(i) The certified doctors whose main task is to provide his/her patients with the best suited cure at that point of time. He/she can do that by viewing and analyzing all the previous medical symptoms and medications as taken by the patient keeping in mind the current scenario. He/she has the choice of adding the prescription and diet plans either in the digital typed format or by uploading or scanning a hardcopy of the same.
(ii) The patients or general users of the app. These users are provided with the basic facilities such as searching for the nearest hospitals, looking for medical details, first aid tips, simple home remedies, setting medicine reminders, chatting with

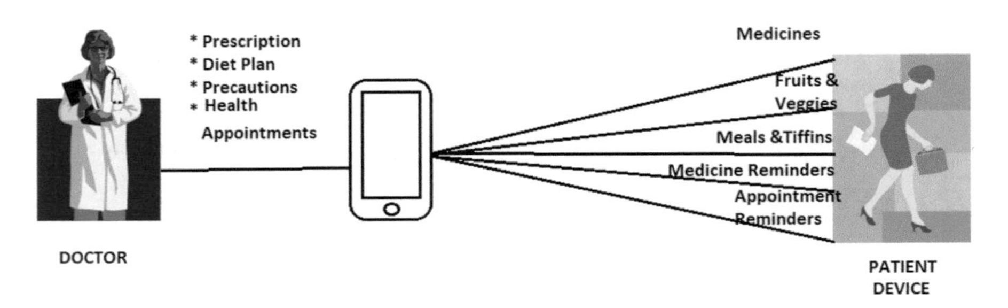

Fig. 16.2 Basic functionalities of a healthcare application

the chatbot, and so on. For the patient, especially for the elderly, there would be an option for voice consultation instead of merely chatting, and the voice could be one cloned in a way similar to the patient's close relative (emotional support for better and faster recovery). Before visiting a doctor in the case of mild symptoms, a user of the application can make use of the symptom checker module and get an estimate of the probable disorder he/she might stand a chance of suffering. Not only that, a regular patient (say someone suffering from a chronic disease who needs to go for check-ups and doctor visits on a regular basis) can view his/her medical records, prescription, diet charts, appointment, and medicine reminders as well as upload their latest reports too.

Figure 16.3 depicts a use case showing the role of the two actors and their way of interaction with the system, as proposed for the healthcare system.

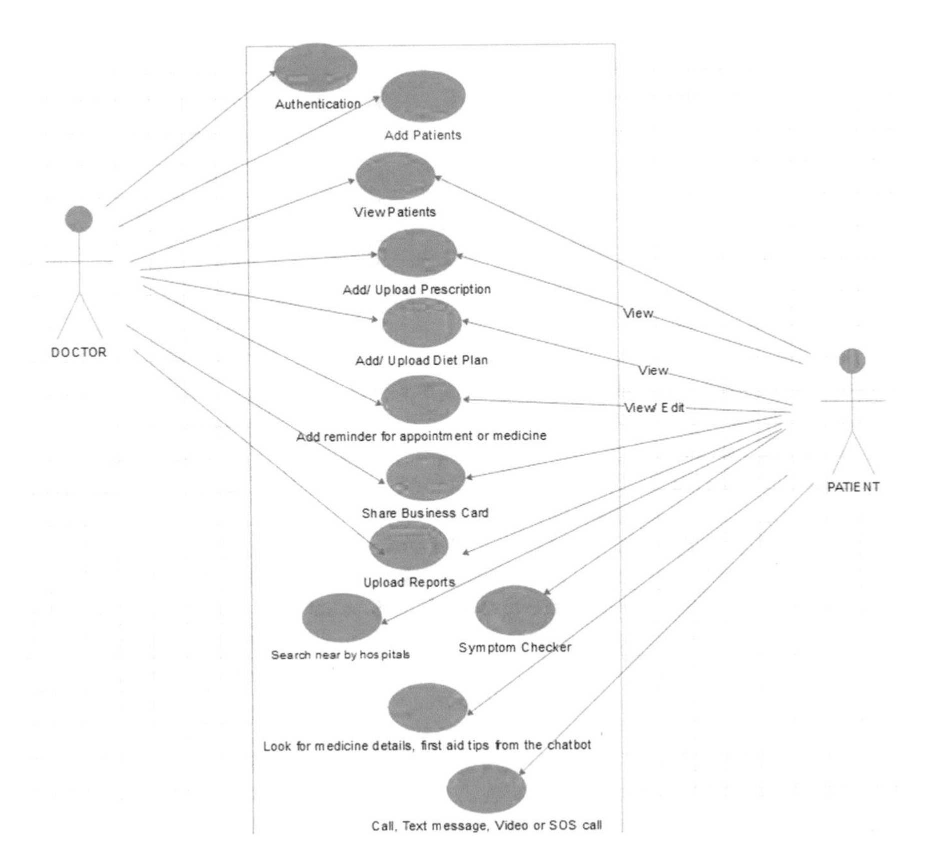

Fig. 16.3 A use case depicting the role of the two actors and their way of interaction with the system

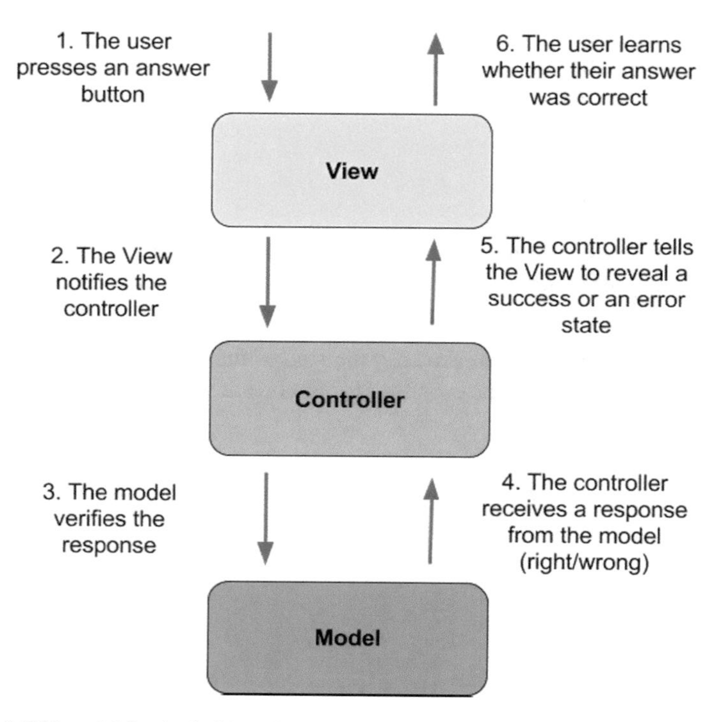

Fig. 16.4 MVC model for Android application development

3.1.2 The Application Development Approach Used

The Android application is designed keeping in view the MVC (Model-View-Controller) architecture of software development. MVC model for Android application development is depicted in Fig. 16.4.

Model signifies the data taking part in the entire application flow and is modified as and when required by the controller itself. The database handlers and the utilities all comprise the model layer of the application. This layer is usually known as the data layer. In Android separate classes are created for database handling tasks. The view determines the kind of interface that the end users would be interacting to. This presentation layer hence deals with the layout files, resources, and other UWE components and specifications (here mainly the XML files). As regards the kind of look, feel, placement of app components, color, theme, and font size, everything is decided in this layer.

In Android, the activities, fragments, and view components form the view layer. The third layer is the controller which works as the monitor or the mastermind of the system. It decides the exact workflow of the application, and how the different components would be called and connected is all done at this layer. This layer provides the actual business logic for the application, and whenever the user interacts with the view, it is the controller that triggers the model in order to make any changes to it. In Android, the classes are created separately for each task such that the object-oriented paradigms are followed.

3.1.3 Application's Workflow Through Data Flow Diagrams

This E-medical assistance Android application consists of two major interfaces – one for the user (or the patient) and the other for the doctor's site.

Model Design and Data Flow for the Doctor Site's Functionality

Functionalities at the doctor's side are depicted in Fig. 16.5 and can be summarized as:

(i) The doctor would be allowed access to the app only after his verification of being a licensed doctor is successfully done. This would be done by checking for the presence of his entry (mainly, registration ID and state) in the licensed doctor's database and its absence in the blacklisted doctor's database.

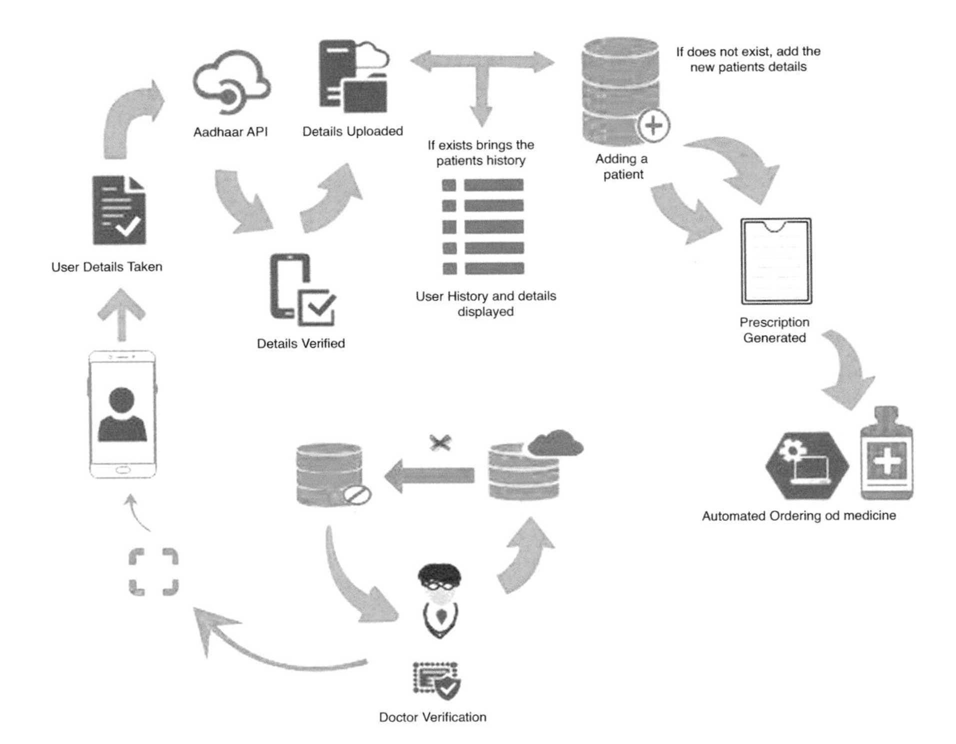

Fig. 16.5 Functionalities provided for the doctor's site interface

(ii) After getting the access, the doctor can either view the list of all his previous patients and contact or view their history, or the doctor on the visit of a new patient can simply scan his/her (patient's) Aadhaar card (or any other card used for the unique identification of patients).
(iii) Upon scanning the patient's unique identity card, the patient's entire medical record would be made available to the doctor if it exists; otherwise, it would add a new entry upon scanning.
(iv) The doctor can then add a new prescription, diet plan/precautions or reminders for food, medicines, etc. if any based on the patients' medical grounds.
(v) A price comparison module would then compare the price of the medicines from two or more sites and generate the minimum bill for the customer to order or not the medicines.
(vi) The doctor also has an additional functionality of sharing their business card with the users/patients both within and outside of his/her clinic.

Model Design and Data Flow for the User Site's Functionality

User end functionalities are depicted in Fig. 16.6 which are explained as:

(i) Any new user would be required to first of all register to the app, or an already existing user could simply login into the app to use its features.
(ii) Once a registered user, the user can search for the nearest available doctors and pathologies through the app or can even search them through the doctor's specialization, the symptoms entered, etc.
(iii) The patient if using the app since quite a long time can also view his/her data of the previous visits, the prescriptions, and the reminders and the diet plans given then.
(iv) Also the app could provide the facility of tracking their recovery just by entering their disease and stage. Also the relevant tests and nearest pathologies would be immediately displayed from the fog layer itself.
(v) The user has certain functionalities for fetching the medicine details, first aid methods, simple home remedies, and possible disease check all through a user-friendly chatbot.
(vi) The patient would also have the feasibility of contacting their current or recently visited doctor in case of help through a call or text message. In case of emergency, an SOS option would also be available which would only be allowed at times of grieve injuries or health emergencies.
(vii) The patient can upload their pathology reports before visiting a doctor or even in the case of regular check-ups for data record's completeness and accuracy for any unforeseen disease trends.
(viii) At last, an option for feedback given is to each patient after every visit to the doctor.

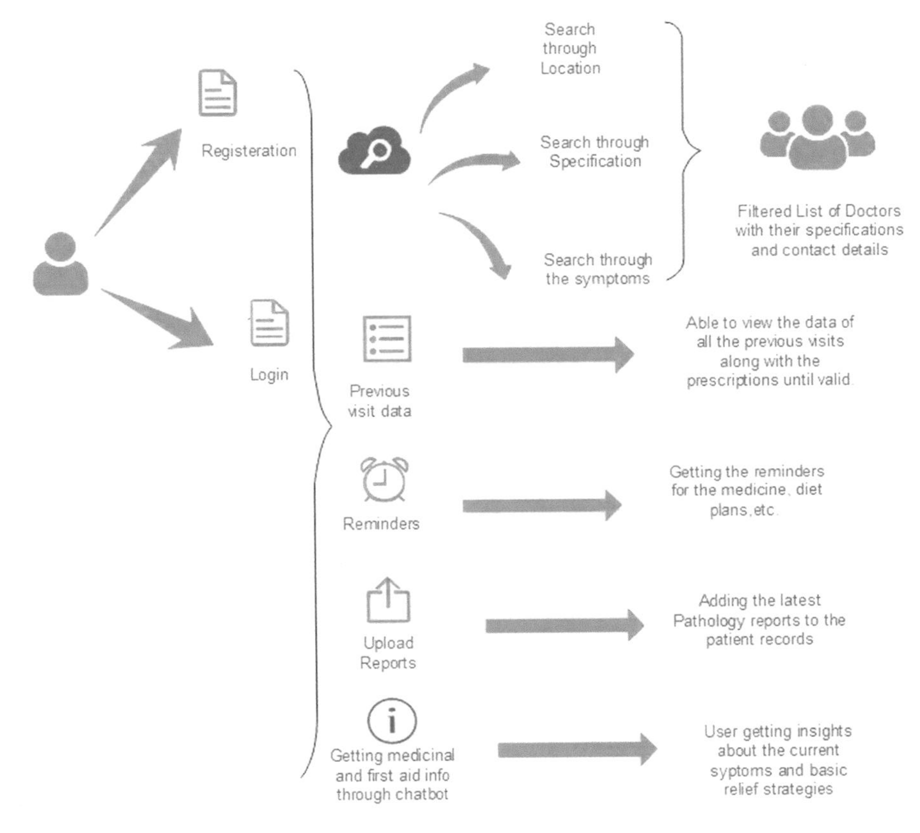

Fig. 16.6 The functionalities provided for the user's site app

4 Implementation

Functionalities and modalities discussed in Sect. 3 for the doctor's and patient's ends are implemented in steps and categorically can be summarized as follows.

4.1 Steps for Doctor's Site App Implementation

Step 1: *Doctor's verification from the authentic database* (*stored as JSON file on* the real-time database Firebase) and then granting access based on the OTPsent to the registered number only (Firebase's authentication service)

Step 2: *Option for viewing and contacting the patient's, already visited in the past, verified from the database*

Step 3: *Aadhaar QR code scanning and retrieval of primary user data*

Step 4: *Storage of that data to the cloud database (Firestore cloud database used here) and retrieval of the existing data if present*
Step 5: *Displaying the medical records timestamp-wise*
Step 6: *Adding prescription manually or by image text recognition*
Step 7: *Adding diet plans or precautions*
Step 8: *Adding and setting reminders for regular intake of medicines and future doctor appointments if compulsory*
Step 9: *Price comparison module*
Step 10: *Business card sharing*

4.2 Steps for User's Site App Implementation

Step 1: *Registration and login for users of the app*
Step 2: *Search for the nearest doctors and pathologies*
Step 3: *Searching doctors based on specialization or symptoms too*
Step 4: *Viewing previous visit data, prescriptions, and reminders*
Step 5: *Suggestion of precautions and the recovery time on the basis of the diseaseand stage entered*
Step 6: *Feedback collection module*
Step 7: *Searching for medicine details, first aid, and relief remedies through achatbot*
Step 8: *Chatbot implementing text to speech conversion along with a voice cloning module for emotional health recovery*
Step 9: *Option for upload of the test reports*
Step 10: *Feature for emergency calling and SOS service available for mainly the elderly or the frequently sick patients*

4.3 Pseudo Code/Steps/Algorithm

Module functionalities for doctors and patients are categorized as follows.

1. Doctor's Verification
Steps involved are:

(i) Creating a new project in Firebase.
(ii) Adding the package of our app to this new project created on Firebase.
(iii) Using the real-time database.
(iv) The *google-services.json* file is downloaded from the Firebase and added to the src folder in our project.
(v) Adding the Firebase dependency to our app *build.gradle* file.
(vi) Uploading the authentic data to the Firebase in the form of *JSON* file.
(vii) In the Android application code:

(a) We create a *Firebase* object.
(b) We get the reference of the table (or the parent node).
(c) We parse the entire table for the presence of the node with the registration ID and state as entered by the doctor.
(d) If record exists, the next activity shows up; otherwise, a message is displayed showing "Access Denied!".

2, 3 & 4 Option for Viewing and Contacting the Patients Visited in the Past
Following options are provided for viewing and contacting patients already visited:

(i) For viewing the patients the doctor has examined in the past (each doctor entry having several unique IDs of patients using which the patient's details are fetched).
(ii) For every new visit providing a scanner that could scan the QR code and retrieve the details from the Aadhaar card in XML format which is then parsed and where information is extracted based on the tags.
(iii) These retrieved details are then compared to the already existing entries (stored in the Firestore cloud database). This is done by taking the reference of the file present on the cloud database where it needs to be stored. If it is present, the entire previous medical record of that patient is displayed on the next activity. These medical records are displayed according to the timestamps, i.e., the date of a visit. The symptoms patient showed then and the medications that were given are all displayed to be viewed by the doctor for better judgment (these records are displayed on an *expandibleListView*).
(iv) In case the patient has no previous record, his entry would be added to the Firestore, and the symptoms and medications given would be added along with the personal details.

5. Displaying the Medical Records in the Timestamp Order
(i) An activity having the expandable lists for display of personal details along with the previous medical records of the patient, linked with his/her Aadhar UID.
(ii) These lists when expanded show the medical records in a chronological order based on the date-time when that patient's record was added. On clicking on any of these date-time values, details of that date would be displayed.
(iii) From this activity, the doctor can also add a new prescription, diet plan, or reminders by clicking on the corresponding button.

6. Adding Prescription
(i) The prescription activity uses the *MultiAutoCompleteTextView* for entering more than one symptom and the medications being prescribed to the patient. This addition is made simpler since the symptoms and the medications are predicted once two or more characters are typed in the space provided there.
(ii) An option for uploading a handwritten prescription is also there. The prescription or diet plan can also be uploaded in image format by clicking a picture of a neat and clear hardcopy which would then be processed by the ML kit *Text*

Vision library provided by Google's Firebase. After processing the image to text data, the data would be stored explicitly to the *Firestore*'s patient medical record.

(iii) All these details when submitted are stored along with the patient's personal details to the cloud database (here *Firestore*).

7. Adding Diet Plans or Precautions
This activity provides an option for adding diet plans for the patients as suggested by the doctor or to add any of the precautions, which the user has to take during his illness period. These suggestions or the diet plan can be easily viewed by the user from his/her profile in user app by the unique Aadhaar ID.

8. Sharing of the Business Cards

(i) A doctor or a practitioner usually holds a business card of their clinic or place of work, and in this era of ultimate hygiene and distant socializing, sharing of these digital business cards is the best way to share info without human contact as well as avoiding the hassle of meeting in person. Doctors too can share these business cards.

(ii) All it takes is a one-step phone number login and OTP verification in order to see the cards stored under that number (Firebase phone authentication is enabled and the dependency added to the app *build.gradle* file).

(iii) Any of those cards can be easily shared by long pressing (*OnLongClickListener* is used); on that card, another activity would pop up asking for the number on which it needs to be shared to.

(iv) The card details are simply copied to the number the card is shared to and the card is visible to the person when he/she logins with their number. (Firebase real-time database was used for this purpose so that the changes are reflected immediately.)

4.4 User's Site Activity Flow

User's site activity can be summarized as:

(i) *Registration and user's login*

Any existing user is given an option to login into his account in order to view his details and history and his appointments or any of the activity provided by the app. New users are given an option to get registered and are able to use all the facilities provided by the app. For the authentication purpose, the *Google Firebase* plays the whole role here, and only authenticated users are able to visit the app.

(ii) *Search for the nearest doctors and pathologies*

The authenticated users are able to search the nearest doctor in their locality and are able to get their contact details. In order to get the doctors near the user in their locality, the *Google Maps APWE* was used.

(iii) *Searching doctors based on specialization too*
The doctors can be easily searched by just mentioning their specialization and the user's location making it convenient for the user.
(iv) *Viewing previous visit data, prescriptions, and reminders*
The verified user who has already visited a doctor can easily view all his details, precautions, and the medicines suggested by the doctor to him and can also look into his whole medical record till now.
   (a) All he/she needs to do is scan the QR code present on their Aadhaar card, and the corresponding entry would be fetched from the Firestore and displayed to the user patient.
   (b) The user also has an option for setting a reminder for the intake of medicines, drinking water every 2 hours in a day, taking a fixed diet, and future doctor appointments or check-ups.
   (c) Suggestion regarding precautions for the infection/disease and the recovery time on the basis of the level of infection or stage of disease.
   (d) The system takes the disease and stage as input and shows the recovery time and precautions to be taken as output. System does real-time analysis of the disease by the details entered by the medical expert.
(v) *Feedback collection module*
Facilitates the doctor visiting the patient with details of previous doctor visits, diagnosis and medication.
(vi) *Searching for medicine details, first aid, and relief remedies through a chatbot*
The user login starts the app with the options for searching the nearby hospitals, setting a medicine intake reminder or so (done using the *BroadcastReceiver* class), looking for any medicine-related information or the option to chat (*HealthOS APWE* used), and getting simple remedies with the help of the bot (the chatbot makes use of the Dialogflow APWE provided by Google).
(vii) *Chatbot implementing text to speech conversion along with a voice cloning module for emotional health recovery*
The chatbot can not only carry out general conversations but also help predict the possibility of a disease by simply asking about the symptoms and a few more follow-up questions (the chatbot makes use of the *Dialogflow* service provided by Google for the basic or service-based interactive interfaces for the user). The bot can also be asked about the home remedies or first aids in case of mild symptoms or discomfort (for the symptom checker module, the *Infermedica APWE* is used).

The text to speech service allows oral interaction with the bot as well. An additional solution for the suffering patient could be the chatbot conversing to the person in a close relative's cloned voice. After all recovery is high paced when a person is emotionally fit (recently in June 2019, an AWE system on voice cloning was created and made open source which would serve the purpose here).

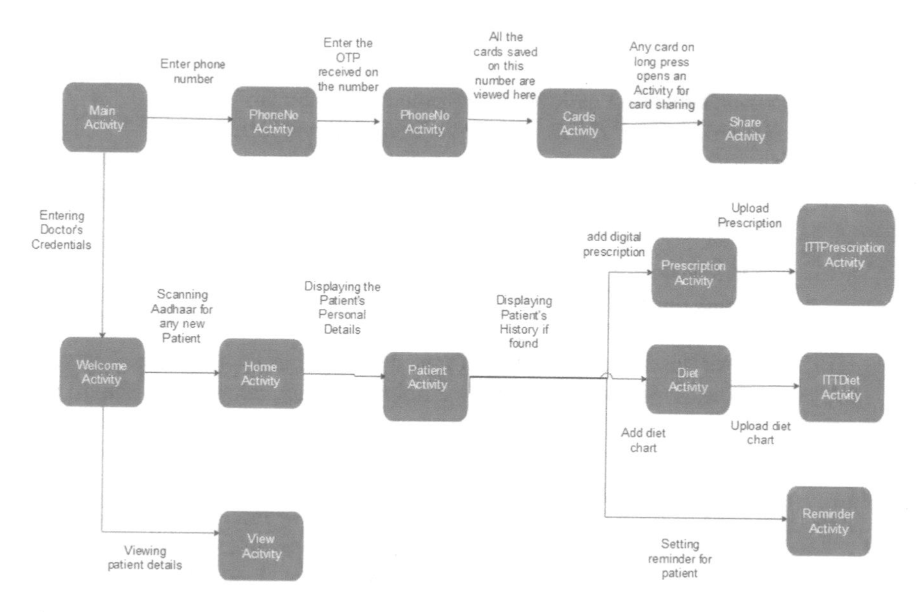

Fig. 16.7 Activity to activity flow for doctor's login

Figure 16.7 depicts activity to activity flow for doctor's login, and Fig. 16.8 depicts activity to activity flow for patient's login, and Fig. 16.9 presents database table structure as stored on the cloud storage.

4.5 *Option for Upload of the Test Reports by the Patient*

The option for uploading the regular test reports by the patient from time to time would all add to the medical records on the *Firestore*. This module uses the text recognition technique for extracting the details in text format from the report image uploaded, and then updations are being made to the cloud database.

4.6 *Feature for Emergency Calling and SOS Service Available for Mainly the Elderly or the Frequently Sick Patients*

Among the important services provided at the user's end, some other essential services such as emergency calling facility, SOS alarm, and chat interaction facility with the recently visited doctor are also required. These services are implemented by simply extending the *BroadcastReceiver* in the Java class or with the help of

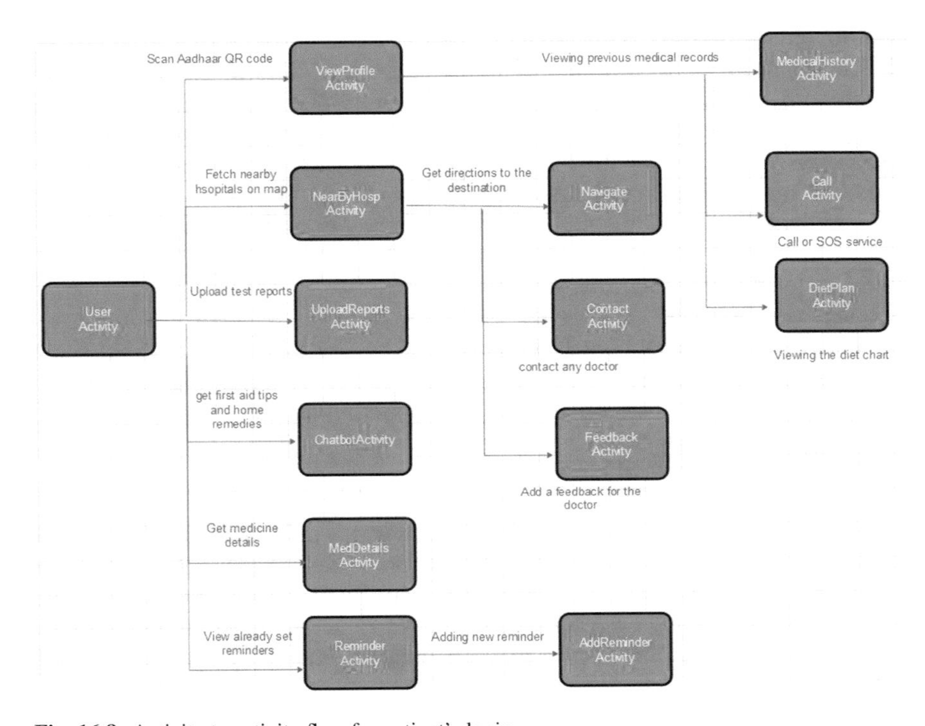

Fig. 16.8 Activity to activity flow for patient's login

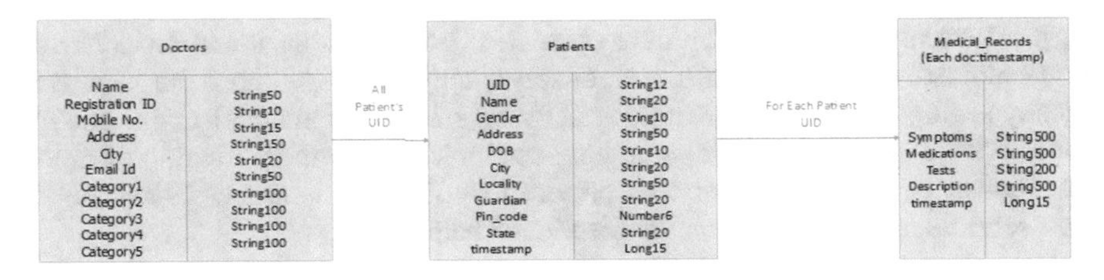

Fig. 16.9 Database table structure as stored on the cloud storage

Intent objects. SOS would either send an alarm to the added contacts or the doctor the patient is currently in touch with.

4.7 Choice of Tech Stack

The choice of our tech stack greatly influences the system's quality and popularity among its target audience and its maintainability. Some of the major technologies and services used and the reason for choosing the same are listed below:

(i) Android Mobile Application

Mobile applications are the most convenient form of digital access. Moreover, almost 95% of the world's population uses Android phones, wearable, and other devices that make use of the Android OS. Hence, designing and working on an application that could be used and reviewed by many has a higher chance of improvement and reaching all strata of people.

Development of Android applications also hardly costs anything, and it's open source; therefore, for innovation purposes, the sky is the limit, and the UIs are all easily customizable in Android.

Language used: Java

Though recently Kotlin has been introduced by Google for the Android application development because of its easy syntax and short format, majority of the applications in Android are till date built and working on Java. This language has proved itself as quite robust.

Toolset used: Android Studio IDE

Android Studio and IntelliJ are the two most popularly used Android application development platforms. Previously, MyEclipse was widely used, but because of the ease of development, developers moved on to the trending IDEs. Android Studio provides an entire environment for the systematic development and testing of programs without the developer having to worry about the file storage, library installations, etc. The IDE's smart suggestion capabilities make it the high in use development platform.

(ii) Firebase

Firebase is a service provided by Google mainly for mobile development platforms but it is also used for Web App development. Firebase is a back-end service that serves almost all the needs of a remote server applications without the need for actually setting up a server or writing an APWE or any other authentication code or cloud server database set up or commonly used machine learning training models. The number of full-fledged services provided by Firebase is nearly 19. Some of these services used in the project are mentioned below.

(a) Real-time Firebase storage: Normally the databases that are set up need to be asked for data whenever changes are being made, and the language used for doing the same itself makes use of another coding language or rather querying language. Here, in Firebase, databases used are NoSQL databases, and therefore the insertion, retrieval, and deletion of the entries is a task of only a few minutes. These databases are accessed by creating the connection of the project with the Firebase and then getting the reference of the database and its table for doing the manipulations. Real-time databases are the popular one and the most widely used. These databases are synced with all the clients connected to it. Any changes made by one client to the database are immediately reflected to all the other clients within a fraction of time. Hence real-time databases are commonly used for building chat applications.
(b) Firestore cloud storage: Another Firebase feature is the cloud storage facility provided for storing the binary files, images, and any other kind of files

directly from the user to the storage. These Firestore are again NoSQL and are referenced using their object for use.

(c) Firebase authentication: Authentication is the identification of the genuine users of our app and then allowing of access to only the authenticated ones. This feature is mostly needed in almost all the applications and can be done either by email ID and password, Google, and Facebook login or with the help of phone number authentication. Firebase provides the complete solution either as the FirebaseUWE drop-in authentication service or in the form of an SDK for custom integration into app as per the usage and need. The database security rules are also declared there and can be easily changed to be accessible only to the authenticated users or so.

(d) ML kit: This SDK tool was an add-on to the Firebase's cloud backend services. This simplified machine learning development toolkit made the lives of the mobile app developers a lot easier. The integration and usage through the available APIs was easy for all to make use of. Currently it has in total 11 functionalities available for use.

(iii) Speech to Text Conversion

Speech synthesis is the conversion of text into oral form spoken in an artificially generated human voice. Both speech to text and text to speech conversion mark a key role in making our application user-friendly. Their main usage is within the chatbots for interacting with them not like a human computer interaction but rather as a human to human interaction. Android applications implement it using the STT (speech to text) and TTS (text to speech) engines. The TTS and STT engines need to be told beforehand about the language it would be converting. They support nearly 14 languages.

(iv) Text Recognition (OCR)

Optical character recognition (OCR) is the electronic or mechanical conversion of the printed, typed, or handwritten text into a format that could be stored and manipulated on a machine. The process of conversion of the text in an image to machine-encoded data is done in three steps of filter application (for differentiating between the characters and their background), of contour detection for recognizing each character separately, and by image classification or pattern matching techniques.

Text recognition was required to be implemented in order to save time for the doctors who may not have the time to type each prescription; rather, they can simply upload an image of the hardcopy that they produce which would then be converted into text and stored on the database or displayed as needed.

Moreover, the health reports obtained by the patients for their regular check-ups or tests again cannot be entered by the patient from the keyboard every time. At this time, text recognition module works wonders, and just by clicking an image of the report, the report content is automatically converted into text data format and stored without any manual interference.

(v) Dialogflow

Dialogflow is a free computer and human interaction service provided by the Google Cloud Platform. It does all the natural language processing and the other difficult tasks on its own and provides the developers with a much refined and simplified APWE for integration of chatbot services.

The training of the model for the chatbot begins with providing it with the sample phrases which would all denote the same "intent" or intention of the user interacting.

The second part that is picked is the "Entity" or the object that is being discussed or talked about. Sometimes certain "parameters" are also associated with the entities, and if proper answer is not obtained, the bot might ask certain follow-up questions to clarify that.

Next come the responses which the bot sends in reply to the user. It does that based on the intent of the user. A bot's response can also be triggered on the occurrence of a particular "event."

Though many chatbot services are available today, Dialogflow is one of the easiest services to work with.

(vi) APIs

An APWE (application programming interface) provides a way of gaining limited yet useful information which could provide us an answer in yes or no and give us an up-to-date correct data for providing it to the user or for authentication purposes. The data such obtained can be either directly displayed to the app or first processed and then shown to the end users in some other form. The app usually interacts by sending a request to the web service, and the web server sends back a response in the form of either an XML or mostly a JSON file. This interaction is mainly http based, or in the case of web socket connectivity, the response is received faster. Figure 16.10 *depicts the usage of an APWE (application programming interface).*

The use of an APWE saves immense of our time spent on coding for the services from scratch. Moreover, the tasks may still not be up to the expectations required to be met by the client application.

The APIs used in this application are:

(a) Infermedica: A disease prediction service based on the symptoms entered by the user
(b) Dialogflow: A chatbot dialog service provided by Google for building simple yet intriguing chatbot with ease

Fig. 16.10 APWE (application programming interface)

(c) Google Maps API: Used for the integration of maps in our Android application (in the module for searching for the nearby hospitals)
(d) HealthOS: An APWE service providing all the information about the medicines, food and nutrition, etc.

(vii) QR Code Scanner

Embedding a QR (quick response) code in our Android application is the smartest, fastest, and best way of extracting data within a fraction of a second. These QR codes can store quite sufficient amount of data including the links, text, images, geographical coordinates, etc. unlike the barcodes that were meant for storing only a limited data. QR codes provide local small-scale data storage for quick retrieval. Nowadays, every smartphone comes with an in-built QR code scanner whether it is for making payments or for transferring any file or contacts to someone.

The integration of QR code scanner into an Android app is made easier using the Google Mobile Vision API. Using the QR code present on the Aadhaar card of the citizens of India, the person's basic personal information is obtained in XML format which is then parsed, and the relevant details are extracted and stored to the database and displayed on the UI.

(viii) XML and JSON Parsers

JSON stands for JavaScript Object Notation which was initially used in JavaScript for coupling the programs by sending objects to them in this JSON format. Applications make use of this format for sending and receiving data from a server or a service API. Even Firebase connectivity needs a JSON file (google-services.json) to be downloaded and added to the project that needs to be connected.

```
A json file looks like this:
"Employee"=    {
"eid" = 0123 ,
"name" = "Sheetal" ,
"Department" = "IT" ,
}
```

In order to parse the JSON string, a JSONObject class object is created, and the received string is passed into it. Then the JSON values are extracted from the key-value pairs using the getString(), getLong(), getInt(), etc. methods.

Sometimes this data can also be fetched from the source in the form of an XML file (like here in case of QR code scanning). XML stands for Extensible Markup Language. In fact, the layouts in Android are created as .xml files. XML is also more secure and does not support array for data storage.

```
XML file content looks somewhat like:
<Employee>
<id> 0123 </id>
<name> Sheetal </name>
<Department> IT </Department>
</Emloyee>
```

The XML files can be parsed by any of the DOM, SAX, and XMLPullParser, but Android mainly makes use of the XMLPullParser. Its object is created firstly by creating the object of the XMLPullParserFactory and then calling the getPull-Parser() function on it. The values are retrieved based on the attribute names used in the XML file.

(ix) SQLite Database

Android app or any other software sometimes have a need to store the data being processed by that software or app. This processed information or data is stored in a database. This database can be on the device itself (SQLite database) or on some other device or cloud-based service (Firebase) as well. Firebase is a cloud service in which the data is stored and processed in a cloud. It is a no SQL, real-time distributed database that is apt for working on some data that needs frequent updates and changes. Only its limited access is free, and an Internet connection is required for accessing it. Whereas a SQLite database is absolutely free, independent, and an open-source relational database that isn't actually present on the Android IDE (Android Studio) but is available on the Android device on which the app is being used, SQLite thus does not always require an Internet connection and is quite handy to use.

(x) Voice Cloning Module

In June 2019, an open-source code for the AWE Toolbox and developed by Corentin Jemine was made available on the GitHub repository. The system was an implementation from the research paper, *Transfer Learning from Speaker Verification to Multispeaker Text-To-Speech Synthesis (SV2TTS).*

The AWE Toolbox works on a text and sample audio as input and generates an augmented audio of the given text in the voice of the sample audio.

The system encodes the given voice waveform sample into vector representation of a particular fixed dimension; similarly, the sample text to be spoken is also converted into its vector representation. These two vectors of voice and text are combined and then decoded. This is represented in the form of mel spectrogram. The vocoder then transforms the spectrogram into an audio waveform file which is audible and can be played. Figure 16.11 *explains about the working of the voice cloner AWE Toolbox.*

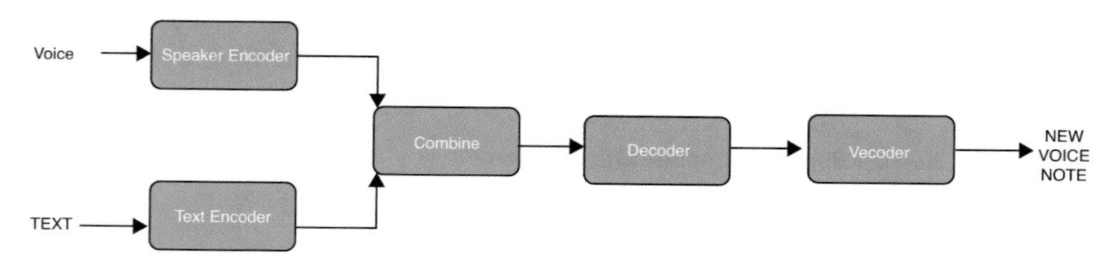

Fig. 16.11 Functional blocks of voice cloner AWE Toolbox

4.8 Generalized Block Diagram

The work can be sub-divided into some modules as portrayed in Figs. 16.12 and 16.13.

Being classified into two main user accessibility modes from the initial phases itself, the modules are also divided distinctly as:

(i) Doctor's login
(ii) User's (patient's) login

Modules for the doctor's login in the block diagram:

(a) Doctor verification
(b) Entry of a visiting patient
(c) Viewing/contacting previously visited patients
(d) Prescription generation and data collection
(e) Setting reminders, routines, and diet plans
(f) Bill generation and price comparison module
(g) Push notifications for any new pathology reports being updated
(h) Business card sharing with the patients/users

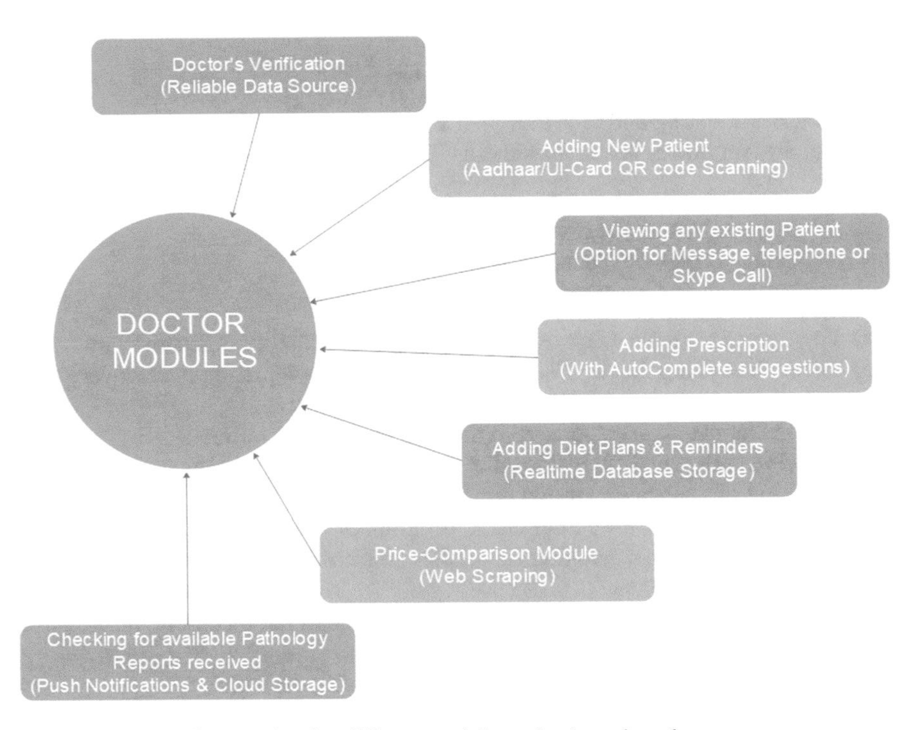

Fig. 16.12 Block diagram showing different modules at the doctor's end

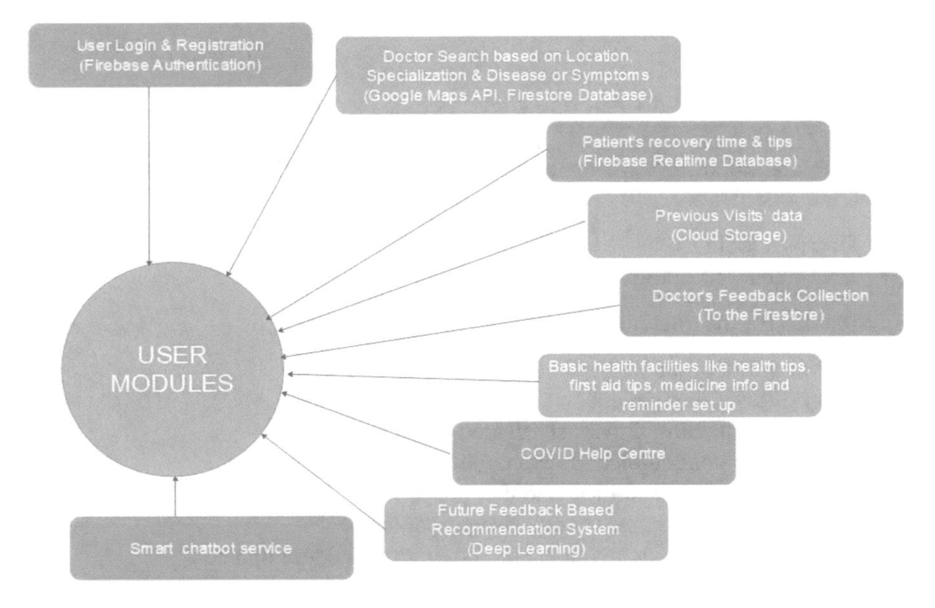

Fig. 16.13 Block diagram showing the main functionalities provided to the users

Modules for the user's login block diagram:

(a) Login of existing and registration of a new user
(b) Facility for doctor search based on:

- Location
- Symptoms/specific disease
- Doctor's area of specialization
- Name, region, contact no., etc.

(c) Fog layer for medical test suggestions along with the nearest pathology
(d) Gantt chart using the *AnyChart* Android library for depicting the recovery time and tips
(e) Uploading check-up reports with the help of text recognition
(f) Viewing entire medical history/diet plan if present, by the patient
(g) Addition of a medicine or diet reminder
(h) A friendly chatbot for casual chats, information about the medicines, first aid tips, home remedies, and disease prediction from the symptoms
(i) A COVID help center
(j) Getting immediate first aid and healthcare tips and the medicinal information at ease
(k) Option for feedback collection for the doctor after each visit
(l) (Future module) doctor's recommendation system

4.9 Flowcharts and Workflow

The flowcharts here describe the functionalities aimed at providing the best of medical health and support to its users and patients as well as an ease of usage and data maintenance for the doctors as well. Flowchart is depicted in Fig. 16.14.

Flowchart steps depicted in Fig. 16.15 can be summarized as:

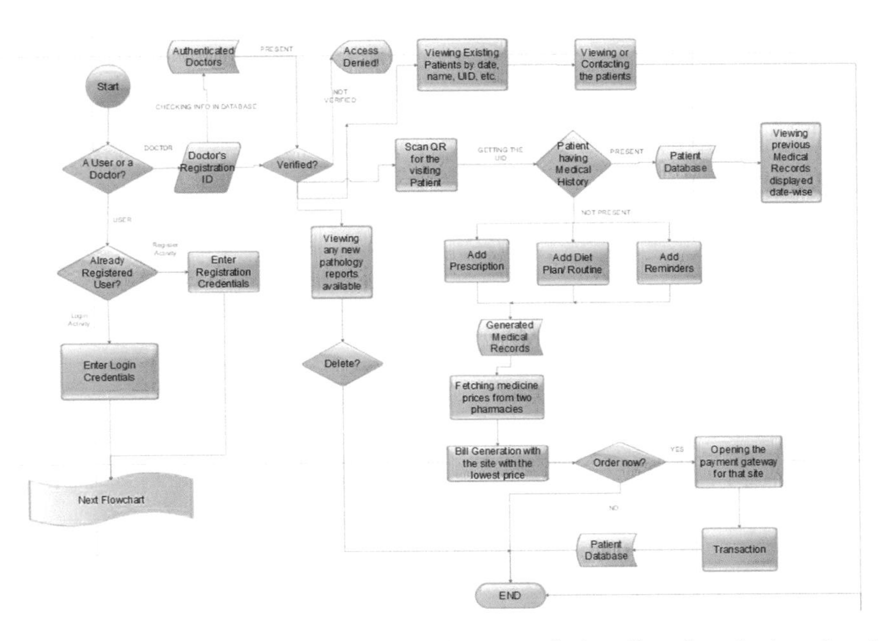

Fig. 16.14 Flowchart depicting the functioning of the "e-medical care" app from the doctor's end

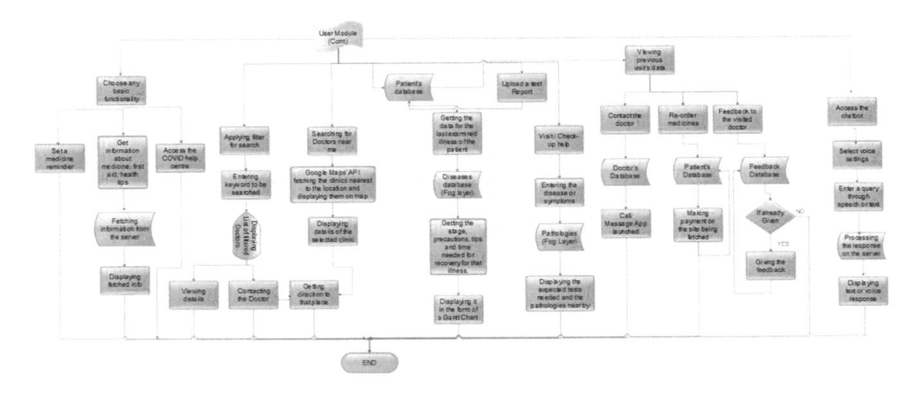

Fig. 16.15 Flowchart depicting the functioning of the "e-medical care" app from the user's end

(i) *Step 1: The very first app activity allows login for a user or a login for a doctor as well.*
(ii) *Step 2: In the case of a user, the successive activities are explained in the next flowchart,* Fig. 16.4.
(iii) *Step 3: In case of a doctor, he enters his registration ID and state (together they form a unique identification key) and further an OTP response sent by the system proposed on the doctor's registered contact number ensures enhanced security and avoids any breech or unauthorized access by a non-doctor.*
(iv) *Step 4: These credentials are all verified from a database of authenticated doctors (provided the list does not contain any blacklisted doctor); if found, the doctor is allowed access to the application; otherwise, the access is denied and message displayed.*
(v) *Step 5: Next, the doctor can either check or see a notification about any new pathology report being updated or view any existing patient's records or contacts (usually for patients he/she is currently seeing) or simply scan the QR code on the Aadhaar card of the patient.*
(vi) *Step 6: The Aadhaar card scanning process would bring all the existing details and medical records of the patient if present, and if not, the new patient's personal credentials would be added to the database.*
(vii) *Step 7: In the case of a doctor adding a prescription or any other reminder or detail, it will all be stored along with the patient's records.*
(viii) *Step 8: Once a prescription is submitted, a bill would be generated there and then for the medications, drugs, and vaccinations available online (minimum bill total after comparison of their price from the two different online pharmaceuticals).*
(ix) *Step 9: Placing or not placing an order is optional, but the entire data would be stored to the patient database in either case.*

Flowchart steps depicted in Fig. 16.15 are explained below:

(i) *Step 1: As discussed in the previous flowchart, when a user is trying to access the application for his/her usage, he/she can either login or register into the app if using it for the very first time.*
(ii) *Step 2: After getting the access, the user is free to explore into the app.*
(iii) *Step 3: He/she can choose to search for a doctor nearest to him/her.*
(iv) *Step 4: They can even search the doctors by name, region, specialization,* etc. *(all by filtering through the doctor's database) or by symptoms or disease name as well (from the diseases database).*
(v) *Step 5: Searching for a doctor would not only allow viewing the general details about a doctor or search for their location but also help the users contact any doctor available nearby.*
(vi) *Step 6: In case of an emergency or otherwise, a user can simply enter his/her illness details (the disease he/she is suffering from) and get a list of all the tests one needs to go through and the pathologies nearby, all from the data stored in the disease database at the fog layer, thus giving the results to its users almost immediately without any latency.*

(vii) *Step 7: A patient also has an option of uploading his/her pathology report by himself/herself as a scanned picture or can ask the pathology center to upload it in softcopy format to be able to be viewed by the visiting doctor or for any future reference.*

(viii) *Step 8: The app also has an option for viewing his/her recovery status in the app itself in the form of Gantt charts (a form of data visualization) that would automatically collect the data from the last doctor's visit (patient database) and accordingly build a chart showing the stages of the disease (diseases database) one might be going through and the steps and the precautions that could be taken in order for early recovery.*

(ix) *Step 9: Another option provided here is the option for viewing the previous doctor's visit data by the patient himself if there are any records existing. The patient can then either contact the concerned doctor (doctor's database), re-order the prescribed medicines for the second or third time in a month and not beyond that (doctor's database), or simply give an already visited doctor his/her honest feedback (feedback database linked to the doctor's database through a unique foreign key).*

(x) *Step 10: Other functionalities are getting the first aid tips, health tips, and medicinal information on one click.*

(xi) *Step 11: The COVID help center provides the option for donation to the PM CARES Fund; an easy prediction of the chances a patient stands of suffering from the deadly coronavirus; a corona tracker for displaying the number of corona cases till date; currently active, recovered, and deceased corona patient count; an option for sending a note of appreciation to all the health professionals having their entry in the standard database; and lastly a feature for reading about the updates, symptoms, and precautions related to COVID-19.*

(xii) *Step 12: Last is a chatbot service for getting the home remedies, small talks, reminders, sentiment analysis, and integration of voice cloning module allowing the chatbot to converse in a known voice, thus promoting early recovery of the patient.*

4.10 Application Algorithm

The algorithm explaining the application's working is elaborated below:

Step 1: *Start.*

Step 2: *If an already existing user enters the credentials or in case of a new user, click on registration option and continue. If you're a doctor, select DoctorLogin from the menu and continue with Step 4.*

Step 3: *After logging into the app, the user can perform any of the 11 activities as:*

(a) *Searching doctor by any field*

(b) *Option for doctors near me*
(c) *Viewing the Gantt chart based on the last illness examined*
(d) *Finding the list of medical tests/check-ups needed and the pathologies nearby, based on the disease/symptoms entered (these details are fetched from the fog layer reducing latency and improving user experience)*
(e) *Uploading a test report with date (added to the patient's record)*
(f) *Viewing previous medical records and the visits of the patient (if needed, the patient can even contact the doctor in any of the three ways,* i.e., *through messaging, telephone calling, or video chat)*
(g) *Setting a reminder for medicine or doctor's appointments,* etc.
(h) *Fetching immediate first aid/health tips or medicine details*
(i) *Accessing the COVID help center*
(j) *Accessing the chatbot for friendly chats and learning about the home remedies for mild symptoms and the possibility of the disease based on the symptoms entered*
(k) *Submitting a feedback for every doctor visited*

Step 4: *On the doctor's part, the first step that a doctor needs to do in order to get access to the app is to get verified. For this, he needs to enter his registration number along with the state he got himself registered.*
Step 5: *If verification is done, he's allowed access to the app.*
Step 6: *The doctor can then either view the list of his/her patients that visited him/her or check on/add a new patient by simply scanning the patient's Aadhaar card.*
Step 7: *The next activity then has the basic personal information (such as the name, age, gender,* etc.*) along with the medical history if found.*
Step 8: *Option for adding a prescription, a diet plan, or a reminder is also provided (this could either be entered manually or through text recognition).*
Step 9: *On submitting the prescription, a generated bill with the lowest priced medicines is generated.*
Step 10: *If selected, the payment gateway for that particular online pharmacy is opened for the payment transaction; otherwise, the prescription is stored along with the patient's records.*
Step 11: *The diet plans are sent to the patient's login and reminders for medicines are set.*
Step 12: *End (exit).*

5 Enhanced Features: Security and Reduced Latency

5.1 Methodology Employed: Security Issue

Any robust system works on a set of data (which may be regular or critically sensitive). The system stores the data, manages it, and provides a secure, easy, and fast means of accessibility to it. Data as a part of any system brings with itself a number of security threats and exploitable loopholes. Same goes with the healthcare systems which store and operate on huge amounts of patient and medical data. It may include the patient's private information and their health information and medical history on a timeline along with the details about the healthcare providers. The security and maintenance of such data units becomes of utmost importance keeping in mind the threats and information breaches that occur which have been seen in the past.

5.1.1 Need for Security and Privacy in Healthcare

Here, since we are dealing with the idea of maintaining a common platform for all of the patients' medical records at one place irrespective of the place of treatment, this centrally stored data would prove to be an easy target for the attackers. Security of healthcare data is one of the primary concerns with use of technology and if breached could expose lot of sensitive/personal data to the world within no time. Data on any network brings with itself the dire necessity to be secured especially when it can easily act as a threat to the privacy and integrity of its participants involved. Any healthcare system including our very own E-healthcare application that deals with the electronic health records (EHRs) provides a pool of information for the hackers and intruders. The private data of the patients including their name, date of birth, contact details, address, etc. obtained from the EHRs can prove to be a cause of the identity thefts. Several perfectly fine individuals can then claim access to the prescribed rare or expensive medications at the cost of the genuine patient buyers. Not only that, an immoral agent or hospitals can claim reimbursement from the health insurance companies based on false treatment and documentations.

Apart from these, the vast medical history of a patient suffering from a chronic disorder or some rare medical condition is a rich source of study and research for the scientists and the laboratories. The attacks on the healthcare system are hard to detect (unlike the banking systems); as these attacks are for data retrieval, attackers are not identifiable, and hence, it proves to be the most common and easy form of cyberattack. Majorly because of these reasons, the security of the healthcare data for the IT health professionals becomes of utmost importance.

5.1.2 Ingredients for a Perfectly Secure System on Healthcare

To make the system robust, the current study identified the major security issues and provided here possible solutions to overcome these issues. They are discussed as below:

(i) Securing the network: No matter how advanced the technology becomes, attacks would still be inevitable. But what one can do is to make the attack either very difficult or very time-consuming to break through the system. Possible ways of doing this are through data encryption, by generating hash codes (using hashing algorithms such as SHA-256) for the data, or by the addition of a multilevel verification.

(ii) Replacement for a centralized storage: A locally available database system is the foundation of a weak system in general. In order to avoid this, the data can be distributed and so will the points of vulnerability.

(iii) A secluded network: Another important step would be the availability of the private network only available to the key participants of the healthcare network. These key participants could be the doctors and other healthcare professionals, the hospital and healthcare units, pharmacies, laboratories/pathologies, health insurers, and patients.

(iv) Transparency to some extent: This doesn't mean that the insertion and retrieval of health records or any other information needs to be public but with a security mechanism should be maintained to indicate any unauthorized intrusion to the network or an attempt to make changes in data.

(v) Limited access: No doubt each participant who is also a part of the network is authorized to access and modify the data available there. But the extent to which each of them can do that can be controlled or the permission to do so would be required every time some need comes up.

5.1.3 Blockchain: Solution to Problems

As discussed above, the challenges one faces in setting up an all-roundly secured system are numerous, but with the implementation of blockchain, all of the security issues discussed above can be managed. Management and security of the health records and other private data can be efficiently implemented through the latest blockchain technology. There are reasons for the same:

(i) It provides complete data security based on the data encryption and hash code generation.

(ii) It makes use of a decentralized storage mechanism.

(iii) It allows knowledge of the network and data only to the participants of the network.

(iv) Any transactions, insertion, deletion, and modification of the stored records are visible to all (not the data which is actually in the encrypted form but the information or notice about the same).

(v) No transaction, accessibility, or alteration in record can be done without the approval of both, medical expert and the patient concerned.

Blockchain can be defined as a distributed, decentralized, and encrypted network data storage and accessibility service. The various stakeholders that are a part of the blockchain control the retrieval of records, and only and only they have the authority to access the data. The blocks that are linked tend to have a hash code derived from its previous block, and that block derives it from its previous block and so on. Hence, whenever an intruder or someone unauthorized to the network tries to access or make changes to the ledger, the entire chain changes its hash code, and the participants get to know. Moreover, the two-way key encryption ensures that even if an intruder gains access to the chain, he/she should not be able to access the data thereafter as well.

5.1.4 Healthcare Record Maintenance in a Blockchain Network

As already discussed above, blockchain technology allows the patients to control the accessibility of their records in the distributed system by changing the access rules for the various participants, for example, allowing a practitioner or researcher to access the records only for a fixed period of time. The medical records are added as ordered records arranged in a block structure. These blocks are connected to each other in a peer to peer network forming a decentralized network. Every new block of information is validated by any of the participant nodes with the help of a consensus algorithm (proof of work (PoW)) and added only if validated successfully. The process is termed as mining, and the node that successfully performs the validation is the miner.

Blocks are uniquely identified by a hash code which acts as a digital fingerprint for each block. The hash of a new block is generated taking into account the hash of the previous block. This logic makes it nearly impossible to modify the data in the midst of the block chain since changes made to any one block would alter not only the hash of the current block but also the hash of all the blocks following it at the same time, thus allowing the participants to be aware of the intruders in the network.

Blockchain implements smart contracts which are a kind of computer protocols that execute automatically as and when the defined prerequisites are met. This ensures the transactions only between the trusted parties and eliminates the need for a central authority for verification. These are not a direct part of the blockchain network but work rather separately.

Besides everything, a system that was originally designed for storing the money transaction details can be deduced to be secure and reliable enough to be trusted with any other type of information too. The medical record management would become efficient and the insurance claim process simpler. Interconnection of the hospitals, clinics, and pathologies in terms of data storage and retrieval would deal with the mishandling of patient's medical records. Acceleration of the clinical and biomedical research would lead to improved inference outcomes. Safety and accuracy of the treatment are provided to the patients based on the medical history, hence making the patients and data owners aware of their ownership rights, thus allowing the system to be patient-centric in interoperability unlike the conventional institution-centric system.

5.1.5 Challenges to Be Overcome in a Blockchain

A robust blockchain technology which has nearly overcome the major security barriers faces challenges of its own when it comes to its implementation.

(i) The more the number of participants there would be in a blockchained network, the more secure it would be. But this security comes with a cost. The system becomes more complex with each added node (here, a new healthcare professional, pathology, doctor, researcher, etc.).
(ii) Data standards are set for the network which needs to be fulfilled every time a new data is added to the blockchain in the form of a new block.
(iii) With the increasing size of the blockchain, scalability and speed become a problem if there isn't an infrastructure to support it. Since we know every transaction that happens needs a validation and after that a copy is maintained at each node, this requires not only immense computation power but also storage for its increasing demand which asks for a huge sum for supporting the robust blockchain infrastructure.
(iv) Another loophole or possible threat in a small blockchain network was what Satoshi Nakamoto, the inventor of Bitcoin, suggested. According to him, if at any time or in any case a minor or a group of minors can take control of over more than 50% of the blockchain network, they can have the majority of the mining power using which the entire network's transaction history can be deleted and not only that new transaction history can also be added in that case.

5.2 *Latency Issue Addressed*

A network's latency is measured by either the speed at which the transaction is updated to all the nodes (in a blockchain) or the number of data packets the network can transfer per second from the source node to the destination node (termed as the node capacity). Thus by altering the capacity or increasing the speed of transfer, the overall latency can be reduced.

As a blockchain and our application user, a client would always prefer to have a flawless, bug-free, no-wait user experience. This makes it even more important to look into the performance, adaptability, and responsiveness of our app. With time, the requirements for high-volume workload and data exchanges increase leading to decreased tolerance for latency. The problem is seemed to be proportionally related to the underlying issue, which was the distance between cloud services, users, and value chain partners. However, placing the IT traffic exchange points in those dense locations that are in proximity to the customers and third-party business partners can significantly reduce the latency caused between the users and the cloud services.

Minimizing the distance also reduces the latency between the interconnected digital consumers (enterprises) and producers (service providers) in dense ecosystems of network, cloud, SaaS, and supply chain partners. Similar to what high-frequency trading, digital content, and online advertising industries do, one can overcome the problem of latency to some extent by placing the applications in some co-location data center that privately connects users, supply chains, and workloads within the same facility or campus, thus providing the shortest distance to the largest number of counterparties.

Usually the most difficult and tricky part can be the identification of the factors causing the latency in the application and the elimination of those factors. Alternatively, if they cannot be removed entirely, they can still be reduced and/or managed to some extent.

5.2.1 Possible Ways to Reduce Latency

Before, during, and after computing the response, there are a number of areas that can add unwanted latency. These are basically related to the network, communication media, server-side processors, etc. Some of the common sources and their means of reduction can be:

(a) Network I/O
Majority of the applications make use of the network today in some manner or the other. It could be either between the client and the server or between server-side processes and applications. The important thing is that the closer the client is to the server, the lower the network latency would be. For instance, a round-trip latency between nodes within the same data center can cost 500 microseconds, while it can be an additional 50 milliseconds for nodes in California and New York. This can be taken care of by:

(i) Using faster networking, such as better network interface cards and drivers and 10GigE networking (5G Internet, the upcoming era of Internet speed).
(ii) The elimination of the network hops. If somehow data can be scaled horizontally, unnecessary round-trip connections can be avoided and so would the cost in the form of time taken to fetch a response for a request reduce.
(iii) Trying to keep the client and server processes as close as possible and if not bulky enough both shall be maintained on the same system itself.
(iv) Seeing to it that the processes happening on the cloud are within the same availability zone and not distributed far away or in a segregated fashion.

(b) Disk I/O
Sometimes the real-time applications require a database server in order to service the real-time incoming requests. These databases have to make the data durable by storing it to the persistent storage. This often leads to unwanted delay and query response, and disk I/O, like network I/O, is costly.

Steps that can be taken to reduce this latency can be by:

(i) Avoiding writing to the disk; instead, use write-through caches or in-memory databases or grids (modern in-memory data stores are optimized for low latency and high read/write performance).
(ii) Combining writes where possible in case of disk writes. The goal is to optimize algorithms to minimize the impact of disk I/O. Consider asynchronous durability as a way to avoid stalling main line processing with blocking I/O.
(iii) Making use of fast storage systems, such as SSDs or spinning disks with battery-backed caches.

(c) The operating environment
Another important factor that contributes to an app's latency is the operating environment in which we run our real-time application. It could be on a shared hardware, in containers, in virtual machines, or in the cloud.

Ways to manage this kind of latency can be by:

(i) Running our application on dedicated hardware so other applications can't inadvertently consume system resources and impact our application's performance
(ii) Avoiding virtualization since it is quite evident that our application is running in a shared environment and may be impacted by other applications on the physical hardware
(iii) Smart selection of the programming language used to build the operating system environment, since some languages like Java and Go that use automatic memory management and use periodic garbage collection to reclaim unused memory introduce unwanted latency at seemingly random times

(d) Optimized coding practices
Sometimes the use of unnecessary loops and inefficient algorithms can prove to be the major cause of latency. These should be checked and replaced with improvised code snippets if overlooked.

(i) Inefficient algorithms are the major cause of latency in code. Wherever possible, look for unnecessary loops or nested expensive operations in code and restructure the loops. Dynamic programming where results of the subproblems are stored (or memorized) for future reference must be implemented where needed.
(ii) Making use of concurrent execution and multithreading.
(iii) Blocking operations cause long wait times, so use an asynchronous (non-blocking) programming model to better utilize hardware resources, such as network and disk I/O.
(iv) Unbounded queues may sound counterintuitive, but these lead to unbounded hardware resource usage, which no computer has. Limiting the queue depths and providing back pressure typically leads to less wait time resulting in more predictable latencies.

(e) Combating the enemy

Building real-time applications requires that the application developer not only writes efficient code but also understands the operating environment and hardware constraints of the system. Provisioning the fastest networking equipment and the fastest CPUs won't singularly solve your real-time latency requirements.

Thoughtful application architecture, efficient software algorithms, and optimal hardware operating environment are all key considerations for reducing latency.

5.2.2 Benefits It Will Provide

Client's feedback and user experience are what matters the most at the end of the day. Moreover, the importance of high application responsiveness and minimized latency in the healthcare sector cannot be quantified. Rather the concepts of telemedicine and emergency medicine would stand meaningless without the immediate care response to its patients in time of need. Therefore, latency reduction is of significant importance in the field of healthcare. Some of the advantages can be listed as:

Just in time delivery of care – seamless integration of data and analytics. Although the data is fetched and stored into the databases and blockchain ledgers, the information that can be derived from the data shall be processed within microseconds and provide with the results. Reduced latency shall hence provide with immediate alerts in case of abnormal changes or meaningful insights in data.

(i) Quick diagnostic results would be available which normally take minutes or hours to process conventionally. These results could also be made available onto the network remotely.
(ii) Smart pharmaceuticals would be a trend. Based on the patient's current health condition and observations, the appropriate medication or emergency care unit could be called, and guaranteed help would reach the sick/old patient within time.
(iii) The exercise routines, after surgery, or pre-pregnancy, during pregnancy, and post-pregnancy routines shall be available at all times on the application which could be downloaded, viewed, or shared, thus promoting not an incremental but an exponential growth in healthcare.
(iv) Remotely monitored surgeries would be a trend and prove to be evolutionary for the countries and rural areas that lack the resources as well as trained professionals to carry out a critical operation in case of an emergency. At these times, fast network can prove to be a blessing as the trained professionals or female doctors shall be able to monitor and guide the nurses and other caretakers through. This practice shall not only provide the helpers with the authority to carry out the emergency procedure but also give them the confidence and practice they lack in the professionals' absence [24–30].

6 Future Scope

The healthcare and medical facilities play a major role in people's lives, and their scope of improvement is vast. Some of the future improvisations can be listed as:

(i) A doctor recommendation system based on the huge data collected with time through the feedbacks provided by the patients. The data cannot be more genuine than what would be comprised of only and only the honest reviews.
(ii) A patient's medical records through time can indicate certain trends or signs which might go unnoticed otherwise. A system that could analyze and denote such absurd or unusual signs could notify the patient directly without the manual involvement of a doctor or physician.
(iii) So many enhancements can be made to the chatbots already available. These chatbots would not only respond but tackle every incoming request as new and rethink (or reprocess) before each response is sent. They would even be capable of counseling their patient from the knowledge acquired and even advice based on their past experiences with them.
(iv) The more interactive and pleasing the application UWE is, the more it attracts the user's attention. Use of animation and graphs can take the involvement of the user to a whole new level without making the user feel overwhelmed by the complexities of the app UI.
(v) A module for guided meditation and yoga tracks for its general users could help them add value to their health and lives.
(vi) Things get a lot simpler when they are easier to interact with, i.e., if every feature of the application could be operated with the help of speech itself, similar to the Google Assistant. This would be easier if the application is molded and made goal-oriented.
(vii) Emotional support providing chatbot making use of the voice cloning technology in order to make it more patient-friendly [31–35].

7 Open Issues and Challenges

Though technological innovations are now exploring new domain of latency-free medical support solutions, as the process involves sensitiveness to human life, a number of open issues and challenges are there, which need to be addressed. Issues are diversified in nature and revolve around quality of portal, security, and privacy issues, privacy of patients and security of their data, usability of portal, safety of platform, financial implications, administrative and ethical issues, and negative effects of social distancing, and Fig. 16.16 depicts major open issues with employing mobile application for medical support [36–43].

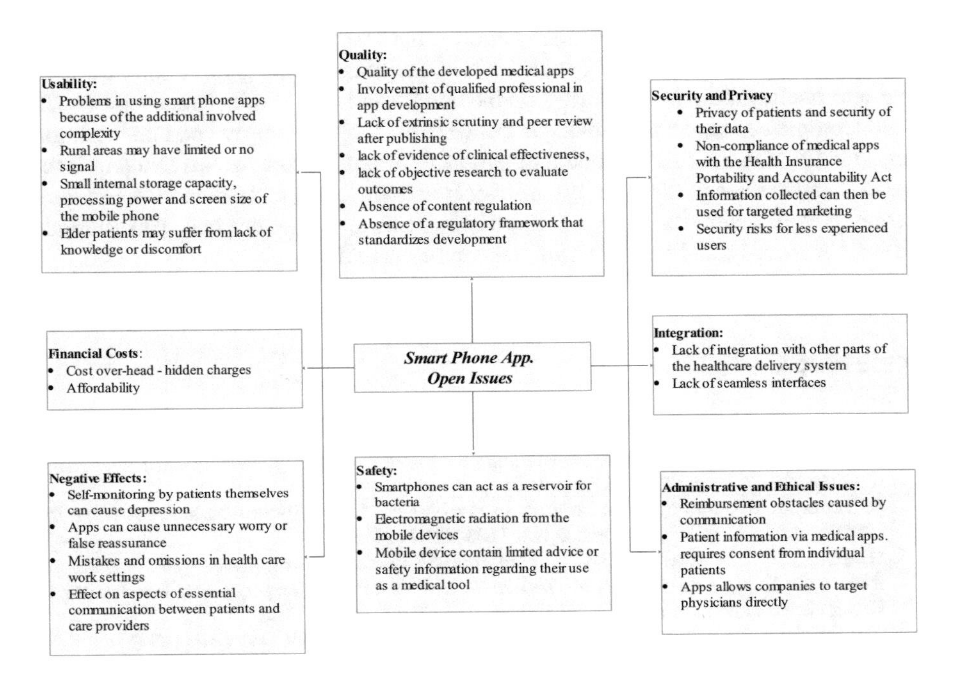

Fig. 16.16 Open issues and challenges

8 Conclusion

Percentage share in population in India above the age of 60 years has almost tripled in the last 50 years and is expected to rise relentlessly in the near future. Percentage projections are almost 133.32 million by year 2021, 178.59 million by year 2031, 236.01 million by year 2041, and 300.96 million by year 2051. Owing to the change in percentage share of age group in population, changes in medical structure are highly recommended. Technological innovations have given the platform for the much-needed digitized medical care. The proposed algorithm offers many benefits as compared to already existing solutions; some differentiating characteristics include:

(i) Electronic prescription and report generation
(ii) Finding directions through the app itself
(iii) Automated ordering of medicines and online transaction
(iv) Maintaining huge amount of patient data at one place

Algorithm proposed for app-based medical support for the elderly people offers a number of advantages as compared to already existing solutions summarized in Table 16.1. The algorithm supports automated generation of electronic prescription and report generation; bill and pathology reports make the management of data far easier. Entry of symptoms, medications, and tests is made easy with the help of AutoComplete suggestions, the app also facilitates finding directions through the app itself, automated ordering of medicines and easy online transaction with AutoComplete suggestions make it convenient for the elderly, and in addition algorithm maintains a huge amount of patient data at one place for faster and quicker access, and since a patient's entire history from symptoms to medications and tests required is all maintained at one place through this app, any reference can be made to the old records, and further treatment can be carried out.

References

1. R. Yan, C. Lwe, D. Chu, The design and implementation of the elderly healthcare information mining platform, in *2017 IEEE International Conference on Bioinformatics and Biomedicine (BIBM)*, Kansas City, MO, 2017, pp. 1501–1506
2. R. Verma, P. Khanna, National Program of Health-Care for the Elderly in India: A Hope for Healthy Ageing. Int. J. Prev. Med. **4**(10), 1103–1107 (2013)
3. H. Jiang, W. Xu, How to find your appropriate doctor: An integrated recommendation framework in big data context, in *2014 IEEE Symposium on Computational Intelligence in Healthcare and e-Health (CICARE)*, Orlando, FL, 2014, pp. 154–158
4. C.J. Salisbury, How do people choose their doctor. Br. Med. J. **299**, 608–610 (1989)
5. J. Sun, J. Ma, X. Cheng, Z. Liu, X. Cao, Finding an expert: A model recommendation system, in *International Conference on Information Systems*, 2013, pp. 1–10
6. J.M. Pierre, On the automated classification of Web Sites, in *Electronic Transactions on Artificial Intelligence*, 2001, p. 6
7. H.J. La, M.K. Kim, S.D. Kim, A Personal Healthcare System with Inference-as-a-Service, in *IEEE International Conference on Services Computing*, 2015, pp. 249–255
8. P. Zhang, S. Hu, J. Hu, Y. Zhang, G. Zhang, J. Zhang, Building cloud-based healthcare data mining services, in *IEEE International Conference on Services Computing*, 2016, pp. 459–466
9. A. Mohindra, D.M. Dias, H. Lei, Health cloud: An enabler for healthcare transformation, in *IEEE International Conference on Services Computing*, 2016, pp. 451–458
10. M. Pearce, Public health information systems: Priorities and practices for successful deployments. Stud. Health Technol. Inform. **225**, 680 (2016)
11. D.E. Elsaied, H.U. Khan, Implementation of health information system-A case study of Magrabi Hospitals. J. Comput. Sci **13**(4), 91–104 (2017)
12. R. Nambiar, R. Bhardwaj, A. Sethi, R. Vargheese, A look at challenges and opportunities of Big Data analytics in healthcare, in *2013 IEEE international conference on Big Data*, IEEE, 2013, pp. 17–22. https://doi.org/10.1109/BigData.2013.6691753
13. C.R. Keenan, H.H. Nguyen, M. Srinivasan, Electronic medical records and their impact on residents and medical student education. Acad. Psychiat. **30**(6), 522–527 (2006). https://doi.org/10.1176/appi.ap.30.6.522
14. M.J. Deen, Information and communications technologies for elderly ubiquitous healthcare in a smart home. Pers. Ubiquitous Comput. **19**(3–4), 573–599 (2015)
15. C. Chekuri, M. Goldwasser, P. Raghavan, E. Upfal, Web Search Using Automatic Classification, 1999, pp. 1–11
16. www.dontmesswithmama.com/steadymd/
17. www.plushcare.com/blog/plushcare-joins-livelys-hsa-marketplace/
18. www.nuemd.com/blog/21-million-doctor-demand-means-providers
19. www.apa.org/monitor/2017/02/online-therapy
20. www.trueventures.com/blog/welcoming-first-opinion

21. www.citymomsblog.com/blog/virtual-doctor-visits-yes-please-amwell-telehealth-services/
22. www.regpacks.com/blog/online-registration-software/
23. www.blog.practo.com/
24. S. Tanwar, Q. Bhatia, P. Patel, A. Kumari, P.K. Singh, W.C. Hong, Machine learning adoption in blockchain-based smart applications: The challenges, and a way forward. IEEE Access **8**, 474–488 (2020)
25. S. Tanwar, J. Vora, S. Kanriya, S. Tyagi, N. Kumar, V. Sharma, I. You, Human arthritis analysis in Fog computing environment using Bayesian network classifier and thread protocol. IEEE Consum. Electron. Mag. **9**(1), 88–94 (2020)
26. J. Vora, S. Tyagi, N. Kumar, M.S. Obaidat, A systematic review on security issues in VANET. Secur. Priv. J. Wiley **1**(5), 1–27 (2018)
27. S. Tanwar, K. Parekh, R. Evans, Blockchain-based Electronic Healthcare Record System for Healthcare 4.0 Applications. J. Inf. Secur. Appl. **50**, 1–14 (2019)
28. J. Hathaliya, S. Tanwar, An exhaustive survey on security and privacy issues in Healthcare 4.0. Comput. Commun. **153**, 311–335 (2020)
29. J. Hathaliya, S. Tanwar, R. Evans, Securing electronic healthcare records: A mobile-based biometric authentication approach. J. Inf. Secur. Appl. **53**, 1–14 (2020)
30. P. Khanna, S. Kumar, Engineering 4.0: Future with disruptive technologies, in *Blockchain Technology for Industry*, ed. by R. Rosa Righi, A. Alberti, M. Singh, vol. 4, (Blockchain Technologies. Springer, Singapore, 2020), p. 0. https://doi.org/10.1007/978-981-15-1137-0_7
31. P. Singh, P. Khanna, S. Kumar, Communication architecture for vehicular Ad Hoc networks, with blockchain security, in *2020 International Conference on Computation, Automation and Knowledge Management (ICCAKM)*, Dubai, United Arab Emirates, 2020, pp. 68–72. https://doi.org/10.1109/ICCAKM46823.2020.9051499
32. S. Tanwar, S. Kaneriya, N. Kumar, S. Zeadally, ElectroBlocks: Blockchain-based energy trading scheme for smart grid systems. Int. J. Commun. Syst. Wiley **33**(15), 1–16 (2020)
33. J.-H. Yoo, The meaning of information technology (IT)mobile devices to me, the infectious disease physician. Infect. Chemother. **45**(2), 244–251 (2013)
34. T.D. Aungst, Medical applications for pharmacists using mobile devices. Ann. Pharmacother **47**(7–8), 1088–1095 (2013)
35. M.R. Slaper, K. Conkol, mHealth tools for the pediatric patient-centered medical home. Pediatr. Ann. **43**(2), e39–e43 (2014)
36. M. Arnhold, M. Quade, W. Kirch, Mobile applications for diabetics: A systematic review and expert-based usability evaluation considering the special requirements of diabetes patients age 50 years or older. J. Med. Internet Res. **16**(4), 50–61 (2014)
37. B.L. Elias, S.A. Fogger, T.M. McGuinness, K.R. D'Alessandro, Mobile apps for psychiatric nurses. J. Psychosoc. Nurs. Ment. Health Serv. **52**(4), 42–47 (2014)
38. K.E. Muessig, E.C. Pike, S. LeGrand, L.B. Hightow-Weidman, Mobile phone applications for the care and prevention of HIV and other sexually transmitted diseases: A review. J. Med. Internet Res. **15**(1), 65–81 (2013)
39. A.C. Nwosu, S. Mason, Palliative medicine and smartphones: An opportunity for innovation? BMJ Support. Palliat. Care **2**(1), 75–77 (2012)
40. A.D. Workman, S.C. Gupta, A plastic surgeon's guide to applying smartphone technology in patient care. Aesthet. Surg. J **33**(2), 275–280 (2013). https://doi.org/10.1177/1090820X12472338. D.S. Eng, J.M. Lee, The promise and peril of mobile health applications for diabetes and endocrinology. Pediatr. Diabetes **14**(4), 231–238 (2013)
41. A. Moodley, J.E. Mangino, D.A. Goff, Review of infectious diseases applications for iPhone/iPad and Android: From pocket to patient. Clin. Infect. Dis. **57**(8), 1145–1154 (2013)
42. F. Haffey, R.R. Brady, S. Maxwell, A comparison of the reliability of smartphone apps for opioid conversion. Drug Saf. **36**(2), 111–117 (2013)
43. D. Dubey, A. Amritphale, A. Sawhney, N. Amritphale, P. Dubey, A. Pandey, Smart phone applications as a source of information on stroke. J. Stroke **16**(2), 86–90 (2014)

Chapter 17
Blockchain- and Deep Learning-Empowered Resource Optimization in Future Cellular Networks, Edge Computing, and IoT: Open Challenges and Current Solutions

Upinder Kaur and Shalu

1 Introduction

In this, we proposed the vision of a deep learning approach with the integration of blockchain technology in various applications. The blockchain is a distributed database technique that provides immutable, transparent, more secure, and reliable decentralized computing services. In [2–3], the author presented a cost-effective administration, dynamic access, and spectrum services application using blockchain for next-generation cellular networks. Some other authors [63] utilize the blockchain technology to increase the performance of the vehicular edge networks. They used smart contract blockchain framework design for data sharing techniques in vehicular networks. They used proof of work based on mining and required huge resources and energy consumption. In [6], the author discussed the mobile edge computing solutions for boosting the communication speed and provided seamless communication in heterogeneous networks and device-to-device communication. The authors [8–10] proposed caching schemes to resolve the backhaul congestion problems in mobile edge computing. They also proposed some offloading strategies to minimize task duration in the case of heterogeneous networks. There is a big challenge for optimizing the computing resources and caching issue on the mobile edge servers due to versatility in the time variant, channels, requirements, and heterogeneity in emerging 5G cellular networks. Some authors worked on high-performance computing algorithms to overcome the demands of upcoming cellular networks.

U. Kaur (✉)
Department of Computer Science, Akal University, Talwandi Saboo, Punjab, India
e-mail: upinder_cs@auts.ac.in

Shalu
Department of Computer Science, Baba Farid College, Bathinda, Punjab, India
e-mail: drshalu@babafaridgroup.edu.in

S. Tanwar (ed.), *Blockchain for 5G-Enabled IoT*,
https://doi.org/10.1007/978-3-030-67490-8_17

The artificial intelligence branch has great potential in handling the issues in 5G cellular networks, blockchain, IoT, and other emerging technologies. Some researchers worked on the integration of blockchain with artificial intelligence (AI) in wireless networks to create intelligent networks. AI can be integrated with cloud computing, cellular networks, mobile edge computing networks, and IoT networks to facilitate intelligent and secure resource management. Reinforcement learning is the branch of AI that can be more generic for problem solving approach and explicit programming. RL can be processed sequentially and automatically adjust the policies by observing the result and behavior. In [11], the author focused on the DRL approach for handling the caching issue in mobile edge computing. Further, the author [12] designed an application to address the expansion of the network scale and in-depth feature discovery. The author [13] presented a DL-based algorithm for traffic control and potentially reduced the offloading time. Some others worked on a secure and intelligent framework for future generation networks integrated with blockchain.

The vogue of machine learning triggers the colossal engrossment application approach deep reinforcement learning in various research fields. The authors [21, 24–26] presented deep learning development models for computer vision, resource optimization, pattern recognition, and speech synthesis. Some of them presented a comprehensive report on application and open issues in the deep learning approach. Others summarized the principles, evolutionary methods, architecture, and core algorithms for deep reinforcement learning. The authors [29] highlight the remarkable achievement of deep neural networks in blockchain, IoT, and cellular networks. Further [91] presented a survey on more application areas of deep learning. The author [35] shed more light on the deep learning potential in mobile edge computing and popular application in IoT and blockchain.

1.1 Aim

There were numerous research articles published in the scope of machine learning integrated with blockchain to date on different aspects of emerging technology. Some researchers presented comprehensive surveys with a focus on the integration of machine learning with blockchain on specific fields and application areas. The proposed chapter covers the details of fundamental aspects of a deep learning-based approach in the integration of blockchain with IoT and 5G cellular networks.

There are several survey articles available that highlight the benefits of the amalgamation of blockchain to 5G networks, and the authors [8] provide adoption of blockchain for secure 5G network resource management. The author [7] focused on the potential of blockchain in Industry 4.0. And further [10] presented the analysis of blockchain application in handling the privacy issues for smart technology like IoT, smart grid, healthcare, etc. The author [9] presented blockchain for 5G IoT applications. Some other researchers presented a systematic survey on D2D caching techniques for content sharing and 5G networks. Our work is different from others as we did research mainly focused on the deep learning techniques for resource management in the integration with blockchain to address research challenges and future prospective for cellular networks, edge computing, and IoT networks. The comparison of this chapter with the existing paper is given in Table 17.1.

1.2 Research Contribution

The proposed work covers all the concepts of deep learning to be applied in blockchain applications with IoT and 5G cellular networks. Following are the research highlights of this chapter.

- We presented a review of the existing survey on blockchain-empowered IoT and 5G cellular network services.
- We presented a systematic discussion on the perspective of blockchain using machine learning and deep learning for resource management in 5G and IoT networks.
- Based on the study, we outline the open research issues, research challenges, case study, and further future research directions.

1.3 Organization

This chapter structure is presented as below. Section 1 covers the introduction of the chapter. Section 2 highlights the overview of the DL existing techniques and provides a brief introduction to the technologies blockchain, IoT, and 5G. The generalized architecture and taxonomy for resource management are detailed in Sect. 3. Section 4 presented the discussion on the open issues, challenges, and further prospects. In Sect. 5, we presented the case study and finally concluded the chapter.

2 Background

In this chapter, we illustrate the analysis of existing surveys and provide the current deep learning solution and application integrated with blockchain in future cellular networks and IoT. The author in [14] explained the blockchain technology with their key requirements, consensus algorithms, and blockchain platforms with their pros and cons. The author [50] provided a detailed study of blockchain in decentralized blockchain consensus mechanism with BFT (Byzantine fault tolerance) strategies, other mining protocols, and hybrid protocols. In [13], the author gave a complete survey on the permissionless blockchain technology and observed the system

Table 17.1 Summarized chart of available surveys in blockchain and machine learning

Reference	Title	Technology	Major contributions	Limitations
[1]	Deep learning in mobile and wireless networking: A survey	IoT, ML, blockchain	The author presented a detailed study of deep learning techniques in mobile and wireless networking, IoT, and signal processing	Discussed the deep learning approach but not covered blockchain application
[2]	Machine learning adoption in blockchain-based smart applications	ML, blockchain, smart applications	The author focused on the use of machine learning integrated with blockchain and smart applications	They missed the deep learning approach
[3]	Blockchain for 5G-enabled IoT for industrial automation	Blockchain, 5G, IoT	The author presented the comprehensive view of blockchain for 5G-enabled IoT in various applications like smart city, smart home, and industry 4.0	Machine learning and deep learning are not considered
[4]	Machine learning for resource management for cellular and IoT networks	Machine learning, deep learning, IoT, 5G networks	In this, the authors provided the taxonomy and machine learning approaches in IoT and 5G networks	They didn't cover it with blockchain technology
[5]	Deep learning-based IoT for secure smart city	Deep learning, IoT, blockchain, smart cities	In this, the author presented the methodological flow of SDN with blockchain technologies for smart cities	They covered blockchain and IoT technologies only
[6]	Adaptive resource allocation in future networks with blockchain	Mobile edge computing, blockchain, deep reinforcement learning	In this, the author provides the systematic framework of the future wireless networks and open issues and future directions	They missed the IoT technologies
[7]	Blockchain for industrial 5G networks and IoT. it only covers the concepts	Blockchain, 5G, and IoT networks	The author provides the content for the applications of blockchain in 5G networks and industry 4.0	Machine learning, deep learning, and specific use cases are missing
[8]	Blockchain for 5G network management	Blockchain, deep reinforcement learning	In this, the author presented the adoption of blockchain in the 5G network resource management and deep learning techniques	They missed focusing on other 5G technologies, D2D communications, IoT
[9]	Blockchain for UAV	Blockchain technology, UAV	The author presented a viewpoint for use of blockchain in solving UAV network issues	They missed the focus on blockchain applications in 5G network domain
[10]	Blockchain and 5G networking	Blockchain technology, D2D communication	They provide complete survey on D2D communication in 5G	They missed focus on the role of blockchain in other 5G services
Our paper	Blockchain for resource management in 5G cellular and IoT networks and deep learning techniques	Blockchain, IoT, deep learning, 5G cellular networks	Here we are presenting a comprehensive survey of integration of blockchain with deep learning in 5G cellular and IoT networks	

performance, cost of participation, and topologies adopted with design to improve the efficiency of blockchain technology. Further in [51], the author investigates the security factors, authentication, privacy, access control, resource utilization, and quality assurance in blockchain technologies. The authors in papers [52–53, 56, 57] provided the comprehensive view of te application of blockchain technology in smart cities, edge computing and IoT with increased efficiency, trace ability, transparency and security of the system at low cost.

In the [59–62], the authors provided the details of the ML approach in WSN, optical communication networks, SDN, and CRN in 5G networks. Some of the authors in [53, 58, 64] scrutinize the applications of the DL to provide intelligence in communication networks and IoT networks. They also provided a review of the ML techniques for traffic analysis and network controls in communication networks. In [79–81, 116], the authors addressed the security issues, threats, and machine learning approach in VANETs, data mining, and detection systems. Further in [90], the author presented the Leech framework for secure wireless sensor networks. They also covered critical security issues in monitoring security threats. Further in [48], the author presented the more secure anonymity-preserving authentication protocol for blockchain-envisioned global mobility networks. Further other worked on vision in [54] blockchain to empower more security with accuracy attack detection and prevention in manets. Nowadays, the machine learning technique is considered propitious technologies in integration with other technologies. Both machine learning and blockchain are used with integration in IoT, future cellular networks, and cloud computing to improve the overall performance. The author in [71] provides the integration of machine learning for resource allocation in WSN. Further, we explore the current deep learning techniques and its application and afterward provide a brief introduction to blockchain, IoT, and future generation networks.

We aggregated all the acronyms used throughout this chapter in Table 17.2.

2.1 *Deep Learning Techniques*

DL is the sub branch of ML, where ML is a broader term, categorized into supervised, unsupervised, and reinforcement learning techniques. Supervised learning is a technique where we can estimate and then do prediction based on unknown parameters. The existing techniques used for supervised learning are Bayes classifier, vector support method, decision tree, random forest, and many more. The unsupervised learning is opposite to the previous one; they are used in a heuristic manner. In unsupervised learning data is unlabelled and used to train the data. Reinforcement learning is a technique that learns from the environment. Q learning is the most popularly used deep learning technique. Machine learning is used in several applications: load balancing, spectrum sharing, sensor device management, channel utilization, energy trading, etc. In [94], the author presented the machine learning models for the data analytics and a threat model. In [109],

Table 17.2 List of acronyms

Acronym	Description	Acronym	Description
ML	Machine learning	DBM	Deep Boltzmann machine
DL	Deep learning	PoW	Proof of work
DRL	Deep reinforcement learning	D2D	Device-to-device
DNN	Deep neural networks	IoT	Internet of Things
ANN	Artificial neural network	SDN	Software-defined network
DBN	Deep belief network	UAV	Unmanned aerial vehicle
AEDL	Auto-encoder deep learning	NOMA	Non-orthogonal multiple access
CNN	Convolution neural network	MIMO	Multi-input multi-output
RNN	Recurrent neural network	PoS	Proof of stake
GAN	Generative adversarial network	FL	Federated learning
eMBB	Enhanced mobile broadband	CRN	Cognitive radio network
V2V	Vehicle-to-vehicle	WSN	Wireless sensor network
QoE	Quality of experience	LoRA	Long Range
BaaI	Blockchain as an Infrastructure	NVM	Network virtualization management
FCN	Future cellular networks 5G and beyond	URLLC	Ultra-low latency communication

the author presented the machine learning-based blockchain framework for smart applications – smart grid, smart cities, healthcare, UAV, etc. – with open research challenges and future scope. The current accessible deep learning techniques are summarized in Fig. 17.1 (a, b, and c).

Deep Reinforcement Learning (DRL)
The DRL is the sub-branch of ML that can solve complex many problems. It is a sequential learning technique that can adjust to the policies and gives rewards accordingly. They mainly focused on the Markov decision process that consists of an environment and has a set of agents [14, 15]. Machine learning followed through deep learning has basic paradigm learning. The agents perform an important role by applying meaning actions. Further, the action gives feedback to improve the output action because when the agent interacts with the environmental setup, it gets rewards as the feedback signal. The main target is to achieve an optimized solution. Recently this approach gained popularity in a wide range of applications. The deep reinforcement learning techniques have two main categories: model-based and model-free [16].

In a model-based deep learning approach, the agent can get access to the model environment. The environment function helps the agent to predict the probabilities and transaction states. The agent can predict better choices by understanding the environment. This approach helps plan policy. The author in [18] presented the model-based predictive control, where the agent predicts the environment and gives an optimal solution. The agent learns the environment and prepares new plans after every interaction in the system. In other [19] research areas, the author used model-based deep learning to train the agent and use agent experience to give probable solutions.

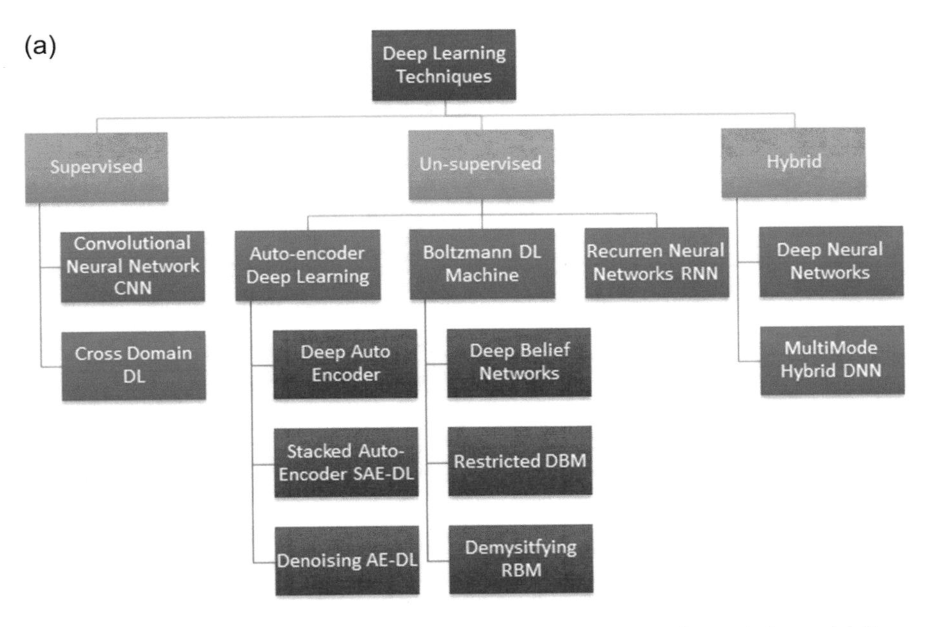

Fig. 17.1 (**a**) Taxonomy of deep learning techniques. (**b**) Deep learning techniques. (**c**) Deep learning libraries

In a model-free deep learning approach, the agent learns the ground truth and uses different aspects of the environment. This is the best approach as it is complex to train the model that can be exploited by agents. Thus, it is more feasible and can incubate multiple factors for rewards. The main agenda is to train the agent with the best optimal solution. The deep learning approach attracted researchers where a large number of possible states can be considered as a feedback output. The model-free approaches are deep Q networks, Asynchronous Advantage Actor Critic, deep deterministic policy gradient, and distributed proximal policy optimization. In [125–127], the authors used a deep learning approach in AI gaming, automatic driving, and robotics. In [19], the author gave DeepMind solutions for Atari video games. Further [93] worked on actor-centric mechanism, where different actors were deployed with different threads of CPU. Thus, it improves the training of agents in CPU processing. In [84], the author presented distributed proximal policy optimization. They implemented this approach on multi-threading CPU and apply different strategies to train the agents. Many other researchers applied the deep learning approach in solving more complex problems. In paper [128], author presented a game based agent system, for detecting defeats in the real time multi agent in game environment. Some authors in [49] presented the deep neural network for the diagnosis of lung cancer. The deep learning techniques also have a scope in a variety of applications. In [38, 39], the author presented the applications of deep learning in price prediction and facial sentiment analysis, respectively. In [128–130], the authors focused on the Markov decision process to train the switching

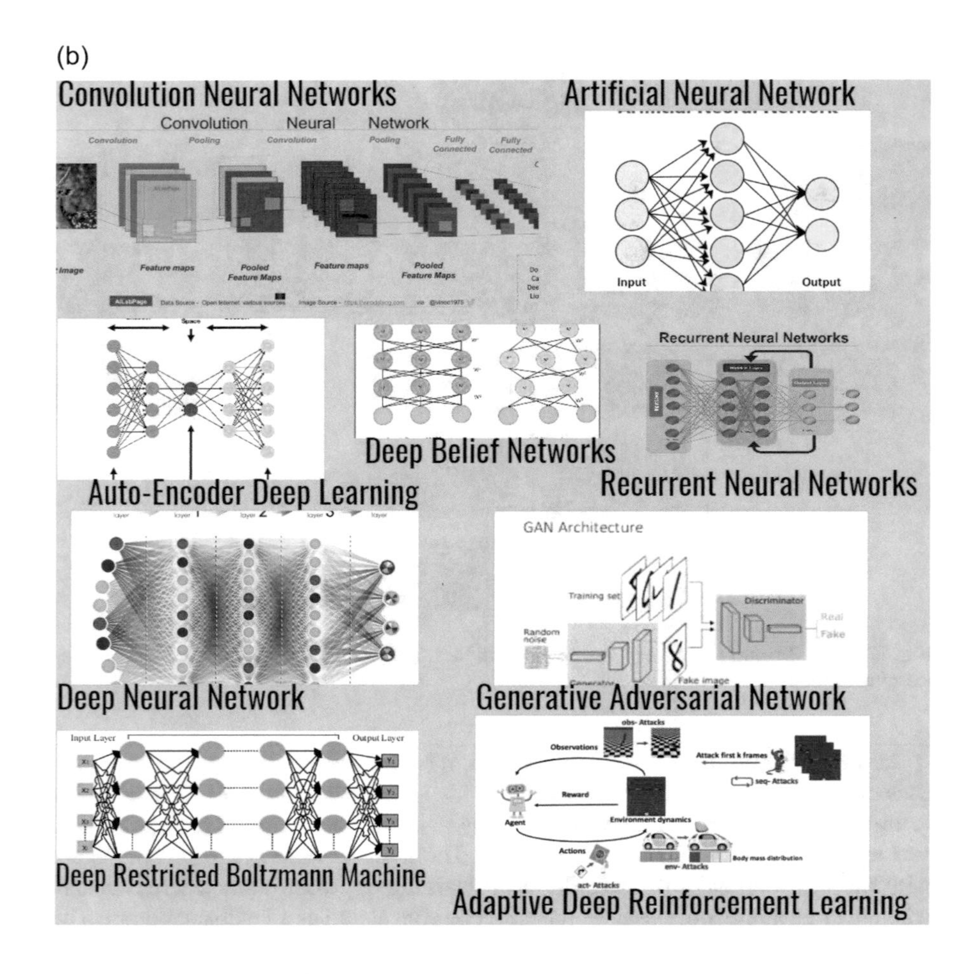

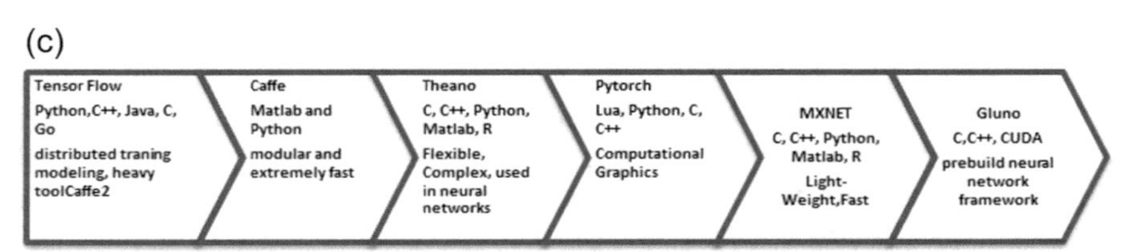

Fig. 17.1 (continued)

policies, routing in networks, tracking control systems, and others in [68] presented the comparison of RL based algorithms for application scheduling. The ability of deep learning to work efficiently in high dimensionality makes this more useful in next-generation computing.

Deep learning is the more propitious technique in machine learning. Due to its versatile feature, it became the most demanding machine learning approach in the current scenario. DL also envisioned a huge success rate in implementing a variety of applications – natural language processing, blockchain, IoT, future generation networks, image processing, voice synthesis, smart devices, healthcare, and many more. The present existing DL techniques are deep neural networks, artificial neural networks, recurrent neural networks, auto-encoder network, convolution neural network, deep reinforcement learning, generative adversarial network, and deep Boltzmann machine network. It is an improved ML approach that requires more training costs than other techniques. Further, we provide brief details of deep learning techniques.

Convolution Neural Network CNN can be employed supervised, unsupervised, and reinforcement learning models. This technique was initially developed for digital image processing; later on, researchers focus on multiple applications of CNN. Some presented CNN for the traffic analysis, image classification, network analysis, resource management, intrusion detection system, TCP protocols, etc. The CNN model worked efficiently with an in-depth feature of the given problem. It uses the transfer learning to reduce the training cost and classification time. In [49], the author proposed the CNN model for deep cooperative spectrum sensing. The CNN algorithms have interrelated features to learn and train the patterns in the convolution layers.

Deep Restricted Boltzmann Machine It is an unsupervised deep learning technique. It consists of visible and hidden layers, where visible layers are conditional independent of hidden layers. The data input will be given to the visible layer and trained algorithm. The training model uses the backpropagation model to achieve gradient contrastive divergence algorithms. The performance of DRBN models envisions superior performance in various applications – network performance, robotics, spatial representation, forecasting, etc.

Auto-encoder Deep Learning This technique is an unsupervised DL model. The working of the auto-encoder depends on the twin-pair model, encoder, and decoder neural networks. The input data is provided to the encoder and generates output as mean and variation functions. Auto-encoder DRL model minimizes the error effect and automatically generates new data samples based on the trained model. It is more applicable in data security-related models, intrusion detection model, prediction-related problems, mobile activity tracking, etc.

Recurrent Neural Network This technique is supervised, unsupervised, and reinforcement learning. This model is based on sequential data modeling because they work on the principle of the feed-forward network model. This model reported gradient vanishing problems. Some researchers got success in accomplishment of RNN model framework for forecasting model in voice precipitation, data flow analysis in mobile networks, massive data analysis etc.

Generative Adversarial Network This is an unsupervised DL technique. It consists of two training models – generative and discriminative model – that target the data distribution in the trained model and finally estimate the sample and output the estimated probability. This training model has a higher learning rate and supports higher parameters. These models are more suitable for migration management in mobile wireless networks, resource allocation problems, and real-time data analysis.

2.2 *Blockchain*

Blockchain technology is a distributed ledger technology that can provide an immutable set of transactions. It is a more secure, reliable, and transparent technology. All the transactions in blockchain are managed in a tamper-proof ledger. It also has a consensus mechanism for attaching a new block to the chain. It offers ample of opportunities to manage untruthful parties by decentralized strategies of transaction governance. The key features of this technology are decentralization, privacy, transparency, auditability, and immutability. Due to its potential in various applications, it gained much attention for mobile networks, cloud computing, IoT, mobile edge computing, and infrastructure commissioners. Distributed ledger-based technology [24] used distributed databases for organizing hash tree, with irreversible and tamper-proof transaction management. Its consensus mechanism ensures the integrity of the transaction and guarantees the consistency of the transaction. The consensus mechanism performance parameters are transaction throughput, security, and scalability that also depend upon application scenarios. PoS, Byzantine fault tolerance, and PoW are the commonly used consensus mechanisms.

The distributed ledger technology can be benefited for all the upcoming and emerging technologies. The transparency in the blockchain is the value for both end-users and the service providers. This is an ideal technology to track the history of all the transactions that occurred. These features help in a revolutionary change in improving the current system efficiency and reliability [35]. In the paper [36], the author presented blockchain as the infrastructure as a service on the cloud with the feasibility that trade can be done in a distributed manner without being centrally managed. Further, they incubate BaaI (Blockchain as an Infrastructure) which managed a tough bond between the service provider and end-user. Thus, blockchain acts as the backbone in a distributed resource management technique and keeps the transparent transaction record. In [95], the author provided the evidence for the use of blockchain and ML amalgamation for efficient energy management in future generation networks. Few stressed the application of blockchain in the healthcare sector. In [28, 55, 109], the authors presented habits as a blockchain-based framework for healthcare services. Some other authors in [115] presented a blockchain-based automated secure tamper-proof system for cheque clearance chain: MudraChain.

2.3 Internet of Things (IoT)

IoT is the emerging technology that integrates Internet with objects, called as Internet of Things. It is a powerful technique that can sense, actuate, dynamic behavior, better networking. The IoT devices are found everywhere – smart homes, smart vehicles, smart transport systems, intelligent grids, smart E-health, etc. the versatility in IoT devices provides numerous applications of IoT in heterogeneous networks of IoT devices. It has several key features that is highly recommended for dynamic environment, high power computing devices, and federated system for secure resource management in IoT networks. The dynamic nature of IoT helps to work across a large geographical area, but needs more scalable models and low-latency response time algorithms. It can be integrated with the other technologies – cloud computing, blockchain, and mobile edge computing – for the deployment of ingestion bandwidth, processing, and better resource management services. In [47], the author focused on the workability of IoT in the big data for secure streaming, data validation, and verification. In [22], the author presented the application IoT in big data computing. The IoT devices provide a large-scale geographic locality and can scale up the resources to empower the real-time services of IoT devices. They can work in distributed decentralized networks and promising technology for future electronic devices, micro-grids, etc.

2.4 Future Cellular Networks

The 5G is popularly known as the next-generation cellular communication networks. The 5G technologies have three broad services: eMBB, URLLC, and m-MIMO. High data rate at the speed of 10 Gbps, high scalability for massive machines communication, network silicing, software defined networks, better network virtualization, device to device communication are the key features of 5G networks. The integration of blockchain and machine learning in 5G cellular networks provides a great perspective to manage the research challenges in implementation. In this, we covered the 5G cellular resource management with deep learning and blockchain technology. In [23], the author presented the systematic use of AI and ML for 6G communications networks. In [89], the author provided the expanded scope of 5G revolution for future generation smart applications.

2.5 Generalized Architecture of Blockchain and Deep Learning

The generalized architecture of amalgamation of blockchain and deep learning for different application consists of four layers: cloud layer, 5G cellular network and mobile edge computing layer, blockchain technology with collaborative deep reinforcement learning, and at the bottom Internet of Things device layer. All layers operate with the collaboration of each other. The cloud layer provides the service function chain and deployed the secure cloud platform. The next layer is the 5G cellular network and mobile edge computing layer. The service offered is the covenant distribution of edge clouds, providing blockchain reliable resource management. It consists of multiple modules – blockchain, controller, network virtualization module, and multiple edge nodes. It consist of multiple modules:- blockchain controller, network virtualization module, multiple edge nodes and blockchain collaborated deep reinforcement learning model. This layer governs major services like the smart contract, covenant resource reservation, user authentication and secure transaction. This provides the credible and reliable services to the device and upper layers. The smart device layer consists of the IoT smart device networks. The device layer handles the multiple device requests and sends request to the upper layers. This DRL-based architecture manages resource optimization, network virtualization, and server nodes at different layers.

3 Deep Learning for Resource Management in Blockchain-Empowered Cellular and IoT Networks

This section presented the current solutions of a deep learning approach in the integration with blockchain for 5G cellular networks and IoT networks. Deep learning is the machine learning technique based on artificial neural networks that have multiple layers between the input and output layers. The deep learning network trained the data set based on the environment and performs an action based on this. This has huge potential in problem-solving – prediction, categorization, resource allocation, classification, speech synthesis, natural language processing, etc. The process used in deep learning is to perform feature extraction and then do further classification in various layers. It consist of multiple layers; first layers works on finding the patterns, second layer works on recognising complex problems then further layer deal with the optimization solutions. A large amount of data can be arranged and managed using deep learning. As this is a powerful technique for high-level feature extraction, it is suitable in blockchain, IoT, and 5G networks because it can exploit the large amount of unlabeled data using the different deep learning models. Some existing deep learning models are shown in Fig. 17.1. Deep Boltzmann network has two layers of neurons, and in [50], the author used this for network communications. They suggest that the deep learning reduces the complexity of the complex task. It has great potential with technologies like blockchain, 5G networks, SDN, IoT, and cloud applications. In deep learning, the agents learn by themselves and experience the solutions by maximizing the effort in generating long-term rewards. Its major disadvantage is the large set of data needed to train and test before use in any application. In [51, 52], the authors presented the applications of deep neural networks in various domains. The author in [53] presented the deep learning for the prediction analysis in IoT networks. Some other researchers in [56, 57] presented the advanced level deep Q learning techniques. The further section provides the summarized research done in the blockchain, future generation networks, and IoT networks using machine learning and deep learning techniques.

3.1 Resource Management Using Blockchain and Deep Learning Techniques for Future Cellular Networks

Blockchain has three different types of blockchain: private blockchain, public blockchain, and consortium blockchain. Table 17.3 showed the blockchain services in future generation communication networks. A public blockchain is a conventional approach that allows everyone to participate. The summarised architectural view shown in Fig. 17.2 provides the integration of Blockchain, IoT, 5G networks and deep learning techniques. Private blockchain is acquired by some peculiar conglomerate, and the consortium-based blockchain used permission nodes to create new blocks. Generally, blockchain has three components – transactions, block records, and a consensus algorithm. When a transaction occurred, it is encrypted and signed digitally. The block is packed with a cryptographically tamper-proof node block. The consensus algorithm helps invalidate the block in terms of consistency and order. In the future generation, communication network security is a big issue. So the application of blockchain is considered for secure and privacy-preserving future networks. Blockchain has the potential to improve the 5G cellular network services. In this chapter, we summarized the detail of the services offered by blockchain for 5G cellular networks. Table 17.3 provides the details of the key services of blockchain in 5G cellular networks. The integration of blockchain in future generation networking communication promises better services including data sharing, NVM, resource optimization and management, FL, privacy services, and spectrum management. The main foucs is to provide the immutable decentralised transparent services of blockchain in the future generation communication networks.

This section provides the resource in the future cellular networks and discusses in detail the prospective of blockchain and ML and DL techniques in resource management in FCN communication.

Table 17.3 Summarized taxonomy for resource management using deep reinforcement learning/machine learning blockchain approach in future cellular networks

Resources	References	Technology	Findings
Data collection and sharing	[11–14, 37]	Blockchain, IoT, DRL, 5G	Presented storage and sharing mechanism for decentralized storage management A secure sharing approach was presented for industrial IoT Integration of blockchain and fog for data sharing They presented energy-efficient framework for secure data collection and sharing DRL-based approach in which blockchain-enabled efficient improve data sharing and collection scheme proposed with Ethereum blockchain and DRL for 5G networks and created the same environment
Device-to-device communication	[33, 34]	Blockchain, DRL, 5G networks	Presented caching content at mobile devices and reduces data traffic using ML Presented DRL approach to address the resource management using blockchain and 5G networks
Network virtualization management	[15–17, 42]	Blockchain and 5G cellular networks	A blockchain approach to support MVNOs with wireless network virtualization They presented a secure virtual machine orchestration system The architecture proposed for blockchain-based network virtualization DL-based blockchain framework for network resource management was presented
Federation learning	[21, 24, 30, 41, 133]	Blockchain and 5G cellular networks, mobile edge computing	Blockchain-based federated learning architecture proposed Proposed blockchain-based nonrepudiation and tamper resistance federation learning methodology Proposed deep learning-based deep chain framework FDC – Federated DRL approach for data collaboration in blockchain applications Presented and DRL and FL mechanism for intelligent resource management in edge computing
Spectrum management	[25–27, 29, 33]	Blockchain and 5G cellular networks	Proposed secure spectrum sharing in CRN Blockchain-based secure spectrum sensing platform proposed Unlicensed spectrum sharing scheme proposed with blockchain Presented DRL approach to address the resource management using blockchain and 5G networks

Resource management	[18–20, 31, 32, 44, 45, 88, 132]	Blockchain, IoT, deep learning, cloud computing, mobile edge computing	Presented architecture for mobile edge computing Blockchain-based resource allocation strategy for IoT Presented blockchain radio access network Proposed network resource management using deep learning Presented adapted resource allocation in future generation networks using a deep learning approach Secure spectrum-based incentive scheme presented An intelligent resource management scheme with DRL and blockchain was presented Blockchain empowers resource management and sharing for future generation networks Presented the blockchain-envisioned UAV communication in 6G networks Presented mobile edge computing using blockchain as a service for UAV
Content caching	[33, 43]	DRL, blockchain, 5G networks	Presented DRL approach to address the resource management using blockchain and 5G networks DRL blockchain approach for addressing open challenges
Energy trading	[35, 36, 40, 44, 46]	DRL, ML, blockchain, 5G networks	They presented the ML approach that is used for smart vehicles and mechanisms for energy trading with blockchain They proposed V2V energy trading framework with ML and blockchain techniques DeepCoin blockchain and DRL-based approach for energy trading in smart grids An intelligent resource management scheme with DRL and blockchain was presented Proof of deep learning is proposed for energy recycling blockchain

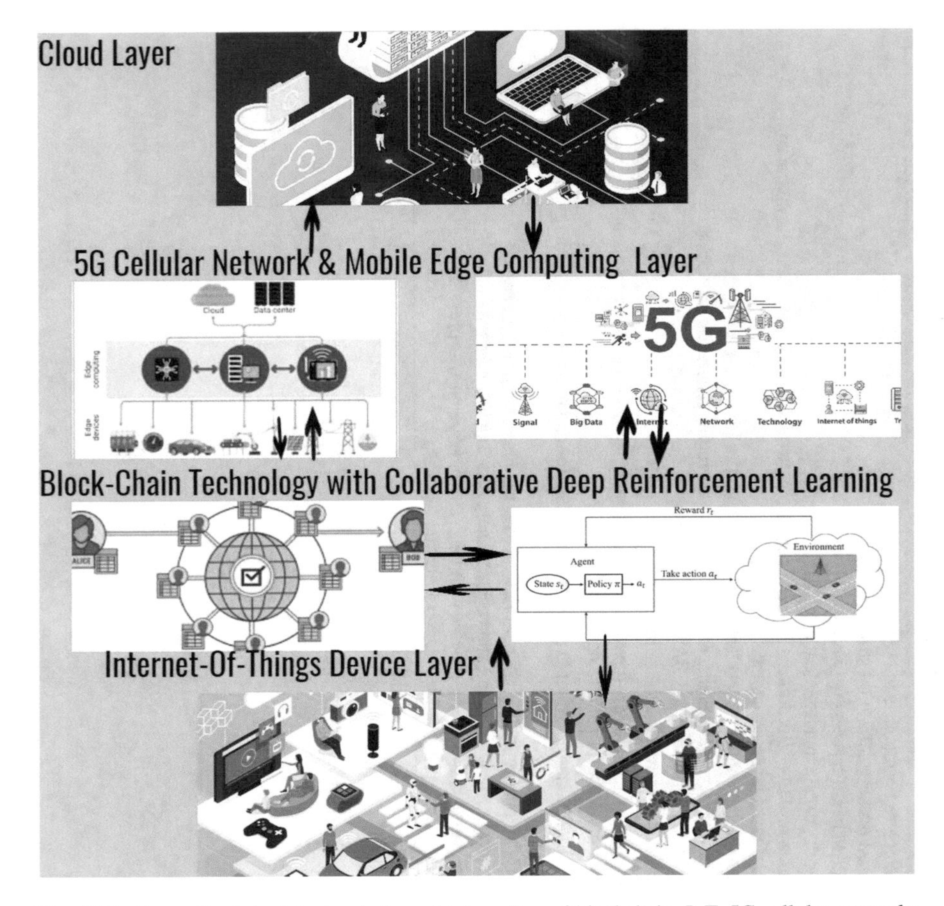

Fig. 17.2 Generalised Architectural view of integration of blockchain, IoT, 5G cellular networks and deep learning techniques

3.2 Blockchain and Deep Learning Application for Resource Management in IoT Networks

The preeminence of deep learning techniques encourages researchers to develop more deep learning in upcoming technologies. The author in [48] proposed the cognitive IoT network with the implementation of Q learning approach of deep learning. It helped in the optimization of packets transmitted in multiple channels. This enhances the overall network efficiency in IoT networks. In this paper [58], author presented DQL based transmission scheduling algorithm for cognitive IoT and estimated good accuracy. Some authors [59–61] worked on the deep learning-based approach in spectrum sharing and sensing. Some others researchers worked on the deep learning techniques for MIMO and NOMA technologies. Similarly, others framed the deep neural network for load prediction and balancing in the IoT networks in [62]. In [63], the author proposed a deep learning technique for spectrum management by using primary and secondary users and created an automated learning platform. Further in [64], a deep recurrent neural network approach was proposed for channel selection, spectrum access, and carrier sensing. They claimed the maximization of the network throughput using a deep learning approach. In Table 17.4, we summarized taxonomy and the application of blockchain-based deep learning in resource management in IoT networks.

4 Future Research Challenges

The open issues in adopting the ML and DRL approach for blockchain empowered resource management in 5G future generation networks and IoT networks are discussed below. The main open challenges are related to the emerging blockchain technologies and then issues related to the 5G and IoT interoperability; after that, there is the issue of using ML and DRL techniques with blockchain technology. The research challenges related to this are given in Fig. 17.3.

4.1 Blockchain Technology

The future generation promising technology blockchain provides a limited number of nodes in the blockchain networks. So scalability is a major issue in servicing the blockchain-empowered 5G and IoT networks. We also need consensus protocols for the validity of blockchain technology. Due to the development in blockchain, the availability of the standard platform and experts is very less. So we need more skills to expertise the benefits of blockchain in various application domains. All the other research issues are discussed below.

The *scalability of blockchain in IoT and 5G cellular networks* becomes a major issue, where the lakhs of users need to be served and need to be scaled up. The blockchain integration with the IoT and 5G cellular future generation networks needs to be proceeding with high computational capabilities for handling a large amount of data and transactions. In the paper [124], author addresses the scalability problem in maintaining resource lanes for sharing lanes and multiple transaction chain that were also maintained in parallel. So scalability needs to handle the fast processing of all the transactions in distributed decentralized networks. The computational power is limited to the resources available in the application domain. So the scalability of transactions is appreciated when the growing demand of the user needs independent lanes for transactions. Further, the processing of scalability in these technologies is still a challenging task. The scalability in hybrid blockchain, IoT networks, resource lanes, and the hybrid consensus in the emerging technologies is an open challenge. The various ML models are trained to address the scalability

Table 17.4 Summarized taxonomy for resource management using deep reinforcement learning/machine learning blockchain approach in IoT networks

Resources	References	Technology	Findings
Resource allocation in vehicular networks	[65–67]	IoT, blockchain, machine learning, DRL	Proposed relocation mapping with DRL approach for vehicle-to-vehicle communication DRL technique for resource allocation in vehicular networks Resource allocation-based DRL algorithm for orchestration in vehicular networks
Spectrum sharing	[63, 65]	IoT, blockchain, DRL	Deep Q-based learning for spectrum sharing Proposed power allocation for multicellular technology
Content caching		IoT, DRL, blockchain	DRL framework for cache replacement technique in multi-timescale resource management
Traffic scheduling and duty cycle	[79, 80, 97]	IoT, DRL, ML	Energy-efficient resource scheduling is presented using machine learning Smart home resource scheduling technique is presented using deep Q learning in IoT networks ML-based traffic scheduling is proposed in a real-time scale
Resource allocation for QoS	[70–73]	IoT, ML, wireless networks	Machine learning and cloud used together for resource optimization using beam management in wireless IoT networks Present a hybrid approach using both supervised/unsupervised machine learning approach to optimised the QoE ML scheme for energy efficiency in IoT networks DRL algorithm-based approach to sense inference and noise in IoT networks
Power allocation	[74]	IoT, blockchain, machine learning	Presented ML-based approach for low-power transmission in IoT devices using LoRA3
Inference management	[75–79]	IoT, 5G networks	Q learning algorithm for power and inference management in IoT and 5G networks Proposed a mechanism for inference level detection for radio channels Presented inference detection and identification model in real-time traffic in IoT networks

Device-to-device communication	[82–87, 90]	ML, IoT, blockchain	Worked on improving the spectral and energy efficiency Presented D2D problems – caching, security, and privacy ML-based approach for resource management in the distributed D2D system Q learning-based technique presented for ideal resource management in D2D networks DRL-based approach to improve and provide optimal resource allocation to the device-to-device communication networks Q learning-based approach for power resource management and to reduce energy consumption in IoT networks
Clustering and data aggregation	[86, 87, 90–93]	ML DRL, IoT	Presented algorithm for decreasing clustering time and to increase efficiency based on k-means clustering algorithm Presented hybrid ML approach for active learning SVM for IDS management Multilayer clustering using ML for intrusion detection was proposed SAX ML technique for handling density-based clustering Presented IoT device access control mechanism using ML (INSTRUCT) Deep Q learning proposal for resource allocation and cluster head selection and resource allocation mechanism in vehicular networks
Resource and cell selection	[96–98, 100, 101]	ML, DRL, blockchain, IoT	DRL-based algorithm proposed for multi-band spectrum sensing Extreme ML algorithm for secure sensing scheme for multi-user environment and to provide more accurate channel complexity states Bayesian ML technique for heterogeneous spectrum states in the ML process Proposed intelligent spectrum mobility management model for IoT networks Presented better spectrum motility between two different modes using the TACL algorithm
Traffic and mobility patterns [69]	[102–106]	DRL, blockchain, IoT	Presented IoT traffic classification using the ML approach Presented analyzed details of real-time packet header using ML SVM and BP approach DRL-based framework for traffic forecast in IoT networks Presented OFDM channel estimation and signal detection using the DRL approach DRL-based traffic learning model and capturing traffic path detection and forwarding

(continued)

Resources	References	Technology	Findings
Heterogeneous networks	[72, 107, 108, 110]	ML, blockchain, IoT	The ML-based scheme proposed for mobility management in heterogeneous networks Presented fuzzy logic-based approach for reducing energy consumption ML framework for radio access in heterogeneous networks Resource allocation technique proposed for cognitive radio networks using ML
MIMO	[111–114, 117–119]	ML, DRL, blockchain, IoT	ML-based Bayesian learning approach presented to address channel estimation issues Ml-based artificial neural network approach proposed for MIMO channel estimation problems DRL-based approach proposed to address MIMO pilot contamination issues ML-based algorithm presented for link adaption in the MIMO system The ML-based algorithm proposed for the modulation-based scheme and to improve coding rate in the MIMO system ML-based approach for handling grouping problem in m-MIMO systems ML-based solution for beam allotment in MIMO networks
NOMA	[120–124, 131]	ML, DRL, blockchain, IoT	ML-based approach for spectral performance and handling multi-user for NOMA techniques Presented DRL-based approach for solving NOMA network problem Presented NOMA techniques for handling multi-users in BS with DRL approach They used ML-based online adaptive filters for NOMA transmissions Q learning-based technique for transmission and dynamic downloading in NOMA transmissions They presented NOMA-based management of multi-cell networks with ML approach NOMA technique to address heterogeneous IoT network with DRL-based RNN model

issue. The improper handling of the scalability issue leads to the delay in processing transactions. The present ML and DRL techniques are trained with the limited capacity data sets that have significantly difficult-to-handle scalability issues. Thus the more diversified algorithms need to be a frame in this versatile environment to address the scalability difficulties.

Efficient Consensus Protocols In the blockchain environment, we are having blockchain consensus protocols in the middle layer of the blockchain environment. In [125–127], several consensus protocols like proof of learning, PoW, proof of useful work, proof of training quality, and proof of DL were proposed. The validity of the assigned work is validated by the validator committee in the blockchain consensus protocol. Further, the consensus grouping and consensus block will be sent to the validator to reach a consensus after some iteration. Thus the complex computation is required to solve these huge hash-based calculations in the blockchain. In [125], the author proposed the proof of DL to validate the blockchain transaction and train the data set using the DRL approach. The valid proof of the

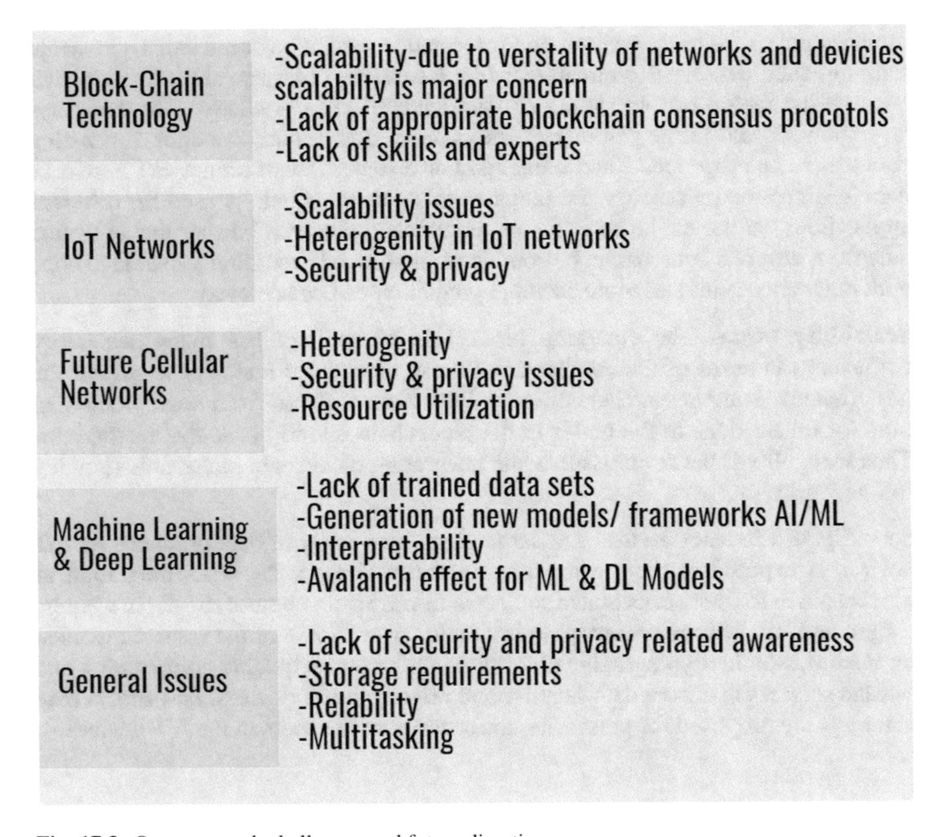

Fig. 17.3 Open research challenges and future directions

transaction block is created only if the DRL model is produced for the same. Further in [127], the author presented the proof of useful work protocol to validate the blockchain cryptocurrency using the DRL deployment model. In this, a new block is mined with a trained model and checked its threshold limit. Thus proper consensus mechanisms are required to address these growing issues.

Lack of Skills and Experts In the blockchain technology, we lacked the number of skilled personnel and the researcher worked on various models, architectures and new frameworks for blockchain application development in different domains. Thus this affects the popularity of this technology.

4.2 IoT Networks

Heterogeneity of IoT and 5G Data The heterogeneous nature of both networks has multiple compatibility issues while servicing various applications. The data generated in IoT networks and 5G cellular networks are heterogeneous and multi-dimensional so the exact features and information cannot be predicted. Thus, while training, such data need pretreatment like cleansing, ordering, and preprocessing because the fusion of this variety of data leads to false prediction. So this plays an important role in the training of the data set and feature selection for testing data sets in heterogeneous environments. For instance, smart homes IoT networks data is generated by sensors and humans; this generated data is used for different applications, so it must be collected at certain central server. Further that collected data from different sources can be combined and trained accordingly to able to cope with other anomalies and more accurate prediction can be achieved.

Scalability Issues The emerging blockchain technology has many scalability bottlenecks in terms of the number of nodes in blockchain that also constraint the performance. Some researchers presented the work suffered from huge processing time for transactions to the nodes in the blockchain due to block size restrictions. Thus scalability is the major issue while integrating blockchain in versatile growing IoT networks.

Security and Privacy Issues The large gigantic quantity of data produced by IoT devices is exposed to the greater threat security and privacy. The integration of blockchain to IoT networks is also subjected to verification before storing in a public ledger, and still privacy preservation is a major issue. Some of the research worked on the lightweight cryptographic algorithms for resource-specific computation and storage services to ensure data security and privacy in IoT devices. Few blockchain gateways are proposed for preserving the security of the users in the IoT devices.

4.3 5G Future Cellular Networks

Heterogeneity The data generated in 5G cellular networks are heterogeneous and multidimensional so the exact features and information cannot be predicted. Thus, while training, such data need o pretreatment like cleansing, ordering, and preprocessing because the fusion of this variety of data leads to false prediction. So this plays an important role in the training of the data set and feature selection for testing data sets in heterogeneous environments. The researchers focused on this issue, and some proposed solutions with a machine learning approach to optimize heterogeneity in blockchain-enabled 5G networks.

Security and Privacy In 5G cellular networks, utilizing blockchain solution has major security and privacy issues. The existing solution was badly suffering from serious security issues. The acceptance of blockchain has various security and privacy issues; the consensus protocol development needs significant testing before deployment in real-time 5G cellular networks. Researchers worked on the consensus protocols that suffered from serious security attacks and also faced critical analysis of smart contract due to improper code. Thus, security and privacy are open challenges in the integration of blockchain for 5G solutions against security threats.

Resource Utilization The researcher emphasizes on the efficient uses of resource scheduling and optimization in the future generation 5G networks. To manage multiple heterogeneous resources in 5G cellular networks, researches provided a blockchain framework with machine learning. Few of them gave solutions for PoW blockchain consensus protocols. The resource management that are assigned with blockchain empowered 5G networks like device to device communication, SDN, network virtualization etc needs efficient solutions for coordination between them. Only a few researchers explored the consensus protocol in this regard. Thus the quantitative analysis of resource utilization with blockchain-empowered 5G networks is a sensitive research issue.

4.4 Machine Learning and Deep Learning

The trends of building an intelligent system have made a drastic change in machine learning and deep learning techniques. Due to the emergence of new IoT networks, it is really difficult to train machine learning and deep learning technology for a greater level of abstraction in real-time analysis. Research put their efforts to build a new system for deep learning to use in blockchain and emerging 5G and IoT networks.

Interpretability Implementation of DRL models needs proper interpretability to get a deep insight of particular machine learning model. They need to be trained as per the policies of the blockchain for IoT and future generation networks to take all the accurate decisions and predictions. The training of deep reinforcement learning techniques needs proper insight into deep learning models for proper prediction. Deep learning models like ANN, DNN, CNN, and RNN have more difficult-to-interpret results. New ML and DRL models need to be framed to interpret greater accuracy. The overall performance of deep reinforcement learning models can be quantified on parameters like computation power, learning capability, reliability, and more accuracy. The existing algorithms can be improved with repeated experimental testing data and training sets.

Training Cost for ML and DRL Techniques The data sets for the experimental models are required to be trained, but the cost to train the ML and DRL model is really high in real time data applications. The training is required to keep the data set up to mark for accurate prediction. The overall retraining of the data set is the continuous learning process in real-time data environments. The system model framework needs to be taken for the initial data training process and keep the learning process in continuous mode for maintaining the accuracy level. Thus the cost of providing the initial training depends on the complexity of the deep learning model used for the process as a sufficient amount of data is needed to train for the testing purpose also.

Lack of Availability of Data Sets The ML and DRL models/algorithms are completely based on the amount of training data set. The emerging technologies are subject to the collection of real-time data sets. The artificial synthesis of data sets is a challenging task as compared to the training of the actual data sets. The lack of availability of proper data sets for a particular application domain also gives rise to another major issue that is the uniformity of data sets in heterogeneous environments.

Avalanche Effect for DRL Approach This is the desirable characteristic of the secure algorithms that the minor change in the input reflected in output also. The use of ML and DRL approaches helped to train the system from these vulnerable adverse security threats. So preprocessing is required before the actual training of the data. In [132], the author discusses the guarantee to the integrity of input data set data. Thus the necessary measures need to be taken for the actual data count and trained input data set to maintain the integrity.

4.5 General Issues

Security and Privacy Issues In the emerging technologies due to the availability of large data sets, the communication has security and privacy issues. The critical aspect of this is that the secure blockchain technology is still suffering from attacks in various application domains e.g. security of ethereum platform breached and resulted $50 million worth of ether as presented in paper [128]. The consensus blockchain brought thus serious issues that led to Ethereum platform. The numerous ML approaches were proposed by the researchers to solve these issues. Some focused on intrusion detection and developed DRL-based mechanism to significantly solve the security and privacy issue.

Storage Blockchain technology has open research areas because a large amount of data will be generated by IoT device networks and 5G cellular networks. So the integration of blockchain in IoT and 5G cellular networks cannot directly use distributed ledgers. The particular information needs to be stored separately before transaction verification. The concept to handle the storage requirement is to combine blockchain with better storage distributed databases that can accommodate a large amount of data in blockchain block nodes. Some research also proposed blockchain storage as a service. The block in the blockchain is replicated and needs more storage space to complete the transaction. That leads to the major concern and burden on blockchain-empowered IoT and 5G technologies to limit participation in the blockchain system.

Reliability The nature of ML and DRL algorithms is very sensitive to the change of data. The prediction mechanism used in adversarial learning changes action and rewards. Thus the reliability of any model and algorithm depends upon the data set, and changes lead to change in output also. Even a few changes may lead to a drastic change in the output prediction. Therefore, the reliability of any model/algorithm referred to the input of learning algorithms and the further processing of data; their classification policies change the outcomes. In [129], the author predicts that adding noise in the trained data set can manipulate the outcome as well. The attackers can target the ML and DRL algorithms and models based on their feature selection. They provided that attackers manipulate the input data set with additional features and they get their desired consequence outcome. In IoT the reliability depends on the resource allocation policies and the data sets used by particular model.

Multitasking Approach This approach depends upon the complexity of the specific problem domain. The ML and DRL models need to be trained according to the specific type of data required. If the preferences in the problem domain changes continuously, then the solution must contain the reassessment and retain in the required data model. Sometimes the data model needs to be restructured for real-time applications like traffic patterns in IoT and 5G cellular networks; the difficulty arises in training the data and preserving the changes. Therefore, the proposed model or algorithm must be tested for their multitasking approach according to the application domain. The complexity of problem domain varies as per the level of depth might take versatile decisions of specific problem in different applications.

5 Case Study: Blockchain and Deep Learning Technique for Resource Management in 5G Networks

In this, the author [33] proposed a secure and intelligent framework for next-generation wireless networks and secure resource sharing and also presented a caching scheme. In this, they explore the deep learning to establish secure content caching scheme. In this, they presented blockchain and deep reinforcement approach to achieve secure content caching scheme. They illustrate the blockchain framework for intelligent resource management in 5G networks. They consider the caching process for D2D by blockchain consortium. In 5G networks, content caching scheme MBSs are used to collect the transaction records. All local transaction records are sent to the nearest MBSs and stored after completing the transaction. The DRL technique is used for caching supply and mapping caching resources. The mapping of caching resources helps in providing bandwidth management between the devices and MBS centers. The blockchain provides a secure transaction by guaranteeing the authentication of all the records. All the transaction records are maintained in the blocks with blockchain and their cryptographic hash to the previous block. The MBS center has access to all transaction records and is able to broadcast its audit results. The DRL approach formulated the optimized solution to the content caching problem in D2D communication. The DRL approach has three key elements – system, action, and reward. The content caching requests and content size can be collected by agents that can match the caching pairs and managed resource mapping. The state in the DRL consists of the <D, C, B> where D represents the content, C represents the caching resource, and B represents the bandwidth of the service provider. The action in DRL matches the state space to the action. After that, the agent received the reward function based on the station and provides a reward function. The DRL frameworks use deep neural network as a primary network and utilize the actor to explore the policies and critic network to estimate the network performance. Thus the DRL helps in training the networks based on the previous experience and uses the current state and maps the appropriate state, action, and reward. This DRL approach empowers the resource management and D2D communication effectively using secure content caching pairing scheme and maximizing caching capabilities. The working model is shown in Fig. 17.4.

6 Discussion and Conclusion

The current scenario perception provides the evidence that deep learning becomes the predominant technology. In this chapter, we acquire the extensive view of deep learning applications in future cellular networks, edge computing, and IoT networks. We epitomize the deep learning current techniques, deep learning in resource management, and advanced approaches in future cellular and IoT networks. The amalgamation of deep learning with blockchain emerging technology for future applications has a significant effect. In this chapter, we explore the opportunities available with the emergence of ML and DRL in blockchain to empower the FCN and IoT networks. The work is persuaded by the extensive growth of ML and DRL to future intelligent technologies. In this chapter, we presented the comparative study of the existing survey in the area of deep learning with blockchain-empowered 5G and IoT networks. We also provided the background

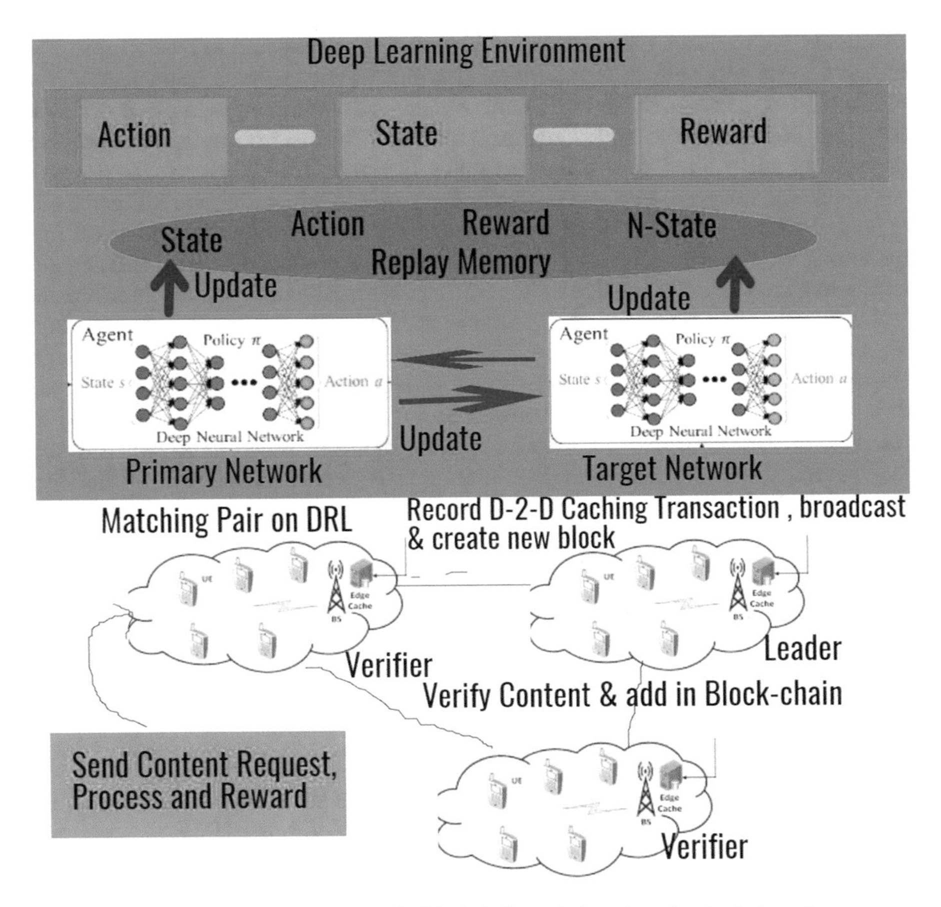

Fig. 17.4 Working model of case study blockchain and deep learning technique for resource management in 5G networks

knowledge of machine learning, deep learning, blockchain, resource management, IoT, and FCN. Then we provide the taxonomy of the resource management services utilizing the machine learning and deep learning techniques with blockchain. Here we focused on intelligent resource management instead of the traditional approaches. The resource management taxonomy in the FCN is summarized in Table 17.3 and provides the existing solution with the technology used. The resource concerned is network virtualization, resource management, data collection and sharing services, federation learning, and spectrum management. The major concern is utilizing the issue focused on machine learning and deep learning. Similarly, the resources that emanate in IoT networks are resource allocation in vehicular networks, spectrum sharing, content caching, traffic scheduling, QoS, power allocation, inference management, device-to-device communication, cell selection, MIMO, NOMA, and traffic mobility patterns. Table 17.4 provides the taxonomy for resource management in blockchain-empowered IoT networks. The integration of ML and DRL techniques for intelligent resource management for blockchain empowered 5G and IoT networks helps in estimating optimised decision for large scale versatile and complex networks. The heterogeneity of emerging technology needs intelligent scalability and efficient, reliable networks. In the end, we conclude the open issues and research challenges in integrating the ML and DRL approach with blockchain to empower the 5G future generation networks and IoT networks. We categorize the open research challenges wrt technological and general issues in Fig. 17.4. We provide detailed future research challenges in blockchain, IoT networks, 5G networks, and utilizing machine learning and deep learning. This study concludes the prospective of ML and DRL techniques in blockchain empower large scale fully heterogeneous 5G and IoT networks. We also discuss the case of using blockchain and deep learning approach for content caching in resource management in 5G networks. Finally, blockchain can be accepted in the empowerment of blockchain-enabled 5G and IoT network services. The availability of blockchain uplifts the shape and future of emerging technologies. This chapter will be helpful for those who study the resource management services in 5G and IoT blockchain-empowered networks utilizing machine and deep learning approach.

References

1. C. Zhang, H. Haddadi, P. Paul, DL in Mobile and wireless networking: A survey. J. IEEE Commun. Surv. Tutor. **21**(3), 1–67 (2019)
2. Q. Bhatia, S. Tanwar, A. Kumari, P. Patel, ML adoption in blockchain-based smart applications: The challenges, and a way forward. J. IEEE Access **8**, 474–489 (2019)
3. I. Mistry, S. Tanwar, S. Tyagi, N. Kumar, Blockchain for 5G-enabled IoT for industrial automation: A systematic review, solutions, and challenges. J. Mech. Syst. Signal Process. Elsevier **135**, 1–21 (2020)
4. F. Hussain, S.A. Hassan, R. Hussain, E. Hossain, Machine learning for resource management in cellular and IoT networks: Potentials, current solutions, and open challenge. J. IEEE Commun. Surv. Tutor. **22**(2), 1–26 (2019)
5. S.K. Singh, Y.S. Jeong, J.-H. Park, A deep learning-based IoT-oriented infrastructure for secure smart city. J. Sustain. Cities Soc. **60**, 1–22 (2020)
6. F. Guo, F. Richard Yu, H. Ji, M. Liu, Adaptive resource allocation in future wireless networks with blockchain and mobile edge computing. J. Trans. Wireless Commun. **19**(3), 1689–1703 (2019)
7. I. Jovovi, S. Macek, I. Forenbacher, S. Husnjak, Innovative application of 5G and blockchain technology in industry 4.0. J. Ind. Trans. Netw. Intell. syst. **18**(6), 1–6 (2019)
8. Y. Dai, D. Xu, S. Mahrajan, Blockchain and DRL empowered intelligent 5G beyond. J. IEEE Netw. **33**(3), 10–17 (2019)
9. P. Mehta, R. Gupta, S. Tanwar, Blockchain envisioned UAV networks: Challenges, solutions, and comparisons. J. Comput. Commun. **151**, 518–538 (2020)
10. D. Prerna, N. Kumar, R. Tekchandani, D-2-D content caching techniques in 5G: A taxonomy, solutions, and challenges. J. Comput. Commun. **153**, 48–84 (2020)
11. Y. Zhang, S. Wang, Z. Yaling, A blockchain-based framework for data sharing with fine-grained access control in decentralized storage systems. J. IEEE Access **6**, 38437–38450 (2018)

12. U.R. Krieger, H.L. Cech, G. Marcel, A fog comp. architecture to share sensor data by means of blockchain functionality, in *Int. Conference of Fog Comp. IEEE ICFC*, pp. 31–40, 2019
13. X. Zhang, K. Cjao Liu, K. Chai, Adaptive blockchain-based electric vehicle participation scheme in smart grid platform. J. IEEE Access **6**, 25657–25665 (2018)
14. D.C. Nguyen, N.P. Pubudu, D. Ming, A. Seneviratne, Secrecy performance of the UAV enabled cognitive relay network, in *IEEE Third International Conference on ICCIS*, pp. 117–121, 2018
15. A. Alshammari, B.R. Danda, M.S. Parwez, Edge computing enabled resilient wireless network virtualization for internet of things, in *Proceedings of IEEE Third International Conference on Collaborative Computing: Networking, Applications and Worksharing*, pp. 155–162, 2017
16. G. Pujolle, N. Bozic, S. Stefano, Securing virtual machine orchestration with blockchains, in *The Proceeding of the International Conference of First Cyber Security in Networking Conference*, pp. 1–8, 2017
17. G.A. Rebello, I.D. Alvarenga, O. Carlos, Securing configuration management and migration of VNF using blockchain, in *Proceeding of IEEE Sym. on Network Operations and Management (IEEE/IFIP – NOMS)*, pp. 1–9, 2018
18. L. Chen, C. Xia, X. Liu, J. Wu, Etra: Efficient three-stage resource allocation auction for mobile blockchain in edge computing, in *Proceeding of 24th – IEEE International Conference on Parallel & Distributed System-ICPADS*, IEEE, pp. 701–705, 2018
19. V.C.M. Leung, L. Yiming, X. Li, F. Richard, Resource allocation for video transcoding and delivery based on mobile edge computing and blockchain, in *Proceeding of IEEE Global Communications Conference*, pp. 1–6, 2018
20. J. Wang, Y. Le, L. Xintong, D. Zhi, Prototype design and test of blockchain RAN, in *The Proceeding of IEEE – International Conference on Communications Workshops – (ICC Work.)*, pp. 1–6, 2019
21. M. Bennis, K. Hyesung, S.L. Kim, Block-chained on-device federated learning. J. IEEE Commun. Lett. **24**(6), 1279–1283 (2019)
22. N. Kumar, S. Tanwar, S. Tyagi, Multimedia big data computing for internet-of-things applications: Concepts, paradigms and solutions, in *Intelligent Systems Reference Library*, (Springer Nature Singapore Pvt Ltd, Singapore, 2019), pp. 1–425
23. H. Shah, K. Sheth, N. Kumar, S. Tanwar, R. Gupta, K. Patel, A taxonomy of AI techniques for 6G communication networks. J. Comput. Commun. **161**, 279–303 (2020)
24. D. Niyato, K. Jiawen, Z. Junshan, X. Shengli, Incentive mechanism for reliable federated learning: A joint optimization approach to combining reputation and contract theory. J. IEEE Internet Things **6**(6), 10700–10714 (2019)
25. S.G. Bilén, K. Kotobi, Blockchain-enabled spectrum access in CRN, in *Proceeding of (WTS) Wireless Telecomm. Symposium*, pp. 1–6, 2017
26. A. Zubow, B. Suzan, W. Adam, G. Piotr, Smart contracts for spectrum sensing as a service. J. IEEE Trans. Cogn. Commun. Netw. **5**(3), 648–660 (2019)
27. J Gazda, M. Taras, J. Minho, H. Longzhe, Blockchain-based intelligent network management for 5G and beyond, in *Proceeding of Third International Conference on (AICT) Advance Information and Communications Technologies*, pp. 36–39, 2019
28. J. Vora, S. Tyagi, N. Kumar, J.P.C. Rodrigues, Home-based exercise system for patients using IoT enabled smart speaker, in *IEEE 19th International Conference on e-Health Networking, Applications and Services (Healthcom-2017)*, Dalian University, Dalian, China, 12–15 October 2017, pp. 1–6
29. K. Samdanis et al., On multi-access edge computing: A survey of the emerging 5G network edge cloud architecture and orchestration. IEEE Commun. Surv. Tutor. **19**(3), 1657–1681 (2017)
30. J. Zhang, L. Weiqi, W. Jian, Z. Yue, Deepchain: Auditable and privacy-preserving DL with blockchain-based incentive, Archive, Report No. 2018/679, p. 1, 2018
31. G. Shaoyong, X. Siya, Q. Feng, D. Yao, Trusted cloud-edge network resource management: DRL-driven service function chain orchestration for IoT. J. IEEE Internet Things **7**(7), 1–13 (2019)

32. G. Fengxian, Y.F. Richard, Z. Heli, C.M.L. Victor, Adaptive resource allocation in FWN with blockchain and mobile edge computing. J. IEEE Trans. Wireless Commun. **19**(3), 1–15 (2019)
33. Y. Dai, X. Du, C. Zhuang, Z. Yan, M. Sabita, Blockchain and deep reinforcement learning empowered intelligent 5G beyond. J. IEEE Intell. Netw. Cogn. Comput. ML **33**(3), 1–10 (2019)
34. J.J. Kang, W. Maogiang, Z. Yan, H. Xumin, Y. Rong, S. Maharjan, Blockchain for secure and efficient data sharing in vehicular edge computing & networks. J. IEEE Internet Things **99**(1), 1–13 (2018)
35. J. Kang, Y. Rong, S. Maharjan, Z. Yan, H. Ekram, Enabling localized P2P electricity trading among plug-in HEV using consortium block-chains. J. IEEE Trans. Ind. Inform. **13**(6), 3154–3164 (2017)
36. Z. Li, J. Kang, Y. Zhang, Y. Rong, Consortium block-chain for secure energy trading in industrial internet-of-things. J. IEEE Trans. Ind. Inform. **14**(8), 3690–3700 (2017)
37. C.H. Liu, W. Shilin, L. Qiuxia, Block-chain-enabled data collection and sharing for industrial IoT with deep reinforcement learning. J. IEEE Trans. Ind. Inform. **15**(6), 3516–3526 (2018)
38. S. Tanwar, M. Patel, N. Kumar, R. Gupta, A DL-based cryptocurrency price prediction scheme for financial institutions. J. Inf. Secur. App. **55**, 102583 (2020)
39. K. Patel, D. Mehta, C. Mistry, R. Gupta, S. Tanwar, N. Kumar, M. Alazab, Facial sentiment analysis using AI techniques: State-of-the-art, taxonomies, and challenges. IEEE Access **8**, 90495–90519 (2020)
40. A.F. Mohamed, M. Leandros, DeepCoin: A novel deep learning and blockchain-based energy exchange framework for smart grids. J. IEEE Trans. Eng. Manag. (Early Access) **67**(4), 1285–1297 (2019)
41. H. Yin, J. Zexun, W. Yulei, B. Yin, FDC: A secure federated DL mechanism for data collaborations in the internet-of-things. J. IEEE Internet Things **7**(7), 6348–6359 (2020)
42. S. Garg, M.-P. Singh, N. Kumar, G.-S. Aujla, A. Singh, DL based Blockchain Framework for Secure SDN. J. IEEE Trans. Ind. Inform. (Early Access), 1 (2020)
43. F. Jameel, W.U. Khan, U. Javaid, R. Jantti, RL in block-chain-enabled IIoT networks: A survey of recent advances and open challenges. J. Sustain. MDPI **12**(12), 1–12 (2020)
44. C. Xu, W. Kun, G. Mingyi, Intelligent resource management in block-chain-based cloud datacenters. J. IEEE Cloud Comput. **4**(6), 50–59 (2017)
45. H. Xu, V.K. Paulo, O. Onireti, B. Cao, Block-chain-enabled resource management and sharing for 6G communications. J. Digital Commun. Netw. **6**(3), 261–269 (2020)
46. C.C. Chenl, L. Boyang, J. Taeho, S. Yiju, Energy-recycling block-chain with proof-of-deep-learning, in *Proceeding of IEEE International Conference on Block-Chain and Cryptocurrency, IEEE*, 14–17 May 2019
47. A. Kumari, S. Tyagi, N. Kumar, Verification and validation techniques for streaming big data analytics in internet of things environment. IET Netw. **8**(3), 155–163 (2019)
48. S. Banarjee, V. Odelu, A.K. Das, S. Chattopadhyay, N. Kumar, Y. Park, Design of an Anonymity-Preserving Group Formation Based Authentication Protocol in global mobility networks. IEEE Access **6**(1), 20673–20693 (2018)
49. J.A. Alzubi, B. Bharathikannan, R. Manikandan, A. Khanna, C. Thaventhiran, Boosted neural network ensemble classification for lung cancer disease diagnosis. Appl. Soft Comput. **80**, 579–591 (2019)
50. Z. Chen, Q. Zhang, T.Y. Laurence, L. Peng, A survey on deep-learning for big data. J. Inf. Fusion **42**, 146–157 (2018)
51. S. Vassanelli, A. Hussain, M.S. Kaiser, M. Mahmud, Applications of deep-reinforcement-learning to biological data. J. IEEE Trans. Neural Netw. Learning-Syst. **29**(6), 2063–2079 (2018)
52. W. Saad, P. Taehyeun, N. Abuzainab, Learning how to communicate in the internet-of-things: Finite resources and heterogeneity. J. IEEE Access **4**, 7063–7073 (2016)
53. B.B. Budinska, B. Frankovic, Advantage and disadvantage of heuristic and multi agents approaches to the solution of scheduling problem. Proc. IFAC **33**(13), 367–372 (2000)

54. S. Doss, A. Nayyar, G. Suseendran, A. Khanna, L.H. Son, P.H. Thong, APDJFAD: Accurate prevention and detection of jelly fish attack in MANET. IEEE Access **6**, 56954–56965 (2018)
55. R. Gupta, S. Tanwar, S. Tyagi, N. Kumar, M.S. Obaidat, B. Sadoun, HaBiTs: Blockchain-based Telesurgery Framework for Healthcare 4.0, in *International Conference on Computer, Information and Telecommunication Systems (IEEE CITS-2019)*, Beijing, China, August 28–31, 2019, pp. 6–10
56. M. Debbah, A. Zappone, M. Di-Renzo, Wireless networks design in the era of DL: Model-based, AI-based, or both? arXiv:1902.02647 **1**, 1–43 (2019)
57. S. Gong, D. Niyato, P. Wang, L.C. Ying, D.T. Hoang, Applications of deep reinforcement learning in communications and networking: A survey. J. IEEE Commun. Surv. Tutor **21**(4), 3133–3174 (2019)
58. Y. Song, S. Houbing, D. Jiang, Z. Jiang, A new DQL based transmission scheduling mechanism for the cognitive internet of things. J. IEEE Internet Things **5**(4), 2375–2385 (2018)
59. D. Cho, K. Minhoe, L. Woongsup, Deep cooperative sensing: Cooperative spectrum sensing based on CNN. IEEE Trans. Veh. Technol. **68**(3), 3005–3009 (2019)
60. M. Kim, D.H. Cho, W. Lee, Deep cooperative sensing: Cooperative, spectrum sensing based on CNN. IEEE Trans. Veh. Technol. **68**(3), 3005–3009 (2019)
61. D. Niyato, L. Zhang, G. Feng, J. Tan, Y.-C Liang, DRL based modulation and coding scheme selection in cognitive Heterogeneous-Networks. arXiv preprint – arXiv:1811.02868, pp. 1–12, 2018
62. J.M. Kim, H-Y. Kim, A load balancing scheme based on DL in Internet-of-Things. J. Cluster Comput. Springer, **20**, 873–878 (2017)
63. X. Du, L. Yujie, G. Mohsen, L. Huang, Resource management for future mobile networks: Architecture and technologies. J. Comput. Netw. Elsevier, **129**(2), 392–398 (2017)
64. W. Saad, U. Chalita, L. Dong, Proactive resource management in LTE-U systems: A DL perspective. arXiv-1702.07031, pp. 1–15, Feb 2017
65. E. Hossain, K.I. Ahmed, A DQL method for downlink power allocation in multi-cell networks (Apr 2019), https://arxiv.org/pdf/1904.13032v1.pdf
66. E. Hossain, K.I. Ahmed, H. Tabassum, DL for radio resource allocation in multi-cell networks. J. IEEE Netw. **33**(6), 188–195 (2018)
67. H. Yin, N. Zhao, Y. Hongxi, Integrated networking, caching, and computing for connected vehicles: A DRL approach. IEEE Trans. Technol. **67**(1), 44–55 (2017)
68. N. K. George, N. Chauhan, N. Choudhury, A comparison of RL based approaches to appliance scheduling, in *Proceeding of Second International Conference on Contemporary Comp. and Informatics – (IC3I)*, pp. 253–258, 2016
69. P. Hu, S. Katti, S. Chinchali, Cellular network traffic scheduling with DRL, in *The Proceeding of International Conference on Artificial-Intelligence (AAAI)*, June 2018
70. T. Verbelen, S. Bohex, B. Dhoedt, S. Pieter, Discrete event simulation for efficient and stable resource allocation in collaborative mobile cloudlets. J. Simul. Model. Pract. Theory **50**, 109–129 (2015)
71. M.M. Zorzi, M. De Filippo, A. Testolin, A ML approach to QoE-based video admission control and resource allocation in wireless systems, in *Proceeding of 13th Annual Mediterranean Ad Hoc Networking Workshop*, IEEE, pp. 31–38, 2014
72. B. Shihada, I. Alqerm, Enhanced ML scheme for energy efficient resource allocation in 5G heterogeneous cloud RCN, in *IEEE 28th Annual International Sym. on Personal, Indoor, and Mobile Radio Comm. (PIMRC)*, pp. 1–7, 2017
73. M. Alizadeh, M. Hongzi, S. Kandula, I. Menache, Resource management with DRL, in *Proceedings of the 15th ACM Workshop on Hot Topics in Networks (HotNets)*, ACM, pp. 50–56, 2016
74. R. Sidhu, A. Patil, L. Zhang, A. Basu, V. M. Suresh, Powering the IoT through embedded ML and lora, in *Proceeding of IEEE 4th World Forum on Internet-of-Things*, pp. 349–354, 2018

75. M. Chen, Z. Fan, X. Gu, S. Nie, D-2-D power control based on supervised and unsupervised learning, in *Third IEEE International Conference on Computer and Comm. (IEEE – ICCC)*, pp. 558–563, 2017
76. P. Lynggaard, Using ML for adaptive interference suppression in WSN. IEEE Sensors **18**(21), 8820–8826 (2018)
77. J. Vora, S. Kanriya, S. Tyagi, N. Kumar, V. Sharma, I. You, Human arthritis analysis in fog computing environment using Bayesian network classifier and thread protocol. IEEE Consum. Electron. Mag. **9**(1), 88–94 (2020)
78. M. Gidlund, S. Grimaldi, A. Mahmood, Real-time interference identification supervised learning: Embedded. Coexistence awareness in Internet-of-Things devices. IEEE Access **7**, 835–850 (2018)
79. X. Wang, S-K. Sharma, Collaborative distributed Q-learning for RACH congestion minimization in MIoT networks. IEEE Commun. Lett. **23**(4), 600–603 (2019)
80. R.I. Ansari, M. Guizani, C. Chrysostomou, M. Shahid, 5G device—2—device networks: Techniques, challenges, and future prospects. IEEE Syst. J. **12**(4), 3970–3984 (2018)
81. B. Vucetic, M. Ding, Y. Li, P. Cheng, M. Chuan, H. Yongjun, Localized small cell caching: A ML approach based on rating data. IEEE Trans. Commun. **67**(2), 1663–1676 (2018)
82. S. Maghsudi, S. Stanczak, Channel selection for network-assisted device-2-device communication via no-regret bandit learning with calibrated forecasting. IEEE Trans. Wireless Commun. **14**(3), 1309–1322 (2015)
83. Y. Miyanaga, A. Asheralieva, An autonomous learning-based algorithm for joint channel and power level selection by device-to-device pairs in HNets cellular networks. IEEE Trans. Commun. **64**(9), 3996–4012 (2016)
84. Y. Luo, S. Zhiping, Y. Qicong, Dynamic resource allocations based on Q-learning for device-to-device communication in CN, in *Proceeding of 11th International Comp. Conference on Wavelet Active Media Tech. Info. Processing*, pp. 385–388, 2014
85. I. AlQerm, S. Basem, Energy-efficient power allocation in multitier 5G networks using enhanced online learning. IEEE Trans. Veh. Technol. **66**(12), 11086–11097 (2017)
86. A. Thakral, S. Gupta, S. Sharma, Novel technique for prediction analysis using normalization for an improvement in k-means clustering, in *Proceeding of 2016 International Conference on IT. – The Next Generation IT Summit on the Theme-IoT: Connect your Worlds*, pp. 32–36, 2016
87. V.V. Kumari, P. Ravi, K. Verma, A semi-supervised IDS using active learning svm and fuzzy c-means clustering, in *Proceeding of International Conference on I-SMAC (IoT in Social, Mobile, Analytics and Cloud) (I-SMAC)*, pp. 481–485, 2017
88. R. Gupta, A. Kumari, S. Tanwar, N. Kumar, Blockchain-envisioned Softwarized multi-swarming UAVs to tackle COVID-19 situations. IEEE Netw., 1–7 (2020)
89. R. Gupta, S. Tanwar, S. Tyagi, N. Kumar, Tactile internet and its applications in 5G era: A comprehensive review. Int. J. Commun. Syst. **32**(14), 1–49 (2019)
90. A. Khan, S. Anwar, M.A. Khan, M.N. Khan, A novel learning method to classify data streams in the IoT, in *Proceedings of IEEE National Software Engg. Conference*, pp. 61–66, 2014
91. H. Ochiai, M.S. Jamal, K.S. Venkata, E. Hiroshi, K. Kataoka, Instruct: A clustering based identification of valid comm. in internet of things networks, in *5th Int. Conference on Internet-of-Things: Systems, Management and Security*, pp. 228–233, 2018
92. Y.J. Omar, D.Y. Paul, Y. Al-Hammadi, M. Al-Qutayri, S. Muhaidat, Semi-supervised multi-layered clustering model for ID. J. Sci. Direct Dig. Commun. Netw. **4**(4), 227–286 (2018)
93. A. Pourkhalili, H.R. Arkian, R.E. Atani, A cluster-based vehicular cloud architecture with learning-based resource management. J. Supercomput. **71**(4), 1401–1426 (2015)
94. R. Gupta, S. Tanwar, S. Tyagi, N. Kumar, Machine learning models for secure data analytics: A taxonomy and threat model. Comput. Commun. **153**, 406–440 (2020)
95. A. Kumari, R. Gupta, S. Tanwar, N. Kumar, Blockchain and AI amalgamation for energy cloud management: Challenges, solutions, and future directions. J. Parallel Distrib. Comput. **143**, 148–166 (2020)

96. V. Koivunen, J. Oksanen, J. Lunden, RL method for energy efficient cooperative multiband spectrum sensing, in *Proceeding of IEEE International Workshop on ML for Signal Processing*, pp. 59–64, 2010
97. H. Kuang, M. Xiaolin, S. Ning, Y. Hong, X. Liu, Co-operative spectrum sensing using extreme ML for CRN with multiple primary users, in *Third IEEE Advanced IT, Electronic and Automation Control Conference (IEEE-IAEAC)*, pp. 536–540, 2018
98. B. Vucetic, Y. Xu, Z. Chen, Y. Li, P. Cheng, Mobile collaborative spectrum sensing for HNets: A bayesian ML approach. IEEE Trans. Signal Process. **66**(21), 5634–5647 (2018)
99. R. Thakkar, P. Thakor, K. Singh, M-Tesla-Based Security Assessment in Wireless Sensor Network, in *International Conference on Computational Intelligence and Data Science (ICCIDS 2018)*, NorthCap University, Gururgram, 7–8 April 2018, Procedia, Computer Science, Elsevier, pp. 1154–1162
100. Z. Li, W. Wu, X. Liu, P. Qi, Improved cooperative spectrum sensing model based on ML for CRN. IET Commun. **12**(19), 2485–2492 (2018)
101. S. Kumar, F. Hu, A.M. Koushik, Intelligent spectrum management based on transfer actor-critic learning for rateless transmissions in CRN. IEEE Trans. Mobile Comput. **17**(5), 1204–1215 (2018)
102. T. Otoshi, M. Murata, Y. Takahashi, K. Shimoto, Y. Ohsita, K. Ishibashi, Traffic prediction for dynamic traffic engineering. J. Elsevier Comput. Netw. **85**(5), 34–60 (2015)
103. X. Xu, C. Han, J. Qin, T. Zhou, G. Han, A learning-based multi-model integrated framework for dynamic traffic flow forecasting. J. Springer Neural Process. Lett. **85**, 407–430 (2018)
104. M. Qiao, M. Yanqing, J. Liu, Y. Blan, Real-time multi-application network traffic identification based on ML. J. Adv. Neural Netw. **9377**, 473–480 (2015)
105. Y. Zuo, Y. Wu, C. Laizhong, G. Min, Learning-based network path planning for traffic engineering. J. Elsevier Future Gen. Comput. Syst. **92**, 59–67 (2018)
106. B.H. Juang, X. Zhang, Power of DL for channel estimation and signal detection in OFDM systems. IEEE Wireless Commun. Lett. **7**(1), 114–117 (2018)
107. M.S. Omar, S. Ali-Hassan, N. Qiang, S. Mumtaz, L. Musavian, H. Pervaiz, Multi-objective optimization in 5G hybrid networks. J. IEEE Internet Things **5**(3), 1588–1597 (2018)
108. K. Vasudeva et al., Fuzzy logic game-theoretic approach for energy efficient operation in H-Nets, in *Proceeding of IEEE International Conference on Communication Workshops*, pp. 552–557, 2017
109. S. Tanwar, Q. Bhatia, P. Patel, A. Kumari, P.K. Singh, W.C. Hong, Machine learning adoption in Blockchain-based smart applications: The challenges, and a way forward, IEEE Access **8**, 474–488 (2020)
110. J.S. Perez, S. Lane, S.K. Jayaweera, ML aided cognitive rat selection for 5G HNets, in *International Black Sea Conference on Communications and Networking – IEEE*, pp. 1–5, June 2017
111. P. Ting, J.C. Chen, K-K. Wong, Channel estimation for massive MIMO using gaussian-mixture bayesian learning. J. IEEE Trans. Wireless Commun. **14**(3), 1356–1368 (2015)
112. Y. Zhao, J. Zhang, Y. Zhao, Mobile location based on SVM in MIMO communication systems, in *Proceeding of IEEE International Conference on Information, Networking and Automation*, vol. 2, (IEEE, 2010), pp. 352–360
113. J. Choi, K. Kim, J. Lee, DL based pilot allocation scheme for 5G massive MIMO system. IEEE Commun. Lett. **22**(4), 828–831 (2018)
114. X. Gao, Z. Dong, W. Wang, S. Junchao, ML based link adaptation method for MIMO system, in *29th Annual International Symposium. on Personal, Indoor and Mobile Radio Comm. IEEE*, pp. 1226–1231, Sept 2018
115. N. Kabra, P. Bhattacharya, S. Tyagi, MudraChain: Blockchain-based framework for automated cheque clearance in financial institutions. Futur. Gener. Comput. Syst. **102**, 574–587 (2020)
116. J. Vora, S. Tyagi, N. Kumar, M.S. Obaidat, A systematic review on security issues in VANET. Secur. Priv. J. Wiley, **1**(5), 1–27 (2018)

117. R.W. Heath, R.C. Daniels, C.M. Caramanis, Adaptation in convolutionally coded MIMO-OFDM wireless systems through supervised learning and subcarrier ordering. J. IEEE Trans. Veh. Technol. **59**(1), 114–126 (2010)
118. T.F. Maciel, V.F. Weskley, F.C. Hugo Neto, F.M. Rafael Lima, D.C. Araujo, A low complexity solution for resource allocation and SDMA grouping in massive MIMO systems, in *15th IEEE International Symposium on Wireless Communication System (IEEE-ISWCS)*, pp. 1–6, Aug 2018
119. J. Wang et al., A ML framework for resource allocation assisted by CC. IEEE Netw. **32**(2), 144–151 (2018)
120. K. Higuchi, Y. Saito, Y. Kishiyama, T. Nakamura, A. Benjebbour, A. Li, NOMA for cellular future radio access, in *77th International Conference in Vehicular Technology – IEEE (VTC Spring)*, pp. 1–5, June 2013
121. Z. Ding, Z. Wang, NOMA: Common myths and critical questions. arXiv preprint arXiv: 1809.07224, 2018
122. H.V. Poor, L. Xiao, C. Dai, L. Yanda, RL based NOMA power allocation in the presence of smart jamming. J. IEEE Trans. Veh. Technol. **67**(4), 3377–3389 (2017)
123. A. Mohajer, F. Nikjoo, A. Mirzaei, A novel approach to efficient resource allocation in NOMA HNets: Multi-criteria green resource management. J. Appl. Artif. Intell. **32**(7–8), 583–612 (2018)
124. M. Liu, G. Gui, T. Song, Deep cognitive perspective: Resource allocation for NOMA based HetNets internet-of-things with imperfect SIC. IEEE Internet Things J. **6**(2), 2885–2894 (2018)
125. B. Li, C. Chenli, T. Jung, Y. Shi, Energy-recycling blockchain with PoDL, in *IEEE International Conference on Block-Chain & Cryptocurrency*, pp. 1–10, 2019
126. S. Maharjan, Y. Lu, X. Huang, Y. Zhang, K. Zhang, Blockchain and FL for privacy preserved data sharing in industrial internet of things. J. IEEE Trans. Veh. Technol. **69**(4), 4298–4311 (2019)
127. Y. Saez, A. Baldominos. Coin. AI: A PoUW scheme for block-chain-based distributed DL. arXiv preprint arXiv:1903.09800, pp. 1–17, 2019
128. D. Siegel, Understanding the DAO attack (2016). Web: http://www.coindesk.com/understanding-dao-hack-journalists
129. R. Agrawal, A. Swami, T. Lmielinski, Mining association rules between sets of items in large DB. J. ACM Sigmod Record **22**(2), 207–216 (1993)
130. R.K. Raman, N.K. Bore, I.M. Markus, Promoting distributed trust in ML and computational simulation via a block-chain network. arXiv preprint arXiv:1810.11126, 2018
131. H. Huang, H. Sari, G. Gui, Y. Song, DL for an effective non-orthogonal multiple access scheme. IEEE Trans. Veh. Technol. **67**(9), 8440–8450 (2018)
132. D. Niyato, A. Asheralieva, Distributed dynamic resource management and pricing in the IoT systems with blockchain-as-a-service and UAV-enabled mobile edge computing. IEEE Internet Things J. **7**(3), 1974–1993 (2020)
133. N. Shan, X. Cui, Z. Gao, "DRL + FL": An intelligent resource allocation model based on deep reinforcement learning for mobile edge computing. Comput. Commun. **160**, 14–24, (2020)

Chapter 18
Importance of 5G-Enabled IoT for Industrial Automation

Arpit Verma, Sharif Nawaz, Shubham Kumar Singh, and Prateek Pandey

1 Introduction

The term automation can be described as a technology involved with performing a process with the use of programmed instructions combined with automated feedback control to make certain right execution of the instructions. The ensuing device can run without human intervention [1]. The growth of various industrial sectors with the usage of the Internet of things (IoT) is at an exponential rate for the past couple of years [2–5]. Ever since the IoT has become part of this world, it ensured that the variety of linked devices could be in billions. According to Gartner, the number of IoT-enabled devices might also additionally attain 24 billion by 2020 [6], and Statista.com additionally gave a figure that, through 2025, there could be at least 75 billion devices connected to the Internet. Managing one's gadgets and ensuring that everyone has the specified pace and data bandwidth, telecommunication agencies and other companies have begun investing in 5G, the brand-new wireless communications era. As we discussed that IoT and 5G would change many things, now we are discussing what they are.

IoT is a system of interlinked computing devices, mechanical and virtual machines supplied with specific identifiers, and it can transfer data over a network without requiring human-to-human or human-to-computer interactivity. IoT is a concept in which devices are interconnected via various mediums like the Internet and transfer data among themselves, without human intervention [7].

While in previous generations of mobile technology (consisting of 4G LTE) centered on ensuring connectivity, 5G takes connectivity to the subsequent degree by handing over connected experiences from clients' cloud, it would enhance

A. Verma · S. Nawaz · S. K. Singh · P. Pandey (✉)
Jaypee University of Engineering & Technology, Guna, Madhya Pradesh, India
e-mail: 181b049@juetguna.in; 181b193@juetguna.in; 181b211@juetguna.in; prateek.pandey@juet.ac.in

S. Tanwar (ed.), *Blockchain for 5G-Enabled IoT*,
https://doi.org/10.1007/978-3-030-67490-8_18

connectivity in underserved rural regions and towns wherein demand can outstrip today's 4G technology ability. Mobile app development is also facing various challenges, and newer development models are proposed for fast growth and a better experience [8–11]. The industrial world is going through many technological adjustments that grow the urgent demand for excellent services and products, which could only be provided with the aid of using an excessive degree of productivity. This process requires engineering systems, computerized manufacturing, and business automation. Competitiveness is the whole thing to manufacturers, and much-wanted profits in performance and profitability will be accomplished through an advanced process of innovations; 5G and IoT, these two technologies play a key role in implementing this automation. A combination of the above two technologies, 5G-enabled IoT, is formed, and this technology can take the level of automation to very higher levels, and it would be very helpful in making drastic changes in various industrial sectors.

The Internet of Things (IoT) has not been around us for very long. However, there were visions of machines communicating with each other since the early 1800s. IoT, as a concept, wasn't named formally till 1999. By 2013, it had developed right into a device using more than one technology, starting from the Internet to wireless communication and from micro-electromechanical systems to embedded systems. The conventional fields of automation (together with the automation of city, building, and homes), wireless sensor networks, GPS, manipulate systems, and others all support the IoT [12]. 5G technology is expected to change the future of technology; from augmented reality and autonomous vehicles to smart cities and distant surgery, the ability for 5G uses is sort of unfathomable. Verizon led the manner in growing and deploying 5G and accelerating 5G innovation [13]. It may even effectively connect billions of low-complexity IoT gadgets, which include environmental sensors and utility meters. These IoT devices are commonly delay-tolerant with a focus on being low-complexity and energy-efficient. To deal with this IoT segment, 3GPP (Third Generation Partnership Project) keeps conforming to NB-IoT (200 kHz bandwidth) and eMTC (1.4 MHz bandwidth) to bring extra abilities and efficiencies. 3GPP Release 16 added support for these technologies to perform with a 5G core network and in-band with 5G NR deployments – making NB-IoT (Narrowband Internet of Things) and eMTC (enhanced Machine-Type Communication) the preliminary solutions for 5G large IoT [14]. This chapter focuses on why these ultra-high-level technologies are required and why it is a matter of importance for various industrial sectors. It covers various topics such as 5G-enabled IoT with industrial automation along with their applications. Every technology has both advantages and disadvantages; this chapter will describe them and also about the rumors around the public for these technologies. Lastly, it will cover the scope of future work with the help of 5G-enabled IoT.

1.1 Research Contribution

In today's era, we're surrounded by the Internet and technology. According to Statista, in July 2020, there were 4.57 billion active users on the Internet. We know that 4G is sufficient for everyday use of the Internet, but if we compare it with 5G, 4G has more latency than 5G, and it is not enough for the automation, which we are going to discuss in this chapter. Seeing this, Verizon is the first company to introduce 5G. It is the most significant contribution in the field of cellular networks because it will open all the doors of those significant advancements using IoT. IoT devices connect with the Internet, and if connected with 5G, it can achieve an excellent data transfer speed. There has been significant work on integrating 5G with various domains like smart city Phan et al. [15]. Chan et al. [16], provided an excellent resource for future works on the field.

1.2 Motivation

After seeing the possibilities of such significant advancement using 5G and IoT, we're highly motivated to write this chapter. Many network and communication industries, including some research institutions, indulge in research activities in 5G and IoT. This encourages us to provide a research contribution toward it. 5G and IoT, these two technologies can change the entire world in the field of industrial automation. The scale by which human lives are going to be made more accessible, productive, secure, and healthy and the need for this to be brought forward by the masses are what serve as the motivation for this chapter.

1.3 Organization

This chapter is organized in a way so that the reader can get a better understanding of 5G, IoT, and 5G-enabled IoT and its importance in the upcoming industrial automation. Firstly, there is a section on 5G-enabled IoT. In this section, it briefly described 5G, and its benefits, IoT, and 5G-enabled IoT. After this industrial automation is discussed, in this section, it is explained why industrial automation is required and what will happen if these would get implemented. After giving a brief overview of these topics, applications of 5G-enabled IoT are discussed, such as advanced healthcare, smart homes, autonomous vehicles, smart city, smart agriculture, and smart supply chain management. Every technology has some drawbacks; likewise, 5G and IoT have it as well. After all the applications, disadvantages, and rumors are discussed, and lastly the conclusion, in this section, briefly explained about the crux of this chapter and scope in the future for these advancements.

2 Literature Review

5G-enabled IoT and its impact on industrial automation are undergoing outstanding research all over the world. We have introduced its effect on some of the industries, along with the ill effects. To put forward concrete facts and figures, we went through many websites, blogs, and research articles on Google Scholars, IEEE, ScienceDirect, etc. We found some significant research going on in this field [4]. Comparison of existing literature is presented in Table 18.1.

In an article by Mistry et al. [2], he gives a systematic review and solutions to the challenges in integrating blockchain with 5G-enabled IoT. He mentions that the centralized architecture of access control mechanism protocols, etc. are the challenges to which he provides solutions.

In a paper by Khurpade et al. [17], the author brings forward the requirements of IoT, which can be fulfilled by the introduction of 5G. It also talks in brief about the architecture of 5G and its merits and demerits.

There has been significant work on the integration of 5G with various industries like by Phan et al. [15] in a smart city, autonomous vehicles and Chan et al. [16], which are currently providing an excellent resource for future works on the field.

3 5G-Enabled IoT

The Internet of Things (IoT) is a system of interrelated computing devices, mechanical and digital machines, supplied with particular identifiers (UIDs) and the capacity to transfer information over a network without requiring human-to-human or human-to-computer interaction. IoT is a concept in which devices are interconnected via various mediums like the Internet and share data among themselves, without human intervention.

Though IoT is a relatively new concept and has gained much popularity in recent years, its primary impression on our daily lives is expected to come after integration with 5G. IoT devices rely heavily on the connection between them, as they need to have high standards of data transfer to achieve the best results. Moreover, connections that are good enough can even open new realms in the world of IoT, for example, allowing events to occur in real time, etc.

To understand the upcoming upgrades in IoT much better, we should first understand what 5G means.

5G is the fifth generation of wireless technology, which is meant to deliver higher multi-Gbps peak data speeds, ultra-low latency, more reliability, massive network capacity, increased availability, and more uniform user experience to more users. From the definition, 5G promises a great solution to the current problems of evolving IoT.

In a paper named "A survey on IoT and 5G network," Khurpade et al. [17] state that with the development in wireless technology, we estimate that IoT and 5G can

Table 18.1 Comparison of existing literature

Author	Year	Description	Merit	Demerits	1	2	3	4
Skouby et al. [18]	2014	Put forward a 4-layer model bringing together the concepts of IoT, smart home, and smart city	Took into account latest technologies like 5G, AI, etc.	The working of the model lacks details	5G, IoT, and AI	Model	Smart city	✓
Dewey et al. [19]	2018	Explains the uses of 5G, IoT, and blockchain in trade, finance, and supply chain	A detailed explanation of integrating blockchain with 5G and IoT	Practical uses and experiences could have been explained in more detail	Blockchain, IoT, and 5G	Review	Supply chain	✓
Sushanth et al. [20]	2018	Presented a smart agriculture system comprising WSN and IoT	The efficient end-to-end algorithm proposed for smart agriculture systems	The model requires a continuous supply of high-speed Internet, which is not always feasible, especially in rural areas	IoT and WSN	Model	Agriculture	✓
Fan et al. [21]	2018	Explained the use of high-end IoT devices in the realization of a smart city	Used Ethereum (blockchain) to manage data from IoT sensors	It lacked proper summarization of the framework	IoT	Framework	Smart city	✗

1 used technology, 2 model/review/survey/framework, 3 type of industry, 4 open issue and challenges. Notation: ✓ considered, ✗ not weighed

combine to shape great surroundings that may satisfy the cutting-edge call for IoT devices. 5G can alternate the conduct and could help to flourish the growth of IoT devices. Though it has much to offer to IoT, most of the excitement is around two specific promises of 5G; they are data transfer speed and greater network reliability. These are discussed below in brief:

- *High data transfer speeds*
  Centrally, the success of IoT commercially depends highly on its performance. However, the performance of any IoT system itself depends on the speeds with which it can send or receive information from other IoT devices, wearables, mobile devices, tablets, and software in the form of its mobile application, web portals, or other platforms, and the list continues. With a significant increase in data transfer speeds by the introduction of 5G, these information exchanges would be much faster and, thus, would drastically increase the performance of IoT systems and their commercial success.
- *Great network reliability*
  5G is expected to operate more reliably, creating more stable connections. Having a reliable and stable network condition will be extremely beneficial to IoT-based systems. Web applications and other software are highly affected by the speed of the Internet; that's why selecting an appropriate development method is highly recommended. The categories of IoT systems that seem likely to be most advantageous are the ones designed especially for connected security and healthcare devices like locks, security cameras, and other monitoring systems that depend on real-time updates.

4 Industrial Automation

According to the current scenario of competition among various industries, buyers, as well as sellers, both want to buy and sell the product at the cheapest rate with the best quality, respectively. To tackle this challenge, industries are manufacturing products using automated machines. Industrial automation is the use of such systems as robots, computers, and information technology for managing different processes to replace humans and errors made by them. The requirement for automation is to increase productivity, and for a human, it is quite not possible to work 24 hours a day, but for a machine, it is possible, and also it reduces the cost associated with a human worker, i.e., wages. As of the current scenario, almost every industry is shifting toward automation for affordability and high performance and to maximize their productivity. As you read about 5G-enabled IoT, wireless communication systems are being commonly used in industrial automation [22].

4.1 Why Industrial Automation?

There are many advantages of industrial automation as explained earlier; some of them are listed below:

- *Increase in productivity*
  Automation is the process of manufacturing that improves the production rate through better control. For mass production, it reduces the time per product massively with great production quality. It allows the company to run the manufacturing plant to work for 24 hours a day, 7 days a week, and 12 months a year. So, for the given labor input, the amount of output is extremely large.
- *Minimizes the cost*
  In almost every type of industry or business, buyers, as well as sellers, both want to get cheaper deals for the product. So automated machinery minimizes the effort and cycle times, and it reduces human labor. Thus, the maximum investment, which was for the payment of employees, gets saved by automation.
- *Quality products*
  When automation is involved in any industry, then the involvement of humans gets reduced, and the error also associated with them gets alleviated. So, with the use of automated machines, the uniformity of the product quality is maintained at any point of the time. As quality performs a major role in any type of industry, therefore with machines, defective materials will also get easily sorted.
- *Increase in safety*
  Industrial automation can increase the safety level for the employees by substituting them with automated machines from working in precarious conditions. Safety measures get more increased when it comes to alarming situations like fire or any other disaster. In this situation, it would be easier with machines to stop the work immediately with less loss of goods.

So, with these facts, we can say that industrial automation is highly suggested to the industries for their better growth in the upcoming situation. There are many examples associated with industrial automation and 5G-enabled IoT.

5 Industrial Automation with 5G-Enabled IoT

The 5G-ACIA (5G Alliance for Connected Industries and Automation) states that the fourth level of the Industrial Revolution, additionally termed "Industry 4.0," is another level of upgrade in commercial manufacturing which is planned to extensively enhance the flexibility, versatility, usability, and performance of the smart factories expected soon. Industry 4.0 integrates the Internet of Things (IoT) and related services in commercial manufacturing and can provide seamless vertical and horizontal integration down the whole value chain and throughout all layers of the automation pyramid. There exist a lot of issues in today's most popular software variant, which is mobile apps and its development. 5G allows software and mobile development companies to develop enhanced user experiences. This ultimately increases mobile app efficiency, and users can more easily decide to retain with an app or to delete it altogether. Connectivity is a key aspect of Industry 4.0 and could guide the continuing developments by supplying effective and pervasive connectivity among machines, people, and objects. Moreover, wireless communication, and especially 5G, is a vital method of attaining the desired flexibility of production, helping new superior mobile applications for workers, and permitting mobile robots and smart automobiles to collaborate on the shop floor – those being only a few examples. The following are some examples on which the effects of 5G-enabled IoT have been explained in brief.

The fourth level of the Industrial Revolution, additionally termed "Industry 4.0," is another level of upgrade in commercial manufacturing, which is planned to extensively enhance the flexibility, versatility, usability, and performance of the smart factories expected shortly. Industry 4.0 integrates the Internet of Things (IoT) and related services in commercial manufacturing and can provide seamless vertical and horizontal integration down the whole value chain and throughout all layers of the automation pyramid.

5.1 Advanced Healthcare

Corresponding to the exponential growth in population around the world, there is an increasing requirement for treatment and assistance in the healthcare industry. This has emphasized the need for technological improvements in healthcare to offer efficient and low-cost solutions to increasing patients [23]. Ericsson predicts a USD 76 billion revenue opportunity for operators addressing healthcare transformation with 5G in 2026. As we know that healthcare services are costlier than ever, and medical diagnosis consumes a large portion of hospital bills, technology might flow the exercises of clinical tests from a hospital-centric to the home-centric. The Internet of Things in conjunction with state-of-the-art blockchain technology can transform the healthcare sector [24–27]. It has considerable applicability in lots of areas, and healthcare is one of them, but IoT devices used for healthcare applications require high data speed, long battery life, and a network with very low transmission delay, and all these necessities of IoT devices may be fulfilled through 5G networks. With the usage of 5G-enabled IoT devices, there are incomparable benefits that will enhance the quality and treatment efficiency and decrease cost, time, and labor. Some of the benefits are as follows:

- *Real-Time Remote Monitoring*

  Chronic disease sufferers experience highs and lows of the body's parameters, which include blood sugar levels, heart rate, and blood pressure. Those numbers are liable to rise to dangerous levels with little warning. These patients typically visit hospitals between one and four times per year, and this costs lots of money and time. According to Anthem, 86% of doctors say wearables, which might be a normal form of remote monitoring, increase affected person engagement with their health. Additionally, hospital expenses are predicted to lower by 16% in the subsequent 5 years through wearables. By using IoT-enabled devices, healthcare providers can get real-time data of patient's body parameter continuously, and the device also compares the previous data of patients with the current data, and if something abnormal detected, then it immediately sends an alert to the healthcare providers so that they can review the report and respond accordingly in real time without physically moving a patient.

- *Remote Surgery*

  Remote surgery, also called telesurgery, is the technology that uses various technologies like 5G-enabled IoT, artificial intelligence, robotics, etc. that enables doctors to perform surgeries on patients regardless of location. The geographical barrier between surgeons and patients is thus removed. The major benefit is that patients no longer need to leave their local hospital to get benefit from specialized surgeons around the world, thus reduced cost and time.
- *Connected Ambulance*

  An ambulance is called when an emergency arises; the paramedics on the ambulance have little information about the patient's body parameters. The addition of technology like 5G-enabled IoT to the ambulance, enables hospitals to receive information about the patient medical records like X-ray, ECG, glucose level, heart rate, and blood pressure from the ambulance so that doctors can give suggestions and are well prepared for the patient. A connected ambulance is very useful when patients require emergency support in very little time, and during a traffic jam, it can be a life savior. Connected ambulances have already become a reality in Barcelona, Spain.
- *Elderly Monitoring*

  5G-enabled IoT solutions enable an aging population to live independently. According to the United Nations, the number of older people is projected to double to 1.5 billion in 2050. An assortment of solutions is available for elderly care through technology intervention [28–32], but 5G will take them all more effectively. These statistics are enough reason to build a discreet way to take care of the aging population, and 5G-enabled IoT already makes it less challenging. Various 5G-enabled devices allow relatives and caretakers to watch over the elderly and monitor their movements. It also helps analyze their behavioral changes, order groceries and medicines, give them a reminder to take medicines, and enable quick alert to the nearest hospitals in case of emergencies.

5.2 Smart Homes

A smart home is a residence that uses Internet-connected devices to enable the remote monitoring and management of appliances and systems inside the building. Being a part of IoT, smart home systems and devices often operate together, sharing consumer usage data and automating actions based on the homeowners' preferences. They are equipped with sensors, actuators, and/or biomedical monitors. The gadgets function in a network connected to a remote center for records collection and processing. The remote center diagnoses the ongoing scenario and initiates assistance procedures as required [16]. The technology can be extended to wearable and implantable gadgets to monitor humans 24 hours a day, both outside and inside the house. This system relies heavily on the amounts and speed of data transfer between devices. The expected capabilities of 5G can make smart homes' whole infrastructure more reliable, efficient, and overall smarter. These are briefly discussed below:

- *High data transfer rates and low latency*

  In a complex ecosystem of devices like that of a smart home, huge amounts of data must be shared between devices with a small delay as possible to achieve real-time responses and automation. This is possible with the integration of 5G. The huge amounts of data that 5G can carry at once, with such low latencies, can revolutionize the way these IoT devices interact with each other.
- *More reliability*

  The connection between devices must be reliable enough, or else it may take down the whole system. If data from one device does not reach the other at a time, the entire system can collapse and may lead to loss of critical data to such a point where recovery may be impossible, as these devices are extremely simple and have very small storage and thus no backup systems. This can be illustrated with an example. Suppose some implanted device inside a patient recognizes a critical condition. It sends this data to a device responsible for alerting the owner. However, due to an unreliable network, this data never reaches the other device and thus gets lost. Such circumstances can lead to extreme conditions. Smart homes demand highly reliable networks to avoid such situations, where 5G comes as a great option.
- *Connects to a great number of devices*

  5G can easily connect to a large number of devices at once without losing its critical features. Therefore, more IoT devices can be brought together to create great ecosystems for smart homes.
- *Cloud-based architecture*

  As mentioned earlier, these IoT devices are very simple, which have extremely low computing and storage capacities. However, the amount of data created by such devices and sensors at a point in time is huge, even referred to by many as big data. Therefore, as a solution to such data computing and storage demands, cloud architecture is used. The data is stored as well as processed in the cloud. The 5G system itself is also anticipated to run on the cloud. Thus, 5G blends in well with the smart home systems.

Thus, it can be concluded that 5G-enabled IoT is the future of smart homes. Though smart homes have become a reality recently, their real power is yet to be unleashed. 5G-enabled IoT is the way to utilize the benefits of smart homes to the fullest.

5.3 Autonomous Vehicles

These days, the vehicle enterprise is experiencing an exponential boom of driverless features, and this fashion is expected to continue in the future. The autonomous vehicle is a vehicle that can operate itself and performs all the necessary actions required without any human interference through its ability and senses. Many factors make these vehicles a necessary transformation in technology for the automobile industry. Above all these factors, a major problem is the slow speed of the wireless connection. Although 4G is rapidly sufficient to share updates or request rides, it cannot supply automobiles the human-like reflexes. Also, Forbes published an article in 2018 titled "Autonomous Cars will not work – Until we have 5G." Autonomous test vehicles of many companies like Uber, Waymo, Tesla, and Toyota are on roadways in some of the places like Pittsburgh, Phoenix, and Boston. A recent accident by Uber's autonomous vehicle raised questions about whether these cars would be ready for the roads. Yes, it would be possible, and they can become road-ready only with 5G. The fifth-generation wireless data will be the most significant data network advancement to date. It is also anticipated to connect nearly everything around us with an ultra-fast, completely responsive, and exceedingly reliable network. It may also permit us to leverage the whole capacity of superior technology, which includes virtual reality, artificial intelligence, and the Internet of Things (IoT). There are some more benefits of using autonomous vehicles:

- *Advantages of Autonomous Vehicles*

  Autonomous vehicles could provide certain advantages compared to human-driven vehicles. According to WHO, approximately 1.5 million people die each year due to road traffic crashes; in this case, the autonomous vehicle can work as an advantage by increasing safety on the roads. Autonomous vehicles could probably lower the number of casualties because the software program utilized in them is in all likelihood to make fewer mistakes in contrast to humans. A lower in the variety of injuries may also lessen traffic-associated issues, which is likewise a bonus of autonomous vehicles. Another benefit is removing using fatigue and being able to sleep for the duration of an overnight journey. It can also help those who cannot drive due to certain factors like age and disabilities (if any) for them to use automated cars as a more convenient transport system.

5.4 Smart City

There are many applications of 5G and 5G-like technologies that give features like ultra-high speeds and ultra-low latency and connect almost every device. Implementation of smart cities is considered as one of the high-priority application domains for 5G. Smart cities' operation relies enormously on IoT, and 5G wireless networks are a great IoT enabler. A smart city is an urban area connected with numerous technologies consisting of the Internet of Things and might improve

Table 18.2 City challenges in developing nations [33]

Governance	Economy	Mobility	Environment	People	Living
Less metropolis institutional capacities	Higher shortage of infrastructure	Lack of public conveyance	Lack of resources	Poverty in urban areas and inequality	Increase of slum
Uncertainty in governance	The deficit in access to technology	Higher shortage of infrastructure	Scarcity of water	The deficit in access to technology	Violence and insecurity in urban areas
The gap between government and governed	Weaknesses with the economy and lack of compet-itiveness	Pollution	Effects of climate change	Certain problems of urban youth	Fast growth and urban sprawl
Uneven geographical development	Certain problems of urban youth	Fast growth	Pollution	Menaces to cultural identity	Lack of social services
Lack of social services	A finite number of urban-based industries		Fast growth and urban sprawl	Low level of education	Menaces to cultural identity
	Uneven geographical development				Poverty in urban areas and inequality

people's lifestyle. It generally occupies six dimensions: technology, infrastructure, management, people, economy, and government. It may sound very complex, but it is possible; there are many sensors and connected devices which allow cities to manage and monitor infrastructure, financial systems and transport, postal service, and many more. Life cycles will be changed once these systems will be implemented. There would be several benefits, which include transportation, healthcare services, and smart homes. As discussed, autonomous vehicles, apart from that, would be an effective traffic management system. As 5G gives high-pace connections among devices, it might be great for home automation too. One of the most crucial applications and advantages of the 5G effect on smart cities is the betterment of a metropolis' healthcare and security structures. Remote monitoring or home caring systems can be used to decrease hospitalization rates. There are more advantages, including social and economic ones, in implementing the concept of a smart city with 5G wireless technology; they are economic growth, better quality of life, better environment, excessive job market, safe environment, and better communication [15]. Table 18.2 describes the challenges faced by developing nations.

- *Privacy and Security Concerns*

  The large use of smart applications has induced many safety and privacy issues. The development of extra superior safety models and frameworks is important and distinctly demanded in each business and educational field. In smart city scenarios, a few common harms, which include packet interception in communication, malware in cellular gadgets and applications, sensitive information leakage, hacking on servers, and falsification permission, whether intentional or unintentional, are the principal purpose of privacy breaches [34–36]. To keep away from misuse with the aid of using unauthorized persons, good enough and powerful countermeasures, which include encryption techniques and nameless mechanisms, and a few novel techniques, different privacy [38], have to be applied [35].

5.5 *Smart Agriculture*

Who wouldn't agree that if there is a sector that human beings need to become more efficient and advance using all the possible technologies such as IoT, 5G, AI, etc. it's the agriculture sector. According to the UN Food and Agriculture Organization, to feed the ever-increasing population, farmers across the world have to grow 70% more food in 2050 than they did in 2006. This statistic seems to be nearly an impossible task for farmers due to the hurdles they are facing like water shortage, climate change, soil degradation, new species of pests, and various new crop diseases. To solve all the hurdles that farmers are facing and to fulfill the demand for foods, farmers have to switch to smart agriculture that uses modern technologies like 5G-enabled IoT to enhance, automate, improve, and monitor agricultural processes. 5G-enabled IoT sensors can measure the moisture of soil and nutrition level, predict the weather, monitor livestock maturity, and generate lots of data, and this can be used to predict susceptibility to crop diseases so that farmers act accordingly. Hence farmers who adopt smart agriculture can optimize traditional farming and increase their productivity and speed of operation. Krishna et al. [38] proposed a wireless robot that is geared up with numerous sensors for measuring distinctive environmental parameters. The main features of this novel intelligent wireless robot are it can execute duties such as scaring birds and animals, moisture sensing, spraying pesticides, moving forward or backward, and switching *ON/OFF* motors. The robot is equipped with a wireless camera to screen the activities in real time. Smart agriculture has numbers of benefits; let's understand how it will improve agriculture and in which step of agriculture it can be implemented:

- *Irrigation System*
  It is one of the most important steps in agriculture because crops need a controlled amount of water at different intervals. To optimize it, 5G-enabled IoT devices and sensors can get data like soil moisture, humidity, predicting the weather, etc. and use it to analyze how much water to supply and at which interval. According to researchers at Chile's UCSC University, the integration of remote sensors around farming areas has reduced the volume of water used by 70%. Rao et al. [33] proposed an embedded based automated irrigation machine that observes the moisture and temperature variations across the crop area and

automatically switches ON the motor, and after the required volume of water supplies, it turns OFF the motor.

- *Drilling, Seeding, and Spraying*

  In traditional farming, drilling and seeding are done by hand, which causes a poor distribution of seed and low productivity compared to using 5G-enabled IoT devices that use sensors to detect specific depth before seeding and ensure that seeds are covered properly by soil. This saves them from birds or being dried up due to exposure to the sun and improves crop yield. Lorain et al. [39] proposed a model of vehicle that will sow the seed by moving around the field. After finishing the seed sowing process, the vehicle will return; it provides the water pumping when the soil seems to be dry. The vehicle uses a soil sensor to monitor the humidity and temperature of the soil. Side by side, it provides good-quality pesticide, spraying, and fertilizer spraying. To prevent the plants from over fertilizers, the time delays are set using the relays. It also provides a special function, which picks out the unwanted grass from the soil.
- *Fertigation*

  It is the injection of fertilizers into the irrigation system used for water amendments and soil amendments. With the help of 5G-enabled IoT sensors, farmers can remotely decide volume and how many fertilizers are injected. Farmers can also monitor soil parameters like fertilizer concentrations, pH value, nutrition, etc. and adjust them using sensors.
- *Crop Monitoring*

  Crop monitoring is a very challenging and costly step, but with advanced technology devices like drones, which are equipped with many sensors and cameras to expose issues like plant disease, fungus, and soil variation in real time, farmers can manage their crop for better yields. Decision support systems for the prediction of various crop disease and solid waste management can also be benefitted from fast sensors and the Internet. Reinecke et al. [40] have proposed the usage of drones for the improvement of crop quality. This might facilitate the farmers to increase their production by sleuthing the loopholes in advance. The crops can be managed by cameras connected to the drones to observe water shortages and harmful pests.

5.6 Smart Supply Chain Management

Supply chain management (SCM) is a process and entities consisting of factories, suppliers, distributors, and retailers concerned with fulfilling a customer order and maximizing customer value. Supply chain management's function is to maximize the surplus that is the price paid by the end customer minus all of the expenses incurred during the supply chain [41]. Internet of Things offers a wide range of connected reality gadgets, benefiting all types of businesses and services.

Benefits of 5G and IoT in supply chain are depicted in Fig. 18.1. Supply chain management is one of the most important beneficiaries of this interconnected

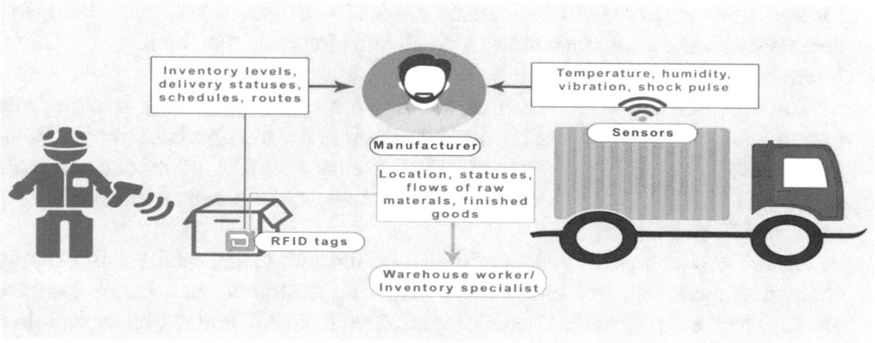

Fig. 18.1 Benefits of 5G-enabled IoT in SCM. (Credits: https://www.scnsoft.com/blog/connected-supply-chain-top-questions-answered)

system. In supply chain management, 5G-enabled IoT offers many benefits like tracking and authenticating products and shipment in real time, monitoring storage conditions like temperature and humidity, and adjusting them according to the product requirement, which will enhance management throughout the whole chain. To utilize these benefits, Arumugam et al. [42] propose a smart logistics solution that encapsulates logistics planners, smart contracts, and condition tracking of the assets within the supply chain management area. Some of the benefits that are provided by 5G-enabled IoT in supply chain management are:

- *Real-Time Monitoring*
  To reduce spoilage of some goods like medicines, chemicals, and foods that require ideal temperature, various IoT devices are used to monitor environmental elements like temperature, exposure to an atmosphere, light intensity, and humidity and automatically adjust to the ideal environment condition required by the products or even trigger an alarm so that employees can act accordingly. It can also keep track of the availability of the product in warehouses as well as their locations in warehouses or transit.
- *Unparallel Transparency*
  Transparency of the supply chain and the enterprise method enables us to put up the belief and reliability of the trademark and creates a unique image of the trademark. With the use of 5G-enabled IoT devices, the clients can without problems get a clear vision of how a product is produced, is processed, and gets to the market. Security and transparency can be achieved in different domains through efficient deployment of blockchain technology [26]. To demonstrate traceability and transparency in supply chain management, Caro et al. [43] presents AgriBlockIoT, a blockchain-based traceability solution

for agri-food supply chain management, capable of seamlessly combining IoT devices generating and consuming virtual data alongside the chain.

- *Improve Contingency Planning*

  The efficient use of 5G-enabled IoT devices can be made for tracking and route planning to easily identify the delay points for the goods in transit. Real-time data transfer technology will make it easy to make plans and alternate solutions for unexpected situations that may come up in practical situations.
- *Improved Segmentation*

  An IoT device generates lots of data, and through these, retailers and supply chain managers segment products by targeting customers and hence increase profit. Thanks to advanced technologies like IoT, AI, and cloud computing, through which company stakeholders will have a good understanding of what types of customers are attracted to what type of product, this will ensure the main motivation of supply chain management, i.e., customer satisfaction, and maximize surplus.

6 Disadvantages and Rumors

Any revolutionary technology has to face wild rumors and incorrect facts, and 5G-enabled IoT is no different. There has been concern growing about 5G being highly hazardous to nature as well as human health. However, those facts are mostly invalid and void. They will be briefly discussed in this section. However, we do not state that this technology comes with absolutely no faults or disadvantages. There exist some disadvantages currently, but they may be resolved in the future. Firstly, talking about the existing disadvantages, they can be summarized in the following two points.

6.1 Expenditure on 5G

The current bandwidth in which 4G operates (3 kHz–6 GHz) is getting saturated as more and more devices are coming online, and thus, 5G plans to use the shorter wavelength bandwidth of 30–300 GHz. However, these high-frequency waves tend to have very low piercing power and, therefore, get absorbed quickly in the surroundings. The solution to this is to build smaller network transmitters, in large numbers, placed very close to each other around the city. The model is depicted in the image below.

However, this demands a whole new infrastructure in the cities. Building a large number of smaller cell towers around the city will be very expensive and time-consuming. The time expected to change the infrastructures of current cities at such massive levels throughout the world can be easily over a decade. Moreover, the longer the project, the higher the investments. To add to it, most of the devices we currently use, and which can use wireless technologies, would be incompetent to

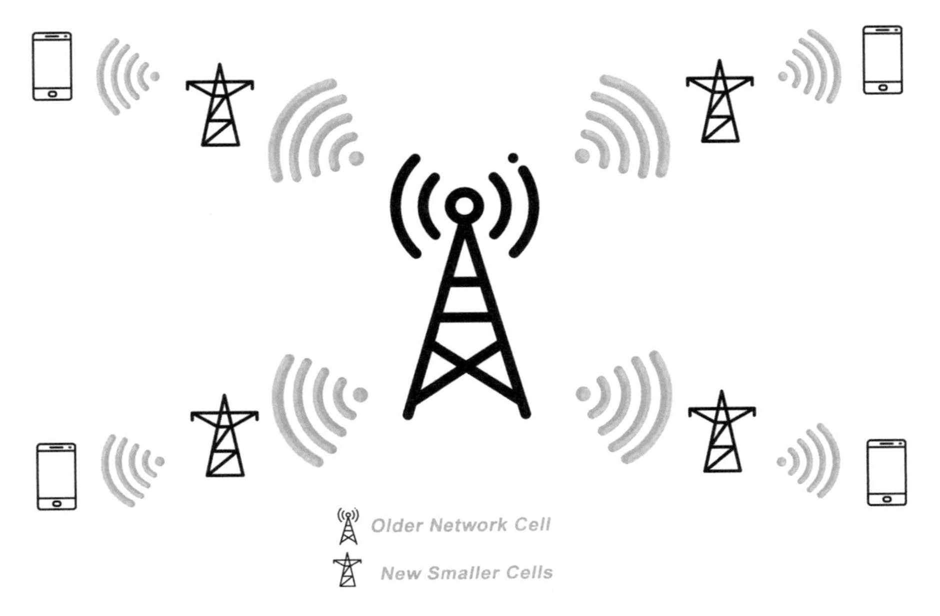

Fig. 18.2 Expected 5G infrastructure

5G, and thus, all of them need to be replaced with new ones, hence adding up to the already high costs of this technology. Expected 5G infrastructure is shown in Fig. 18.2.

6.2 Security Concerns

Since 5G and 5G-enabled IoT are newly emerging ideas, the security and privacy concerns have not yet been resolved. 5G requires a convoluted infrastructure. When combined with IoT, it demands a highly complex digital ecosystem. However, a complex system results in greater security challenges. The high-speed data transfer abilities of 5G can easily be exploited by malicious attackers. The strength of 5G-enabled IoT can be abused by attackers to pose security threats. There are chances that distributed denial-of-service (DDoS) assaults will possibly increase, as 5G will enhance IoT's participation in real-time company systems. And IoT is developed at the old client/server model, with old reliability mechanisms. 5G technology is built upon the extensive use of virtualization. It also uses the containerized systems (as it is expected to run on the cloud). This combination highly widens the gamut of options that attackers can exploit.

6.3 Rumors

There have been many rumors about the ill effects of 5G. Most of them arise because 5G will use very-high-frequency carrier waves, with a significant increase in the number of network transmitters expected around the cities. Another fact that leads to fake rumors in the case of 5G is that the frequency at which it will operate is similar to the frequency which microwave ovens use to cook food. This has become a cause of discomfort for the end-users as they think that these waves will cook up their organs, similar to how microwave ovens cook food inside it. However, there is no truth in such statements. A basic understanding of radiation is required to make such fears void. Though the frequency range of both the waves is similar, a microwave oven gets its ability to cook food from the power output it can generate. A modern-day microwave oven can generate power anywhere between 500 and 1000 watts. This comparatively massive power is what enables the microwaves to boil up the water molecules inside food particles. On the other hand, our smartphone or other modern-day IoT-enabled devices can barely generate power anywhere near 1 watt. Thus, the fearful fact about our organs getting cooked up has no validity.

7 Case Study: Cloud VR Services

Virtual reality (VR) refers to software-generated simulation in which anyone can experience and interact with an artificial 3-D environment with the help of electronic gadgets. This technology can be integrated with many other industries to bring the most out of each, some examples being, in healthcare, allowing tasks like remote surgery; in gaming, giving an immersive experience to the players; and in the automobile industry, allowing engineers to experiment easily with the design and build of any automobile before commissioning expensive prototypes, to name a few. However, to make the aforementioned tasks a possibility, there is a need for cloud-based VR services. The following is a case study on the launch of cloud VR services, a possibility using 5G-enabled IoT. *[The following facts and figures were produced by Huawei at the 10th Global Mobile Broadband Forum. Source].*

The operators who shall provide cloud services for VR can enjoy great opportunities in the market. Carrier serviceable addressable market (SAM) is expected to reach more than 93 billion USD globally by 2025. It is estimated that cloud-based infrastructure can reduce the cost of VR equipment by 70–80%. This will considerably increase the demands for VR hardware as their cost is a major disadvantage in their widespread use among common masses. Required network performance is shown in Table 18.3.

The network architecture for cloud-based VR services demands end-to-end public networks connected to cloud-based GPUs (graphics processing units). The latency allowed for the 4 k resolution (H 265) in VR is less than 16 milliseconds. Such low latency is almost impossible to achieve with the previous generation

Table 18.3 Required network performance

	2 K resolution (H 264)	4 K resolution (H 265)	4 K resolution (H 266)
Latency	<20 ms	<16 ms	<10 ms
Speed	20–50 Mbps	50–200 Mbps	200–600 Mbps

of wireless technologies, as the average latency of a 4G network is around 50 milliseconds and the best it can do is up to 20–30 milliseconds. Thus, the architecture highly demands a network with ultra-low latencies. This is where 5G-enabled devices come into the picture as even 5G-enhanced mobile broadband can reach latencies as low as 4–5 milliseconds and the 5G-URLLC (Ultra-Reliable Low-Latency Communications) systems can achieve latencies as low as 1 millisecond. Hence, the latency requirements of cloud-based VR services can only be fulfilled by 5G-enabled devices.

Talking about the data transfer rates required for such a system to work flawlessly, gadgets on previous generations of connecting technology cannot be satisfactory. This is because, for 4 k resolution (H265) in VR, the required data rates range from 50 to 200 Mbps. However, 4G has average download speeds between 5 and 12 Mbps, and the peak of the transfer rate possible is 50 Mbps. Therefore, there is a call for 5G's high data transfer speeds. The peak speed of 5G can reach up to 20 Gbps, and even the average data transfer speeds can easily reach 100+ Mbps. Thus, this requirement of the network architecture for cloud-based VR services can only be met by 5G.

8 Conclusion and Scope of Future Work

In this chapter, we had briefly introduced the impact 5G-enabled IoT will put on industries and industrial automation. Industrial growth has known no limits in the last few years, but we have only included those categories which are expected to go through a major revolution with the advent of 5G and its integration with IoT. 5G-enabled IoT is expected to make healthcare much more advanced, bring usable self-driving vehicles closer to reality, and evolve many typical industrial products and systems into their smarter versions, like smart homes, smart cities, smart agriculture, smart supply chain management, etc. All these mentioned changes have been briefly discussed in the chapter. The capability to carry large amounts of data with ultra-low latency, which 5G boasts of, is what increases the expectations and hype around the industrial growth in the upcoming years. 5G promises to bring a greater number of devices, including the IoT devices on board, which has created a hype around automation in the future.

However, not only the positives but we have also talked about the negatives of the upcoming technology. It requires a serious input of money to make 5G even a reality. The fact that most current devices will have to be replaced itself makes the very technology time-consuming and something to expect only in the distant future. Moreover, the benefits of 5G are very easy to be abused, and malicious attacks are certainly expected to increase manyfolds. Nevertheless, the evolution in the industries which is expected is worth the troubles needed to fix or compensate for the disadvantages. Rumor has it; this technology is going to have severe health issues for humans as well as other living beings; however, we have discussed some of them, which are fake and just the result of lack of scientific knowledge in common folks.

As 5G is still a concept on paper, there is a huge scope for further research to bring it to reality. The fact that current IoT devices will need to undergo modifications to become compatible with 5G opens many doors for further research. Moreover, it is not practically feasible to discuss all the industries that will undergo major evolution in a single chapter. Thus, some more sections need to be explored. There is a major need for research to compensate for the disadvantages of 5G architecture. There can be papers proposing cheaper or more efficient models for the infrastructure of 5G cell towers in cities. The security of the 5G-based architecture currently demands lots of research to make it practically usable.

Through this chapter, we have tried to bring together and discuss all the major developments in the aforementioned technologies, as predicted by the rigorous research going on all over the world. The readers can get an estimate or a peek-look at the revolution that is to come due to the advancements in wireless technology (5G) and its integration with the Internet of Things (IoT). The chapter introduces the reader to the world of 5G, IoT, as well as 5G-enabled IoT and achieves its primary goal, to bring out the "importance of 5G-enabled IoT for industrial automation."

References

1. M.P. Groover, Automation. *Britannica* [Online]. Available: https://www.britannica.com/technology/automation. Accessed 10 Aug 2020
2. I. Mistry, S. Tanwar, S. Tyagi, N. Kumar, Blockchain for 5G-enabled IoT for industrial automation: A systematic review, solutions, and challenges. Mech. Syst. Signal Process. **135**, 106382 (2020)
3. P. Pandey, R. Litoriya, An IoT assisted system for generating emergency alerts using routine analysis. Wirel. Pers. Commun. **112**(1), 607–630 (2020)
4. S. Kaneriya, J. Vora, S. Tanwar, S. Tyagi, Standardising the use of duplex channels in 5G-WiFi networking for ambient assisted living, in *2019 IEEE International Conference on Communications Workshops (ICC Workshops)*, 2019, pp. 1–6
5. U. Bodkhe, P. Bhattacharya, S. Tanwar, S. Tyagi, N. Kumar, M.S. Obaidat, BloHosT: Blockchain-enabled smart tourism and hospitality management, in *2019 International Conference on Computer, Information and Telecommunication Systems (CITS)*, 2019, pp. 237–241
6. M. Hung, Leading the IoT, 2017
7. M. Mittal, S. Tanwar, B. Agrawal, L. M. Goyal (eds.), *Energy Conservation for IoT Devices Concepts, Paradigms and Solutions* (Springer Nature Singapore Pte Ltd, Singapore, 2019)
8. M. Pandey, R. Litoriya, P. Pandey, Novel approach for mobile-based app development incorporating MAAF. Wirel. Pers. Commun. **107**(4), 1687–1708 (2019)
9. M. Pandey, R. Litoriya, P. Pandey, Mobile app development based on agility function. Ingénierie des systèmes d'information RSTI série ISI **23**(6), 19–44 (2018)

10. M. Pandey, R. Litoriya, P. Pandey, The applicability of machine learning methods on mobile app effort estimation: Validation and performance evaluation. Int. J. Softw. Eng. Knowl. Eng. **30**(1), 23–41 (2020)
11. M. Pandey, R. Litoriya, P. Pandey, Validation of existing software effort estimation techniques in context with mobile software applications. Wirel. Pers. Commun. **110**, 1659–1677 (2019)
12. K.D. Foote, A brief history of the Internet of Things. *Dataversity* (2016) [Online]. Available: https://www.dataversity.net/brief-history-internet-things/
13. Personal Tech, When was 5G introduced? *Verizon* (2019) [Online]. Available: https://www.verizon.com/about/our-company/5G/when-was-5G-introduced. Accessed 10 July 2020
14. OnQ Blog, With 5G here, what's next for the Internet of Things? *Qualcomm* (2020) [Online]. Available: https://www.qualcomm.com/news/onq/2020/05/12/5g-here-whats-next-internet-things. Accessed 10 June 2020
15. A. Phan, S.T. Qureshi, 5G impact On Smart Cities, 2017
16. M. Chan, E. Campo, D. Estève, J.-Y. Fourniols, Smart homes — Current features and future perspectives. Maturitas **64**(2), 90–97 (2009)
17. J.M. Khurpade, D. Rao, P.D. Sanghavi, A survey on IoT and 5G network, in *2018 International Conference on Smart City and Emerging Technology (ICSCET)*, 2018, pp. 1–3
18. K.E. Skouby, P. Lynggaard, Smart home and smart city solutions enabled by 5G, IoT, AAI and CoT services, in *2014 International Conference on Contemporary Computing and Informatics (IC3I)*, 2014, pp. 874–878
19. R.S. Hill, J.N. Dewey, R.M. Plasencia, *Blockchain and 5G-Enabled Internet of Things (IoT) Will Redefine Supply Chains and Trade Finance* (Holland & Knight, 2018), pp. 42–45
20. G. Sushanth, S. Sujatha, IOT based smart agriculture system, in *2018 International Conference on Wireless Communications, Signal Processing and Networking (WiSPNET)*, 2018, pp. 1–4
21. L. Fan, J.R. Gil-Garcia, D. Werthmuller, G.B. Burke, X. Hong, Investigating blockchain as a data management tool for IoT devices in smart city initiatives, in *Proceedings of the 19th Annual International Conference on Digital Government Research Governance in the Data Age – dgo '18*, 2018, pp. 1–2
22. F. Jammes, H. Smit, Service-oriented paradigms in industrial automation. IEEE Trans. Ind. Inform. **1**(1), 62–70 (2005)
23. B. Kumar, S.P. Singh, A. Mohan, Emerging mobile communication technologies for health, in *2010 International Conference on Computer and Communication Technology (ICCCT)*, 2010, pp. 828–832
24. P. Pandey, R. Litoriya, Securing E-health networks from counterfeit medicine penetration using Blockchain. Wirel. Pers. Commun. (2020)
25. P. Prateek, L. Ratnesh, Securing and authenticating healthcare records through blockchain technology. Cryptologia **44**(4), 341–356 (2020)
26. J. Vora et al., BHEEM: A Blockchain-based framework for securing electronic health records, in *2018 IEEE Globecom Workshops (GC Wkshps)*, 2018, pp. 1–6
27. P. Pandey, R. Litoriya, Implementing healthcare services on a large scale: Challenges and remedies based on blockchain technology. Health Policy Technol. **9**(1), 69–78 (2020)
28. P. Pandey, R. Litoriya, An activity vigilance system for elderly based on fuzzy probability transformations. J. Intell. Fuzzy Syst. **36**(3), 2481–2494 (2019)
29. S.Y.Y. Tun, S. Madanian, F. Mirza, Internet of things (IoT) applications for elderly care: A reflective review. Aging Clin. Exp. Res. (2020)
30. P. Pandey, R. Litoriya, Elderly care through unusual behavior detection: A disaster management approach using IoT and intelligence. IBM J. Res. Develop. **64**(1), 15:1–15:11 (2019)
31. E. Borelli et al., HABITAT: An IoT solution for independent elderly. Sensors **19**(5), 1258 (2019)
32. P. Pandey, R. Litoriya, Ensuring the elderly well being during COVID-19 like pandemic using IoT. Disaster Med. Public Health Prep., 1–10 (2020)
33. R.N. Rao, B. Sridhar, IoT based smart crop-field monitoring and automation irrigation system, in *2018 2nd International Conference on Inventive Systems and Control (ICISC)*, 2018, pp. 478–483

34. P. Pandey, R. Litoriya, in *Legal/regulatory issues for MMBD in IoT BT - Multimedia big data computing for IoT applications: Concepts, paradigms and solutions*, ed. by S. Tanwar, S. Tyagi, N. Kumar, (Springer Singapore, Singapore, 2020), pp. 367–388
35. L. Cui, G. Xie, Y. Qu, L. Gao, Y. Yang, Security and privacy in smart cities: Challenges and opportunities. IEEE Access **6**, 46134–46145 (2018)
36. R. Litoriya, A. Gulati, M. Yadav, R.S. Ghosh, P. Pandey, Social, ethical, and regulatory issues of fog computing in healthcare 4.0 applications, in *Fog Computing for Healthcare 4.0 Environments*, (Springer, Cham, 2021), pp. 593–609
37. C. Dwork, F. McSherry, K. Nissim, A. Smith, *Calibrating Noise to Sensitivity in Private Data Analysis* (2006), pp. 265–284
38. K.L. Krishna, O. Silver, W.F. Malende, K. Anuradha, Internet of Things application for implementation of smart agriculture system, in *2017 International Conference on I-SMAC (IoT in Social, Mobile, Analytics, and Cloud) (I-SMAC)*, 2017, pp. 54–59
39. A.N. Lorain, U. Ramamalini, R. Veeralakshmi, Smart automated agriculture monitoring and controlling system using Arduino, in *International Conference on Electrical, Information and Communication Technologies*, 2017, pp. 25–29
40. M. Reinecke, T. Prinsloo, The influence of drone monitoring on crop health and harvest size, in *2017 1st International Conference on Next Generation Computing Applications (NextComp)*, 2017, pp. 5–10
41. E. Hassini, Supply chain optimization: Current practices and overview of emerging research opportunities. INFOR Inf. Syst. Oper. Res. **46**(2), 93–96 (2008)
42. S.S. Arumugam et al., IoT enabled smart logistics using smart contracts, in *2018 8th International Conference on Logistics, Informatics and Service Sciences (LISS)*, 2018, pp. 1–6
43. M.P. Caro, M.S. Ali, M. Vecchio, R. Giaffreda, Blockchain-based traceability in Agri-Food supply chain management: A practical implementation, in *2018 IoT Vertical and Topical Summit on Agriculture – Tuscany (IOT Tuscany)*, 2018, pp. 1–4

Chapter 19
Blockchain Based Framework for Document Authentication and Management of Daily Business Records

Prakrut Chauhan, Jai Prakash Verma, Swati Jain, and Rohit Rai

1 Introduction

Documents exist of various kinds such as financial papers, identification documents, historical records, etc. With the advent of digitization, softcopies are replacing the hard copies for most purposes. While digital documents have a lot of advantages, there are some challenges associated with them as well. The issues are that digital documents can easily be modified, forged, altered, and tampered. It becomes difficult for individuals and organizations to ascertain the authenticity of digital documents [10]. The main challenges involved in the authentication of digital documents arise due to the widespread use of document manipulating tools, the need for intermediaries, the opaque authentication process, and the cost involved [12]. Thus, there is a need for a novel approach for authentication and verification of documents.

Manipulation of Digital Assets There are several software and tools that people with basic computer literacy can use to alter or modify digital documents. Individuals with malicious intentions also use such tools to forge and manipulate documents. It is difficult for other people to determine the authenticity of such documents. Sophisticated tools are required to expose the manipulation. Therefore, it is a tough challenge to authenticate and verify digital documents [4].

P. Chauhan · J. P. Verma (✉) · S. Jain
Institute of Technology, Nirma University, Ahmadabad, India
e-mail: 16BCE044@nirmauni.ac.in; jaiprakash.verma@nirmauni.ac.in;
swati.jain@nirmauni.ac.in

R. Rai
Indian Naval Ship (INS) Valsura, Jamnagar, Gujarat, India
e-mail: rohit.rai@navy.gov.in

S. Tanwar (ed.), *Blockchain for 5G-Enabled IoT*,
https://doi.org/10.1007/978-3-030-67490-8_19

Need for an Intermediary As documents are frequently exchanged on the Internet, it becomes challenging for individuals and organizations to determine the ownership, source, and authenticity of a document [17]. Thus, organizations often hire an intermediary that performs the task of authenticating and verifying them for the organization. Without an intermediary, an organization would have to spend considerable resources to perform mundane tasks for confidential documents [22].

Lack of Transparency The authentication and verification process is generally not very transparent. The individual or organization has to trust the intermediary. There is hardly any information about the process and rules that were followed to authenticate and verify the documents [20].

Cost Ineffectiveness Every organization deals with a huge number of digital documents, many of them need to be validated by the organization, and they have to respond to verification requests. These are ordinary tasks, but they are required to be performed frequently due to large numbers of documents that need verification. The organization would either hire a third party to perform these tasks or spend its own resources, both of these methods are cost-ineffective.

The proposed decentralization application addresses these challenges. The issues regarding document authentication have been addressed by leveraging the blockchain technology [3]. The immutability of blockchain prevents the manipulation of documents, and the decentralization of blockchain eliminates the need for an intermediary, hence adding transparency.

The paper contains seven sections. Section 1 gives the introduction to this article (the need, findings, and the proposed system). Section 2 discusses the related work on the domain of blockchain (different approaches, findings, and analysis). Section 3 presents our model and the methodology adopted in our approach, Sects. 4 and 5 go about our implementation strategy for the prescribed blockchain model, while Sect. 6 comes out with the results obtained. Finally, Sect. 7 concludes and proposes future research of the article.

2 Related Work

Blockchain technology has been used for authentication of documents such as educational certificates, birth certificates, and identity documents. As per the needs and use cases, different blockchain platforms and methodologies have been implemented to cater to different needs [15].

Blockchain Based Identity Verification Model Hyperledger Fabric blockchain, a private permissioned blockchain, is used in this paper to facilitate the verification and storage of identity documents issued by the government. The front-end interface is built using HTML, CSS, and PHP. Node.js is used for the middle-tier, and Hyperledger is the underlying blockchain.

Developing Ethereum Blockchain-Based Document Verification Smart Contract for Moodle Learning Management System A smart contract is developed to store certificate details on the Ethereum blockchain. Ganache, a personal Ethereum development tool, is used as the local blockchain. Smart Contract is developed in Remix through Solidity programming language. Using the web3 library, the details of the certificate are directly taken from the Moodle System. The smart contract is deployed on the Ethereum Ropsten Test Network.

Tamper Proof Birth Certificate Using Blockchain Technology This paper proposes the use of BigchainDB to store birth certificates using IPFS. The solution allows the user to login by generating a BigchainDB key pair and RSA key pair and storing the user details on the blockchain. The issuing authority adds the birth certificate based on the user's phone number.

Proposing a Blockchain-Based Solution to Verify the Integrity of Hardcopy Documents The proposed solution to verify hardcopy documents consists of four key components—barcodes, cryptographic hashing, digital signatures, and optical character recognition. The solution design has two main processes—generation and validation. In the generation phase, first, the text in a document is encoded and the metadata is stored on the blockchain. Its unique id and public key are encoded in a barcode, and the barcode is added to the bottom of the document. The hardcopy is generated. During the verification phase, the hardcopy is first scanned and, using the barcode, the metadata is obtained from the blockchain. Using optical character recognition, the text is generated and its metadata is compared with that stored in the blockchain.

A Permissioned Blockchain-Based System for Verification of Academic Records This paper presents an end-to-end solution to transmit and verify academic records. The system architecture has two main components—a web application layer and blockchain application layer. Hyperledger Fabric and Hyperledger Composer are used along with Google Cloud to host the web application. The implementation is divided into two—transcript request and validation process.

Table 19.1 shows the comparative analysis of the different approaches discussed in this section.

3 Blockchain Technology for Document Authentication

Blockchain, as the name suggests, is a "chain of blocks" based on the peer to peer technology. It is an immutable time-series record of data that is not managed by a single entity but a cluster of computers. Each block in the blockchain contains data related to transactions and linked with others through cryptographic rules [5]. In a blockchain network, every participating entity holds an independent copy of the blockchain and updates its copy as per the underlying consensus algorithm of the blockchain network. Blockchain was invented by Satoshi Nakamoto, and the

Table 19.1 Comparison of different approaches

Ref	Objective	Methodology	Pros	Cons
[13]	Blockchain Based Identity Verification Model	Hyperledger blockchain along with IPFS is used to store identity documents	Entire file is stored by using the Inter Planetary File System	Every time n individual or an institution wants to verify the document, they have to be added to the Access Client List
[8]	Developing Ethereum Blockchain-Based Document Verification Smart Contract for Moodle Learning Management System	MOOC Course certificates are generated and the details are stored on Ethereum	The certificates are generated automatically	Other different types of documents cannot be verified by this methodology
[23]	Tamper Proof Birth Certificate Using Blockchain Technology	BigchainDB is used to store birth certificates using IPFS	Every user can control who can view the birth certificate	Document can only be shared via blockchain technology
[16]	Proposing a blockchain-based solution to verify the integrity of hardcopy documents	Documents are verified by using barcodes, hashing, digital signature, and OCR	Even hardcopies can be verified using this method	Use of barcodes limits the size of the documents that can be stored
[2]	A Permissioned Blockchain-Based System for Verification of Academic Records	Academic records are stored on Hyperledger Fabric	Web application allows users to perform transactions without actually joining the network	Students need to request the university every time to send their documents

technology came into light in 2008 when a paper about Bitcoin was circulated. The identity of Santoshi Nakamoto is unknown, and it is speculated that it could be a group of people or an individual. Bitcoin is a cryptocurrency based on blockchain technology. The popularity of Bitcoin led to the widespread use of blockchain technology for various purposes. Over the years, different types of blockchain networks have emerged and this technology is being used to solve different problems across domains due to inherent characteristics of a blockchain network.

3.1 Key Features of Blockchain Technology

Blockchain's success has been primarily due to the three key features of this technology.

3.1.1 Decentralization

The Internet largely works on centralized technologies. Most popular services use the client–server model where data stored in centralized databases is owned by private entities. This centralized model has several drawbacks. Firstly, because the system is centralized, it can be prone to attacks that can lead to the failure of the entire system. Secondly, the owner can shut down the system at their will, and data could be lost or corrupted permanently. The blockchain network is decentralized, and no single entity controls it. All the participating entities on the network are on the same hierarchy, it does not have a single point of failure, and also it cannot be shut down at the will of an individual. Since the data is replicated by each entity in the blockchain network, the risk of data loss or corruption is negligible. A 51% attack is needed to take over a blockchain network, and experts believe that it is virtually impossible in the present scenario [21].

3.1.2 Transparency

With public addresses on blockchain, all transactions are viewable to everyone on the network and it provides a fully auditable and valid ledger of transactions. Blockchain does not compromise security and privacy for transparency. Blockchain provides anonymity by securing a person's identity through private keys. Private keys also ensure security [6, 7]. Encryption and cryptography ensure that blockchain is secure enough to handle financial transactions and also elections [24].

3.1.3 Immutability

The records and transactions on the blockchain network are immutable, which means that an appended block cannot be altered. This feature makes blockchain valuable for multiple applications. If a hacker alters one block, subsequent blocks need to be altered because all the blocks are linked through cryptographic hash functions. Also, changes in one copy of blockchain will not do any damage because all the entities in the blockchain hold a copy of the same. It ensures that once a block has been added to the blockchain and approved using the consensus mechanism, it cannot be altered or modified [9].

These three features have increased the utility, efficiency, and significance of the blockchain technology for a wide variety of applications. The description of blockchain characteristics and their applicability for document authentication and verification is shown in Table 19.2.

Table 19.2 Blockchain features for document authentication and verification

Characteristic	Description	Applicability
Decentralization	No single entity controls the network	Ensures that platform is fault tolerant and less prone to breakdown
Transparency	With public addresses and transactions, all activity is auditable	Leads to trust between the authenticating and verifying entities
Immutability	Anything once written on the blockchain cannot be altered	Ideal for documents of permanent nature such as birth certificates and university transcripts

3.2 Types of Blockchains

Blockchains can be primarily categorized into four categories based on how the nodes can join the network, the need of the application, and the nature of the participating entities. The choice of the type of blockchain to use primarily depends on the application. Conventionally, public blockchains are used by organizations whose key focus is on transparency, while private blockchains are used by private companies. Consortium blockchains are used when multiple organizations are collaborating, and hybrid blockchains are ideal for business models that want to use the features of both private and public blockchains.

3.2.1 Public Blockchain

This type of blockchain is truly decentralized because it is permissionless and unrestricted. Anyone with Internet can access and join the network and perform all the different tasks like mining, auditing, and performing transactions. Public blockchains ensure transparency and thus build trust amongst the participating nodes. These can be used for applications such as fund-raising and voting. The largest and the most popular blockchains—Bitcoin and Ethereum—are public blockchains.

3.2.2 Private Blockchain

This blockchain network is a restricted version. There are network administrators who control participation in it by granting permissions. This type of blockchain network is usually smaller and is used by industries or entities such as banks and private corporations that want a higher level of control. These blockchains are faster and scalable. They are usually used for internal use cases such as supply chain management. Hyperledger Fabric is an example of a private blockchain network [11].

3.2.3 Consortium Blockchain

It is a semi-decentralized network that is managed by multiple organizations. Role and rights related to a node are governed by a group of nodes. These blockchains are ideal when various organizations are collaborating for one task. They can be used by organizations that are collaborating for research, group of banks, and inter-governmental organizations. Corda and Quorum are examples of consortium blockchain [18].

3.2.4 Hybrid Blockchain

It is a combination of a public and private blockchain, combining the features of both types. It provides the flexibility to keep some data public and some data private. This type is gradually getting more popular for use cases that require the features of both private and public blockchains. It can be used to semi-restricted environments such as financial markets. Dragonchain is a popular hybrid blockchain.

These four types of blockchain networks have certain advantages and disadvantages in terms of transparency, scalability, security, and reliability. Restrictions may reduce scalability and fault tolerance while protecting privacy. Public networks offer more scalability and reliability. The choice of the type of blockchain network depends on the nature of the use case and the organization.

3.3 Working of a Proposed Blockchain and Consensus Algorithms for Document Authentication

There are various consensus algorithms that facilitate the coordination of all nodes in a blockchain. The most widely used consensus algorithm is Proof of Work (PoW) that is implemented by Bitcoin and Ethereum. In this algorithm, each entity (node) in the blockchain network preserves a copy of the blockchain network and communicates with other nodes to ensure that all the nodes have the same version of the transactions data. A new block is created periodically to record the transactions. The process of creating new blocks for updating the transactions is termed as mining. Mining is an integral part of every blockchain that is based on hash functions.

Since hash functions are one-way functions, all the nodes try to guess the input to generate the particular output. The node that first solves this hash function adds the new block to the blockchain. Solving this hash function is simply a method of hit and trial; due to the complexity of hash functions, it consumes a lot of computational power and electricity. The newly mined block broadcasts the updates to all the nodes in the blockchain. The other nodes verify the transactions using the hash function for verification of newly mined block. Once verified, all the nodes update their

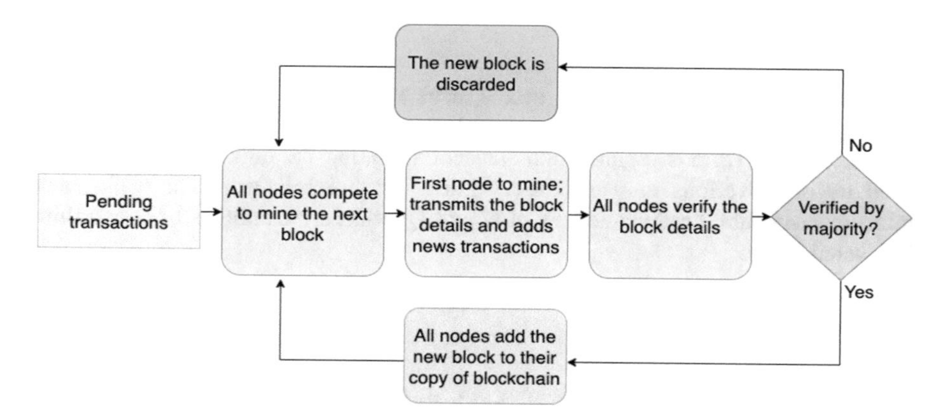

Fig. 19.1 Working of a blockchain

blockchain by adding information about the new block. A node will not be able to add an incorrect block because of the verification by all the nodes in the network. To attack the integrity of such a blockchain network, one would need more than 50% of the network's computing power, which is highly unlikely in actual blockchain networks like Bitcoin and Ethereum. Figure 19.1 shows the working of a blockchain network.

There are other consensus algorithms as such Proof-of-Stake (PoS) and Proof-of-Authority (PoA) [14]. As per the Proof-of-Stake algorithm, nodes lock a certain number of their coins as a stake in the network. The number of locked coins affects the chances of being selected as the node that adds the new block. The larger the stake, more the chances of being selected; however, this is a pseudo-random process, so the wealthiest node will not be selected every single time. Proof-of-Authority works similarly like Proof-of-Stake, but instead of staking coins, the nodes stake their reputation. In this algorithm, there are some nodes that are validators and only these validators can add the new blocks. The validators are incentivized to add the correct transactions because a negative reputation affects their authority in the network. In all the three algorithms, the nodes are incentivized to select the correct transactions because incorrect verification causes a loss of resources in the case of PoW, loss of coin in the case of PoS, and loss of reputation in case of PoA [1].

4 Introduction to Ethereum

Conceptualized in 2014 by Vitalik Buterin, Ethereum is a public, decentralized, programmable blockchain. Its native cryptocurrency is called Ether and is represented as ETH. The Ethereum community is the most active and largest community

Table 19.3 Comparison of different blockchains

Characterisitc	Ethereum	Hyperledger	Corda
Platform	Generic blockchain	Modular blockchain	Specialized for Financial Industry
Governance	Ethereum Developers	Linux Foundation	R3
Consensus	Ethereum Developers	Broad understanding of consensus	Specific understanding of consensus
Mode of operation	Public	Permissioned	Permissioned
Smart Contracts Language	Solidity	Permissioned	Kotlin, Java

in the world. One of the primary reasons for the massive popularity of this open-source platform is that it enables the development of decentralized applications also called dapps. This is facilitated by smart contracts that allow the dapps to be executed on the Ethereum network through Ethereum Virtual Machine. Decentralized applications, as the name suggests, do not execute on a centralized single machine but run on a decentralized blockchain network. These are basically open-source software based on blockchain technology. A comparison of different blockchain platforms is shown in Table 19.3.

A contract has certain conditions that need to be fulfilled by participating parties. A smart contract is executed automatically when the necessary requirements are fulfilled. Smart contracts are basically written as code and committed to the blockchain. The code is executed when the events specified in the smart contract are triggered. Simply put, a smart contract is a piece of code running on Ethereum. These are written in the Solidity programming language. A smart contract allows the dapps to connect to the blockchain.

4.1 Ethereum Virtual Machine

The runtime environment for the smart contracts in Ethereum is called the Ethereum Virtual Machine. It has 140 unique opcodes that allow the execution of different tasks. These opcodes make the Ethereum virtual machine Turing complete. Therefore, all computational problems can be written as smart contracts. Technically, any logic implemented in standard programming languages can be in written in Solidity language. Thus, sophisticated applications can build using smart contracts. It is one of the reasons why Ethereum is leading as the popular blockchain for decentralized applications [19].

Table 19.4 Comparison of different test networks of Ethereum

	Rinkeby	Ropsten	Kovan
Consensus	PoA	PoW	PoA
Security	Immune to spam	Not immune	Immune to spam
Blocktime	15 s	30 s	4 s
Support	Geth	Geth and Parity	Parity
ChainData	>30 GB	>120 GB	>80 GB
Ether	Requested	Mined	Requested

4.2 Ethereum Test Networks

It is important to first deploy all dapps on a test network before deploying them on Ethereum main net because actual assets are involved in the main net. Actual Ether needs to be purchased to run a dapp on the main net of Ethereum, but developers can freely obtain Ether to run the dapps on test networks. The working of test nets is similar to the main net, but there are variations in terms of scale, consensus algorithms, and block-mining time. The three test networks of Ethereum are Rinkeby, Ropsten, and Kovan. Developers may choose a test network according to their preferences. Table 19.4 shows the comparison of the different test networks of Ethereum.

5 Methodology

The proposed solution aims to leverage blockchain technology to solve the existing problems associated with document authentication and verification. The key challenges in the process of authentication of documents can be overcome by using a decentralized public ledger. The inherent characteristics of a blockchain network help in solving these problems. The immutable nature of blockchain ensures that documents cannot be modified once they are written, the decentralization removes the need for an intermediary, open-source decentralized apps that ensure transparency, and the transaction fees associated with the verification process are minuscule compared to hiring a third party. The Ethereum blockchain will be used because of its smooth and efficient platform that supports the use of smart contracts to build decentralized applications.

A smart contract shall be developed that includes the functions of authentication and verification. This will be deployed on the Ethereum blockchain. The fundamental idea is to create a cryptographic hash of every document. There are two main reasons for doing this; firstly, a cryptographic hash is a one-way function; therefore, once the hash is generated, it is impossible to find out the original input document

Algorithm 1 Authenticating the documents

```
1: procedure ADD DOCUMENT(doc.extension)
2:     doc_hash ← hash(doc.extension)        ▷ Hash code generated using inbuilt hash function

3:     BlockchainNetwork ←doc_hash           ▷ Write the hash code on the blockchain network

4:     write_time ← BlockchainNetwork.time
5:     block_no ← BlockchainNetwork.block

       return doc_hash, write_time, block_no
6: end procedure
```

Algorithm 2 Verifying the documents

```
1: procedure SEARCH DOCUMENT(doc.extension)
2:     doc_hash ← hash(doc.extension)        ▷ Hash code generated using inbuilt hash function

3:     Block_no ← search_block(doc_hash)
4:     write_time ← search_time(doc_hash)              ▷ Use blockchain's search function

5:     if block_no! = 0 then
6:         Document verified
   return block_no, write_time
7:     else
8:         Document not verified
9:
```

from the hash, and this ensures privacy. Secondly, the size of the output hash is always fixed irrespective of the input size. This gives the flexibility to use any size or any kind of document. The hash code size is quite small and can be written on the blockchain, while writing the entire document on the blockchain is an undesirable and inefficient alternative because of the issues of cost, privacy, and speed. SHA-256, a strong and presently an uncompromised algorithm, has been used as the hash function. It generates a 256-bit hash code that is a 64-character alpha-numeric code. Once the hash code is written on the blockchain, it is immutable and stored with a timestamp. Later a document can be verified by comparing its hash code with that already stored on the blockchain. Even a tiny change in the document, i.e., a change of one pixel or character, will completely change the hash code due to the "avalanche effect." Therefore, no two documents will ever have the same hash code. Figures 19.2 and 19.3 show the processes of authentication and verification, respectively.

The ability to add the document hash code on the blockchain will be restricted by using access control modifiers of functions in the solidity programming language, but any individual will be able to search for a hash code because it is impossible to generate the document from the hash code. Algorithm 1 shows how the authentication of the documents is to be carried out using the proposed approach and the Algorithm 2 shows how the verification of the documents is to be carried out using the proposed approach.

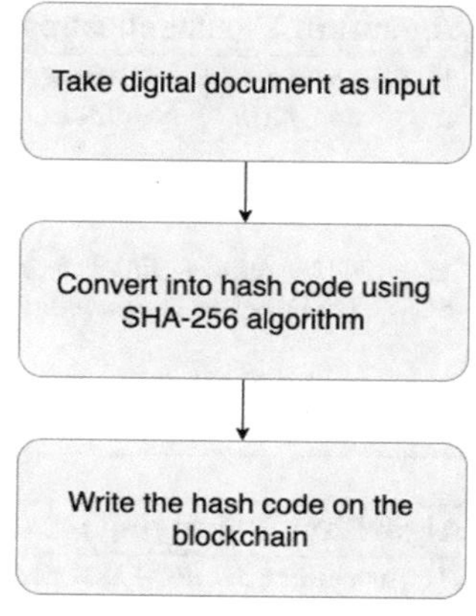

Fig. 19.2 Process of authentication

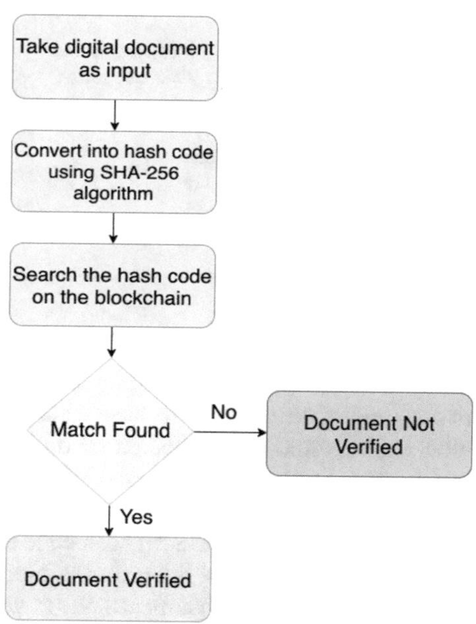

Fig. 19.3 Process of verification

6 Execution and Implementation

6.1 Development of Smart Contract

A basic smart contract is developed with two functions: one adds document hash to the blockchain and other searches for the document hash on the blockchain. The "add document hash" function will return a transaction id, and "find document hash" will return the block number and timestamp of writing that hash on the blockchain. If the hash code is not found on the blockchain, null values will be returned. These functions take a hash code as the input parameter and not the document itself because converting the document into hash code inside the smart contract would increase the transaction fee. The Ethereum smart contracts are written in Solidity programming language.

6.2 Deployment on Local Blockchain

Ganache is a tool that simulates the Ethereum blockchain and runs locally. Thus, it can be used as a personal blockchain for development and testing purposes. The smart contract was deployed on Ganache to test its working. To deploy it, the configuration file of Truffle was modified to set the development network as "localhost." Using the truffle migrate command, the smart contract is deployed. Figure 19.4 shows the Ganache interface.

6.3 Development of Command-Line Interface

A JavaScript file is created that interacts with the smart contract. Firstly, the smart contract is connected to the JavaScript file using the contract's address. Secondly, a function to calculate document hash is created, which takes a document as an input and generates a hash code using the SHA-256 algorithm from the jsSHA library. This approach has been adopted because if the conversion happens inside the smart

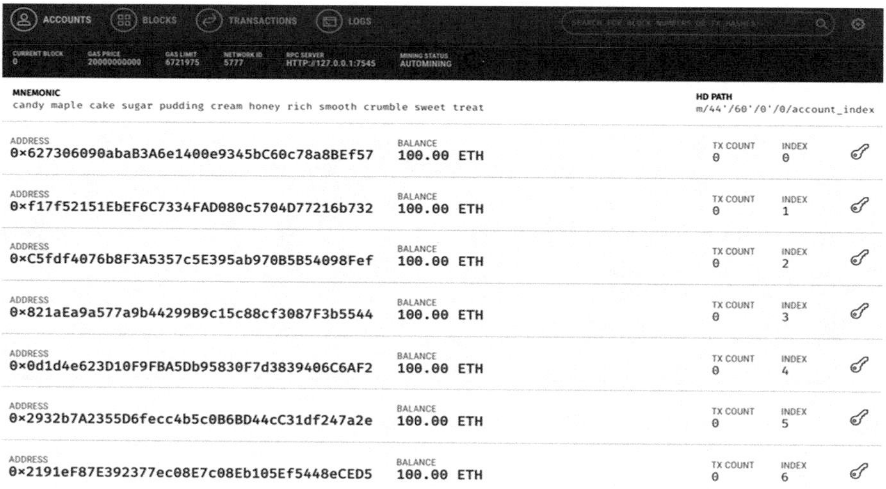

Fig. 19.4 The Ganache interface

contract then the user has to pay the transaction fee for converting the document into the hash code. AddDocumentHash and FindDocumentHash functions in this file send the hash code to the smart contract. The working of the smart contract does not change. So, it can be accessed using a command-line interface. A user can now select the file to either add its hash code or find its hash code on the blockchain.

6.4 Testing the Smart Contract

As seen in Fig. 19.5, the smart contract is tested using a command-line interface. The file is specified, -a is used to add the document hash code, and -f is used to search for the document hash code on the Ganache blockchain. When the document hash code is written on the blockchain, the hash value and transaction id are displayed to the user, and if the document hash code is found on the blockchain after searching, the block number where the hash code was written is displayed along with the time that block was mined. In case the document hash value is not found on the blockchain, a message indicating the same is displayed to the user. Figure 19.5 shows the testing of smart contract in a command-line interface.

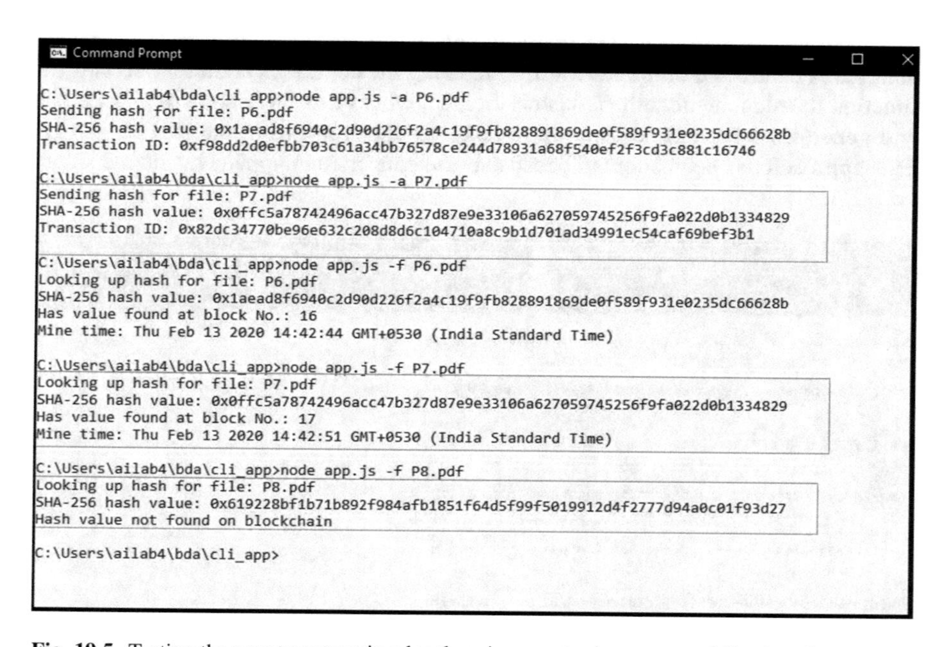

Fig. 19.5 Testing the smart contract in a local environment using command-line interface

6.5 Setting Up of a Rinkeby Node

Before we can deploy our network on the main net, a smart contract must be deployed on Ethereum's test network. The smart contract was deployed on the Rinkeby test network. But before deploying the smart contract on the Rinkeby, the computer needs to be a node of the network, i.e., it needs to be connected to the Rinkeby network. Therefore, the computer must download all the blockchain data and state entries of the network and continue to update the latest data and entries as the blockchain is constantly growing. This process is called syncing. Once syncing is complete, the computer needs to continue to be connected to the network as long as it wants to write to the blockchain. A snapshot of the Rinkeby network can be seen in Fig. 19.6. Since the blockchain is continuously growing, the size of the data increases, and hence the time required to sync the network also increases. In this case, it took nearly 60 h to complete the syncing process. About 125 million state entries were processed and 38GB of data was downloaded. After completing the syncing process, the smart contract was deployed. Figure 19.7 shows the Rinkeby syncing process.

6.6 Requesting Test Ether

All transactions on Ethereum require Ether, the cryptocurrency of Ethereum. Actual Ether needs to be purchased to complete the transactions, but Ethereum provides free Ether that can only be used for test networks. The Rinkeby Ether can only be requested through social media accounts so that malicious users do not get large amounts of Ether to jam the Rinkeby networks through too many transactions. First,

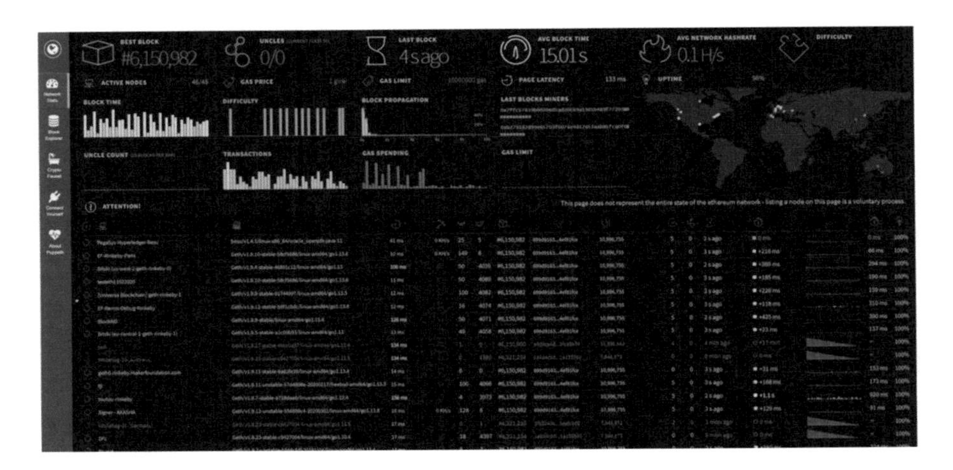

Fig. 19.6 A snapshot of Rinkeby network statistics

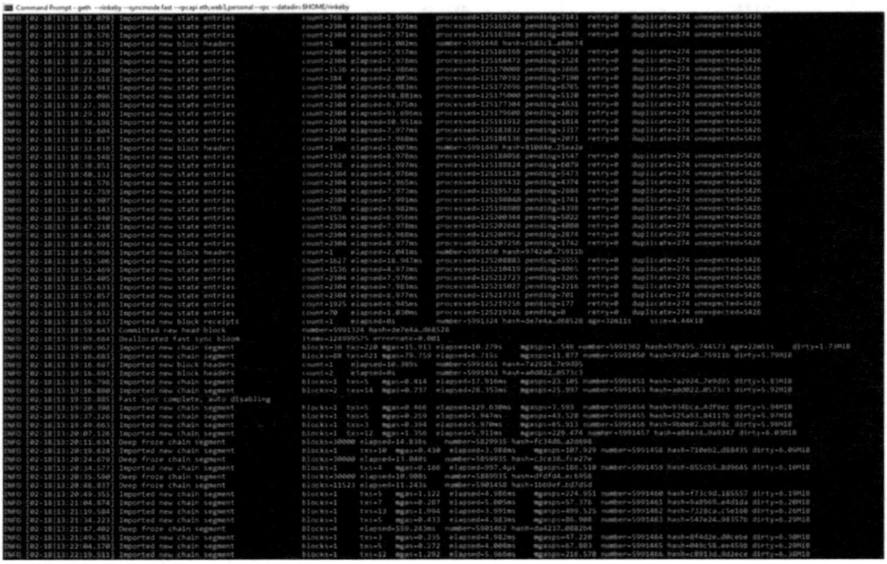

Fig. 19.7 The on-going Rinkeby syncing process

an account was set up using geth. This is a public address to a wallet that can send and receive Ether. The Rinkeby Ether can be requested in two ways: (a) Posting a tweet on Twitter with Ethereum Account Address. (b) Making a public post on Facebook with Ethereum Account Address. The link of either the tweet or the post has to be pasted on the website of Rinkeby Faucet and the Ether can be requested. The same social media account cannot request Ether again for a specified duration.

6.7 Deploying the Smart Contract on Rinkeby

The Truffle configuration file is modified to connect the smart contract to the Rinkeby Test Network. A new Rinkeby network is added to the file by setting the network id parameter to "4," which represents the Rinkeby Network. Using the truffle migrate command, the smart contract is deployed on Rinkeby. In case, the smart contract needs to be deployed on any other network, first, the computer must be in sync with that network, and only the network id needs to be changed in Truffle configuration file. The network ids of different test networks are shown in Table 19.5.

Table 19.5 Network Ids

Network	Network Id
Main net	1
Ropsten	3
Rinkeby	5
Kovan	42

```
Command Prompt
:\Users\ailab4\bda\cli_app>node app.js -a certi1.pdf
;ending hash for file: certi1.pdf
;HA-256 hash value: 0x1aead8f6940c2d90d226f2a4c19f9fb828891869de0f589f931e0235dc66628b
'ransaction ID: 0x5aba3439c9b9650ec60e2b0db8cd6968b476ce0360db1d12af4b6e8bd3d21a85

:\Users\ailab4\bda\cli_app>
```

Fig. 19.8 Adding document hash on Rinkeby

6.8 Testing the Smart Contract on Rinkeby

After deploying the smart contract, it was tested by adding the document hash on the Rinkeby blockchain and finding a document on the same. All transactions can be searched on the Rinkeby Network. Firstly, a document called "certi1.pdf" was used, and its hash was written on the Rinkeby blockchain. The hash value and transaction id can be seen. The command-line interface works in the same manner as it worked while using the Ganache blockchain. The process to add the document may take several seconds because all pending transactions are added into new blocks and each new block is added every 15 s in Rinkeby. In Ethereum main net, each new block takes 10–20 s to be added. Figure 19.8 shows the document being added. Once the new block is generated, all the details related to the transaction of adding the document hash such as block number, transaction id, transaction fee, timestamp, etc. can be publicly viewed. A document hash on the block can be searched on the Rinkeby blockchain. If the document hash was written, its timestamp and block number shall be displayed, else a message is displayed that says that hash value is not found on the blockchain. Figure 19.9 shows the document being searched.

6.9 Designing a Web Application

Blockchain architecture is being used by different organizations for financial transactions, record keeping, digital notary, smart contracts, etc. The end users of these applications need a simple interface to perform the tasks. A web application will be helpful for all the stakeholders of the blockchain model for document authentication and management. The command-line tool is replaced by a web application that can

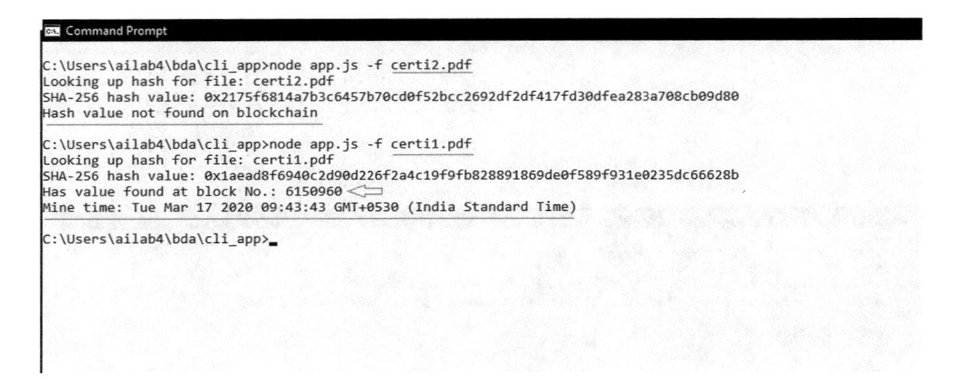

Fig. 19.9 Searching for document hash on Rinkeby

be accessed by other users. HTML-CSS and bootstrap framework is used to create the front end. HTTP-Server was used local server. MetaMask extension needs to be downloaded so that it can inject a web3 provider in the browser. It is available as an extension for Google Chrome. Rinkeby network has to be selected as the network for the wallet so that end users can sign the transactions from their accounts. Multiple accounts can be set up, and the user can deposit or send Ether through this wallet. The javascript file is modified by changing the initialization code. Web3 provider is set so that it can check if a compatible extension such as MetaMask is installed. The smart contract can now be executed using the web browser. If no web3 object is found, the user is alerted through a message. The primary reason for connecting the smart contract with a web application is that it allows users to access our contract without having to run their computer as a node. The user can now access the functions to add and search for hash code on the blockchain from the browser using the web application. The transaction fee to write the document hash will be charged from the MetaMask wallet. The Ether for transactions can be requested from Rinkeby faucet.

6.10 Usability

The decentralized application that has been created is ideal for digital documents that are of permanent nature and require periodic verification such as birth certificates, academic degree and transcripts, purchase receipts, etc. Universities provide a hard copy of academic transcript to their students at graduation. Later, graduates have to carry these hard copies to prove their credentials to employers and other universities. A digital document of degree and transcript would be convenient for students to share their credentials. However, other entities would be skeptical to trust these digital documents as they can be easily modified. In order to verify these digital documents, they have to contact the issuing university and wait for them to validate the degree. Universities have to spend considerable resources for a traditional client–server model to store the digital documents and respond to the entities that request verification.

The proposed decentralized application allows the universities to store the digital document hash on a public blockchain. The students can share the digital document in the normal manner to other entities that can verify the document hash instantly on the blockchain through the application. The need for a database and server is eliminated, and universities do not need to respond to verification requests as it is facilitated by the decentralized application.

The issuing organization has to setup a node to connect to the blockchain network, and it needs to be in sync with it. The verifying organizations have to use MetaMask or similar tools to connect to a decentralized application through Internet. The resources for initial setup are compensated in the longer run as this approach is faster and efficient.

7 Conclusion and Future Scope

A fully functional web application has been created that leverages the blockchain technology to solve the problems in authentication of documents. This solution is cost-effective, scalable, secure, and easy to use. The solution works seamlessly on the Rinkeby test network. An individual can digitally share their issued documents that may be verified by others through the blockchain platform. This solution removes the need for intermediaries and does not rely on a central server, and thus the service has virtually zero downtime. The solution has been thoroughly tested in both local and actual blockchains. Command-line and graphical user interfaces have been developed. The central idea to use document hash codes to authenticate and verify documents can be used in different implementations by changing the underlying blockchain.

There is a scope to update the implementation of this work in the future. Firstly, instead of Rinkeby, it can be deployed on the main net to truly utilize the blockchain properties. However, the cost must be taken into consideration as actual Ether needs to be purchased for the main net. Secondly, the use of private blockchain could be considered if storing the documents is desired and the nature of documents is sensitive. A central administrator is required in this case, who controls the access and grants the permissions.

Acknowledgments This work was done at the campus of INS Valsura under MoU between Nirma University and Indian Navy Represented by INS Valsura. The research behind this chapter would not have been possible without the exceptional support from Nirma University and resources at INS Valsura.

References

1. A. Andrey, C. Petr, Review of existing consensus algorithms blockchain, in *2019 International Conference "Quality Management, Transport and Information Security, Information Technologies" (IT QM IS)* (2019), pp. 124–127
2. A. Badr, L. Rafferty, Q.H. Mahmoud, K. Elgazzar, P.C.K. Hung, A permissioned blockchain-based system for verification of academic records, in *2019 10th IFIP International Conference on New Technologies, Mobility and Security (NTMS)* (2019), pp. 1–5
3. U. Bodkhe, S. Tanwar, K. Parekh, P. Khanpara, S. Tyagi, N. Kumar, M. Alazab, Blockchain for industry 4.0: a comprehensive review. IEEE Access **8**, 79764–79800 (2020)
4. A. Cheddad, J. Condell, K. Curran, P. McKevitt, Combating digital document forgery using new secure information hiding algorithm, in *2008 Third International Conference on Digital Information Management* (2008), pp. 922–924
5. T.M. Fernández-Caramés, P. Fraga-Lamas, A review on the use of blockchain for the internet of things. IEEE Access **6**, 32979–33001 (2018)
6. R. Gupta, S. Tanwar, F. Al-Turjman, P. Italiya, A. Nauman, S.W. Kim, Smart contract privacy protection using AI in cyber-physical systems: tools, techniques and challenges. IEEE Access **8**, 24746–24772 (2020)
7. R. Gupta, S. Tanwar, N. Kumar, S. Tyagi, Blockchain-based security attack resilience schemes for autonomous vehicles in industry 4.0: A systematic review. Comput. Electr. Eng. **86**, 106717 (2020)
8. E. Karataş, Developing Ethereum blockchain-based document verification smart contract for Moodle learning management system. **11**, 399–406 (2018)
9. H.S. Kim, K. Wang, Immutability measure for different blockchain structures, in *2018 IEEE 39th Sarnoff Symposium* (2018), pp. 1–6
10. C. Lakmal, S. Dangalla, C. Herath, C. Wickramarathna, G. Dias, S. Fernando, Idstack – the common protocol for document verification built on digital signatures, in *2017 National Information Technology Conference (NITC)* (2017), pp. 96–99
11. D. Li, W.E. Wong, J. Guo, A survey on blockchain for enterprise using Hyperledger fabric and composer, in *2019 6th International Conference on Dependable Systems and Their Applications (DSA)* (2020), pp. 71–80
12. Y. Liu, D. He, M.S. Obaidat, N. Kumar, M.K. Khan, K.-K. Raymond Choo, Blockchain-based identity management systems: a review. J. Netw. Comput. Appl. **166**, 102731 (2020)
13. G. Malik, K. Parasrampuria, S.P. Reddy, S. Shah, Blockchain based identity verification model, in *2019 International Conference on Vision Towards Emerging Trends in Communication and Networking (ViTECoN)* (2019), pp. 1–6
14. D. Mingxiao, M. Xiaofeng, Z. Zhe, W. Xiangwei, C. Qijun, A review on consensus algorithm of blockchain, in *2017 IEEE International Conference on Systems, Man, and Cybernetics (SMC)* (2017), pp. 2567–2572
15. I. Mistry, S. Tanwar, S. Tyagi, N. Kumar, Blockchain for 5G-enabled IoT for industrial automation: a systematic review, solutions, and challenges. Mech. Syst. Signal Process. **135**, 106382 (2020)
16. S. Mthethwa, N. Dlamini, G. Barbour, Proposing a blockchain-based solution to verify the integrity of hardcopy documents, in *2018 International Conference on Intelligent and Innovative Computing Applications (ICONIC)* (2018), pp. 1–5
17. G. Mwitende, Y. Ye, I. Ali, F. Li, Certificateless authenticated key agreement for blockchain-based WBANs. J. Syst. Architect. **110**, 101777 (2020)
18. R.M. Nadir, Comparative study of permissioned blockchain solutions for enterprises, in *2019 International Conference on Innovative Computing (ICIC)* (2019), pp. 1–6
19. T. Osterland, T. Rose, Model checking smart contracts for Ethereum. Pervas. Mobile Comput. **63**, 101129 (2020)
20. M. Panjwani, M. Jäntti, Data protection security challenges in digital IT services: a case study, in *2017 International Conference on Computer and Applications (ICCA)* (2017), pp. 379–383

21. R. Patel, A. Sethia, S. Patil, Blockchain – future of decentralized systems, in *2018 International Conference on Computing, Power and Communication Technologies (GUCON)* (2018), pp. 369–374
22. R. Rana, R.N. Zaeem, K.S. Barber, An assessment of blockchain identity solutions: minimizing risk and liability of authentication, in *2019 IEEE/WIC/ACM International Conference on Web Intelligence (WI)* (2019), pp. 26–33
23. M. Shah, P. Kumar, Tamper proof birth certificate using blockchain technology. Int. J. Recent Technol. Eng. **7**(5S3), 95–98 (2019)
24. C. Silver, Council post: how the transparency of blockchain drives value, Last assessed on 14 February 2020. (https://www.forbes.com/sites/forbestechcouncil/2020/02/14/how-the-transparency-of-blockchain-drives-value/?sh=6ab3807131a6)

Part V
Next Generation 5G-Enabled IoT for Industrial Automation

Chapter 20
IoT Wearable Devices for Health Issue Monitoring Using 5G Networks' Opportunities and Challenges

Ahmed Ismail, Samir Abdelrazek, and Ibrahim Elhenawy

1 Introduction

There is no doubt that there are many healthcare issues that require technologies and information solutions to make the life of people better and easier [1–3]. Nowadays IoT provides many advantages of using its applications. In healthcare, many technologies and solutions can improve the available solutions such as wearables. In healthcare applications, finding out a suitable solutions based on different factors such as the time interval (latency), battery life, dynamic range, throughput, adaptability, coverage, and organizational model [4, 5] is a challenging issue. There are a lot of applications and health status for remote monitoring available in homes, hospital rooms, and wearable invasive or noninvasive devices which a combination of the short- and long-range communication technologies is necessary [6–9]. Figure 20.1 shows how healthcare devices get the streaming data through inputs from wearable devices and then send the same to the cloud to be analyzed by healthcare services to visualize its overall impact.

The medical devices based on IoT connections provided healthcare with tracking tools to track the health status continuously [4]. Wearables become nowadays more

A. Ismail (✉)
IT Consultant at Nordson, Munich, Germany

Re DI School of Digital Integration Munich, Munich, Germany
e-mail: a.ebada@students.mans.edu.eg

S. Abdelrazek
Information Systems Department, Faculty of Computers, and Information, Mansoura University, Mansoura, Egypt
e-mail: samir.abdelrazek@mans.edu.eg

I. Elhenawy
Faculty of Computer Science and Information Systems, Zagazig University, Zagazig, Egypt
e-mail: Ibrahim.Henawy@zagazig.edu.eg

S. Tanwar (ed.), *Blockchain for 5G-Enabled IoT*,
https://doi.org/10.1007/978-3-030-67490-8_20

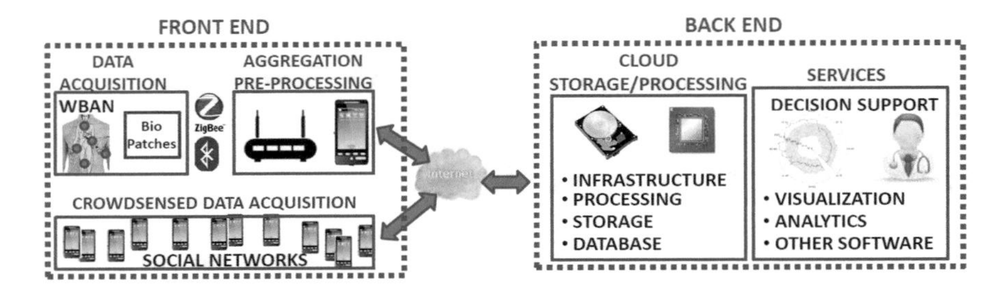

Fig. 20.1 Wearable devices in healthcare solutions [10]

common, and they are used for different objectives in fitness, healthcare, and sports. There is much more interest in employing IoT based on 5G connections and edge computing to be applied in wearable devices' applications [5]. Based on a new study in 2020 that was carried out by the Grand View Research surveys, the healthcare IoT market has considerably reached $330 billion in investments [6].

After the spread of coronavirus known as COVID-19, telemedicine solutions became more popular. Wearable devices are connected to systems to observe patients anywhere and anytime. This technology helped many healthcare providers to provide their services on demand and remotely [7]. The remote health monitoring provided the patients with a new choice to contact their healthcare providers to give them needed care online. These solutions can reduce the time and financial resources allocated at hospitals for senior care and for other patients as well. This technology allows caregivers to help the patients and show them their physical parameters such as blood pressure, oxygen rate, blood pressure, etc. [8]. Most of the available wearables provide a Bluetooth connection with smartphones using iOS or Android and cloud services for edge computing for processing, storage, or analysis [9].

The healthcare devices can send the streaming data to smartphones which can contain mobile applications to collect the data to the cloud or just to show the users the sports activities that impact their calories and health status [10–13].

The accuracy is a very important factor while dealing with the data in the healthcare area. So the target is to minimize false alarms to make the decisions based on that data more accurate. The chapter is presented as follows: Sect. 1 introduces IoT communications for the healthcare area; Sect. 2 introduces solutions for data engineering in healthcare and AI, while Sect. 3 presents edge computing and the structure for healthcare using 5G-connected wearables. Section 4 presents the healthcare data connections. Section 5 presents solutions to collect healthcare data and their applications to make decisions. Section 6 presents edge computing applications in healthcare services. Finally, the conclusions and future solutions are presented in Sect. 7.

2 Healthcare Data-Based Applications

Recently many healthcare providers started using data analysis, data science, and AI to analyze the patient's data to get more insights from the raw data. AI, machine learning, and deep learning provide the healthcare sector many advantages to recognize diseases, predict diseases, and discover the disease state [10]. Wearable devices and medical devices send a lot of health data that can be analyzed and monitored by healthcare providers. The health data coming from the wearables opened a different research area such as biochemistry research, medical engineering, security, commerce, and data science. The medical data analysis is a challenge because there is a need to remove errors, clean the data, and process and analyze the streaming data in a short time [11]. As shown in Fig. 20.2, the data from sensors coming from wearables can be collected to be analyzed to give the healthcare providers particular tools for the healthcare data analysis [14].

Many researchers applied data mining techniques in healthcare on the cloud to use the power of resources in the cloud [12, 15]. Knowledge mining tools are multidisciplinary because it uses machine learning, statistics, and database. We are recently interested in wearable electronic, bio-devices, and bio-patches as a future IT convergence solution. The need for wearable devices and wearable patch systems is gradually increased because they can support user healthcare and mobile environment care with lightweight mobility [13]. This work cultivates the computing platform and information interaction about the wearable bio-patch integrated package. It has also presented the bio-information collection method and bio-data structure of the user through a wearable patch-based biosensing module [14–16].

Wearable bio-patch information system works on gathering bio-information for the wearable patch user as a form available at healthcare and recuperation facilities. In [17], the authors proposed a solution based on edge computing (see Fig. 20.3) to provide more security on the stored data. The data can be transferred to the public cloud only after filtration of what to be kept on the public cloud and what is to be kept on the private cloud at the healthcare provider to keep the data under their control.

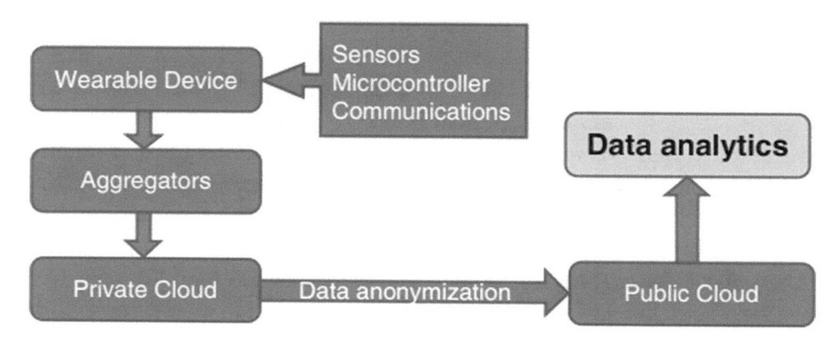

Fig. 20.2 Healthcare data on the cloud

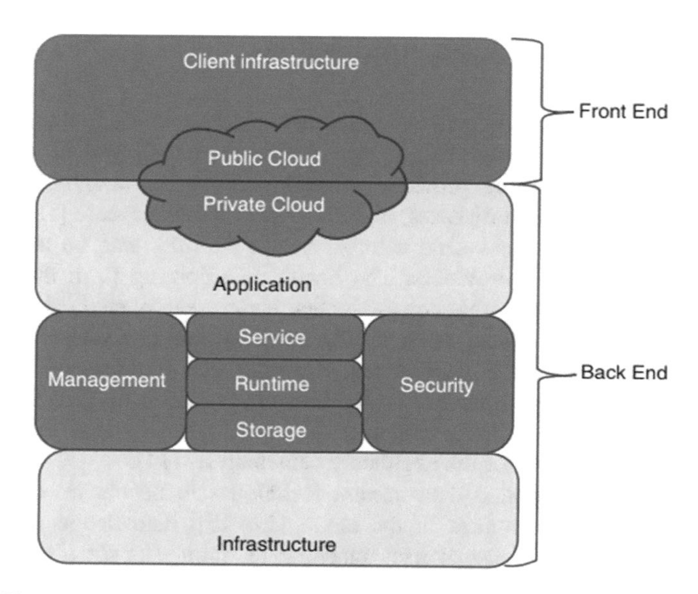

Fig. 20.3 The cloud data structure in healthcare systems

Body condition is associated with an individual's psychological state, and the related cognitive function can control the physiological state. Skin electrical activity may be an indicator to detect the action of the nervous system. Nerve terminal stimulates sweat glands by controlling the physiological activity. This stimulation leads to changes in the conductivity of the skin and is sensed by a physical sensor. To finalize the signal chain of bio-patches and to adjust the bio-patch solution, we consider the signal conditioning impact on the overall power efficiency calculations.

3 Applying 5G in Healthcare Solutions

For modeling of health condition, we define the brink value function and therefore the classified values to classify satisfactory, warning, danger, etc. [18–20]. The health condition management process of biosensing devices should be defined for every user. We define the identification information and also the status information for the digital patch system management [19].

The statistical analysis and evaluation model for analyzing and evaluating the individual user should be defined. Thirdly, they'll configure the knowledge guide process and therefore the control process for wearable patch device users [20]. We can design the data display interface for showing health analysis information of a user. It can show the outbreak of danger and therefore the warning signals on the specific user. It can display a real-time health information analysis, respiration, and health-related signals, and also the physical status analysis information. Besides, it can set up a bio-information device using the learning statistical model of every individual user [21–23]. Finally, they will define the wearable patch device system and wearable system convergence development process. The definition of the data table and process on the bio-patch device development support is required. It presents the definition and guides the important convergence points [24–26]. The structure and components of such a wearable patch system closely interact with the bio-information sensing cycle described in the next chapter. The digital patch bio-information cycle standard model should be proposed and continuously enhanced more and more [27]. A bio-information execution environment for the bio-patch sensing process is provided. More specifically, the information structure for patch bio-information systems is required for more systemized data management.

In the military applications as shown in Fig. 20.4, there are many applications that require a safe, high, and excellent connection. Applications such as wearable computing in defense automation systems require a high-speed network with low

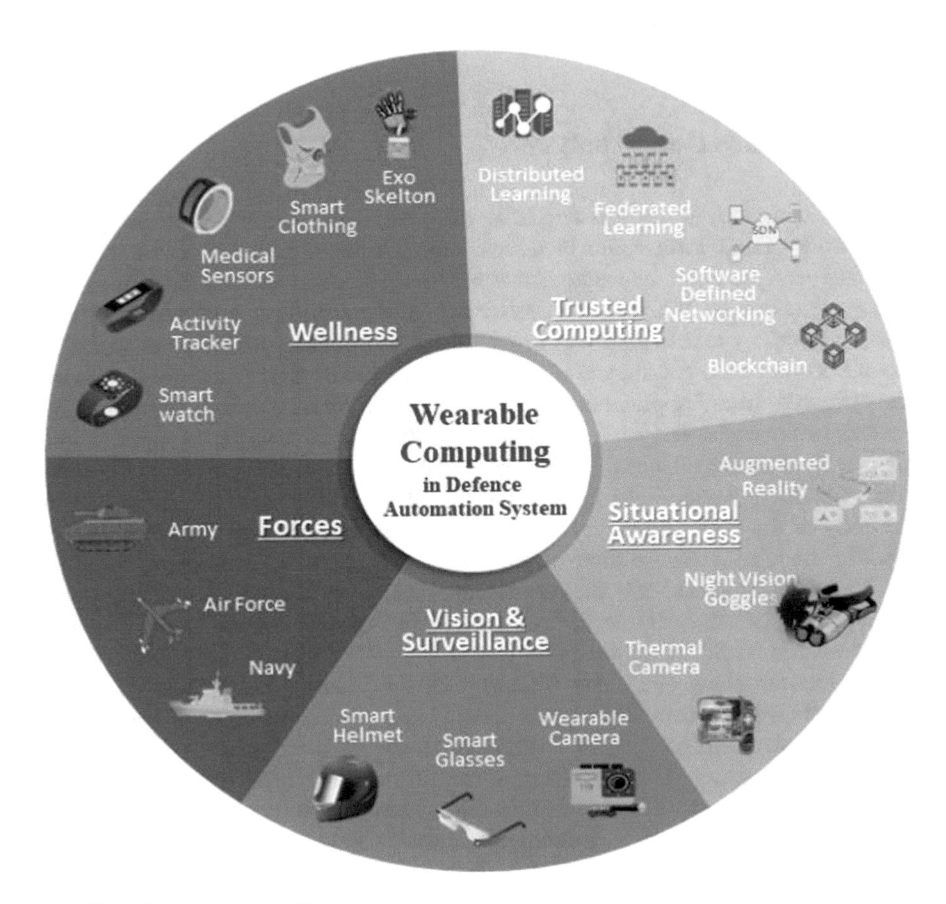

Fig. 20.4 5G applications [10]

latency to keep the data synchronized to the server. There are many connected devices in this area to follow up with the changes in this dangerous environment to watch the environment around the user and keep tracking their health issues, such as smart glasses, smart helmet, and smartwatches. Medical wearables, smart clothing, cameras, and augmented reality becomes essential for the military applications to help them to perform their normal operation in the war situations.

4 The Connection Techniques for Healthcare Solutions

The fifth generation of the cellular network (5G) aims to have ubiquitous communication anytime and anywhere between anyone and anything. In this section, the most important design factors of the IoT paradigm that will significantly affect the performance in terms of security, energy efficiency, and quality of service (QoS) are discussed [29].

5 Healthcare Data Analysis

Healthcare data is useless when it is only raw data without any analysis or transforming it into information because a large amount of data cannot give the users or healthcare providers any output that can be used as a decision [30]. Wearable devices data can be analyzed with many available methods to be presented to the users [31]. Many research studies proved the use of ambient displays of wearable devices to make the parameters more understandable by users more than showing some figures. Many applications work on analyzing data coming from wearable devices or smartwatches such as UbiFit or Fitbit. Those smartphone applications include an ambient, glanceable display to show the physical activity of the users who wear these devices. Health data parameters are collected from biosensors on the body of the users to analyze the changes in parameters during their activities. Most of the users who use these applications in fitness or sport tracking activities use these applications to help them to achieve their goals in training or doing sports.

Analyzing smart wearable devices depends on the application or the target of the applications. Some applications target the motivation of the trainer, so the application can show every day the target of the training to be achieved. Other applications work as a reminder to the patients to take their medicine on time. Other companies provided solutions to detect heart attacks or blood pressure trouble.

Nowadays many people use smart wearable devices to set up their goals and follow a plan to improve their lifestyle. The records of the wearable devices (see Fig. 20.5) can help the users to encourage them to do better on the following day. By using a diary with the training times, sleep rates, and health parameters, the users can get more insights about their lifestyle [28]. Many users of these applications find

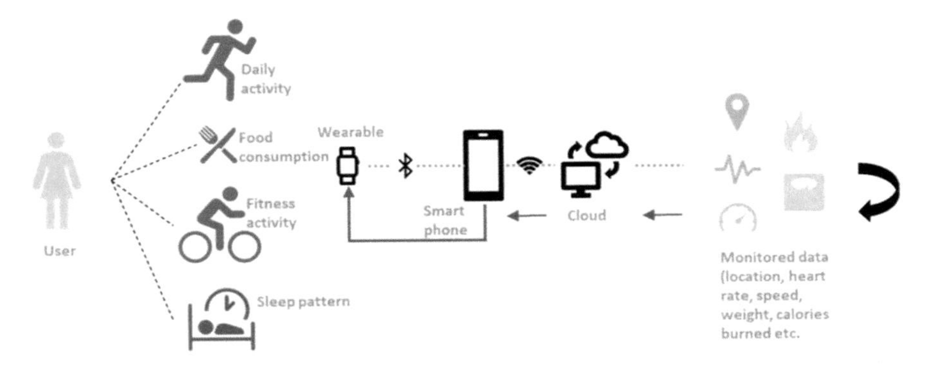

Fig. 20.5 Getting the meaning of lifestyle data

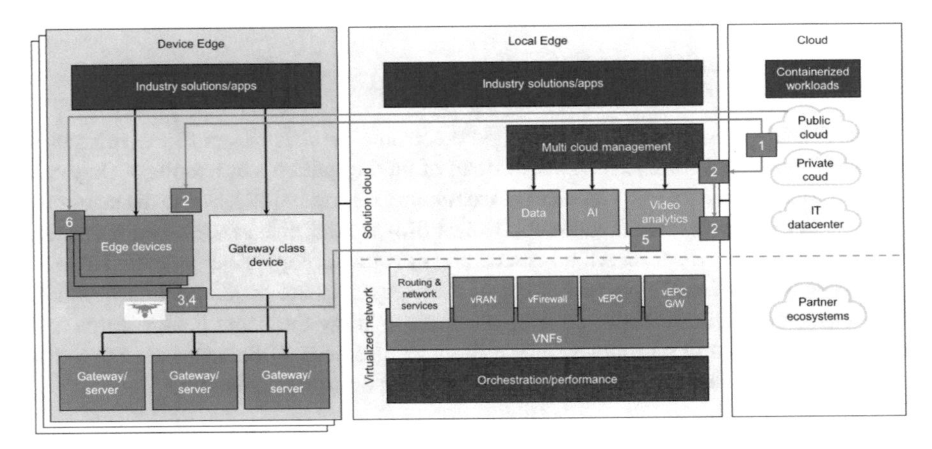

Fig. 20.6 The edge computing architecture

that these applications and the analysis of their activities can make an impact on their lifestyle or can recommend them to see a doctor when it is needed.

6 Edge Computing in the Healthcare

5G networks provide effective network speed better ten times than 4G. Edge computing architecture (see Fig. 20.6) provides an attractive solution to keep the data on local storage and then transfer the data to the cloud based on user preferences. Edge computing can be applied with wearable computing technologies for ensuring the provision of healthcare service for patients at home, at work, or on the move without having to see doctors directly in person [29]. Wearable bio-patch healthcare service technology involves more advanced models than using advanced filtration to share only specific data with healthcare providers. Wearable Patch-Based Bio-information Systems.

Wearable healthcare service draws attention recently, where patients can receive medical results via the healthcare service system based on bio-data collected by a range of medical sensors and devices in living space. As a rule, wearable healthcare service technology is used for ongoing monitoring in daily routines of high-risk groups including pregnant women and patients with chronic diseases and cardiac disorders to determine and predict diseases proactively as part of emergency services. Given the presence of diverse services involving wearable healthcare information, system authority, efficiency, and security issues are considered important. This is a highly sensitive part such as a heartbeat because of patients' life and safety, in which sense the assessment of security and stability is of great importance.

7 Remote Healthcare Services

IoT healthcare solutions allow connected electronic devices to send/receive healthcare data through wearable devices. Instead of having patients at hospitals, they can be monitored at homes and can receive treatments remotely [32]. Some diseases can be tracked using wearable devices such as cardiovascular diseases, diabetes, cancer, and chronic kidney disease. All advanced and new technologies can be called also as mobile health (mHealth) because most of these services are based on smartphones, notebooks, and tablets. These devices are used to show the users health status and their important health parameters as per their requirement. Other services can assist the elderly when they are alone or when they need an immediate care.

8 Conclusion and Future Work

This chapter proposed a wearable biotechnology platform based on 5G networks to show the bio-information methods and biosensing platforms. Such a wearable biosensing cycle and platform configuration carry a plus of supporting the adaptability followed by a change of assorted healthcare system environments like user environment, bio-information gathering type and method, etc. without the invalid or unnecessary system configuration of wearable computing resources. The wearable bio-patch cycle has shown that the bio-information flow supports the bio-information system architecture.

Healthcare solutions based on IoT, AI, and data science can improve healthcare services. Healthcare services based on new technologies can improve the quality of life and provide seniors with a better way to get better healthcare services on demand. Applying new 5G in the future for healthcare applications and wearable devices can solve many issues such as connection issues and can provide more insights from data using artificial intelligence solutions.

References

1. World Health Organization. Available online: http://www.who.int/en/. Accessed on 9 March 2020
2. J. Joonyoung, H. Kiryong, L. Jeonwoo, K. Youngsung, K. Daeyoung, Wireless body area network in a ubiquitous healthcare system for physiological signal monitoring and health consulting. J. Image Process **1**(1), 47–54 (2008)
3. T. Ngoc, D. Phan, Human activities recognition in android smartphone using support vector machine, in *IEEE*, (2016)
4. T. Volkel, R. Kuhn, G. Weber, Mobility impaired pedestrians are not cars: requirements for the annotation of geographical data, in *LNCS*, ed. by K. Miesenberger, J. Klaus, W. Zagler, A. I. Karshmer, vol. 5105, (Springer, Heidelberg, 2008), pp. 1085–1092
5. M. Aziz, T. Owens, U. Zaman, Received signal strength Indicator based localization of Bluetooth devices using trilateration: An improved method for the visually impaired people: World academy of science, engineering and technology, open science index 146. International Journal of Electronics and Communication Engineering **13**(2), 88–93 (2019)
6. K. Loh, D. Ryu, B. Lee, *Bio-Inspired Sensors for Structural Health Monitoring* (Springer International Publishing Switzerland, 2015)
7. G. Honan, A. Page, O. Kocabas, T. Soyata, B. Kantarci, Internet-of-everything oriented implementation of secure digital health (D-health) systems, in *IEEE Symposium on Computers and Communication (ISCC)*, (2016)
8. A. Page et al., Conceptualizing a real-time remote cardiac health monitoring system, in *Medical Imaging: Concepts, Methodologies, Tools, and Applications, Edited by Information Resources Management Association*, (IGI Global, 2017), pp. 160–193. https://doi.org/10.4018/978-1-5225-0571-6.ch007
9. A. Ismail, A. Shehab, L. Osman, M. Elhoseny, I. El-Henawy, Quantified self-using IoT wearable devices, in *International Conference on Advanced Intelligent Systems and Informatics*, (Springer, 2017), pp. 820–831
10. Z. Ma, S. Li, H. Wang, W. Cheng, Y. Li, L. Pan, Y. Shi, Advanced electronic skin devices for healthcare applications. J. Mater. Chem. B **7**(2), 173–197 (2019)
11. P. Mukherjee, A. Mukherjee, Advanced processing techniques and secure architecture for sensor networks in ubiquitous healthcare systems, in *Sensors for Health Monitoring*, (2019), pp. 3–29
12. T. Adhikary, A. Jana, A. Chakrabarty, S. Jana, The Internet of Things (IoT) augmentation in healthcare: an application Analytics, in *International Conference on Intelligent Computing and Communication Technologies*, (Springer, 2019), pp. 576–583
13. R. Ferrero, M. Rebaudengo, F. Rosique, Personal assistance and monitoring devices applications, in *Advances in Human-Computer Interaction*, vol. 2019., Article ID 8916796, 2 pages, (2019). https://doi.org/10.1155/2019/8916796
14. J. Winter, E. Davidson, Governance of artificial intelligence and personal health information. Digital Policy, Regulation and Governance (DPRG) **21**(3), 280–290. Special issue on "Artificial Intelligence: Beyond the hype?" (2019). https://doi.org/10.1108/DPRG-08-2018-0048
15. N. ElAboudi, L. Benhlima, Big data management for healthcare systems: Architecture, requirements, and implementation. Adv. Bioinforma. **2018**., Article ID 4059018, 10 pages (2018). https://doi.org/10.1155/2018/4059018
16. G. Appelboom, E. Camacho, M. Abraham, S. Bruce, E. Dumont, B. Zacharia, E. Connolly, Smart wearable body sensors for patient self-assessment and monitoring. Arch. Public Health **72**(1), 28 (2014)
17. A. Witte, R. Zarnekow, Transforming personal healthcare through technology-a systematic literature review of wearable sensors for medical application, in *In Proceedings of the 52nd Hawaii International Conference on System Sciences*, (2019)
18. J. José, M. Farrajota, J.M. Rodrues, J.M.H. du Buf, The SmartVision local navigation aid for blind and visually impaired persons. Int. J. Digit. Content Technol. Appl. **5**, 362–375 (2011)

19. A.J. Ramadhan, Wearable smart system for visually impaired people. Sensors **18**, 843 (2018). https://doi.org/10.3390/s18030843
20. J. Ding, C. Yen, O. Cheng, A method to integrate GMM, SVM, and DTW for speaker recognition. Int. J. Eng. Technol. Inno **4**, 1–38 (2014)
21. N.A. Giudice, W.E. Whalen, T.H. Riehle, S.M. Anderson, S.A. Doore, Evaluation of an accessible, real-time, and infrastructure-free indoor navigation system by users who are blind in the Mall of America. Journal of Visual Impairment & Blindness. **113**(2), 140–155 (2019). https://doi.org/10.1177/0145482X19840918
22. M. McMullan, Patients using the Internet to obtain health information: How this affects the patient health professional relationship. Patient Educ. Couns. **63**(1), 24–28 (2006)
23. Y. Mittal, S. Sharma, S.P. Toshniwal, D. Singhal, R. Gupta, V. Mittal, A voice-controlled multi-functional smart home automation system, in *Proceedings of the Annual IEEE India Conference (INDICON)*, (New Delhi, India, 2015), pp. 17–20
24. A. Ismail, A. Shehab, I. El-Henawy, Healthcare analysis in smart big data analytics: reviews, challenges, and recommendations, in *Security in Smart Cities: Models, Applications, and Challenges*, (Springer, Cham, 2019), pp. 27–45
25. S. Tyagi, N. Kumar, *Multimedia Big Data Computing for IoT Applications: Concepts, Paradigms, and Solutions, Intelligent Systems Reference Library* (Springer Nature Singapore Pte Ltd, Singapore, 2019), pp. 1–425
26. R. Gupta, S. Tanwar, F. Al-Turjman, P. Italiya, A. Nauman, S.W. Kim, Smart contract privacy protection using AI in cyber-physical systems: tools, techniques and challenges. IEEE Access **8**, 24746–24772 (2020)
27. Q. Bhatia, P. Patel, A. Kumari, P.K. Singh, W.C. Hong, Machine learning adoption in Blockchain-based smart applications: the challenges, and a way forward. IEEE Access **8**, 474–488 (2020)
28. J. Vora, S. Kanriya, S. Tyagi, N. Kumar, V. Sharma, I. You, Human arthritis analysis in fog computing environment using Bayesian network classifier and thread protocol. IEEE Consum. Electron. Mag. **9**(1), 88–94 (2020)
29. S. Tanwar, K. Parekh, R. Evans, Blockchain-based electronic healthcare record system for healthcare 4.0 applications. J. Inform. Secur. Appl. **50**, 1–14 (2019)
30. J. Hathaliya, S. Tanwar, An exhaustive survey on security and privacy issues in Healthcare 4.0. Comput. Commun. **153**, 311–335 (2020)
31. J. Hathaliya, S. Tanwar, R. Evans, Securing electronic healthcare records: a mobile-based biometric authentication approach. J. Inform. Secur. Appl. **53**, 1–14 (2020)
32. J. Vora, S. Tanwar, S. Tyagi, N. Kumar, J.P.C. Rodrigues, Home-based exercise system for patients using IoT enabled smart speaker, in *IEEE 19th International Conference on e-Health Networking, Applications and Services (Healthcom-2017)*, (Dalian University, Dalian, China, 12–15 October 2017), pp. 1–6

Chapter 21
Blockchain and 5G-Enabled Industrial Internet of Things: Application-Specific Analysis

D. K. Sreekantha, R. V. Kulkarni, and Xiao-Zhi Gao

1 Introduction

This chapter has discussed the intelligent and autonomous IIoT systems designed using AI, blockchain and 5G technologies. IIoT systems are the backbone in building a smart city, smart business and smart industries. The survey and analysis of IIoT market trends, methods, technology components and applications in different industries are studied and presented in five following sub-sections.

1.1 Motivation: International Status and Market Potential

There is a huge demand for the development of intelligent systems using disruptive technologies such as AI, IIoT, 5G and blockchain. These intelligent systems are going to create an immense impact by automating business and industry operations, enhancing productivity and reducing the cost. The report from IndustryARC Analysis and Expert Insights predicted that market size for IIoT will reach US$ 123.89 billion by the year 2021. IoT market has an economic value of US$ 11 trillion by 2025. The global blockchain market is expected to reach US$ 23.3

D. K. Sreekantha (✉)
NMAM Institute of Technology, Nitte, Karnataka, India
e-mail: sreekantha@nitte.edu.in

R. V. Kulkarni
CSIBER, Kolhapur, Maharashtra, India
e-mail: drrvkulkarni@siberindia.edu.in

X.-Z. Gao
University of Eastern Finland, Kuopio, Finland
e-mail: xiao-zhi.gao@uef.fi

S. Tanwar (ed.), *Blockchain for 5G-Enabled IoT*,
https://doi.org/10.1007/978-3-030-67490-8_21

billion by 2023. A report estimated about 64 billion IoT-connected components exists worldwide by 2025. At present Europe and Germany are implementing many blockchain-based pilot projects. The world's first peer-to-peer imperativeness trade occurred in New York City in 2016. In Japan, Marubeni company is conducting Bitcoin business for its customers. In China, Wanxiang company was expected to invest US$ 30 billion in a blockchain-driven smart city project. Power Ledger, a start-up company in Perth of Australia, is working on blockchain-based projects.

1.2 Overview of Blockchain, 5G and IIoT Technologies

The objective of blockchain technology is to create a trusted group in computing systems for addressing the challenges such as security, scalability, mutual trust and collaboration. Blockchain technology applies the concepts of mathematics, peer-to-peer networking, diverse algorithms and cryptography to resolve the problems of synchronization and security in open distributed databases. The time-stamps of transactions are recorded in the blockchain network nodes to provide additional security for avoiding any data infringement in transactions.

5G technology enables high-speed communication between billions of wearable devices, Androids, iPhones, tablets and IoT devices. These devices are executing a large number of diverse applications and also generate huge network traffic. 5G technology can manage this huge network traffic at the lightning speeds and prevent network congestion. 5G communication has many advantages of higher throughput, high reliability, low latency and decreased costs of operations. 5G can be customized for lower latency and higher bandwidth for edge computing devices. IIoT processes promote self-organizing and self-healing capabilities. The deployment of 5G networks exponentially boosts the amount of data generated with enhanced interactions among the users and devices. 5G enabled systems are designed to capture data close to the edge devices with lower latency.

Machine-to-machine (M2M) communication has become an important part of IIoT. Fog computing systems are deployed on diverse hardware and software platforms, process huge amounts of data locally, work on-premise and are completely portable. The IIoT systems permit businesses and industries to gather a huge quantity of data to enhance the efficiency of operations. Assuring security at the edge devices in IIoT systems is a very difficult task. Intelligent data processing at the edge devices in IIoT systems delivers smart and cost-effective cloud services in the production environment.

Blockchain technology would revolutionize the present and future applications in various sectors by decentralized access. Automation of business operations leverages better quality and faster decision-making. Application of blockchain technology prevents the intruders from physically attacking the IIoT devices. The data stored at the nodes of blockchain is immutable for intruders to alter the data.

Hackers would not be able to match the hash value at a specific block level in real time. So blockchain ensures trusted data sharing in IIoT systems.

1.3 Chapter Organization

This chapter is organized into four sections. Section 1 introduces the concepts and subsection 1.1 the motivation, status and market potential of disruptive technologies. Sub Section 1.2 presented overview technologies. The review of recent literature is discussed the second section. Section 2 is further divided into eleven subsections. Each subsection describes papers reviewed from one area of applications. The highlights of each paper are presented in a table format also: Section 2.1, finance; sub section 2.2, business; sub section 2.3, smart cities; sub section 2.4, manufacturing; sub section 2.5, Industry Applications; 2.6 healthcare; sub section 2.7, communication; sub section 2.8, food supply chain; sub section 2.9, cyber security; sub section 2.10, military and civil applications; sub section 2.11, agriculture; and sub section 2.12, research. Comparative Analysis, Section 2.4 Research Issues and Challenges in Implementation Section 4 presents the tool manufacturing industry case study. The section 5 concludes the findings of this chapter. The last section deals with references to cited papers.

1.4 Research Contribution

- Authors have curated papers from highly reputed research journals from IEEE, Elsevier, Science Direct and Springer publications. About 78 papers covering blockchain, 5G and IIoT concepts and applications are reviewed. This study has revealed the various technologies, systems and algorithms applied for IIoT systems across the world. These papers have been further classified industry-wise and presented as a pie chart.
- Presented a taxonomy of concepts applied in 5G, blockchain and IIoT techniques in various business and industries. For each concept in the taxonomy, the existing literature has been mapped to handle several issues.
- A tool manufacturing industry case study is presented to demonstrate the usage of these disruptive techniques.
- The set of curated survey papers are compared with specific parameters such as architecture, applications, open issues, challenges, taxonomy and security to understand the scope of coverage of each survey paper and to understand the research gaps.
- A table of acronyms was developed.

1.5 Taxonomy and Acronyms

This sub section shows the number of papers reviewed industry-wise as pie chart Fig. 21.1. Figure 21.2 shows the taxonomy of various concepts of IIoT technology discussed. Table 21.1 presented all the list of acronyms used in this chapter for easy reference.

Figure 21.3 shows the organization of various sections and topics in the chapter.

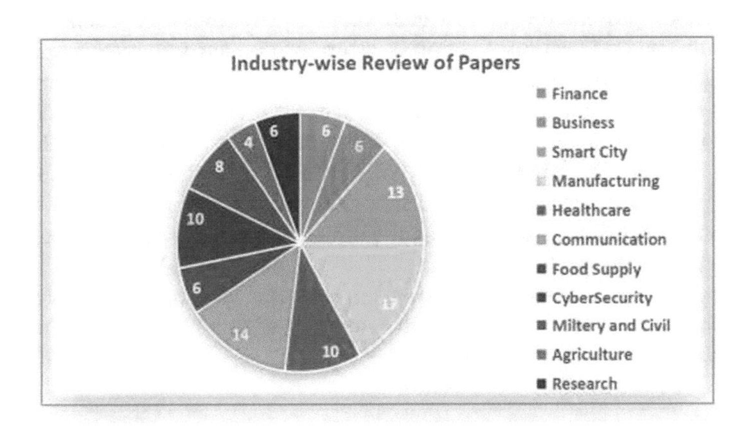

Fig. 21.1 Summary of papers reviewed on applications of 5G, blockchain and IIoT

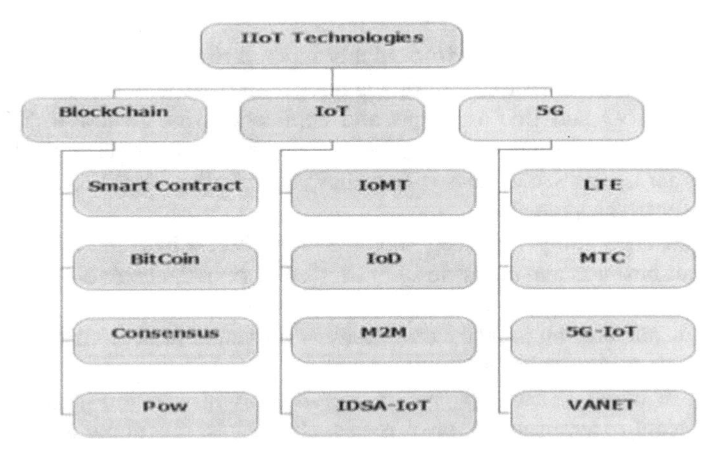

Fig. 21.2 Taxonomy of IIoT technologies and concepts

Table 21.1 List of acronyms used in this chapter (*Acm, Acronym)

Acronym	Description
AI	Artificial intelligence
IoT	Internet of things
IIoT	Industrial internet of things
M2M	Machine-to-machine
PoW	Proof of work
ICT	Information and communications technology
P2P	Peer-to-peer
DER	Distributed energy resource
SHA	Secure hash algorithm
MIMO-UCC	Multiple-input multiple-output unequal concentric chain clustering
IoMT	Internet of medical things
VIM	Virtual infrastructure manager
UDN	Ultra-dense networks
MEC	Multi-access edge computing
RAN	Radio access networks
ML	Machine learning
SUE	Spectrum utilization efficiency
CRN	Cognitive radio networks
CAC	Call admission control
CCN	Content-centric networking
RA	Reference architecture
TFT	Tit-for-tat
SDN	Software-defined networks
PFBT	Practical byzantine fault tolerance
PL	Production logistics
CPS	Cyber-physical system
ABAC	Attribute-based access control
MTC	Machine-type communications
URLLC	Ultra-reliable low-latency communications
LTE	Long-term evolution
5G IIoT	5G-enabled intelligent IoT
QoS	Quality of service
TAM	Technology acceptance model
SEM	Structural equation modelling
IDSA-IoT	Intrusion detection system architecture for IoT
CIA	Confidentiality, integrity, availability
UAV	Unmanned aerial vehicles
IoD	Internet of drones
ECC	Elliptic-curve cryptography
HaBiTs	Blockchain-based secure and flawless inter-operable telesurgery
DSP	Digital smart publication
VANET	Vehicular ad hoc networks
GSBCP	Golden seed breeding cloud platform
BPS	Blockchain-enhanced pub/sub communication
NOMA	Non-orthogonal multiple access
CDETA	Credit-differentiated edge transaction approval
MHT	Merkle hash tree

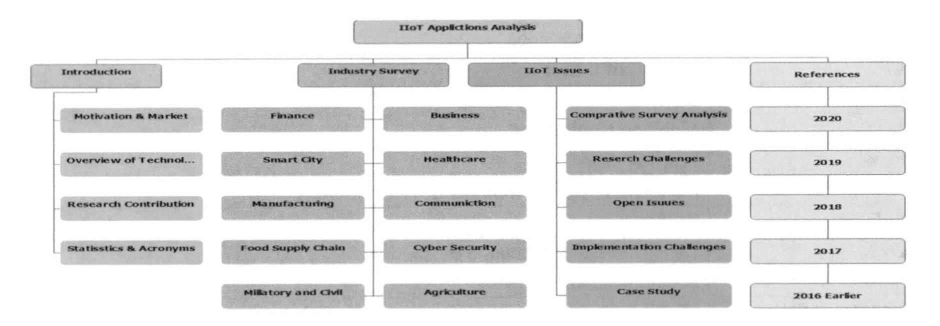

Fig. 21.3 Taxonomy for the organization of the chapter

2 Review of Recent Literature

This section presents the findings and highlights from each of the reviewed paper organized sector-wise.

2.1 Finance Sector

Satoshi Nakamoto conceived electronic cash system because of the financial slowdown in the year 2008 as an alternative to physical currency. The blockchain technology was designed to introduce electronic currency or crypto-currency system called Bitcoin. The objective was to create a secure and decentralized digital currency that can be used as a medium of exchange of currency across the globe. Authors discovered that the unique features of blockchain such as immutable, decentralized, peer-to-peer, open ledger system have shown considerable influence on every industry in the recent decade. The source code of Bitcoin was released as an open-source, leveraging developers and researchers to use this code for developing diverse applications. Researchers began to realize the potential of blockchain and started exploring this blockchain technology applications in other industries also. Blockchain is the underlying technology for the revolutionary crypto-currency and Bitcoin systems [1].

The history and the applications of Bitcoin using blockchain concept are presented. Authors explained the working principles of Bitcoin and legal issues involved in Bitcoin transactions. The real-time applications and challenges in implementing Bitcoin are also covered [2]. The blockchain technology ensures trust in processing transactions by implementing smart contracts, consensus mechanism and avoids the third-party centralized controls. The reduced complexity in solving proof of work (PoW) puzzle also optimizes blockchain platform performance. Decentralization and group trust reduces the cost of transaction processing in

Bitcoin-based solutions. The successful implementation of blockchain in the crypto-currency domain has inspired its extended applications in other domains such as healthcare, banking, industry, etc. The implementation of blockchain technology in IIoT systems also enhances security mechanism [3].

The rapid growth of cyber-attacks on the IoT systems in the Industry 4.0 sector is a major concern. A qualitative and quantitative analysis of cyber-security frameworks, impact assessment on major Industry 4.0 technology and business trends was carried out. Authors described the standards and regulations in the Industry 4.0 and their financial effect on the evaluation methods. The mathematical methods and risk parameters for computing financial loss from cyber-attacks were presented [4].

Table 21.2 presents a summary of highlights from the finance sector review.

2.2 *Business Sector*

An ICT architecture based on blockchain technology for efficient project management was designed. This architecture facilitates the project team to carry out all project-related financial transactions in real time securely. The project execution consists of three types of transactions related to reimbursement of bills, profits and cost saving; these bills are processed using smart contracts. Authors have discussed the proof of concept and implementation of Hyperledger network using IBM® Blockchain Cloud Beta platform. This architecture applies smart contracts for efficient financial management and its automation. Security and privacy are ensured using Hyperledger Fabric in this blockchain architecture [5]. A global crowd-funding platform called BitFund provides an optimum project cost alloca-

Table 21.2 Finance sector review summary

Ref. no.	Methods used	Results	Applications
1	Smart contracts PoW puzzle	Decentralization and group trust. Reduces the cost of transaction processing	Crypto-currency
2	Bitcoin and legal issues	Presented principles of bitcoin and legal issues involved in bitcoin transactions	Implementing bitcoin applications
3	Decentralized peer-to-peer electronic cash system	Secure and decentralized currency	Secure crypto currency Applications
4	The financial effect on evaluation methods of industry 4.0	Cyber-attacks on the IoT systems in the industry 4.0 sector	Industry 4.0 applications

tion. Investors and developers are represented as the nodes of a blockchain. The capitalists would publish the requirements for a specific project or concept by quoting their first bid price. This first bid price includes the cost of development, maintenance and time duration available for the project. The development team analyses the merits and demerits to claim the project implementation responsibility and bids for the project. The investors and developers will have several rounds of discussions to reach a consensus on project implementation. The developers and investors would negotiate and sign a smart contract by expressing their consent on project terms and conditions. The BitFund platform revealed better performance when compared to other similar platforms. This platform would enhance investor's profits and the developer's project execution prospects [6].

IoT networks drive all the smart services in offices, homes, utilities, transport and healthcare in a smart city. These smart services have a great impact on financial and social wellbeing of all citizens. At the same time, these IoT networks are vulnerable to attacks, prone to critical risks, threats to people, industries and businesses. This paper presented different types of attacks on safety, privacy, problems and dangers hidden in hardware and networks of IoT systems. The summary of IoT security algorithms, security attacks and trends in research are presented [7].

Modern enterprises are delivering smart services by deploying IoT ecosystems. These ecosystems comprise six major stakeholders such as network technology developers, software platform developers, application/solution developers, hardware platform developers, users and customers. The enterprise IoT architecture consists of a service management layer, application layer, processing layer, perception layer and network layer. There are operational, analytical and collaborative enterprise IoT systems. There are four features in IoT business models based on networking activities, value proposition, sustainability and resources. Authors developed a case study of an IoT business model for a smart hotel room management. The survey conducted by Business Insider revealed that about 50% of IoT service providers have planned to invest and implement 5G-enabled IoT solutions at the earliest [8].

Blockchain technology was applied to promoting e-tourism. Tourists who are constantly on the move would face payment failures while accessing their account data from the centralized cloud server. This paper proposed a Blockchain-Enabled Smart Tourism and Hospitality Management (BloHosT) system based on distributed technology. BloHosT ensures trust and reputation among different stakeholders in the tourism industry such as airports, taxis, banks, railways, hotel, restaurant and travel agents. BloHosT enables tourist interactions and seem-less integration with all the stakeholders of the tourism industry using a single e-wallet authentication and crypto-currency. BloHosT implements an immutable transaction without demanding any proof of identity while travelling. Tourists will have hassle-free experience using BloHosT system. BloHosT also facilitates smart contracts and browsing through the experience of tourists who have already visited the spots earlier [9].

Table 21.3 Business sector review summary

Ref. no.	Methodology	Results	Applications
5	ICT-based solution architecture for efficient project management	Efficiency in the automation of project transactions ensures security and privacy using Hyperledger fabric with permissioned blockchain	Hyperledger network using IBM® Blockchain cloud Beta Version
6	Global crowd-funding platform	This platform would enhance investor's revenue and also the developer's project assignment prospects	On the line project management
7	IoT drives the framework of all smart entities such as cities, homes, utilities, transport and health	Summary of potential IoT security algorithms and latest IoT security trends in research	Safety and privacy applications of IoT systems
8	Enterprises delivering IoT services by using IoT ecosystem	The survey was conducted by business Insider's points	IoT service providers' solutions
9	Blockchain-enabled smart tourism and hospitality management (BloHosT)	BloHosT enables tourist interactions and seem-less integration with all the stakeholders of the tourism industry using a single e-wallet authentication and crypto-currency	Applications to stakeholders of tourism industry

Table 21.3 shows the highlights of the paper reviewed in the business sector.

2.3 *Smart City Applications*

The blockchain technology implementation in smart city solutions ensures security and trust in data exchange. Smart contracts can be incorporated in smart city applications such as financial budget management, environment. Monitoring government, dwelling and mobility planning. Energy generation and management is a significant factor for a smart city to be self-sustainable. The inexhaustible sources of power such as solar and wind energies are to be best exploited in a smart city. Smart contracts with consumers would facilitate only the necessary power to be transmitted to consumer devices based on their current power needs. Smart contracts ensure automated and dynamic management of power supply and pricing policies. Smart contracts also leverage consumers to select the best services offers provided by different power supply companies leading to healthy competition in the energy market. This innovative power distribution system works on the concept of just in time power or power on demand to the consumer devices. Sharp contracts leverage robust, online financial and operational data recording of power delivered to the consumer. Blockchain implementation proposes a new concept called essentialness business, promoting optimum balance in peer-to-peer (P2P) controlled trading. Small-scale distributed energy resource (DER) suppliers can sell their excess power to the energy grid. Consumers who need power can consume power for a price from this energy grid [10].

Smart platform was developed for managing the big data in Hefei smart city project implementation. Blockchain AI and big data technologies are integrated to build a smart platform. Smart city services are developed by tightly integrating disruptive technologies such as IoT, cloud computing, big data, mobile apps and AI. Smart solutions are developed for urban planning, construction, resource management and operations. The smart big platform is the big data source ensuring secure data sharing in smart cities. The smart big platform supports data sharing and intelligent processing across the departments, organizations and enterprises in a smart city. The open distributed ledger technology and intelligent contract structures ensure security in the smart platform for big data [11].

An intelligent cyber-physical system for waste management keeps the records on the usage of the trash bins and assures the proper working of the system even in adverse conditions. This proposed system consists of the database server, blockchain server, embedded system-based clients and android apps. The economic viability and power consumption analysis of this system was carried out. This proposed system was developed and tested. The collected data was stored in a public blockchain server and used to discover a better strategy for garbage collection. Active participation of local people and support from the government policies such as Swachh Bharat Abhiyan are essential for clean city management (Clean India Movement) [12].

This paper appealed for optimum utilization of energy in smart homes. Every smart home shall send the estimated energy requirements to the power control centre. This control centre transmits energy and the operating instructions to smart homes. The two-way communication between the smart home and power control centre makes them vulnerable to security attacks. The smart meters in homes gather energy utilization data and transmit it to the control centre for further processing. This data also contains consumer identification information leading to breach of consumer privacy, if smart meters are attacked. The present procedures ensure users' privacy by centralization, encryption and data aggregation. This paper proposed decentralized processing of smart meter data based on a consortium blockchain. The security analysis proved that this scheme assures users' privacy and promotes confidentiality and unforgeability. The experiments also showed that this scheme has less computing and communication overheads. The SHA2 encryption algorithm was employed to resolve confidentiality and authentication issues [13].

An application to record weather data values such as moisture, pressure, temperature changes and location changes has been developed by integrating IoT and blockchain technology. Authors studied a huge number of IoT devices with low bandwidth for communication and less computing power for better performance of blockchain systems. This weather data was stored in tamper-proof, open, distributed ledgers which can only be accessed by authorized users. The safety and privacy of data was enhanced by using blockchain by the reduction in computation power, for higher throughput, and increased the strength of the key up to 99% [14]. This paper proposes a new central energy balancing multiple-input and multiple-output unequal concentric chain clustering (MIMO-UCC) protocol for energy management for 5G-enabled IoT systems. MIMO-UCC protocol uses the principal vector projection approach, which builds an unequal concentric chain cluster around the base station. This protocol has four characteristics such as creating a useful blockchain topology, energy depletion balancing at heads of the cluster, protecting the consumption of energy among the members of the cluster and curating appropriate communicating interfaces deployed for IoT sensors. The experimental results revealed that this protocol enhanced the lifetime of the network and optimized the energy consumption at the heads of the cluster comparable with UN-LEACH in the 5G environment [15].

The BackCom fundamental principles, architecture and advanced techniques and its applications in the IoT domain were studied. IoT system inter-connects many millions of smart nodes consuming a large amount of energy in networks, challenging from the perspective of consumption of energy. IoT applications in the future are going to implement BackCom technology extensively to provide truly ubiquitous and pervasive computing advantages [16]. The vision of a smart city is to develop intelligent services for connected society using disruptive technologies such as AI, IoT, 5G and blockchain for building novel smart, urban, efficient, scalable and reliable infrastructure. The smart services are provided in the areas of energy, transportation, pollution, healthcare, sanitation, etc. People from different professions, business and classes should be involved for the successful implementation of smart city systems. All the services shall be citizen-oriented for a better quality of life. Most of the smart city projects today are in the primitive phases of implementation. The quality of services in a smart city depends on the network infrastructure and device capabilities [17].

The advanced smart transportation introduces novel opportunities and challenges for vehicular adhoc networks (VANETs). VANET demands improved network efficiency and highly reliable and secure systems to manage trust and privacy. The disruptive 5G technologies are going to have a huge impact on mobile communication systems and wireless services which are ultra-reliable and has minimum latency time advantages. The integration of software-defined network architecture and 5G-VANET enables data collection and control of the network from any part of the world. The observation and control of vehicular traffic can be enabled in real time using IoT services. These technologies would drive novel vehicular security policies. Authors incorporated blockchain technology and SDN-enabled 5G-VANET to provide a secure transportation system. The experimental results revealed that unreliable vehicles or communication could be discovered to assure safe and reliable vehicular 5G-VANET [18]. A study on applications and challenges in machine learning techniques for designing more robust biotechnology-based smart applications was conducted. Authors explored the implementation of various machine learning algorithms for smart applications such as controlling unmanned aerial vehicles, power grid and healthcare systems in a smart city. A case study on energy trading in smart city demonstrating the machine learning applications was discussed [19].

Table 21.4 Smart city application review summary

Ref. no.	Methodology	Results	Applications
10,11	Smart contracts	Automated and dynamic management of supply and pricing policy	Smart environment, smart government, smart dwelling and smart mobility services
12	IoT, cloud computing, big data, mobile apps, and AI	Smart platform for managing the big data	Data sharing across the departments, organizations and enterprises in a smart city for intelligent data processing
13	Cyber-physical system for waste management	Public server, local server and client interface interactions	Swachh Bharat Abhiyan (clean India movement)
14	Optimum usage of energy in a smart home	Less computing and communication overheads	Smart energy meters in smart homes
15	Distributed ledgers which are tamper-proof and can only be accessed by authorized users	Safety and privacy of the data is enhanced by blockchain to reduce computation power, for higher throughput	Weather forecasting
16	Principal vector projection approach	Enhanced the lifetime of the network and optimized the energy consumption at the heads of the cluster	Protocol for energy management for 5G-enabled IoT systems
17	BackCom fundamental principles, architecture and advanced techniques and its applications	BackCom technology provides true ubiquitous and pervasive computing advantages	The management of consumption of energy
18	Using disruptive technologies such as IoT, 5G and blockchain	The vision of a smart city is to develop an intelligent connected society	The smart services are provided in the areas of energy, transportation, pollution Management, etc.
19	The advanced smart transportation introduces novel opportunities and challenges	This design assures a safe and reliable vehicular 5G-VANET	Software-defined network architecture and 5G-VANET

Table 21.4 shows the summary of papers reviewed in smart city applications.

2.4 *Manufacturing Industry*

The blockchain technology application case study in iron and steel enterprise logistic transaction processing was discussed. The enhanced Practical Byzantine Fault Tolerance (PFBT) algorithm was applied to ensure decentralization and security in transactions. Authors described the role of different layers in blockchain architecture to configure the steel logistics and managing resources. This architecture monitors the logistic data sources in the supply chain. The outcome of these experiments revealed that this enhanced PFBT algorithm improves the performance of the logistics chain and also ensures secure and reliable transactions. An architecture for a production logistics (PL) application using IoT and cyber-physical system (CPS) was designed. The experiments conducted on the test-bed revealed that this architecture performed better than other models based on cost, user-friendliness and efficiency factors. This architecture helps an enterprise in promoting the flexibility and efficiency to be successful in a dynamic and competing environment [20].

Authors introduced a real-time platform driven by IoT, 5G and blockchain for measuring the air pollution index. This platform gathers real-time data from the sensors. This data was protected from tampering and forgery by encryption. The cloud storage space was used to store weather data and to measure the index of air pollution in real-time. Weather information services are provided through technology-intensive edge and cloud computing techniques. This research work contributes to the secure sharing of air pollution data and enables us to control air pollution. This platform can be applied to control the environment in a smart factory [21]. A review of the current state-of-the-art smart contract vulnerabilities that affect the applications of AI techniques in designing better smart contracts was carried out. The problems and challenges for implementing AI and designing smart contracts were examined. A case study of retail marketing that integrates AI technique for designing smart contracts to enhance security and privacy was presented [22].

Table 21.5 shows the highlights of this sector.

Table 21.5 Manufacturing industry review summary

Ref.no.	Methodology	Results	Applications
20	PFBT algorithm	Ensures more secure and reliable transactions to speeds up the processes in the steel industry	Iron and steel enterprise logistic transactions processing
	IoT and CPS	Performed better than other models and presented a barcode system which is cost effective, user-friendly and efficient	PL application
21	Measuring the air pollution index	Gathers real-time data from sensors in IoT-enabled 5G networks	This platform can be applied to control the environment in a smart factory
22	AI techniques in designing better smart contracts	Enhances security and privacy	Retail marketing

2.5 *Industrial Applications*

The intelligent data processing can be enabled at the edge devices using blockchain and 5G technologies in IIoT systems to deliver secure and manageable services. Authors have presented a cross-domain sharing and edge resources scheduling policy with a credit-differentiated edge transaction acceptance technique. The outcomes of this technique revealed a significant enhancement in service cost and capabilities of edge devices [23]. IIoT devices are characterized by mobility, constrained resources and operating capabilities in dispersed environments. It is difficult to access and control these devices from a centralized point in real-time IoT deployments. This paper proposed a method to access and control these IoT systems through Fabric IoT. Fabric IoT works on a blockchain framework based on Hyperledger Fabric called attribute-based access control (ABAC). This framework supports three types of contracts; they are access contract (AC), device contract (DC) and policy contract (PC). DC saves the URL of data resources captured by IoT devices and facilitates inquiry processing. Administrative users can manipulate ABAC through PC. Normal users enforce access control through AC. IoT systems are empowered with dynamic access control management and a decentralized, fine-grained platform for the implementation of Fabric IoT. Authors have simulated two sets of experiments. Their outcome revealed that Fabric IoT provides high output per unit time and ensures consistency of data and efficient consensus in distributed real-time systems [24].

The evolution of mobile network generations, pros and cons are analysed. 5G was designed to provide the optimal connectivity for machine-type communications (MTC) and IoT services. The summary of standards, critical and massive MTC and Ultra-Reliable Low-Latency Communications (URLLC) events are covered. 5G supports massive MTC which is an extension of cellular IoT standards Long-Term Evolution (LTE)-MTC and Narrowband (NB) IoT. Both NB-IoT and LTE-MTC together satisfy massive MTC 5G demands [25].

This paper presented the analysis and applications of blockchain Slice Leasing Ledger in future smart factories. The blockchain system deployed using 5G Network Slice Broker technique decreases the time in creating new services. This new concept enables the machines in the manufacturing industry to effectively get the slice required using autonomous and dynamic approaches. Network slice business can be carried out using slice broker method for implementing smart contracts and slice orchestration strategy. The entire procedure can be carried out automatically without any human involvement [26]. Authors discussed the design of a virtualization platform for supporting various functions of IoT systems. An integrated solution for supporting all interactions of IIoT/IoT elements was presented. The analysis of the modes for transmitting a small quantity of data to large distances in IIoT/IoT systems was carried out in Russia. This paper discussed the Russian communication standards for IoT NB-FI and cyber-security issues in IIoT/IoT systems [27].

The blockchain technology was applied in establishing a decentralized vitality commercial centre. A case study of a start-up called Grid Singularity that applied blockchain to investigate "pay-as-you-go" sun-oriented solar power generation systems was discussed. IoT systems and electric vehicles can utilize the solar power generated from housetop of sun-oriented houses. The customers who need power publish their demand for power, and housetop sun-oriented suppliers can supply power. The power supply transactions are recorded using smart meters in distributed ledgers of the blockchain. Houses with sun-oriented housetop would operate like small-scale power-generating business enterprises. Purchasers could consequently contact suppliers stating their capacity needs. By implementing a blockchain-based record-keeping, business houses can enter into smart contracts with housetop suppliers using distributed ledgers [16].

This paper has presented an overview of blockchain, IoT, 5G and their industrial applications. The open problems, threats and limitations in the implementations of these technologies in industries are also discussed [28]. Authors proposed the design of a single virtualization data platform by integrating all the physical data centres to deliver smart services in IoT systems. The type of data services depends on the geographic position, financial and data centre requirements specified by Federation of Russia. An investigation was carried out on wireless networks for transferring a data to distant places based on the applications of IIoT systems. IoT and IIoT systems generate a huge amount of digital data, and hence providing data security to prevent cyber-attacks is a big challenge [29].

5G-enabled intelligent IoT (5G IIoT) manages a huge quantity of data ensuring optimal communication channels. 5G IIoT has three integrated technologies such as reinforcement learning, data mining and deep learning. The results derived from the experiments revealed that 5G IIoT enhances quality of service (QoS) to a large extent. This paper has also provided a new orientation to the IoT domain in which smart systems are integrated into the 5G's communications and IoT [30].

Table 21.6 shows the review summary of the industrial sector.

2.6 Healthcare Services

A novel abstract model for representing the future of mobile services was proposed. This model applies blockchain technology to ensure transparency and authorization in delivering mobile services. The mobile operators are empowered with this model for billing activities, separating it from network management functionality. Micro-services paradigm integrated with blockchain and 5G infrastructure was explored. Different use-cases of 5G applications such as information transfer, healthcare emergency services and user data privacy management are discussed.

The systems which are not so critical in latency and those that require less throughput can use distributed ledgers in wide area networks [31]. Today's modern lifestyle has been influenced by many wearable devices and wireless body sensor networks. These devices capture data and communicate this data to online service providers to avail their services. Authors presented the use cases of implants to monitor the heartbeat of the patient, transponders with bio-chips and electric clams

Table 21.6 Industrial application review summary

Ref. no.	Methodology	Results	Applications
23	Credit-differentiated edge transaction approval	Significant enhancement in service cost and capabilities of edge devices	IIoT domain
24	Fabric IoT works on Hyperledger fabric blockchain framework and ABAC	This system supports three kinds of smart contracts; they are DC, PC and AC	Control from the centralized point in real-time IoT deployments
25	MTC and IoT services	5G supports massive MTC which is an extension of cellular IoT standards LTE-MTC	Satisfy massive MTC 5G demands
26	5G network slice broker in a blockchain system	The process is automated and there is no need for human intervention	Blockchain slice leasing ledger in future factories
27	Single virtualization platform	Transmission of small amounts of data over long distances in IIoT/IoT systems	An integrated solution for supporting all interaction of IIoT/IoT elements
28	Decentralized vitality commercial Centre	Smart contracts with housetop suppliers using distributed ledgers	Pay-as-you-go applications
29	Industrial application's blockchain integrated with 5G and IoT	Blockchain, IoT and 5G	Deals with open problems, threats and limitations in implementations of these technologies in industries
30	Single virtualization platform by integrating and interlinking all the physical data centres to enable all smart services based on IoT systems	IoT and IIoT systems generated a huge amount of digital, and providing data security and preventing cyber-attacks is a big challenge	Integrated solution for interfacing all IIoT/IoT components
31	Managing a huge quantity of data using intelligent way ensuring optimal communication channels	5G IIoT	5G IIoT enhances QoS to a large extent

in water bikes with embedded sensors. A case study on food distribution systems was also discussed to predict the shelf-life and dynamic distribution patterns of food using IoT technology [32]. The e-healthcare applications powered by 5G and IoT are catching up more attention of the public day by day. The data shared between doctor and patient should be protected from misuse and tampering. There is a need to secure and encrypt medical data which is exchanged between patients and doctors using advanced cryptographic methods. Authors presented the cryptographic techniques such as quantum cryptography, ECC, block cipher

PRESENT and stream cipher Espresso and compared these techniques based on performance and security analysis [33].

Patient healthcare data is the core component of the healthcare service industry. This sensitive data has to be managed with high security as a statutory and ethical requirement. The healthcare industry is influenced by the developments in medical equipment, hardware, software, network infrastructure and emerging technologies such as AI, IoT, 5G and blockchain. The systems built with these technologies assure efficiency, security and transparency in managing the transactions and also open up many new opportunities in the healthcare ecosystem. Sharing of medical data across the stakeholders of the healthcare domain would enhance the accuracy of diagnosis, quality of services and compliance with regulations. The blockchain applications in the healthcare sector would increase data security management. The healthcare data is very sensitive and demands online processing, in case of emergencies, so the selection of correct consensus algorithm, the deployment platform and the type of blockchain has to be carried out with utmost care. The implementation of hybrid clouds, Internet of Medical Things (IoMT) and blockchain technologies are essential for building a robust healthcare system [34].

Technology acceptance model (TAM) discovers the impact of advanced technology in the healthcare services industry. The behavioural attributes (privacy concerns and trust) and cognitive factors (perceived usefulness and perceived ease of use) of patients are studied to assess the acceptance of advanced technology in healthcare services delivery. The objective of this paper is to discover the associations in estimating patients' acceptance of technologies in healthcare services. This model was tested on 416 patients who are randomly selected from primary healthcare centres at different locations in Chennai and Delhi in India. Structural equation modelling (SEM) was applied to conceive the model and to evaluate nine hypotheses covering key constructs. The results revealed that perceived value, perceived user-friendliness, trust and privacy concern are the primary predictors of patient's behaviour to evaluate technology acceptance in healthcare services [35, 36]. The emerging technologies are having a tremendous impact on the healthcare industry and patient care. Healthcare 4.0 standards are enforcing highly technology-driven solutions for patient-centric healthcare. These technical solutions are assisting the medical professionals and patients very well, but these solutions cannot substitute the human touch and relationships between doctor, nurse and patient.

Too much dependence on these technical tools takes away the human touch. Disruptive technologies such as blockchain systems, talking digital assistants and chatbots cannot replace a kind physician. The kind physician placing his or her hand on a patient's shoulder to assure him a speedy recovery and a busy nurse who spares an extra time to listen to a patient's feelings are more valuable to a patient. These modern tools would reduce the cost of medical care. The ratio of nurse to patients in many US hospitals is less than the standards, and there is an acute deficiency in physicians, causing less personal attention to patients. These above facts demand the need for technical solutions to answer frequent questions in hospital systems round the clock. There is an urgent need to shift from fee-for-service philosophy into value-based health care philosophy [37].

The demand for IoT systems and mobile apps that interact with and make payment to each other ensuring secure, safe operations and services is estimated to increase soon. For instance, IoMT devices, wearables and mobile apps that make payment for autonomous connected devices, vehicles, public transportation and emergency/disaster services like drone defibrillator, or a drone for the delivery of medicine ordered, or a self-driving ambulance are in great demand. The blockchain-driven open distributed, peer-to-peer ledger, smart devices, drones and vehicles would reduce the cost of operations and eliminate third-party centralized service providers [38]. This paper presented a blockchain-based secure and flawless interoperable telesurgery (HaBiTs). Smart contracts are designed to provide immutability and interoperability. Authors studied the challenges in existing telesurgery framework and presented the ways to overcome using HaBiTs. HaBiTs implements Hyperledger Fabric or consortium blockchain to ensure the security of data related to patients, doctors and medical data. HaBiTs framework eliminates the role of intermedicate agents and facilitates interoperability between blockchains [39].

This paper proposed a physical exercise based smart support system for patients at home. This is an IoT and cloud technology-based system consisting of smart speaker, a speech learning unit and an exercise database at the edge. This system monitors the patient exercises and generates reports. The progress in doing exercises is compared with standards suggested by physiotherapist. These reports are also stored as electronic medical records. This system enables the patient to workout from home to reduce the stress levels and prevents patients personally visiting doctors frequently. This system can be further improved by incorporating AI components to enhance the efficiency of physical-occupational therapy by proper monitoring and record-keeping [40]. Table 21.7 shows the healthcare service review summary.

2.7 Communication Industry

A novel communication architecture to satisfy the needs of 5G mobile networks was presented. This architecture integrates virtual infrastructure manager (VIM), ultra-dense networks (UDN) and multi-access edge computing (MEC) technologies to enable flexibility for radio access (RA) mobile nodes [41]. The customers and devices are striving for network connection and good data transfer rate before 5G implementation. After implementing 5G technology, the performance of the network and the quality of the connection between the devices have been improved significantly. A network may consist of many drones, users and IoT devices where an autonomous operation is possible by moving 5G radio access networks (RAN) with ML [42]. Understanding the impact of RAN technology on the transport layer of the networks is essential for designing the interfaces in the future. Authors evaluated the 5G scenario-based variations of consumer traffic and the extent of utilization of 4G features in real life to meet the targeted performance. Computed 5G transport network data rate requirements considering per-cell and cumulative cases into account. The statistical multiplexing is an important function in reducing

Table 21.7 Healthcare service review summary

Ref. no.	Methodology	Results	Applications
32	Micro-services-based paradigm	Latency-critical or those that can operate with a lower throughput can use distributed ledgers	User data privacy management, healthcare emergency services
33	Wearable devices and wireless body sensor networks for healthcare	Implants to monitor the heart of the patient, transponders with bio-chips	Improved delivery of healthcare
34	Quantum cryptography, ECC, block cipher PRESENT and stream cipher espresso	e-health applications using 5G-based IoT	Serving the patient's with quality healthcare
35	A consensus algorithm, the deployment platform	Efficiency, security and transparency in managing the patient transactions	The implementation of hybrid clouds, IoMT and blockchain
36, 37	TAM to discover the impact of advanced technology in the healthcare service industry	Tested on 416 patients randomly selected at primary healthcare centres at different locations in New Delhi	Perceived value, user-friendliness, trust and privacy concern are primary predictors of patients' behaviour
38	Talking digital assistants and chatbots	Human touch and relationships between doctor, nurse and patient	Kind physician and busy nurse who gives extra time to listen to a patient's feelings can not be replaced
39	IoMT devices, wearables and mobile apps	Reduce the cost of operations and eliminate third-party centralized service providers	Make payment for autonomous connected devices and vehicles
40	HaBiTs	Ensure the security of data related to patients, doctors and medical data	Supports online medical services to patients
41	Physical exercise smart support system for patients	This system monitors the patient exercises and generates reports	Online support for physical-occupational therapy by proper monitoring and record-keeping

the cost and capacity of 5G networks. The procedure for designing a 5G transport network was discussed. Authors introduced new functional modules to prevent additional costs for increasing the transport capacity as demanded by mmWave and massive MIMO technologies [43]. An investigation on the spectrum utilization efficiency (SUE) in cognitive radio networks (CRNs) for assessing the availability of channels and the probability of completion of service was carried out. Spectrum-efficient CRN (SE-CRN) applies a hybrid underlay-interweave (UI) mode of CRNs to secondary users for collaborative communication. Two types of spectrum utilization with high and low levels of priority are taken into account. Multiple attribute-based methods for service discovery for low-priority users' interruption were proposed to ensure the optimum channel accessibility for high-priority users. The outcome derived from this scheme assured significant enhancement in channel accessibility, reduction in no service probability and service retainability [44].

The applications and case studies for 5G-enabled devices are driven by multi-purpose OPC UA. The requirements, design considerations, advantages and role of OPC UA are identified. The elements, patterns and work-flow description for OPC UA in implementing the information modelling and integration of devices are presented [45]. A new Call Admission Control (CAC) scheme uses fuzzy logic and C-RAN preemption technique. The cloud bursting technique was implemented in this scheme for managing congestion. This congestion can be prevented by preempting few delay-tolerant, low-priority user links of public cloud by charging penalty. The results obtained through the simulation experiments revealed less probability of blocking which was only 5%, higher output/unit-time and less energy consumption, and return on investment was up to 95% [46].

A spectrum sharing system was designed using blockchain technology. The users can effectively share the same frequency spectrum in dense networks. Game theory concepts are used to achieve coordination between the customers who are not participating in sharing the spectrum. Simulation experiments were conducted by applying a tit-for-tat (TFT) strategy and blockchain-based spectrum sharing technique. The results revealed that about 55.1% optimization was achieved through spectrum sharing strategy [47]. A well-organized study of trust-based IoT recommendation techniques was carried out. This study reviewed 59 curated papers from 206 papers published during the period 2011 to 2018. The advantages, disadvantages and open issues of IoT were discussed [48].

Authors conducted a review of applications, challenges of IoT and blockchain technologies. Integrating IoT with blockchain technology has a high potential for designing many innovative and robust applications [49]. A general-purpose smart, contract-based IoT monitoring architecture to enhance the security aspects of public blockchains that reduces the operational cost was proposed. This architecture applies reinforcement learning technique which adjusts to dynamic data submission rates of IoT systems. Implementing high-security features using smart contracts of blockchain is a very expensive process due to the fluctuating Bitcoin prices. Authors discussed cost-effective IoT architecture using agents to control the IoT devices through smart contracts. These agents are trained using reinforcement learning and deep Q-network techniques. The results derived from simulated experiments revealed that the agent's strategy is more cost-effective and satisfies all users' requirements [50].

This paper presented a non-orthogonal multiple access (NOMA)-based communication architecture for tactile Internet for smart communities in 5G.

The objective of 5G communication is to ensure high-speed, low-latency and low-power connectivity for billions of IoT devices. The applications of NOMA for real-time applications such as autonomous vehicles and e-healthcare are presented. A case study on e-healthcare has been discussed in this paper [51]. Table 21.8 shows the communications sector review summary.

Table 21.8 Communications sector review summary

Ref. no.	Methodology	Results	Applications
42	UDN, MEC and VIM	Flexibility for RA in mobile nodes	5G mobile networks
43	NOMA	Enhancing the performance moving 5G RAN with ML	Network with many drones, users and IoT devices
44	Impact analysis RAN technology on the transport network	Prevent additional costs for increasing the transport capacity	MmWave and massive MIMO technologies
45	Investigation on the SUE of cognitive radio networks	Substantial improvement in channel availability, network un-serviceability and service retainability	Cognitive radio networks
46	5G-enabled field devices using the versatile OPC UA	Realizing information modelling and device integration phases	5G networks
47	Fuzzy logic-based CAC scheme with C-RAN preemption	This scheme has less probability of blocking which is less than 5%, higher output/unit time and less energy consumption, and return on revenue is up to 95%	5G, C-RAN
48	TFT strategy for spectrum sharing system designed based on blockchain technology	Enhance spectrum sharing by 55.1% through optimization	Same frequency spectrum in dense networks
49	59 curated papers from 206 papers published during the period 2011 to 2018	Advantages, disadvantages and open issues of each IoT recommendation technique were discussed	Trust-based IoT recommendation techniques
50	Review of applications and challenges of blockchain and IoT technologies	A well-organized study of trust-based IoT techniques	Generation of robust application
51	Smart contract-based IoT monitoring architecture	Considerable cost savings based on users' choices	Applications designed using IoT and public blockchain
52	NOMA	Ensure high-speed, low-latency and low-power connectivity for IoT	Real-time applications such as autonomous vehicles

2.8 Food Supply Chain Management

The unique features of 5G and blockchain technology can complement each other in building a robust system. The constraints in blockchain communication are compensated by integrating it with 5G. The 5G technology privacy and security limitations were compensated with blockchain implementation. Authors have researched on a blockchain, 5G, mmWave communication and MIMO technologies which are suitable for networks with large numbers of nodes having security and load balancing issues. The significant growth of blockchain-enabled 5G and IoT technologies has disruptive potential to develop valuable next-generation robust connected systems. The integration of IoT and micro-grid has applications in the supply chains, smart contracts, decentralized autonomous systems, crypto-currency and proof of services [52].

Network coding technique is very cost-effective and trustworthy for future communications. A high data rate in a device-to-device communication is ensured in small cell environment by network coding. The attacks and pollution are major problems in network coding.

Standard cryptographic techniques make it difficult to secure network by re-encoding of packets at the intermediate nodes in the network. This problem is solved using homomorphic message authentication codes and signatures. The 5G wireless technology is required to implement low energy consumption, ultra-low latency and high data rate. The mobile small cells using network coding can play an important role in assuring the maximal output per unit time in network communication. The optimum performance can be ensured by securing these small cells from pollution attacks [53]. Vehicular communications suffer from security aspects such as dedicated network slice and content-centric networking (CCN). The emerging 5G technology and advances in the automotive industry have ensured safer travel and movement of goods. Advanced networking technologies, artificial intelligence and blockchain are integrated into autonomous vehicles to reduce human errors. Reliability, security and stability are the most critical factors in vehicular communications. Authors highlighted the need for researching on consensus protocol designs for dynamic VANET use-cases. The analysis of information in vehicular communication is very helpful to decide the causes for accidents to help insurance companies and police authorities [54]. Table 21.9 shows the food supply chain management review summary.

2.9 Cyber Security

A novel intrusion detection system architecture for IoT (IDSA-IoT) system was implemented at edge devices and public cloud. IDSA-IoT helps in discovering the new attacker's behaviour efficiently. The classification process is carried out at the edge devices, and high-end computing is deployed at public cloud. IDSA

Table 21.9 Food supply chain management review summary

Ref. no.	Methodology	Results	Applications
53	Researched blockchain, 5G, mmWave communication and MIMO technologies	Potential to develop valuable next-generation robust connected systems	Applications in the food supply chain, crypto-currency, proof of services, smart contracts, decentralized autonomous systems
54	Food distribution systems	IoT and wireless sensor networks	Predict shelf-life and dynamic distribution patterns of food using IoT technology
55	Network coding is a very efficient and reliable network	Assuring the maximal throughput over the network	To implement applications requiring high data rate, ultra-low latency and low energy consumption
56	Network slice and CCN	Artificial intelligence and blockchain integrated into autonomous vehicles would reduce human errors	Reliability, security and stability are the most critical factors in vehicular communications

can be used for various kinds of IoT networks by maintaining control on task distribution. The experimental results revealed that the edge devices with low-end resources can perform the classification and detection tasks at the flow rate of 450Kb per second [56]. Content-centric mobile network models are using blockchain technology to solve the privacy problems in 5G environment. This model is based on developing trust between content providers and content consumers mutually. The blockchain ledger ensures transparency, tamper resistance, control and privacy of the transactions. This model would maintain the public ledger efficiently taking the help of selected miners sharing the required data, network delay and congestion to assure green communication. This model also reduces the complexity of managing the keys. Users are empowered to design a micro-level access policy and enhanced control over data. Minors can significantly bring down the congestion level and cut down the delay in the network [57].

Authors surveyed IoT technology and demands in security aspects in IoT systems. The primary focus was on security issues confidentiality, integrity, availability (CIA) and problems in each layer. Authors studied three emerging technologies such as blockchain, AI and ML for solving security problems in IoT applications. Security is a very significant problem in IoT and demands immediate solutions by integrating with emerging technologies [58]. Every company should regularly monitor the probable harmful events such as violation of controls, tampering of data and illegal activities based on reliable information which can be verified easily. These companies need a strong tool for tracing the links between the materials seized and related proof documents. Authors presented tamper-resistant and time-stamped original ledger by utilizing a public pre-existing trusted blockchain that can be tuned to any configuration. This solution ensures many levels of integrity checking and privacy. This solution is also scalable and supports automation and standards, and it is inter-operable [59].

IoT systems are mainly driven by the internet, which is insecure, lot many hackers are threatening to steal the valuable data, suffers from data loss, dis-continuous data transfer and manual handling of devices. Higher computing capability between the objects and extended network connectivity can be achieved in IoT. The sensors and temporary devices would collect, exchange and process data with minimum human intervention. Implementing expensive security features and extending the computational capabilities on the Internet is not a practical solution. Real-time systems generate a huge amount of data which is stored in the cloud and processed using distributed cloud computing methods. Public cloud services are also insecure, and they are also equally vulnerable to cyber-crime attacks such as data tampering, structured query language injection and node failure. Data integrity and data availability cannot be ensured by cloud services [60]. This chapter illustrated security problems related to blockchain technology and presented different encryption methods to ensure safety, security and prevent vulnerable attacks. Authors presented security attacks of blockchain and their influence on the performance and safety of systems. Authors also discussed rapid growth and associated cyber-attacks on the IoT systems in the Industry 4.0 environment. A qualitative and quantitative analysis of cyber-security frameworks, impact assessment models and major Industry 4.0 technology and business trends was carried-out. Authors made efforts for incorporating standards and administration into Industry 4.0 to present an enhanced perception of financial effect evaluation methods of Industry 4.0. Authors discussed the mathematical background for computing financial losses from cyber-attacks. The risk parameter arrays are designed to compute the financial effect of cyber-attacks on regulations and standards [61, 62].

Every person can be uniquely identified by his specimen signature. Signature recognition and verification are very essential in various business and financial activities of a person. The static and dynamic parameters can be analysed while making a signature to identify the person. Authors have analysed the static parameters such as the writing speed, pressure points and curves in signing. Person signature can be verified by both offline and online Processes. In the offline signature verification process, the person is asked to sign on a paper with a pen. This signature is scanned and stored in a secure digital biometric repository. The person current signature is always verified with the stored signature store in biometric repository. In online signature verification, the electronic gadgets are used to capture dynamically all parameters of signature. Initially, the user uses a stylus for making a signature on the electronic screen which is recorded and stored securely. The static signature verification is very useful for verifying documents such as property, legal records, checks, debit cards, credit cards and contracts by comparing signature with a stored digital version of specimen signature. Dynamic signature verification can be applied in e-commerce and retail applications [63]. Table 21.10 shows the review summary for cyber security domain.

Table 21.10 Cyber security review summary

Ref. no.	Methodology	Results	Applications
57	Intrusion detection system architecture	Edge device with reduced resources can perform the classification and detection tasks at the flow rate of 450Kb per second	IoT networks by maintaining control on task distribution and intruder detection systems
58	Content-centric mobile networks	This model empowers users to design a fine-grained access policy to have better control over data	Privacy issues in content-centric mobile network applications
59	Blockchain, AI and ML for solving security problems in IoT applications	Surveyed IoT security	CIA
60	Monitor the probable harmful events	Strong tool for tracing	Tamper-resistant and time-stamped original ledger
61	Preserving the integrity of the data in IoT systems	Data tampering, structured query language injection and node failure	Data integrity and data availability
62	Security problems related to blockchain	Well-organized literature survey	Helps us to build more safe and robust blockchain systems
63	Cyber-attacks on the IoT systems in the industry 4.0 sector	Well-organized literature on security aspects of blockchain	The risk parameter arrays to compute the financial effect of cyber-attacks on regulations
64	Static and dynamic parameters signature verification	The static signature verification is very useful for verifying documents such as property, legal records, checks, debit cards and credit cards	Dynamic signature verification can be applied in e-commerce and retail applications

2.10 Military and Civil Applications

Unmanned aerial vehicles (UAV) or drones are extensively used in defence as well as civilian applications today because of its cost-effective and efficient operations. The drones are controlled from the control room or moving access points in the sky. The drones can also be connected to the Internet forming a new paradigm called the Internet of Drones (IoD) [55]. Ensuring privacy and security to these Internet of Drones (IoD) on the fly is very difficult and must be handled efficiently. Applying blockchain technologies would be promising in this context due to its unique advantages such as decentralization, immutability and traceability of transactions. The capability to prevent any possible attacks is very essential in the

IoD environment. The implementation of this scheme revealed enhanced security, more functional capabilities and less communication and computation costs when compared to other similar Schemes [64].

The growth in IoT systems facilitates more and more IoT-enabled devices to enter into our daily life, risking our life-sensitive data with more attackers. Authors proposed an ECC (elliptic-curve cryptography) asymmetric algorithm to encode the data. The sensed data is compressed and reconstructed to enhance data storage speed in IoT systems. The results obtained from the experiments revealed that this algorithm is much better when compared to other algorithms [65].

This paper proposed one-way permutation function to reduce the storage on the client side of systems. Authors concentrated on improving the performance of forwarding secure searchable encryption schemes. The state-of-the-art forward private SSE scheme is proposed in CCS 2016 which relies on the asymmetric cryptographic primitive [66]. Authors discussed a method to derive the keys and applied it to plain text blocks to generate the cypher text using keys. The reverse procedure needs to be applied at the receiving end to retrieve the plain text from the cypher text. The application of AES technique has enhanced the randomness in inputs. The security is provided by MAC using irreversible process. CBC concept of blockchain is applied in this method. This method has minimum complexity in architecture and reduces time complexity and vulnerabilities. The analysis of this method ensured the randomness in generated keys [67]. Authors highlighted the unique advantages of blockchain and introduced blockchain-enhanced pub/sub communication model (BPS). BPS is decentralized for multi-tenant edge cloud, to rebuild pub/sub system internal security methods. BPS exploits blockchain technology to detect prohibited operations and actions from both vindictive tenants and unreliable publishers or subscribers. BPS applies Merkle hash tree (MHT) directly to verify the integrity of vital and private meta-data. BPS presents smart contract-based micro-level control on content-based classified messages by saving access control list. BPS also adopts an incentive method for sincere publishers and subscribers. Authors have implemented the BPS model on Kafka and EoS blockchain. The rigorous analysis and experiments conducted revealed that BPS performs better with minimum cost compared to other models [68].

Authors have carried out an extensive review of literature on storage systems based on blockchain technology. A comparative study with cloud-based storage systems and consensus protocols was performed. Introduced a new method for data storage to enhance the privacy and security. The problems and demands in data storage domain such as scalability and flexibility in access are discussed. This blockchain-based storage is a new technology and needs to be still explored and researched [20]. Table 21.11 shows the review summary for military and civil applications.

Table 21.11 Military and civilian application review summary

Ref. no.	Methodology	Results	Applications
65	IoD	The proposed scheme can resist several potential attacks that are essential in IoT-enabled IoD environment	Military to civilian applications
66	ECC asymmetric algorithm to encrypt information	The algorithm is superior to other comparable algorithms in security and storage performance	Secure IoT applications
67	Asymmetric cryptographic primitive	Secure searchable encryption schemes	The state-of-the-art forward private SSE scheme
68	CBC concept of blockchain	AES technique	This method has minimum complexity in architecture and reduces time complexity and vulnerabilities
69	BPS model	MHT	BPS performs better with minimum cost compared to other models
70	Cloud-based storage systems and consensus protocols	Blockchain-based storage	Enhance privacy and security

2.11 Agriculture

Authors presented an open distributed tamper-proof ledger method to securely store the data about the fish farm. Fish farm implemented a custom-made version of blockchain using a RESTful interface which is scalable, with a higher rate of output, offline storage and privacy features. The Hyperledger Fabric version of blockchain implementation was applied in the legacy fish farm system, which reduces the risk of a user deliberately inserting malicious code through the smart contract. All the users know each other, and all their actions are saved on the blockchain nodes based on access control rules for a given network and type of transactions. The services offered by Hyperledger Fabric network are evaluated, and transaction processing capability was assessed with various performance metrics. Embedded boards like Raspberry Pi and Arduino are used as a client device to link with the legacy fish farm system and blockchain application [69].

A repute model for consensus protocols is used in the design of IoT systems for ensuring trust. In the repute model, the good conduct would be rewarded, and bad conduct would be punished, to enhance the safety of the protocol. The results from experiments revealed the adequate performance with efficiency and safety of this model [70]. A breeding information management system was developed using enhanced blockchain technology to manage crop breeding data efficiently

Table 21.12 Agriculture sector review summary

Ref. no.	Methodology	Results	Applications
71	Custom-made a version of blockchain using a RESTful interface	Securely store the data about the fish farm	Reduces the risk of a user deliberately inserting malicious code through the smart contract
72	Good conduct was rewarded and bad conduct was punished using repute model	Adequate performance with efficiency and safety	Reputed model for consensus protocols used in IoT systems for ensuring trust in consensus protocols
73	Golden seed breeding cloud platform (GSBCP) for storing breeding data	Proxy encryption technology is applied to ensure data security	Chinese breeding companies and scientific research labs using GSBCP

with higher output per unit time. This paper discussed the Golden Seed Breeding Cloud Platform (GSBCP) which was designed for storing, breeding data efficiently and also to support the faster gathering of breeding data from crops and breeding process. This storage model applies a light blockchain to store the key breeding data. Various types of blockchains are used to store distinct types of breeding data. Proxy encryption technology is applied to ensure data security. Chinese breeding companies and scientific research labs, such as Hunan Longping High Tech Seed Industry Co., Ltd.; Shandong Shengfeng Seed Industry Co., Ltd.; Beidahuang Kenfeng Seed Co., Ltd.; Agricultural Scientific Institute of Yueyang City; Agricultural Academy of Jilin Province; and Agricultural Academy of Henan Province, are using GSBCP [71].

Table 21.12 shows the agriculture sector review summary

2.12 Research Applications

A digital framework called Digital Smart Publication (DSP) based on smart contracts and blockchain can be applied for automating academic research publications. DSP is very secure and also balances the distribution of rewards to all the stakeholders ensuring data integrity and security. The researchers are updated about the benefits whenever their paper is referred to or cited. All the decisions and payments made are transparent, direct in a decentralized manner. DSP ensures that all the stakeholders will be provided with long-lasting benefits in the form of tokens based on predefined conditions. All the incentives to stakeholders are distributed based on their tokens received. The application of smart contracts ensures distribution of tokens [72].

Authors presented the process of evaluating large-scale, multi-tier network's performance using Vienna simulator for 5G systems. This simulation tool works on MATLAB and designed as a separate module that can be accessed on-demand. This simulator can examine any adhoc network and its parameters constellations that can be improved easily. This paper also discussed the unique features, structure and functionality of this simulator in more detail. 5G networks provide enhanced diversity of scenarios, use cases, the large number and combination of parameters used in simulations. Authors designed a simulator for researchers to analyse the significant aspects of multi-tier, heterogeneous networks in future. This simulator source programme code was made available at no cost for academicians to boost progressive participation in this research [73]. A reference architecture (RA) is designed using big data, and machine learning techniques was applied in edge computing systems. This basic edition of RA was implemented on 16 different configurations in edge/distributed computing set-ups. This paper also presented a complete system details for further development and deployment. The main advantages of this RA are reduced development and maintenance costs, minimized risks and enhanced interactions among the stakeholders. The mapping of architectural elements to deployment configurations of different prototypes is discussed [74].

A study on architectures, features, security, efficiency and network problems faced by the systems designed by integrating IoT, 5G and blockchain technology was carried out. IoT has diverse applications in day-to-day life and demands high bandwidth and enhanced connectivity among the devices. Integrating AI with these disruptive technologies enables us to solve some of the issues such as standardization, energy utilization, scalability, regulation and interoperability [75]. Table 21.13 shows research application review summary.

Table 21.13 Research application review summary

Ref. no.	Methodology	Results	Applications
74	DSP	DSP ensures that all the stakeholders will be provided with long-lasting benefits in the form of tokens based on predefined conditions	Smart contracts
75	Vienna simulator for 5G systems	Designed a simulator for researchers to analyse the significant aspects of multi-tier, heterogeneous networks	Evaluating large-scale, multi-tier network's performance
76	Using machine learning techniques for edge computing systems	Reduced development and maintenance costs, minimized risks and enhanced interactions	RA for big data system
77	Systems designed by integrating IoT, 5G and blockchain technology	A study on architectures, features, security, efficiency and network problems	Integrating AI with these disruptive technologies solves standardization, energy utilization, scalability, regulation and interoperability issues

2.13 Comparative Analysis of Survey Papers with Specific Parameters

Authors have selected ten survey research papers for comparative study. The list of specific feature parameters such as 1 architecture, 2 applications, 3 open issues and challenges, 4 taxonomy, 5 security, and 6 future directions are used for analysing these papers. The analysis results are shown in Table 21.14. This chapter is compared with all these survey papers.

3 Research Issues, Implementations, Challenges and Future Directions of IIoT

The major security issues in IoT systems are applications that are deployed in distributed places. The IoT devices are highly diverse in make, capabilities and features. IoT devices have the heavy resource and power constraints. A survey by Hewlett Packard in 2015 revealed about 70% of IoT devices are vulnerable to attacks because of no transport encryption, no robust authentication and insecure web, software and firmware interfaces [77].

3.1 Research Issues in IIoT

The following research issues are identified from the literature [78].

Security Challenges in Designing

- Trusted IoT platforms.
- Less complex encoding techniques.
- Authorization and authentication.
- Data and user privacy.

Distributed Intelligence for

- Data models.
- Virtualization of smart objects.
- Non-intrusive interfaces.
- Multimedia and agent communication.

Leverage Distributed Systems with

- Communication protocols.
- Middleware.
- Context awareness.
- Management and distribution of data.

Table 21.14 Comparative study of survey papers with specific parameters

Author	Year	Study objectives	Advantages	Disadvantages	1	2	3	4	5	6
Khan et al. [6]	2017	To identity common security gaps in similar technologies like edge computing, cloudlets and micro-data centres	Security measures are correctly adhered to overcome the potential limitations identified	The main challenge is to build a system that can efficiently provision security without making eminent sacrifices in performance	√	X	√	X	√	√
Farouk et al. [33]	2020	To combat the production and distribution of counterfeit drugs in partnership with Cisco, IBM and block	Blockchain will improve security, management, and analysis of healthcare big data	Blockchain technology is a very new technology, it is constantly changing and evolving, and there are a limited number of people trained in that area	√	√	√	X	√	√
Mistry et al. [76]	2020	To survey 5G-enabled IoT as a backbone for blockchain-based industrial automation for the smart city applications	Most of the industrial applications have been covered in this work, where blockchain used to maintain the security faster data flow	A great amount of technological research is required to address the specific demands of the collaboration of these technologies	√	√	√	√	√	X
Han et al. [23]	2019	To prevent forgery and tampering of collected data, used in blockchain technology to encrypt, transmit data to cloud	Extracts technology-intensive data through edge computing and cloud computing	Measuring air pollution levels in real industrial sites are not carried out	√	√	X	X	X	X

(continued)

Table 21.14 (continued)

Author	Year	Study objectives	Advantages	Disadvantages	1	2	3	4	5	6
Nehra et al. [14]	2020	Outlines the challenges and research opportunities in IoT domain	Discussed advantages of each paper reviewed	Not comparing the results with similar experiments	√	X	√	X	X	√
Mohanta et al. [66]	2020	A systematic study of ML, AI and blockchain for addressing the security issues in IoT	Critical analysis and research challenges are discussed	Scalability issue of IoT needs to be considered when addressing security protocols	√	√	√	X	√	X
Khan et al. [72]	2020	To discuss the state-of-the-art research efforts in IIoT	Highlighted the enabling technologies for IIoT and recent challenges faced by IIoT	The main challenge is to link and modify the security measures and apply them to fog computing platform	√	X	√	√	X	√
Qin et al. [73]	2020	To survey the recent advances of IIoT including reference architectures, key technologies, relative applications and challenges	Architecture and organization of the paper is good	Comparative analysis of work with others is missing	√	√	√	X	X	X
Wang et al. [30]	2018	Propose a novel paradigm, 5G IIoT, to process big data intelligently and optimize communication channels	Architecture and discussion of open issues is good	Comparative analysis of work with others is missing	√	√	√	X	X	x
Ke Zhang et al. [75]	2019	To propose a cross-domain sharing inspired edge resource scheduling scheme and design a CDETA mechanism	Designed a new CDEETA to reduce the edge service costs and service capacity	An efficient AI processing mechanism to cope up with highly dynamic 5G beyond networks needs to be investigated	√	X	√	X	√	x
Sreekantha D.K. et al. [The proposed approach]	2020	To analyse IIoT applications with a case study	A comprehensive study of various sectors	Quality of presentation can be improved	√	√	√	√	√	√

1. Architecture, 2. Applications, 3. Open Issues and Challenges, 4. Taxonomy, 5. Security, 6. Future Directions

Communication and Computing for

- Energy harvesting.
- Low energy computing.
- Near-field communications.
- Ultra-wideband.

3.2 *Seven Challenges in Implementing IIoT*

1. Security of data and communication is a major concern.
2. Regularly maintaining the visibility of IoT devices.
3. Ensure 100% quality in Internet connectivity at all times and places.
4. Interfacing legacy systems with IIoT infrastructure.
5. Deploying computing resources at the edge of the network.
6. Distributed data storage and management.
7. Selecting the right set of tools for the task in hand [79].

3.3 *IIoT Future Trends*

Microsoft Corporation has conducted a survey and found that 85% of companies have at least one IIoT use-case project implemented. This number is likely to increase, as 94% of respondents would implement IIoT projects by 2021.

1. The implementation of 5G in industries would revolutionize the connectivity for IoT-enabled devices to enhance productivity.
2. Maryville University estimated that by this year 2025 over 180 trillion gigabytes of data will be generated around the world year after year. IIoT-enabled industries would occupy a major chunk of this data segment.
3. The IIoT data analytics emerges as huge potential applications.
4. Predictive maintenance of machinery.
5. IoT platforms can wirelessly patch up and update machinery on time through predictive maintenance enabling a considerable reduction in downtime for machinery.
6. IIoT boosts edge computing applications.
7. IIoT industries transform their computing models to support edge computing. Edge computing facilitates data to be processed near the IoT devices to reduce delay and make best use of available bandwidth.
8. Industrial AI.
9. Applications built using AI and deployed at the edge or on VMs enable predictive analytics.
10. AI-based visual inspection systems.
11. This is an emerging AI technology using deep learning and computer vision for automated visual inspection and quality control [80].

4 IIoT: An Industry Case Study – Stanley Black & Decker Tool Manufacturing Industry

Stanley Black & Decker is a leading global tool manufacturing company. It is one of the largest tool manufacturing plants in Reynosa, Mexico, to serve the North American market. This company manufactures mechanical access solutions, electronic security and monitoring systems, hand tools, power tools and related accessories, products and services for industrial applications. This company upgraded its operations with IIoT solutions from Cisco and AeroScout for visibility of operations and to enhance productivity. Stanley Black & Decker has 40 multi-product manufacturing lines and thousands of employees; this plant produces millions of power tools each year.

The main challenges faced by the company before implementation of IIoT Solution are:

1. Lack of transparency in real-time production to schedule.
2. Lack of visibility to real-time overall equipment effectiveness and line productivity.
3. Reducing the production line changes over time.

The company has implemented AeroScout Real-Time Location System (RTLS) in collaboration with Cisco company in manufacturing plants in Reynosa, Mexico. RTLS is powered by Cisco's robust wireless network and AeroScout industrial leading enterprise visibility solutions. The RTLS includes small and easily deployed Wi-Fi and RFID tags that can be attached to virtually any material to provide real-time location and status to assembly workers, shift supervisors and plant managers for sending notifications of any issues using Wi-Fi infrastructure and plant-wide Ethernet.

The benefits of this successful implementation of RTLS are as follows:

1. Provided 24% increase in overall equipment effectiveness (OEE) on the router production line.
2. Allowed faster decision-making because of immediate increased throughput by about 10%.
3. Reduced inventory of material holding costs by 10%.
4. Empowered employees in the assembly line to notify supervisors of product quality problems.
5. Provided visibility to the line managers immediately to react on line issues.

Mr. Mike Amaya, the plant manager, explains the advantages of RTLS as. "We now have products and machines talking over Internet, and we can monitor and control production almost automatically". Mr. Gary Frederick, the chief information officer, expressed that "With the help of the Cisco and AeroScout Industrial solution, we are on our way towards realizing our vision of a virtual warehouse and fully connected factory, with complete visibility and traceability".

5 Conclusion

Authors are motivated by the market potential and industry attention to IIoT technology as a potential game-changer. An extensive study of literature on IIoT applications deployed in various industries was carried out. The intelligent and autonomous systems are designed using AI, IIoT, 5G and blockchain technologies. All the applications are classified industry-wise in 11 sections from finance and business through research. A summary table highlighting the methodology, results and applications from each paper is explored industry-wise. A taxonomy of concepts from IIoT, 5G and blockchain technologies is presented. A comparative analysis of curated survey papers with this chapter is also presented. A case study of IIoT implementation in world-class tool manufacturing company is analysed. At the outset, the authors realized that AI, blockchain and 5G are empowering IIoT systems by leveraging trust, security, flexibility and efficiency in operations. Authors tried to provide a holistic view of IIoT applications and presented the insight into systems design by discussing the algorithms, techniques and methods applied.

Authors conclude that intelligent systems integrated with IIoT, blockchain and 5G are going to play a vital role in every industry and business to enhance the quality and efficiency of services and products in various industries to reduce the cost of operations. At the end, the research issues, implementation challenges and future trends are a discussed.

References

1. L. Hang, I. Ullah, D.-H. Kim, A secure fish farm platform based on Blockchain for agriculture data integrity. Comput. Electron. Agric.. ISSN 0168-1699 **170** (2020). https://doi.org/10.1016/j.compag.2020.105251
2. R.G. Shukla, A. Agarwal, S. Shukla, Blockchain-powered smart healthcare system. Chapter-10, in *Handbook of Research on Blockchain Technology*, 1st Edition, Academic Press, Paperback ISBN: 9780128198162, eBook ISBN: 9780128204153 (2020), pp. 245–270
3. J.L. Ferrer-Gomila, M.F. Hinarejos, A 2020 perspective on a fair contract signing protocol with Blockchain support. Electron. Commer. Res. Appl. **42**, 100981 (2020)
4. B. Nour, A. Ksentini, N. Herbaut, P.A. Frangoudis, H. Moungla, Blockchain-based network slice broker for 5G services. IEEE Netw. Lett. **1**(1), 99–102 (2019)
5. M. Baniata, H. Ji, Y. Kim, J.-w. Choi, J. Hong, Energy-balancing unequal concentric chain clustering (Mimo-ucc) protocol for IoT system in 5G environment, in *RACS-18: Proceedings of the Conference on Research in Adaptive and Convergent Systems*, (2018), pp. 68–74
6. S. Khan, S. Parkinson, Y. Qin, Fog computing security: a review of current applications and security solutions. J. Cloud Comput. **6**, 19 (2017)
7. Radanliev, P., De Roure, C.., Cannady, S., Montalvo, R.M., Nicolescu, R., Huth, M.: The Economic Impact of IoT Cyber Risk Analysing Past and Present to Predict the Future Developments in IoT Risk Analysis and IoT Cyber Insurance, in Living in the Internet of Things: Cybersecurity of the IoT – 2018. Institution of Engineering and Technology, London (2018)
8. V. Kristiina, B. Jere, Y. Seppo, Blockchain-powered value creation in the 5G and smart grid use cases. IEEE Access **7**, 25690–25707 (2019)

9. U. Bodkhe, P. Bhatacharya, S. Tanwar, S. Tyagi, N. Kumar, M.S. Obaidat, BloHosT: Blockchain-enabled smart tourism and hospitality management, in *International Conference on Computer, Information and Telecommunication Systems (IEEE CITS-2019)*, (Beijing, China, August 28–31, 2019, 2019), pp. 237–241
10. M.A. Afshar, J. Sapna, Blockchain implementation using the smart grid-based smart city. Chapter-6, in *Handbook of Research on Blockchain Technology*, 1st Edition, Academic Press,Paperback ISBN: 9780128198162 eBook ISBN: 9780128204153, Academic Press, (2020), pp. 133–169
11. A. Thada, U.K. Kapur, S. Gazali, N. Sachdeva, S. Shridevi, Custom Blockchain based cyber-physical system for solid waste management, in *International Conference on Recent Trends in Advanced Computing (ICRTAC). Procedia Computer Science*, vol. 165, (2019), pp. 41–49
12. Y. Lee, S. Rathore, J.H. Park, J.-H. Park, A Blockchain-based smart home gateway architecture for preventing data forgery. Hum-Cent. Comput. Inform. Sci. **10**(9) (2020)
13. D.-O. Jaquet-Chiffelle, E. Casey, J. Bourquenoud, Tamper-proof time stamped provenance ledger using Blockchain technology. Forensic Sci. Int. Digit. Investig. ISSN 2666-2817 **33**, pp. 1–15 (2020)
14. V. Nehra, A.K. Sharma, R.K. Tripathi, Blockchain implementation for the internet of things applications. Chapter-5, in *Handbook of Research on Blockchain Technology*, 1st Edition, Academic Press, Paperback ISBN: 9780128198162 eBook ISBN: 9780128204153 (2020), pp. 113–132
15. A. Thakre, F. Thabtah, S.R. Shahamiri, S. Hammoud, A novel Blockchain technology publication model proposal. Appl. Comput. Inform.. ISSN: 2634-1964 (2019). https://doi.org/10.1016/j.aci.2019.10.003
16. W. Liu, K. Huang, X. Zhou, S. Durrani, Next-generation backscatter communication: systems, techniques, and applications. EURASIP J. Wirel. Commun. Netw. **2019**, 69 (2019)
17. J.-h. Noh, H.-y. Kwon, A study on smart city security policy based on blockchain in 5G age, in *International Conference on Platform Technology and Service (PlatCon)*, (Jeju, Korea (South), 2019), pp. 1–4
18. S. Tanwar, Q. Bhatia, P. Patel, A. Kumari, P.K. Singh, W.C. Hong, Machine learning adoption in Blockchain-based smart applications: the challenges, and a way forward. IEEE Access **8**, 474–488 (2020)
19. L. Xie, Y. Ding, H. Yang, X. Wang, Blockchain-based secure and trustworthy internet of things in SDN-enabled 5G-VANETs. IEEE Access **7**, 56656–56666 (2019). https://doi.org/10.1109/access.2019.2913682
20. A. Yang, Y. Li, C. Liu, J. Li, Y. Zhang, J. Wang, Research on the logistics supply chain of Iron and steel enterprises based on Blockchain technology. Futur. Gener. Comput. Syst. **101**, 635–645 (2019)
21. M. Tu, M.K. Lim, M.-F. Yang, IoT-based production logistics and supply chain system – IoT-based cyber-physical system: a framework and evaluation. Ind. Manag. Data Syst. **118**(1), 96–125 (2017)
22. R. Gupta, S. Tanwar, F. Al-Turjman, P. Italiya, A. Nauman, S.W. Kim, Smart contract privacy protection using AI in cyber-physical systems: tools, techniques and challenges. IEEE Access **8**, 24746–24772 (2020)
23. Y. Han, B. Park, J. Jeong, A novel architecture of air pollution measurement platform using 5G and blockchain for industrial IoT applications, in *2nd International Workshop on Smart Manufacturing and Smart Mobility (SMSM). August 19–21, Halifax, Canada. Procedia Computer Science*, vol. 155, (2019), pp. 728–733
24. K. Zhang, Y. Zhu, S. Maharjan, Y. Zhang, Edge intelligence and Blockchain empowered 5G beyond for the industrial internet of things. IEEE Netw. **33**(5), 12–19 (2019)
25. H. Liu, D. Han, D. Li, Fabric-IoT: a Blockchain-based access control system in IoT. IEEE Access **8**, 18207–18218 (2020)
26. Y. Wu, H. Huang, C.X. Wang, Y. Pan, Cellular internet of things, Chapter 10, in *5G-Enabled Internet of Thing*, (2019), CRC Press, DOI: https://doi.org/10.1201/9780429199820, eBook ISBN978042919982. pp. 361–366

27. V. Adat, I. Politis, C. Tselios, S. Kotsopoulos, Blockchain enhanced secret small cells for the 5G environment, in *IEEE 24th International Workshop on Computer-Aided Modeling and Design of Communication Links and Networks (CAMAD)*, (Limassol, Cyprus, 2019), pp. 1–6
28. C. Laroiya, D. Saxena, C. Komalavalli, Applications of Blockchain technology. Chapter-9, in *Handbook of Research on Blockchain Technology*, (2020), pp. 213–243
29. J. Backman, S. Yrjölä, K. Valtanen, O. Mämmelä, Blockchain network slice broker in 5G: slice leasing in the factory of the future use case, in *Internet of Things Business Models, Users, and Networks*, (Copenhagen, 2017, 2017), pp. 1–8
30. D. Wang, D. Chen, B. Song, N. Guizani, X. Yu, X. Du, From IoT to 5G I-IoT: the next generation IoT-based intelligent algorithms and 5G technologies, in *Unlocking 5G Spectrum Potential for Intelligent IoT: Opportunities, Challenges, and Solutions, IEEE Communications Magazine, IEEE Wireless Communications and Networking Conference Workshops (WCNCW)*, (2018), pp. 185–190
31. K. Dilip, A. Bhatnagar, K. Chauhan, D. Bhamrah, S. Srivastava, S. Thakur, S. Bisht, S. Narula, K. Jangid, P. Jundre, A microservices-based virtualized Blockchain framework for emerging 5G data networks, in *2019 IEEE Globecom Workshops (GC Wkshps)*, (Waikoloa, HI, USA, 2019), pp. 1–6
32. R. Fatima, R. Manal, M. Tomader, Cryptography in e-health using 5G based IoT: a comparison study, in *BDIoT'19: Proceedings of the 4th International Conference on Big Data and Internet of Things, 17*, (2019), pp. 1–6
33. A. Farouk, A. Alahmadi, S. Ghose, A. Mashatan, Blockchain platform for industrial healthcare: vision and future opportunities. Comput. Commun. **154**, 223–235 (2020)
34. D. Dhaggara, M. Goswami, G. Kumar, Impact of trust and privacy concerns on technology acceptance in healthcare: an Indian perspective. Int. J. Med. Inform. **141**, 104164 (2020)
35. V. Suresh, K. Prabhakar, K. Santhanalakshmi, K. Maran, Applying technology acceptance (TAM) model to determine the factors of acceptance in out-patient information system in private hospital sectors in Chennai City. J. Pharm. Sci. Res. **8**(12), 1373–1377 (2016)
36. P. Zhang, M.N. Kamel Boulos, Blockchain solutions for healthcare. Chapter 64, in *Precision Medicine for Investigators, Practitioners and Providers*, 1st Edition, Academic Press, Paperback ISBN: 9780128191781 eBook ISBN: 9780128191798 (2020), pp. 519–524
37. M. Sun, J. Zhang, Research on the application of Blockchain big data platform in the construction of a new smart city for low carbon emission and green environment. Comput. Commun. **149**, 332–342 (2020)
38. S. Kiyomoto, A. Basu, M.S. Rahman, S. Ruj, On Blockchain-based authorization architecture for beyond-5G Mobile services, in *12th International Conference for Internet Technology and Secured Transactions (ICITST)*, (Cambridge, 2017), pp. 136–141
39. R. Gupta, S. Tanwar, S. Tyagi, N. Kumar, M.S. Obaidat, B. Sadoun, HaBiTs: Blockchain-based Telesurgery framework for healthcare 4.0, in *International Conference on Computer, Information and Telecommunication Systems (IEEE CITS-2019)*, (Beijing, China, August 28–31, 2019), pp. 6–10
40. J. Vora, S. Tyagi, N. Kumar, J.P.C. Rodrigues, Home-based exercise system for patients using IoT enabled smart speaker, in *IEEE 19th International Conference on e-Health Networking, Applications and Services (Healthco m-2017)*, (Dalian University, Dalian, China, 12–15 October 2017), pp. 1–6
41. N. Nomikos, E.T. Michailidis, P. Trakadas, D. Vouyioukas, H. Karl, J. Martrat, T. Zahariadis, K. Papadopoulos, S. Voliotis, A UAV based moving 5G RAN for massive connectivity of mobile users and IoT devices. Veh. Commun. **25**, 100250, ISSN 2214-2096 (2020)
42. J. Bartelt, N. Vucic, D. Camps-Mur, E. Garcia-Villegas, I. Demirkol, A. Fehske, M. Grieger, A. Tzanakaki, J. Gutiérrez, E. Grass, G. Lyberopoulos, G. Fettweis, 5G transport network requirements for the next generation fronthaul interface. EURASIP J. Wirel. Commun. Netw. **2017**, 89 (2017)
43. A.U. Khan, G. Abbas, Z. Haq, A.M. Waqas, A.K. Hassan, Spectrum utilization efficiency in the cognitive radio enabled 5G-based IoT. J. Netw. Comput. Appl. **164**, 102686, ISSN 1084-8045 (2020)

44. H. Abukwaik, A. Gogolev, C. Gro, M. Aleksy, OPC UA realization for simplified commissioning of adaptive sensing applications for the 5G IIoT. Internet of Things **11**, 1–10 (2020)
45. T. Sigwele, P. Pillai, A.S. Alam, Y.F. Hu, Fuzzy logic-based call admission control in 5G cloud radio access networks with preemption. EURASIP J. Wirel. Commun. Netw. **2017**, 89 (2017)
46. Y. Liu, S. Zhang, H. Zhu, P.-J. Wan, L. Gao, Y. Zhang, Z. Tian, A novel routing verification approach based on Blockchain for inter-domain routing in smart metropolitan area networks. J. Parallel Distrib. Comput. **142**, 77–89 (2020)
47. V. Mohammadi, A.M. Rahmani, A.M. Darwesh, A. Sahafi, Trust-based recommendation systems in the internet of things: a systematic literature review. HCIS **9**, 21 (2019)
48. H.F. Atlama, G.B. Wills, Technical aspects of Blockchain and IoT. Adv. Comput. **115**, 1–39, Elsevier Inc. ISSN 0065-2458 (2019). Doi:10.1016/bs.adcom.2018.10.006
49. B. Dinesh, B. Kavya, D. Sivakumar, M.R. Ahmed, Conforming test of Blockchain for 5G enabled IoT, in *Proceedings of the 3rd International Conference on Trends in Electronics and Informatics (ICOEI 2019)*. IEEE Xplore Part Number: CFP19J32-ART; ISBN: 978-1-5386-9439-8, (2019)
50. Y.J. Choi, I.-G. Lee, Game theoretical approach of Blockchain-based Spectrum sharing for 5G-enabled IoTs in dense networks, in *IEEE 8891144*, (2019)
51. I. Bhudiraja, S. Tyagi, S. Tanwar, N. Kumar, J.J.P.C. Rodrigues, Tactile internet for smart communities in 5G: an insight for NOMA-based solutions. IEEE Trans. Industr. Inform. **15**(5), 3104–3112 (2019)
52. V. Ortega, F. Bouchmal, J.F. Monserrat, Trusted 5G vehicular networks: Blockchains and content-centric networking. IEEE Veh. Technol. Mag. **13**(2), 121–127 (2018)
53. W. Cassells, H. Senger, E.R. de Faria, A. Bifet, IDSA-IoT: an intrusion detection system architecture for IoT networks, in *IEEE Symposium on Computers and Communications (ISCC)*, (Barcelona, Spain, 2019), pp. 1–7
54. K. Fan, Y. Ren, Y. Wang, H. Li, Y. Yang, Blockchain-based efficient privacy-preserving and data sharing scheme of the content-centric network in 5G. IET Commun. **12**(5), 527–532 (2018)
55. B. Bera, S. Saha, A.K. Das, N. Kumar, P. Lorenz, M. Alazab, Blockchain-envisioned secure data delivery and collection scheme for 5G-based IoT-enabled internet of drones environment, in *IEEE 9110761*, (2020)
56. D. Olivier, J. Chiffelle, E. Casey, J. Bourquenoud, Tamper-proof time-stamped provenance ledger using blockchain technology. Forensic Sci. Int. Digit. Investig. **33**, 300977, ISSN 2666-2817 (2020)
57. S. Porkodi, D. Kesavaraja, Integration of Blockchain and internet of things. Chapter 3, in *Handbook of Research on Blockchain Technology*, 1st Edition, Academic Press, Paperback ISBN: 9780128198162 eBook ISBN: 9780128204153 (2020), pp. 133–169
58. N. Gupta, A deep dive into security and privacy issues of blockchain technologies. Chapter 4, in *Research on Blockchain Technology*, (MRIIRS, Faridabad, India, 2020), pp. 95–112. ISBN 9780128198162
59. S. Bhushan, B. Bohara, P. Kumar, V. Sharma, A new approach towards IoT by using health care-IoT and food distribution IoT, in *2nd International Conference on Advances in Computing, Communication, & Automation (ICACCA) (Fall)*, (Bareilly, 2016), pp. 1–7
60. B. Nour, A. Ksentini, N. Herbaut, P.A. Frangoudis, H. Moungla, A Blockchain-based network slice broker for 5G services. IEEE Netw. Lett. **1**(3), 99–102 (2019). https://doi.org/10.1109/LNET.2019.2915117
61. Y. Wei, S. Lv, X. Guo, Z. Liu, Y. Huang, B. Li, FSSE: Forward secure searchable encryption with keyed-block chains. Inf. Sci. **500**, 113–126 (2019)
62. P. Prajapatia, K. Chaudhar, KBC: multiple key generations using key block chaining, in *International Conference on Computatoped, Procedia Computer Science*, vol. 167, (2020), pp. 1960–1969
63. Tanwar S, M S Obaidat, Tyagi S., Kumar N.: Online signature-based biometrics recognition. M. S. Obaidat et al. (eds.), Biometric-Based Physical and Cybersecurity Systems, Springer International Publishing, Springer Nature Switzerland AG 2019, pp.255–285 (2019)

64. B. Huang, R. Zhang, Z. Lua, Y. Zhang, J. Wua, L. Zhan, P.K. Hung, BPS: a reliable and efficient pub/sub communication model with a Blockchain-enhanced paradigm in multi-tenant edge cloud. J. Parallel Distrib. Comput. **143**, 167–178 (2020)
65. N.Z. Benisi, M. Aminian, B. Javadi, Blockchain-based decentralized storage networks: a survey. J. Netw. Comput. Appl. **162**, 102656, ISSN 1084-8045 (2020)
66. B.K. Mohanta, D. Jena, U. Satapathy, S. Patnaik, Survey on IoT security: challenges and solution using machine learning, artificial intelligence and Blockchain technology. Internet of Things **11**, 100227 (2020)
67. E.K. Wang, R.P. Sun, C.-M. Chen, Z. Liang, Proof of X-repute Blockchain consensus protocol for IoT systems. Comput. Secur. **95**, 101871 (2020)
68. Q. Zhang, Y.-y. Han, Z.-b. Su, J.-l. Fang, Z.-q. Liu, K.-y. Wang, A storage architecture for high-throughput crop breeding data based on improved Blockchain technology. Comput. Electron. Agric. **173**, 105395 (2020)
69. M.K. Müller, F. Ademaj, T. Dittrich, A. Fastenbauer, B.R. Elbal, A. Nabavi, L. Nage, S. Schwarz, M.R. Müller, Flexible multi-node simulation of cellular mobile communications: the Vienna 5G system level simulator. EURASIP J. Wirel. Commun. Netw. **227** (2018)
70. P. Pääkkönen, D. Pakkala, Extending reference architecture of big data systems towards machine learning in edge computing environments. J. Big Data **7**, 25 (2020)
71. R.M. Haris, S. Al-Maadeed, Integrating Blockchain technology in 5G enabled IoT: a review, in *IEEE International Conference on Informatics, IoT, and Enabling Technologies (ICIoT)*, (Doha, Qatar, 2020), pp. 367–371
72. Z. Khan, M.H. Rehman, H.M. Zangoti, M.K. Afzal, N. Armi, K. Salah, Industrial internet of things: recent advances, enabling technologies and open challenges. Comput. Electr. Eng. **81**, 106522 (2020)
73. W. Qin, S. Chen, M. Peng, Recent advances in industrial internet: insights and challenges. Digit. Commun. Netw. **6**, 1–13 (2020)
74. D. Wang, D. Chen, B. Song, N. Guizani, X. Yu, X. Du, From IoT to 5G I-IoT: the next generation IoT-based intelligent algorithms and 5G technologies, unlocking 5G Spectrum potential for intelligent IoT: opportunities, challenges, and solutions. IEEE Commun. Mag
75. K. Zhang, Y. Zhu, S. Maharjan, Y. Zhang, Edge intelligence and Blockchain empowered 5G beyond for the industrial internet of things. **33**(5), 12–19, IEEE Network: The Magazine of Global Internet working September 2019. https://doi.org/10.1109/MNET.001.1800526, IEEE Netw. (2019)
76. I. Mistry, S. Tanwar, S. Tyagi, N. Kumar, Blockchain for 5G-enabled IoT for industrial automation: a systematic review, solutions, and challenges. Mech. Syst. Signal Process. **135**, 1–19 (2020)
77. L. Maximilian, E. Markl, A. Mohamed, Cybersecurity management for (industrial) internet of things: challenges and opportunities. J. Inform. Technol. Software Eng. **8**(5), 1000250 (2018)
78. J. Gubbia, R. Buyyab, S. Marusic, M. Palaniswami, Internet of things (IoT): a vision, architectural elements, and future directions. Futur. Gener. Comput. Syst. **29** (September 2013)
79. J. Dillard, 7 challenges for the industrial internet of things adopters, garlandtechnology.com visited on 10th Oct. 2020
80. J. Berry, MobiDev, blog https://mobidev.biz/blog/industrial-iot-internet-of-things-trends visited on 10th Oct. 2020

Chapter 22
Amalgamation of Blockchain Technology and Internet of Things for Healthcare Applications

Dinesh Bhatia, Animesh Mishra, and Moumita Mukherjee

1 Introduction

The integration of blockchain and the Internet of Things (IoT) helps to strengthen and leverage mutual advantages in a variety of industrial domains including healthcare. The healthcare industry is one of the largest and fastest-growing industries involving more than 10% of the gross domestic product (GDP) of several countries globally. The use of blockchain in IoT can help to track multiple devices in the network, avoid duplication of malicious data entries, disallow data tampering, and provide enhanced security and privacy. Several devices are connected in an IoT network for fast-paced 5G-enabled data transfer environment for different industrial applications for collecting information and processing it, and the chances of failure of this centralized system are much higher. The time required to retrieve data from different devices, existing protocols, and access control mechanism employed presently may not work in the future as they are centralized based mechanisms [1, 2]. The deployment of blockchain in IoT allows distributed, decentralized ledger records with high security, privacy, improved efficiency, and authenticity of data [1, 3, 4].

Soon, the blockchain technology would immensely benefit and revolutionize the healthcare sector by ensuring data transparency, reducing costs of service delivery,

D. Bhatia (✉)
North Eastern Hill University, Shillong, Meghalaya, India
e-mail: dbhatia@nehu.ac.in

A. Mishra
North Eastern Indira Gandhi Regional Institute of Health and Medical Sciences, Shillong, Meghalaya, India

M. Mukherjee
Adamas University, Kolkata, West Bengal, India
e-mail: moumita.mukherjee@adamaduniversity.ac.in

S. Tanwar (ed.), *Blockchain for 5G-Enabled IoT*,
https://doi.org/10.1007/978-3-030-67490-8_22

centralized patient data, and hospital services management. It would allow real-time monitoring of the patient data and availability to seek second referral for their medical condition due to the transparency of data and high-speed 5G network services being provided by the healthcare service provider employing blockchain and the IoT [2, 4]. This would enable the patients to learn regarding similar cases being treated or reported in other countries which would help in improving their present medical condition and nature of services being provided by their clinician. Overall the practical utilization of integrated blockchain and IoT technology can benefit patients, clinicians, healthcare service providers, researchers, and biomedical specialists to improve service efficiency, quality of healthcare services, and accurate diagnosis and treatment of diseases and share patient data with enhanced security and privacy. It would prevent loss or modification of patient data being decentralized and transparent in nature [5, 6].

The major concerns for acceptance of this technology are the misuse of data by sharing it with third parties for profit or indirect identification of users through pseudonymous identifiers. With a decentralized database, blockchain can help solve several underlying problems within the healthcare system giving more autonomy and control to patients over their data with selected access rights to researchers relating to the medical records. It is a more patient-centric healthcare model where trust is paramount, and the records cannot be altered or deleted as per convenience. The motivation of writing this chapter is to allow readers to understand the present advancements in the healthcare sector due to integration of the blockchain and IoT technologies which would provide immense benefits to the patients as well as the healthcare service providers. Further, the users must also be aware regarding the technological challenges and limitations of the proposed technology and how it can be overcome to allow optimized usage and maximum potential of the technology to be leveraged to solve the existing healthcare problems, thereby improving quality of life for the patients.

Kumar et al. [1] stated that the industry is under strict regulatory norms, with a complex system of different interconnected entities leading to fragmented patient data and enhanced costs due to system inefficiencies leading to lack of complete transparency in data transactions, data traceability issues, and security concerns for the patient. Tandon et al. [11] wrote to overcome this problem, blockchain technology has gained considerable attention in several domains including the healthcare industry to ensure secure and reliable management of real-time clinical healthcare patient data [1, 11]. Qamar et al. [2] referred that the technology could be employed to support patient data management, hospital supply chain management, pharmaceutical research, online patient monitoring through hospital telemedicine set-up, and hospital billing system. The blockchain follows peer-to-peer integrated network where any transaction request made is broadcasted to all nodes and needs to be approved by all nodes to be added to the blockchain network.

Further, integrated solutions can be built to avoid medical fraud and drug counterfeiting. Priyadarshan et al. [14] referred that using IoT and machine learning, the technology can help in overcoming the deletion of clinical trial record or hiding sources of funding for carrying out medical research. It can also help in saving huge costs in the pharmaceutical industry by having a defined network of supply chains. The contracts incorporated during implementing the blockchain system can be legally binding and enforceable as per law of the land. Thus, they could help in automating the healthcare system such as billing, insurance, diet planning, and pharmacy management, thereby significantly reducing the costs involved. Shakeel et al. [15] and Padmavathi et al. [19] mentioned that the blockchain technology research in healthcare is focused on integrity, integration, and access control of patient-related health records. In the future, work is ongoing on other aspects such as clinical trials, medical research, supply chain management, and medical insurance.

The chapter is organized with introduction to the blockchain and IoT technologies and work reported by different researchers with regard to these technologies in the present industry scenario giving special reference to the healthcare technology. The subsequent section discusses about the blockchain-based IoT architecture and how the data is stored, monitored, and retrieved when required from this system. The remaining sections discuss about the applications of these technologies in the healthcare sector enumerating their salient features and limitations. The chapter also discusses regarding decentralizing IoT networks employing blockchain with the future scope and applications of integrating these technologies for seamless operations and benefit of mankind through improved healthcare services being discussed in the "Discussion and Conclusion" section.

2 Paper Preparation

2.1 Background of Blockchain

Blockchain technology comprises a distributed peer-to-peer network where non-trusting members can interact in a verifiable manner with each other without a trusted intermediary [5]. The blockchain technology is growing rapidly with promising applications in several technological fields. It is a decentralized system where the data entries are immutable and cannot be altered at convenience. The time-stamped chain of blocks are connected in a blockchain framework using cryptographic hashes affecting all industrial domains exponentially even influencing the healthcare sectors through its processes, services, management, education, etc. The blockchain has largely affected the functioning of electronic and personal health records (EHR and PHR) in the healthcare set-up. Cocosila et al. [17] and Lafourcade et al. [44] cited that the blockchain allows better data integrity in all aspects, access control to medical records, interoperability, and smooth transactions. The term blockchain originated from "Bitcoin" or cryptocurrency in the year 2008. This slowly expanded and got acceptance with several million users by 2019, wherein it was felt that its potential could be employed in different applications including healthcare.

Cachin [38]. talked about blockchain, cryptography, and consensus. As per the IBM report published in 2019, several healthcare leaders predicted that blockchain technology will play a major role in the healthcare sector due to its decentralized framework for sharing medical records safely, better management of clinical trials, and regulatory compliance enforceability. Further, it is expected that the global market for blockchain technology in the healthcare sector may soon cross $500 million by 2022 [11, 33, 38, 42]. Still, technology has miles to go and adapt to the existing systems and lend useful value to improve efficiency of the present healthcare industry.

A blockchain comprises a sequential chain of blocks which contains all transaction records. The first block is the genesis block, with successive blocks connected through hash values. Das et al. [31] affirmed that the blockchain concept was first introduced as an integral component of the cryptocurrency Bitcoin. The paper that led to the creation of Bitcoin was published in 2008, and the first genesis block of the blockchain was mined on 3 January 2009. The blockchain system comprises the decentralized database with open access, transparency, and autonomy to all users who are logged into the network to access information and regularly update post approval. Once the records are updated in the blockchain network, it cannot be deleted or removed easily without authentication and approval from more than half (51%) the number of users on the network. Hasselgren et al. [18] stated that blockchain would help to overcome several problems associated with the complex integrated healthcare network including the lack of traceability during different transactions, non-reporting of clinical trials, and issues related to substandard or fake medicines being regularly supplied to patients. This would avoid the lack of confidentiality in handling the patient data and non-sharing with for-profit organizations for commercial benefits or misuse of the information in any form. It would also remove the presence of middleman in the healthcare system by cutting down on costs, thereby enhancing trust and transparency in the system. Drosatos et al. [20] referred that it would reduce dependency on single supplier for procuring resources for the healthcare systems, thereby allowing them to contact different vendors as per requirement without incurring substantial costs or legal constraints. Presently, these blockchain systems are at the proof ofconcept stage; however, their potential is being acknowledged and realized with several projects involving blockchain technology increasing rapidly [20, 21].

The blockchain allows decentralization of records with no central authority controlling the addition of blockchain content. The immutable entries are added to the blockchain network based on consensus through peer-to-peer network. Balandina et al. [7] stated that the blocks are linked to each other through cryptographic protocols which are self-executable smart contracts. Persistence is also a key feature of blockchain as accepted entries cannot be deleted due to a distributed ledger stored across multiple nodes. It allows anonymity in the blockchain set-up with easy traceability due to interconnectivity of different blocks with each other through the hash. The transactions are arranged in the form of a Merkle tree for each block with each leaf value verified to its established root for traceability. Previously, the blockchain was used extensively in the finance sector as it allows security and privacy of data; however, nowadays, it is being applied in several computational fields where voluminous transactions take place. Kumar et al. [3] and Liang et al. [5] mentioned that it is being used in the Internet of Things (IoT) network for recording, analyzing, and transferring of real-time data from nodes or objects, humans, and sensors to automate different assignments or tasks.

The blockchain can be categorized into public, consortium, and private. The public blockchain is referred to as permissionless as anyone can join it to view and contribute to the blockchain and allow changes to its core software. The two largest cryptocurrencies that widely employ public blockchain are Bitcoin and Ethereum. The efficiency is low, but data immutability is high. The consortium of public permission allows a restricted number of identified group entities to access, view, and participate in the consensus protocol. In consortium blockchain, the efficiency is high, but data could be easily tampered with. Priyadarshan et al. [14] and Drostos et al. [20] stated that in the private blockchain, only selected nodes are allowed to participate in the distributed network controlled by a single organization having centralized authority over all transactions in the blockchain. In this blockchain, although the efficiency is high, the data could be tampered with.

All users connected in the block can view ongoing transactions in the network. The record of previous block and transactions is available with each block, thereby making it a secure and immutable ledger record. The chain increases as new blocks are added and arranged in a peer-to-peer network comprising several nodes with each node maintaining a copy of the same. The nodes may comprise "users" or "miners" who check and validate every transaction occurring on the network for its appropriateness. With the "consensus" of users and miners, the data is added to each block. Three majorly employed consensus protocols are proof of work (PoW), proof of stake (PoS), and Practical Byzantine Fault Tolerance (PBFT) protocols. The PoW is the consensus protocol strongly associated and integrated with Bitcoin. The miners are involved in solving some complex computational problems by finding lesser value than the predetermined one in the proposed block. The miner who finds the solution first is rewarded and validates the transaction. However, Kuo et al. [22] claimed that it is limited due to its high computational energy requirements when applied on a large blockchain which may be equivalent to the electricity required for a small country. The stake of each node in the blockchain determines the approval node selection process in the PoS. This may give an unfair advantage to richer nodes and influence the decision-making. Hence, the randomization process is employed to select the approving node. Tandon [11]. and Padmavathi et al. [19] referred that Ethereum, the second-largest cryptocurrency, is in the process to shift from PoW to PoS. The PBFT is based on the Byzantine agreement protocol limiting its usage in the public domain as all nodes must be known in the network. It comprises three phases, namely, as pre-prepared, prepared, and commit. Each node requires two-thirds majority vote approval from all nodes in the network to move to the next phase. The PBFT is currently employed in the Hyperledger Fabric. Saraf et al. [36] stated that till the year 2018, the number of frameworks rolled out by Hyperledger Fabric is equal to five, namely:

1. Burrow.
2. Fabric.
3. Indy.
4. Iroha.
5. Sawtooth.

The Blockchain 3.0 version is based on directed acyclic graph (ADG) and has several advantages over previously employed versions 1.0 and 2.0. Blockchain 1.0 also known as Bitcoin came into existence in 2008 that uses proof of work, whereas Blockchain 2.0 which came into existence in 2014 with Ethereum and smart contracts employs proof of stake concept. Both of these are vertical schemes that have several limitations such as privacy, self-sustainability, scalability, interoperability, and governance. Hasselgren et al. [18] stated that Blockchain 3.0 is a horizontal scheme that addresses several limitations of the previous two versions and hence preferred nowadays.

Ethereum which is a popular blockchain system since 2013 comprises the block header and a set of validation rules that needs to be followed. Hasselgren et al. talked about the Merkle tree root hash comprising the relevant transactions and corresponding hash values in the block. The timestamp records the actual time of transactions as universal time during block creation. The nonce, a 4-byte field, has 0 value initially which gets incremented after the calculation of every hash function. The parent block hash comprises 256-bit values which link to previous blocks in the blockchain [18]. After each transaction in the block, the transaction counter is incremented. The size of the block and transactions decides the maximum number of transactions a block can hold. Asymmetric cryptography techniques are utilized to authorize and validate the transactions using a digital signature technique. Each user in the network uses two keys, namely, private and public keys. The private key is confidential, not shared, and employed for digital signatures. The signed transactions are then shared over the entire network [11, 19, 69]. Blockchains are a new technology built on a relatively simple premise being that a continuously generated stream of information can be commonly shared among all members of a group.

Hashing is a key cryptographic tool which uses the concepts of mathematics for code making and code breaking aspect. Using hashing, the message of any length can be converted to an output of a fixed length of alphanumeric characters. Therefore, when transactions happen, large pieces of information are converted into manageable small outputs which represent the original information. When a large number of transactions happen, then the need for verification and compilation may arise. It is collectively called "validation" and requires the knowledge of mining as the validation of a block leads to the creation of new units of cryptos in the environment. Once the problem is solved by any miner and a new block has been created, everyone on that network must agree that the problem is solved and accept it as the next block. The validation of the work is performed by solving the problem by oneself and matching with the work done by the other successful miner(s). This is a tedious task where proof of work (PoW) comes into play. The miner(s) who solves the problem also broadcasts the PoW to other members, so that they could accept it and move on to solve the next problem and create the block. Niya et al. [58] stated that the proof of stake (PoS) concept involves the validation of block that depends on the number of cryptocurrency tokens that a person already owns. It is said that the number of cryptos is directly proportional to the mining power by the miner(s). Instead of utilizing energy to answer PoW puzzles, a PoS miner is limited to mining a percentage of transactions reflecting the ownership stake.

2.2 *Background of Internet of Things*

Internet of Things (IoT) was introduced in 1999 and is presently a $50 billion industry. It is a network of devices that configure themselves automatically in which the human is not at the center of the system. It allows better understanding of the environment and response to situations wherein the machines are doing better in sensing and reporting conditions. Presently, IoT is involved in home automation devices, road safety and traffic regulation, automatic payments, quality assurance, and remote condition monitoring. In the future, IoT would be involved in every field whether driving of vehicles, wearable devices, scannables, and flyables. Although having good connectivity and sensor technology, it is vulnerable to security threats due to insecure interfaces and unencrypted communications [33, 45].

Cocosila et al. [17] stated that IoT is being employed in several organizations to enhance operational efficiency, deliver better services by acknowledging customers' concerns, and aid in decision-making, thereby enhancing the business outcome. It is believed that the forecasted global revenue growth of IoT products would be around $6.2 trillion with major income from the healthcare sector by 2025. Drosatos et al. [20] showed concern over the potential to transform healthcare and global public health. The IoT business running on a cloud server model would generate global revenues of approximately $ 490 billion soon. The major business tycoons such as Google and Samsung having NetLabs and SmartThings solutions, respectively, are keen on providing innovative IoT technology to their users [71, 79]. The IoT devices are having greater penetration in manufacturing, industrial automation, businesses, wearable devices, and the healthcare sector at large.

The IoT definition states that it is a dynamic global network infrastructure with self-configuring capabilities based on standard and interoperable communication protocols where physical and virtual "things" have identities, physical attributes, and virtual personalities employing intelligent interfaces that are seamlessly integrated into the information network to communicate data associated with users and their environments [6, 9, 11, 23, 59, 81, 82]. The major characteristics of an IoT network are that it is dynamic and self-adapting and self-configuring with interoperable communication protocols and gives unique identity to objects that are integrated into information network [10, 12]. An IoT device may consist of several interfaces such as sensory, actuators, memory, or audio/video for connections to other devices, both wired and wireless. The focus in an IoT network is on configuration, control, and networking via the Internet of Devices or Things that are traditionally not associated with the Internet. Several devices such as thermostats, utility meter, a Bluetooth-connected headset, irrigation pumps, sensors, and control circuits for an electric car's engine can be connected through it [14, 16]. The IoT technology is being driven by advancements in sensor networks, mobile devices, wireless communication, networking, and cloud computing. The scope of IoT is not limited to just connecting things (devices, appliances, machines) to the Internet. IoT allows the things to communicate and exchange data. Data itself does not have a meaning until it is processed into useful information. The information is then organized and structured into knowledge for future use when required [15, 18, 33, 35, 36, 48, 50].

As mentioned by Catarinucci et al. [6] the Internet of Things (IoT) has tremendously transformed and revolutionized the way present healthcare services are being provided to patients. Qamar et al. [3] claimed that the healthcare industry is in a state of great despair with rising healthcare costs, an aging global population, and an increase in new and chronic diseases. This could lead to a state of healthcare services being out of reach of the common man making them unproductive and prone to chronic diseases. The IoT technology allows for remote monitoring in the healthcare sector, thereby contributing in continuous and safe monitoring of the patient's health condition, empowering physicians to deliver quality healthcare at reasonable costs. Fonseca et al. [13] acknowledged that the level of patient-doctor interaction has increased tremendously leading to widespread satisfaction with improved healthcare efficiency among patients due to ease of interaction with the healthcare providers. This decreased the hospital duration stay for the patients and their families showing a deep impact on diminishing healthcare costs and improving treatment outcomes [49]. Several applications in the healthcare sector employing IoT are presently available to benefit the healthcare providers, hospitals, insurance companies, and patients at large. Mendes et al. [16] disclosed that IoT has largely benefitted critically ill patients and the elderly population living alone who require continuous monitoring due to their diseased condition by tracking their health condition on a regular basis and allowing any disturbances or change in routine daily living activities to trigger an alarm to the family member or concerned healthcare providers for an immediate check on their condition and to provide instant care and treatment to avoid any fatalities or long-term damage [15, 16]. However, Priyadharsan et al. [20] projected that several potential challenges exist while designing any IoT-based healthcare system that needs to be addressed such as security and privacy, user identification and authentication, and regular communication and exchange of healthcare data before the adoption of the technology on a mass scale [14, 20, 27].

2.3 Integration of Blockchain and IoT in Healthcare

The wearable or carrying IoT devices employed in the healthcare sector are smart electronic devices that can be worn as clothing or body implants or accessories. These devices in the form of the smartwatch, fitness trackers, or Google applications can collect human physiological data that could be transmitted over the network employing Bluetooth or Wi-Fi setups. Fuqaha [29]. referred that these devices help in reducing costs and improving efficiency as they can monitor remotely vital patient's signs and health status. The wearable devices are playing a vital role in the healthcare industry by monitoring health fitness levels and wellness factor regularly, motivating users for healthy living, and sharing of medical data with healthcare providers. Prasad [32]. established "SMART Asthma Alert using IoT." This device collects the data from the asthmatic patient's environment and understands the pattern of allergen that triggers the attack. Here, the device predicts the threshold value and alerts the patient. To maintain integrity and security, blockchain-Internet of Things concept can be added for the smooth functioning of the device and connectivity with the patient's family and healthcare providers.

Bhatia et al. [21] considered IoT as a boon in the healthcare industry, several difficulties exist in their acceptance such as privacy and security, and since data is being transferred to the cloud server, it may easily fall in the wrong hands and get misused, thereby compromising safety and privacy of the patient. Kuo et al. [21] embarked on data authenticity, integrity, security, and confidentiality which are the major vulnerabilities of an IoT network. Healthcare comprises sensitive patient data, and with digitization, it is easy to transfer and move manual records in the digitized form which helps in easy storage and retrieval. Talpur [4]. cited that the electronic health record (EHR) is digitized that allows easy sharing and analysis of patient data by healthcare providers. Catarinucci et al. [6] and Tandon et al. [11]. brought into light that the digitized record is always under threat of cybercrime to manipulate the confidential patient information. The amount of security, privacy, and confidentiality of the medical data and its interoperability between different healthcare agencies are crucial, and blockchain technology can help to provide robust structure against attacks and system failures. The blockchain also prevents the manipulation and tampering of healthcare data. Besides, it reduces costs, enhances efficiency, and ensures data safety with regulatory compliance with the creation of a secure environment and immutable records [68, 78].

The IoT system operational requirements include dynamic but verifiable group membership with data integrity and authentication, secure transactions again single-node (or small sub-set of nodes), light weight resource operations, data encryption with capability to handle sensor power-off time periods [19, 21]. It should be able to handle diversity of resources comprising data from different sensors and aggregators. For integration of blockchain in IoT, the desirable properties are distributed protocol with verifiable transaction history and dynamic multi-party signature membership [22]. The undesirable properties include requirement of "proof of work," PKI requirement, and size of ledger which could be an issue for miniature devices and anonymous (unverified) join and leave operations. This can be overcome by allowing proof of earlier participation using transaction history, employing hash-based signatures or Merkle tree schemes, and maintaining relevant transaction ledger when device is too resource constrained and group signatures using pre-shared group key(s) to overcome anonymous or unverifiable join and leave operations [24, 26, 28]. Further, since the IoT technologies are increasing exponentially, the IoT devices will always have diverse capabilities and resources.

The use of cryptography is done without clear understanding of the implications, and there are no current standards for lightweight cryptography. Hence, blockchain-inspired protocols combined with new cryptographic primitives might be a way forward to overcome the above limitations [30, 32, 36].

2.4 Challenges Brought by Blockchain and IoT Integration for Healthcare

Several researchers and Dorri et al. [76] mentioned in their research works that they are working on possible solutions to devise blockchain-based IoT (BIoT) architectures that are lightweight to reduce the communication-induced network overhead and are concentrated mostly on smart homes. They cited about distributed time-based consensus (DTC) algorithm which reduces the mining processing overhead and delay. The BIoT architecture is divided into three layers, namely, home, network, and cloud. The home layer comprises interconnected sensing devices that store locally collected data with blockchain features. The network layer connects the devices with the network or the Internet. The cloud is employed for the storage of recorded data. Hasselgren et al. [18] cited that blockchain can be employed with several IoT applications such as healthcare, smart home, smart city, supply chain, e-governance, etc. This fusion technology finds several applications in industrial automation. If big data is combined with BIoT, it could be used to effectively control and monitor the huge amounts of data collected in the IoT ecosystem [18, 20, 26, 28]. Presently, the BIoT consortium is quite less in number, and proper regulations are required to control different factors that presently affect the implementation of BIoT. Padmavathi et al. [19] showed that the issues of prime importance are security, scalability, and cross-platform applications. The commercial products on BIoT presently available include Walton chain, Ambrosus, Power Ledger, Block Mesh, etc.

The different challenges in implementing BIoT include technical integration, interoperability, legal hurdles, and government regulations. Tandon et al. [11] and Fuqaha et al. [29] paved that the technical integration due to scalability and standardized development of products is an important challenge being faced in its smooth implementation. Interoperability can be achieved by having mutual agreement among all stakeholders involved in the implementation of BIoT framework. The government regulations with legal frameworks require real-time product testing to avoid system malfunctioning and seamless integration of different aspects of the combined technology. Drosatos et al. [20] and Bhatia et al. [21] stated that due to the high demand for real-time data by industry and research organizations, it puts them at a larger risk of theft, break-ins, and unauthorized sharing. Further, due to malpractices such as counterfeit drugs, skills of medical staff, etc. in the healthcare system, it may erode public trust in the system [52, 55, 56]. Hence, it is important to implement a system that can overcome such disadvantages and improve the

Table 22.1 Blockchain and IoT characteristics [72, 74]

Characteristics of blockchain	Description of characteristics	Potential of blockchain and IoT in healthcare
Data integrity	Non-repudiation of data; the transactions cannot be denied	Transactions cannot be deleted or manipulated once they are approved
Prior approval or consent	Prior consent or approval	Seeking consent of patients and healthcare service providers
Security and privacy	Less chances of fraud or mismanagement of the secure patient data	Allows privacy and ensures secrecy of medical data and records with
Continuous monitoring	Monitoring of patients may be done in real time with continuous updates regarding the health condition to the patient and the doctors	Continuous and remote patient monitoring and interaction with the system
Reduces paperwork	Reduces storage costs and improves overall efficiency	Effective management of the electronic health records (EHR)
Automates processing	Removal of intermediaries and middlemen	Remove supply of fake and sub-standard drugs in the supply chain

overall management of the system. Talpur et al. [4] leads that blockchain technology leads the way due to its decentralized approach and data integrity provides a potential solution to the potent problems affecting the present-day healthcare system. Catarinucci et al. [6] claimed that blockchain allows greater interoperability, data integrity, sharing of information, and provenance among the stakeholders, thereby enhancing efficiency, system performance, and most importantly mutual trust among parties involved. Xia Qi et al. [34] introduced the indicator-centric schema (ICS), a data model used to simplify the various medical data (like scans, reports, x-rays, etc.) and store them efficiently. Separation of frequently used data from the infrequently used data can be done using this blockchain model [34, 36, 45] (Tables 22.1 and 22.2).

3 Applications in Healthcare and Medical Technology

Blockchain technology has several applications in the healthcare field. It can be employed for patient-based requirements such as seeking consent, allows privacy, and ensures secrecy with continuous monitoring and interaction with the system. The issues that may be associated with its implementation include scalability, interoperability, and patient management. Blockchain technology can be employed for effective management of the electronic health record (EHR) by employing deep learning algorithms. Balandina et al. [7] mentioned the advantages of blockchain technology as it reduces paperwork and storage costs, improves efficiency, automates processing, and lowers the chances of fraud or mismanagement of the secure

Table 22.2 Comparison with similar approaches: advantages and disadvantages [78–82]

References	Technology	Advantage	Disadvantage
[78]	Wellbeing monitoring using wireless sensor network and cloud computing using IoT	This is a cost-efficient technique	Not easy to implement WSN nodes compared to wired networks
[79]	Cipher text policy attribute-based encryption (CP-ABE) for data security	Access-based policy for the need of data protection	This is problematic to implement in non-interactive group of networks
[80]	Health monitoring through wireless body area sensor network (WBASN)	As easy addition to the existing sensors	Sensors should be low in complexity, small in size, lightweight, and easy and configurable
[81]	IOT with smart devices	Real-time data access and intelligent data integration	Constant updation and upgradation of component devices are required
[82]	Technology acceptance model	Recognized technology to use	Adoption of new technologies is difficult to equip with for elders

patient data [19, 23, 27, 80, 82]. The blockchain technology allows the removal of intermediaries giving more control over the information to the patient which is a huge advantage. The technology could aid in biomedical research with the help of smart digital contracts to regulate the functioning of the biomedical databases. It ensures data integrity and non-repudiation. Gong et al. [8] and Amin et al. [10] unveiled different pathways in the healthcare system for privacy protection as it allows for remote patient monitoring with the secure management of medical sensors. The monitoring of patients may be done in real time with continuous updates regarding the health condition to the patient and the doctors [29, 68, 69, 71, 76]. It provides medical interventions with automated notifications to the concerned parties. The only disadvantage being data transfer through open channels from smart devices to blockchain nodes and due to certain delays in block verification, it cannot be used in emergency response situations as the response time would increase comprising patient health. The blockchain technology would help in overcoming fake drug prescription and labelling due to prudent check mechanism. The use of counterfeit drugs leads to adverse drug reactions with over 10 lakhs deaths annually as per the World Health Organization (WHO) data. Padmavathi et al. [19] introduced that to avoid the inclusion of such fake drugs in the supply chain, transparency must exist to avoid tampering, modifications, or stealing of drugs. The pharmacists need to ensure that drugs of correct quantity and strength are available. The blockchain also helps in the medical insurance storage system that allows transactions to be

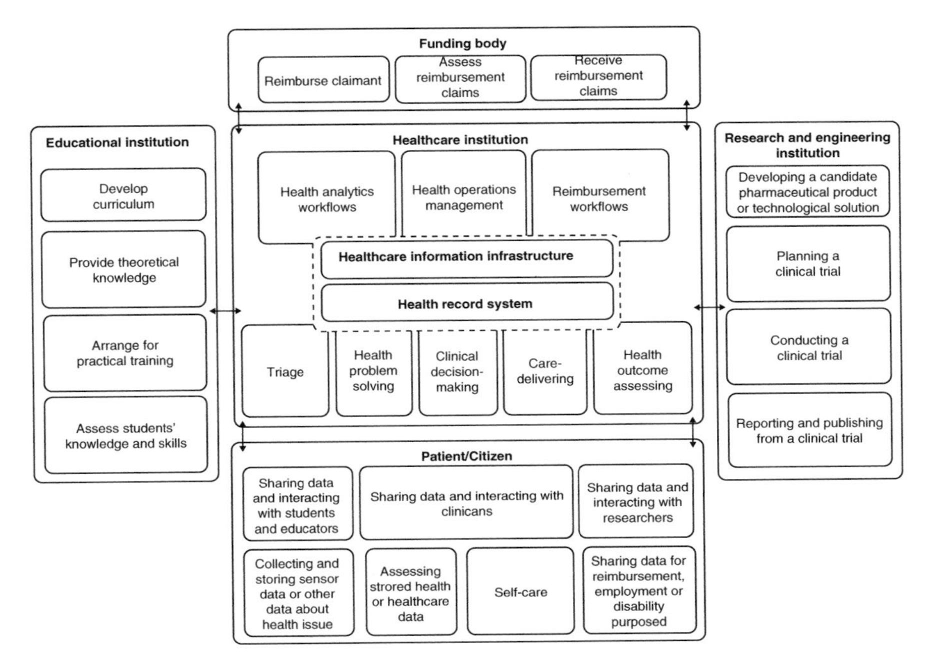

Fig. 22.1 Blockchain and healthcare sector [18]

verified before being added to the blockchain network, thereby saving time, costs, and the transaction verification process for the insurance company [18, 19, 77].

Before blockchain can be employed in the healthcare sector, there is a need to understand the application areas where it has been employed for present applications in biomedical domains along with Internet of Things such as clinical medical records, clinical trials, clinical education, and evidence-based databases as represented in Fig. 22.1. Further, Alhussein et al. [12] and Cocosila et al. [17] clarified that the types of data such as personal clinical records, medical health records, willingness forms, drug supply chain, patient location, and health insurance transactions have been addressed with blockchain technology that need to be elaborated with the level of present expertise available such as architectural design, system component availability of technical tools, etc. [12, 17]. The need for employing blockchain technology in the biomedical healthcare along with its pros and cons needs to be understood. It needs to understand the level of usage of blockchain technology in access control, data integrity, data auditing, and provenance that needs to be studied. Drosatos [20]. referred to the type of blockchain framework such as Ethereum, Hyperledger Fabric, Bitcoin, etc. that would be employed to address the problems encountered in the biomedical domain that needs to be understood [11, 20].

Gong et al. [8] cited that the healthcare data can be broadly classified into two main categories, mainly as medical records focused on medical care produced from healthcare departments or hospitals and the other being personal health records comprising information entered usually by patients willingly. Blockchain technology has shown immense potential in providing privacy, security, and integrity of healthcare records. Alhussein et al. [12] referred that it is being nowadays used in creating patient record ledgers residing at different locations, integrating medical records with healthcare enterprises, and sharing of these resources across different regions and users. The technology also allows patients to have control over their information and can be employed to seek second opinion or referral by sharing if required. Blockchain can lead to integration and unification of personal health records with medical records and allows easy management of one unified healthcare record. The insurance-based health record is encrypted and stored without tampering in the blockchain network for future use. Another area of blockchain application is to safeguard the transactions of researchers with reference biomedical databases such as clinical trials, pharmaceuticals, etc. that are updated regularly. Hasselgren et al. [18] stated that blockchain can also be employed to track the drug supply chain from the manufacturer to the consumer to avoid counterfeit drugs entering the supply chain. Most of the blockchain technologies for medical applications are at the initial stage of maturity or being tested for implementation shortly. However, any technology depicting real-time solutions is not presently available, and more time is required for the blockchain technology to mature in the future [18, 20, 57, 59, 63, 68, 74].

The blockchain technology allows independence to patients to assign rules permitting specific use of their recorded data for research purposes. It also gives freedom to patients to connect to hospitals directly and procure their medical data automatically. Tandon et al. [11] told that with the help of ledger technology in blockchain, it is possible to securely transfer patient medical records without consideration of any boundaries, supply chain management of medicines administered to patients, improved coordination among healthcare organizations, fewer transaction costs and risks involved and help in genetic research or genomics. This can help in streamlining healthcare processes and prevent costly and fatal medical errors. Genomics has been considered to improve human health, and several companies are exploring the possibility of employing this to understand human health for scientific advancement in inpatient treatment and quality healthcare. Bhatia et al. [21], Engelhardt et al. [23], and Randall et al. [24] concluded that the blockchain and IoT technology are found to be perfectly suitable to this area by helping patients to share their encrypted genetic information which allows for wider database creation, thereby helping the scientific community in their research endeavors as represented in Fig. 22.2. The goal of the Internet of Things (IoT) technology is to enable easy, secure, and efficient use of technology for transmission and sharing of data globally. It offers several potential benefits in the dynamic digital world leading to personalized healthcare benefits for patients and care providers which is not a challenging technological problem anymore. With the help of IoT technology for healthcare systems, it is possible to track and monitor anything within the system at any convenient time and place on demand. The present-day advancements in sensor technology and communication networks and the availability of high-performance

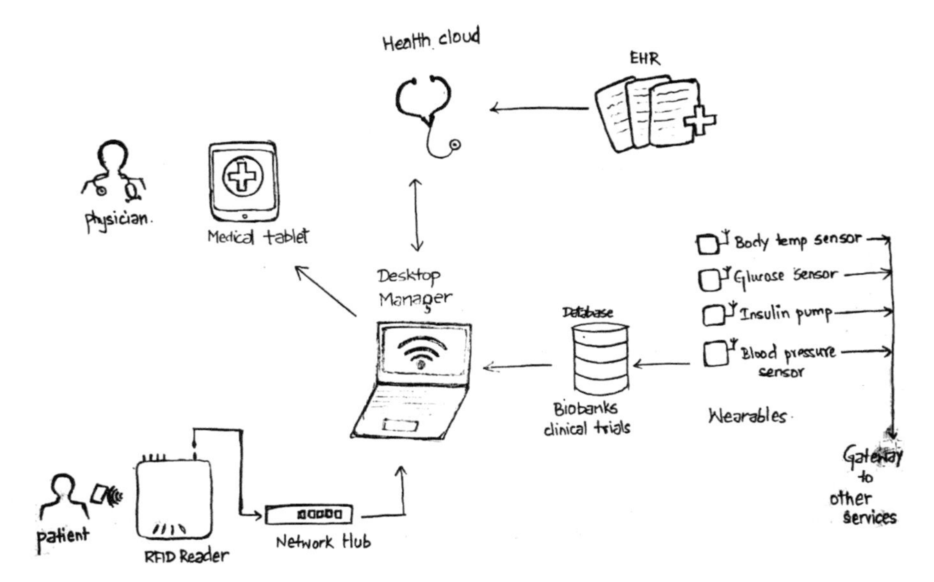

Fig. 22.2 IoT schematic for the healthcare sector [21]

computing systems and IoT along with blockchain technology have been taking rapid strides and coming up as a new potential technology that can be explored to provide numerous healthcare benefits at reduced costs and time. With the help of these technologies, any device can be used to sense its surroundings and do computation employing a wireless network and address. This allows developing real-time mapping of virtual objects to establish communication channels between them using different communication technologies for information about the status of the connected entity as and when required [26, 28, 30, 32, 36].

Blockchain can help the telemedicine field to thrive by developing mutual trust between patients and healthcare professionals. It helps in data integrity, authentication, and identification of parties which ensures transparency by providing incentives in the form of crypto tokens for players to act fairly. Shakeel et al. [15] cited that with the use of regulated artificial intelligence tools, it creates a seamless platform for global healthcare by the effective use of healthcare resources and needs. Mendes et al. [16], Bahga et al. [28], and Reyna et al. [30] showed that the major benefit would be providing remote qualitative diagnostic services especially when a present situation such as COVID-19 lockdowns is prominent across the globe with a large number of people deprived of quality healthcare facilities. A large number of start-ups are coming up in this field, and several are found in the dermatology field due to its close association with the beauty industry [30, 47, 53, 57, 63, 65, 82].

4 Decentralizing IoT Networks Employing Blockchain

Engelhardt et al. [23] and Randall et al. [24] mentioned that the smart devices employed in IoT cannot be directly used in blockchain as they are lightweight with energy and memory constraints. By using the traditional client-server model in IoT, problems related to data synchronization, security, and data privacy may exist. This can be overcome by employing blockchain incorporated within the IoT systems. Jing et al. [26] powered that a blockchain with hypergraphs used in smart homes can maintain security and privacy protection by reducing cyber-attack possibilities. Rothers et al. [25] considered that the system must be robust enough to allow data privacy and integrity which could be managed with a unique identity of devices with little scope of tampering. The system should allow scalability and solve problems of data exchange and trading [40, 41, 46]. With the help of data fusion techniques, the analysis could be done on the cloud, thereby improving the healthcare delivery and reducing operational expenditures. Reyna et al. [30] affirmed that healthcare data privacy is maintained between the cloud nodes and fog with security through blockchain. The advantages include smooth data aggregation, accuracy in the management of data, low-cost IoT software, and less consumption of resources. Sultan et al. [27] cited that for smooth IoT functioning, integration with 5G network is required and the blockchain transactions may be visible to all if secure communication protocols on members are not maintained in the blockchain system [80].

Tandon et al. [11]. stated that employing blockchain in IoT has several distinct advantages. These are *adaptability* as blockchain and IoT integration can be easily done without the requirement of intermediate blocks [13, 37, 39]. Bagha et al. [28] referred that cross-communication can be implemented easily with middleware technologies among blockchain devices. The next is *scalability*, and it is the ability of the blockchain network to accommodate more IoT devices or sensors on the network without any problem. Fuqaha et al. [29] added that the third advantage is *security*, which is of prime importance between the fusion of both blockchain and IoT technologies. Shakeel et al. [15] showed that the security issues in IoT such as denial of services (DoS), authorization, identity management, and data protection can be overcome by using blockchain. The last one is *consistency*, which implies that the data in the system is consistent with the combination of blockchain and IoT. This ensures data integrity and allows the autonomous administration of IoT systems or domains. It allows every domain in the blockchain network to act independently for managing the IoT devices [15, 16]. Hence, with these potential benefits, if integration of blockchain with IoT is achieved successfully, then it would allow better healthcare patient monitoring, improve service efficiency and accurate patient diagnosis, and enhance security with quality services [32, 34, 35, 43]. In their recent study, Kuo et al. [22] presented an overview of blockchain technology to meet the need of futuristic biomedical engineering and e-healthcare. The authors have proposed the pseudonym-based encryption with different authorities (PBEDA) concept to empower a patient to verify, modify, as well as access his/her healthcare data on the IoT-based multi-layered system.

5 Discussion and Conclusion

The blockchain technology allows less dependence on huge servers and reduces redundant work. By providing high security and data privacy for users, the system can become cost-effective with fast computing capabilities. Soon, the blockchain technology would immensely benefit and revolutionize the healthcare sector by ensuring data transparency, reducing costs of service delivery, central server for management of all patient data, and hospital services [14, 19]. The clinicians can monitor their patient's condition in real time without worrying regarding the level of honest information being provided by their patients. Similarly, Tandon [11]. emphasized that the patients need not worry to get a second referral for their medical condition due to the transparency of data and services being provided by the healthcare service provider. Further, Kuo et al. [22] pressed that they can willingly share their data to enable them to learn regarding similar cases being treated or reported in other countries which would help them in improving their present medical condition and nature of services being provided by their clinician. Overall the practical utilization of blockchain technology can benefit patients, clinicians, healthcare service providers, researchers, and biomedical specialists to improve service efficiency, quality of healthcare services, and accurate diagnosis and treatment of diseases and share patient data with enhanced security and privacy [22, 26, 33, 47, 51]. It would prevent loss or modification of patient data being decentralized and transparent in nature. The major concerns for acceptance of this technology are the misuse of data by sharing it with third parties for profit or indirect identification of users through pseudonymous identifiers [67, 70, 73]. Since research is ongoing in this area and the number of publications is available presently, the future for the growth and expansion of this technology looks promising. In the healthcare sector, it is expected that the blockchain technology market may cross $500 million by 2022. Hasselgren et al. [18] informed that due to its data integrity and enhancing mutual trust among users, its demand would be ever-increasing and required by healthcare institutions to perform their tasks with seamless efficiency and precision [20, 22, 43, 57, 69, 74].

The time required to retrieve data from different devices, existing protocols, and access control mechanism employed presently may not work in the future as present IoT-based systems are centralized based mechanisms [54, 60, 61, 62]. The deployment of blockchain in IoT allows distributed, decentralized ledger records with high security, privacy, improved efficiency, and authenticity of data. The healthcare industry is one of the largest and fastest-growing industries globally. The industry is under strict regulatory norms involving complex interconnected entities leading to fragmented patient data and enhanced costs due to system inefficiencies leading to lack of complete transparency in data transactions, data traceability issues, and security concerns for the patient. The blockchain technology has gained wide acceptability in different industrial domains including healthcare to ensure secure and reliable management of real-time clinical healthcare patient data, thereby reducing healthcare costs and accurate disclosure of the clinical trials to overcome supply of duplicate or low-quality medicines to patients and hospitals. The technology could be employed to support patient data management, hospital supply chain management, pharmaceutical research, online patient monitoring through hospital telemedicine set-up, and hospital billing system [75].

This review focuses on a real-time healthcare monitoring system using IoT and cloud computing services which are in various ways important for elders and chronic diseases' patients. Different methods that are available for healthcare services are thoroughly surveyed, and the challenges in the path of realization are described. This review proposes intelligent real-time patient monitoring system to monitor the subject's vital parameters including temperature, pressure, breath activity, as well as ECG using specific prototype model and detects any abnormality with accuracy. In the near future, the research could be carried out on to focus on improving wearing sensor experience by using softer materials and enabling controlled sharing of information among the doctors, the patient, and the patients' family through social networking paradigm. The clinicians can monitor the patient's condition in real time, and the patients could avail second referral for their medical condition due to the transparency of data and services being provided by the healthcare service provider employing blockchain and the Internet of Things [64, 66]. Further, they can willingly share their data with high-speed 5G networking to enable them to learn regarding similar cases being treated or reported in other countries which would help them in improving their present medical condition and nature of services being provided by their clinician. Overall the practical utilization of integrated blockchain and IoT technology can benefit patients, clinicians, healthcare service providers, researchers, and biomedical specialists to improve service efficiency, quality of healthcare services, and accurate diagnosis and treatment of diseases and share patient data with enhanced security and privacy. It would prevent loss or modification of patient data being decentralized and transparent in nature. The major concerns for acceptance of this technology are the misuse of data by sharing it with third parties for profit or indirect identification of users through pseudonymous identifiers.

References

1. P.M. Kumar, S. Lokesh, R. Varatharajan, G.C. Babu, P. Parthasarathy, Cloud and IoT based disease prediction and diagnosis system for healthcare using Fuzzy neural classifier. Future Gener. Comput. Syst. (Elsevier) **86**(1), 527–534 (2018)
2. S. Qamar, A.M. Abdelrehman, H.E.A. Elshafie, K. Mohiuddin, Sensor-based IoT industrial healthcare systems. Int. J. Sci. Eng. Sci. **11**(2), 29–34 (2018)
3. S.M. Kumar, D. Majumder, Healthcare solution based on machine learning applications in IoT and edge computing. Int. J. Pure Appl. Math. **119**(16), 1473–1484 (2018)

4. M.S.H. Talpur, The appliance pervasive of internet of things in healthcare systems. Int. J. Comput. Sci. Iss. (IJCSI) **10**(1), 1–6 (2013)
5. Z. Liang, G. Zhang, J.X. Huang, Q.V. Hu, Deep learning for healthcare decision making with EMRs, in *IEEE International Conference on Bioinformatics and Biomedicine*, (2014), pp. 556–559
6. L. Catarinucci, D. de Donno, L. Mainetti, L. Palano, L. Patrono, M.L. Stefanizzi, L. Tarricone, An IoT-aware architecture for smart healthcare systems. IEEE Internet of Things **2**(6), 515–526 (2015)
7. E. Balandina, S. Balandin, E. Balandina, Y. Koucheryavy, S. Balandin, D. Mouromtsev, IoT use cases in healthcare and tourism, in *IEEE 17th Conference on Business Informatics*, (2015), pp. 37–44
8. T. Gong, H. Huang, P. Li, K. Zhang, H. Jiang, A medical healthcare system for privacy protection based on IoT, in *Seventh International Symposium on Parallel Architectures, Algorithms and Programming*, (2015), pp. 217–222
9. M.M. Alam, H. Malik, M.I. Khan, T. Pardy, A. Kuusik, Y.L. Moullec, A survey on the roles of communication technologies in IoT-based personalized healthcare applications. IEEE Access **6**(1), 36611–36631 (2018)
10. S.U. Amin, M.S. Hossain, G. Muhammad, M. Alhussein, M.D.A. Rahman, Cognitive smart healthcare for pathology detection and monitoring. IEEE Access **7**(1), 10745–10753 (2019)
11. A. Tandon, An empirical analysis of using Blockchain technology with the internet of things and its application. Int. J. Innov. Technol. Expl. Eng. **8**(9S3), 1469–1474 (2019)
12. M. Alhussein, G. Muhammad, M.S. Hossain, S.U. Amin, Cognitive IoT-cloud integration for smart healthcare: case study for epileptic seizure detection and monitoring. Mob. Netw. Appl. **23**(1), 1624–1635 (2018)
13. C. Fonseca, D. Mendes, M. Lopes, A. Romão, P. Parreira, Deep learning and IoT to assist multimorbidity home based healthcare. J Health Med. Inform. **8**(3), 1–4 (2017)
14. D.M.J. Priyadharsan, K.K. Sanjay, S. Kathiresan, K.K. Karthik, K.S. Prasath, Patient health monitoring using IoT with machine learning. Int. Res. J. Eng. Technol. (IRJET) **6**(3), 7514–7520 (2019)
15. P.M. Shakeel, S. Baskar, V.R.S. Dhulipala, S. Mishra, M.M. Jaber, Maintaining security and privacy in health care system using learning-based deep-Q-networks. J. Med. Syst. **42**(186), 1–10 (2018)
16. D.J.M. Mendes, I.P. Rodrigues, C.F. Baeta, C. Solano-Rodriguez, Extended clinical discourse representation structure for controlled natural language clinical decision support systems. Int. J. Rel. Qua. E-Health **4**(1), 1–11 (2015)
17. M. Cocosila, N. Archer, Adoption of mobile ICT for health promotion: an empirical investigation. Elect. Mark **20**(1), 241–250 (2010)
18. A. Hasselgren, K. Kralevska, D. Gligoroski, S.A. Perdersen, A. Faxvaag, Blockchain in healthcare and health sciences-a scoping review. Int. J. Med. Inform. **134**(1), 1–10 (2020)
19. U. Padmavathi, N. Rajagopalan, A research on the impact of Blockchain in healthcare. Int. J. Innov. Technol. Expl. Eng. **8**(9S2), 35–40 (2019)
20. G. Drosatos, E. Kaldoudi, Blockchain applications in the biomedical domain: a scoping review. Comput. Struct. Biotechnol. J. **17**(1), 229–240 (2019)
21. D. Bhatia, S. Bagyaraj, K.S. Arun, A. Mishra, A. Malviya, Role of the Internet of Things (IoT) and deep learning for the growth of healthcare technology, in *Trends in Deep Learning Methodologies Series: Hybrid Computational Intelligence for Pattern Analysis and Understanding*, (Elsevier, Amsterdam, Netherlands, 2020)
22. T. Kuo, H.E. Kim, L.O. Machado, Blockchain distributed ledger technologies for biomedical and health care applications. J. Am. Med. Inform. Assoc. **24**(6), 1211–1220 (2017)
23. M.A. Engelhardt, D. Espinosa, Hitching healthcare to the chain: an introduction to Blockchain Technology in the Healthcare Sector. Technol. Innov. Manag. Rev. **7**(10), 22–35 (2017)
24. D. Randall, P. Goel, R. Abujamra, Blockchain applications and use cases in health information technology. J. Health Med. Inform. **8**(3), 8–11 (2017)

25. A. Rohers, C. André, R. Righi, Omni PHR: a distributed architecture model to integrate personal health records. J. Biomed. Inform. **71**, 70–81 (2017)
26. Q. Jing, A.V. Vasilakos, J. Wan, J. Lu, D. Qiu, Security of the internet of things: perspectives and challenges. Wirel. Netw **20**(8), 2481–2501 (2014)
27. A. Sultan, M. Sheraz, A. Malik, A. Mushtaq, Internet of things security issues and their solutions with Blockchain technology characteristics: a systematic literature review. Am. J. Compt. Sci. Inform. Technol. **6**(3), 1–5 (2018)
28. A. Bahga, V. Madisetti, *Internet of Things: A Hands-On Approach*, 1st edn. (Universities Press, Delhi, 2014)
29. A.A. Fuqaha, M. Guizani, M. Mohammadi, M. Aledhari, M. Ayyash, Internet of things: a survey on enabling technologies, protocols, and applications. IEEE Commun. Surv. Tutor. **17**(4), 2347–2376 (2015)
30. A. Reyna, M. Cristian, J. Chen, E. Soler, M. Díaz, BOn blockchain and its integration with IoT- challenges and opportunities. Futur. Gener. Comput. Syst. **88**(1), 173–190 (2018)
31. Susan Das, M. Debbarma, A. Deka, 2019, "A brief review on Blockchain and distributed ledger technology", International Conference on "Computing for Sustainable Global Development", pp. 2–5, (2019)
32. A.K. Prasad, SMART asthma alert using IoT and predicting threshold values using decision tree classifier, in *Springer Nature, International Conference on Computer Communication and IoT*, (2020), pp. 2–4
33. F. Casino, T.K. Dasaklis, C. Patsakis, A systematic literature review of blockchain-based applications: Current status, classification, and open issues. Telematics Inform. **36**(1), 55–81 (2018)
34. Q. Xia, E.B. Sifah, K.O. Asamoah, J. Gao, X. Du, M. Guizani, MeDShare: trust-less medical data sharing among cloud service providers via Blockchain. IEEE Access **5**(1), 14757–14767 (2017)
35. A. Rosic, What is Blockchain Technology? A Step-by-Step Guide for Beginners. https://blockgeeks.com/guides/what-is-blockchain-technology/, (2016)
36. C. Saraf, S. Sabadra, Blockchain platforms: a compendium, in *IEEE International Conference on Innovative Research and Development (ICIRD)*, (Bangkok, 2018), pp. 1–6
37. S. Nakamoto, Bitcoin: A Peer-to-Peer Electronic Cash System. [Online], (2008)
38. C. Cachin, M. Vukolić, Blockchain consensus protocols in the wild. ArXiv preprint arXiv:1707.01873 [cs. Dc], (2017)
39. K. Fanning, D.P. Centers, Blockchain and its coming impact on financial services. J. Corp. Account. Fin. (Wiley Periodicals) **27**(5), 53–57 (2016)
40. A.S. Thakur, V. Kulkarni, Blockchain and its applications – a detailed survey. Int. J. Comput. Appl. (0975–8887) **180**(3), 29–35 (2017)
41. B. Scott, How Can Cryptocurrency and Blockchain Technology Play a Role in Building Social and Solidarity Finance. 1(1), UNRISD Working Paper, pp. 1–25, (2016)
42. J. Baron, A. O'Mahony, D. Manheim, C. Dion-Schwarz, *National Security Implications of Virtual Currency*", ISBN: 978-0-8330-9183-3 (2015), pp. 1–102
43. A. Roehrs, C.A. da Costa, R.R. da Righi, V.F. da Silva, J.R. Goldim, D.C. Schmidt, Analyzing the performance of a blockchain-based personal health record implementation. J. Biomed. Inform. **92**(1), 103140 (2019)
44. P. Lafourcade, M.L. Platet, About blockchain interoperability. Inf. Process. Lett. **161**(1), 105976 (2020)
45. J.G. Dumas, P. Lafourcade, F. Melemedjian, J.B. Orfila, P. Thoniel, Local PKI: an interoperable and IoT friendly PKI, in *International Conference on E-Business and Telecommunications*, (Springer, 2017), pp. 224–252
46. S. Liu, S. He, Application of block chaining technology in finance and accounting field, in *International Conference on Intelligent Transportation, Big Data and Smart City (ICITBS)*, (2019), pp. 342–344
47. D. Tapscott, A. Tapscott, *Blockchain Revolution: How the Technology behind Bitcoin Is Changing Money, Business, and the World* (Penguin, ISBN-13: 978-1101980149, 2016)

48. K. Korpela, J. Hallikas, T. Dahlberg, Digital supply chain transformation toward blockchain integration, in *Proceedings of the 50th Hawaii International Conference on System Sciences, 50*, (2017), pp. 1–10
49. L.W. Cong, Z. He, Blockchain disruption and smart contracts. Rev. Financ. Stud. **32**(5), 1754–1797 (2019)
50. N. Hackius, M. Petersen, Blockchain in logistics and supply chain: trick or treat? in *In Digitalization in Supply Chain Management and Logistics, Smart and Digital Solutions for an Industry 4.0 Environment. Proceedings of the Hamburg International Conference of Logistics (HICL)*, 23(1), (2017), pp. 3–18
51. M. Milutinovic, W. He, H. Wu, M. Kanwal, Proof of luck: An efficient blockchain consensus protocol, in *Proceedings of the 1st Workshop on System Software for Trusted Execution*, (2016), pp. 1–6
52. M. Conoscenti, A. Vetro, J.C. Martin De, Blockchain for the Internet of Things: a systematic literature review, in *IEEE/ACS 13th International Conference of Computer Systems and Applications (AICCSA)*, (2016), pp. 1–6
53. F. Tian, A supply chain traceability system for food safety based on HACCP, blockchain, and internet of things, in *International Conference on Service Systems and Service Management*, (2017), pp. 1–6
54. H.F. Atlam, A. Alenezi, M.O. Alassafi, G. Wills, Blockchain with the internet of things: benefits, challenges, and future directions. Int. J. Intell. Syst. Appl. **10**(6), 40–48 (2018)
55. B.W. Jo, R.M.A. Khan, Y.S. Lee, Hybrid blockchain and internet-of-things network for underground structure health monitoring. Sensors **18**(12), 4268 (2018)
56. D.E. Kouicem, A. Bouabdallah, H. Lakhlef, Internet of things security: a top-down survey. Comput. Netw. **141**(1), 199–221 (2018)
57. A.A. Siyal, A.Z. Junejo, M. Zawish, K. Ahmed, A. Khalil, G. Soursou, Applications of blockchain technology in medicine and healthcare: challenges and future perspectives. Cryptography **3**(1), 3–6 (2019)
58. S.R. Niya, E. Schiller, I. Cepilov, F. Maddaloni, K. Aydinli, T. Surbeck, T. Thomas Bocek, Adaptation of proof-of-stake-based Blockchains for IoT data streams, in *IEEE International Conference on Blockchain and Cryptocurrency (ICBC)*, (Seoul, Korea (South), 2019), pp. 15–16. https://doi.org/10.1109/BLOC.2019.8751260
59. C.B. Kalis, E. Mitchell, E. Pupo, A. Truscott, Blockchain: securing a new health interoperability experience, in *ONC/NIST Use of Blockchain for Healthcare and Research Workshop*, (2016)
60. M. Crosby, P. Pattanayak, S. Verma, V. Kalyanaraman, Blockchain technology: beyond bitcoin. Appl. Innov. **2**(6–10), 71 (2016)
61. A.M. Antonopoulos, *Mastering Bitcoin: Programming the Open Blockchain* (O'Reilly Media, Sebastopol, California, Inc., 2017)
62. I. Eyal, Blockchain technology: Transforming libertarian cryptocurrency dreams to finance and banking realities. Computer **50**(9), 38–49 (2017)
63. S. Corbet, C. Larkin, B. Lucey, A. Meegan, L. Yarovaya, Cryptocurrency reaction to fomc announcements: evidence of heterogeneity based on blockchain stack position. J. Financ. Stab. **46**(1), 100706 (2020)
64. L.D.K. Chuen, L. Linda, *Inclusive Fintech: Blockchain, Cryptocurrency, and ICO* (World Scientific, Singapore 2018)
65. M.H. Miraz, M. Ali, Applications of blockchain technology beyond cryptocurrency. arXiv preprint arXiv:1801.03528, (2018)
66. J.Y. Lee, A decentralized token economy: how blockchain and cryptocurrency can revolutionize business. Bus. Horiz. **62**(6), 773–784 (2019)
67. J. Crandall, Blockchains and the "chains of empire": contextualizing blockchain, cryptocurrency, and neoliberalism in Puerto Rico. Des. Cult. **11**(3), 279–300 (2019)
68. P. Martino, K.J. Wang, C. Bellavitis, C.M. DaSilva, An introduction to blockchain, cryptocurrency and initial coin offerings. Chapter 7, in *New Frontiers in Entrepreneurial Finance Research*, (World Scientific Publishing Co. Pte. Ltd, Singapore, 2019), pp. 181–206

69. D. Valdeolmillos, Y. Mezquita, A. González-Briones, J. Prieto, J.M. Corchado, Blockchain technology: a review of the current challenges of cryptocurrency, in *International Congress on Blockchain and Applications*, (Springer, Cham, 2019), pp. 153–160
70. A. Manzoor, M. Liyanage, A. Braeke, S.S. Kanhere, M. Ylianttila, Blockchain-based proxy re-encryption scheme for secure IoT data sharing, in *IEEE International Conference on Blockchain and Cryptocurrency (ICBC)*, (2019), pp. 99–103
71. H. Wang, J. Zhang, Blockchain-based data integrity verification for large-scale IoT data. IEEE Access **7**(1), 164996–165006 (2019)
72. O. Novo, Blockchain meets IoT: An architecture for scalable access management in IoT. IEEE Internet Things J. **5**(2), 1184–1195 (2018)
73. G. Sagirlar, B. Carminati, E. Ferrari, J.D. Sheehan, E. Ragnoli, Hybrid-IoT: Hybrid blockchain architecture for the internet of things-pow sub-blockchains, in *IEEE International Conference on Internet of Things (iThings) and IEEE Green Computing and Communications (GreenCom) and IEEE Cyber, Physical and Social Computing (CPSCom) and IEEE Smart Data (SmartData)*, (IEEE, 2018), pp. 1007–1016
74. S. Singh, N. Singh, Blockchain: Future of financial and cybersecurity, in *2nd International Conference on Contemporary Computing and Informatics (IC3I)*, (IEEE, 2016), pp. 463–467
75. A. Dorri, S.S. Kanhere, R. Jurdak, Towards an optimized blockchain for IoT, in *IEEE/ACM Second International Conference on Internet-of-Things Design and Implementation (IoTDI)*, (IEEE, 2017), pp. 173–178
76. A. Dorri, S.S. Kanhere, R. Jurdak, P. Gauravaram, LSB: a lightweight scalable Blockchain for IoT security and anonymity. J. Parallel Distrib. Comput. **134**(1), 180–197 (2019)
77. I. Mistry, S. Tanwar, S. Tyagi, N. Kumar, Blockchain for 5G-enabled IoT for industrial automation: a systematic review, solutions, and challenges. Mech. Syst. Signal Process. **135**(1), 119 (2020)
78. S. Jaiswal, R. Katake, B. Kute, S. Ranjane, P.D. Mehetre, Survey of health monitoring management using Internet of Things (IOT). Int. J. Sci. Res. **5**(11), 2243–2246 (2017)
79. R.S. Pramila, A survey on effective in-home health monitoring system. Int. J. Comput. Appl. **68**(7), 15–19 (2013)
80. D. Kajaree, R. Behera, A survey on healthcare monitoring system using body sensor network. Int. J. Innov. Res. Comput. Commun. Eng. **5**(2), 1302–1309 (2017)
81. S.P. Shinde, N. Phalle Vaibhavi, A survey paper on internet of things based healthcare system. Internet Things Cloud Comput. **4**(4), 131–133 (2017)
82. B. Thaduangta et al., Smart healthcare: basic health check-up and monitoring system for elderly, in *2016 International Computer Science and Engineering Conference (ICSEC)*, (2016), pp. 1–6

Chapter 23
Highly Isolated Self-Multiplexing 5G Antenna for IoT Applications

Sandeep Sharma, Padmini Nigam, Arjuna Muduli, and Amrindra Pal

1 Introduction

The advanced automation in manufacturing is now a reality, so are the advances in health monitoring, fleet management, system maintenance tracking all this and many more thanks to Internet of Things (IoT) [1]. The IoT systems play an essential role in enhancing people's daily lives and establishing the connection between smart home equipment and a smart atmosphere. IoT systems build the concept of smart home, smart cities, smart hospitals and smart building also [2]. The industrial sector is also facing the challenges in the development of the industrial IoT (IIoT) systems for their use, also facing difficulty to establish business model. The major industrial structure, like machines, traffic and components etc., the IIoT is facing many operational and technical difficulties, like trustworthiness, timeless and security of the connection [3]. Nowadays, 3GPP and LTE are the most widely used technologies for communication with IoT systems [4], because these techniques offer wide area cover, low operation cost, more secure, allocated spectrum, and easiness in the management. But the recently available cellular infrastructure is not in the position

S. Sharma
Center for Reliability Sciences & Technologies, Chang Gung University, Taoyuan City, Taiwan

OMKARR Tech, New Delhi, India
e-mail: D000016086@cgu.edu.tw

P. Nigam · A. Pal (✉)
DIT University, Dehradun, India
e-mail: padmini.nigam@dituniversity.edu.in; amrindra.pal@dituniversity.edu.in

A. Muduli
KL Education Foundation, Guntur, Andhra Pradesh, India
e-mail: arjuna@kluniversity.in

S. Tanwar (ed.), *Blockchain for 5G-Enabled IoT*,
https://doi.org/10.1007/978-3-030-67490-8_23

to support the machine-type communications (MTC), which is the heart of the IoT system [5, 6].

The 5G offers the path to the billions of smart devices to automatically interact and share the data. Nowadays, diversified applications of the various instruments are found and create trouble for the IoT systems to identify that devices are capable or not to fulfil the need of the application [7]. Present IoT systems use the Bluetooth Low Energy (BLE), ZigBee, etc. technologies for specific applications. As well as some other data transfer or communication medium like Wi-Fi, low-power wide-area (LP-WA) networks, cellular communications (e.g. MTC using 3GPP, 4G (LTE)), etc. [3]. The IoT systems are changing continuously at a higher rate with the new concepts and technologies and expanding in the new application areas also.

The evolving 5G infrastructure are capable enough to resolve the above-mentioned issues. It offers the high data with low latency and large area of coverage for MTC communication as compared to 4G. In fact, the machine-to-machine (M2M) communication can establish the communication between a large number of smart devices. Figure 23.1 shows the 5G-based IoT application infrastructure [3].

The success of IoT is based not only on wired but equally on wireless communication capabilities. The wireless communication systems have enabled multifold faster development and the implementation of IoT systems as the systems are/now can be mobile. An IoT system essentially requires the data is transmitted in small packets at low transmission rate(s), reducing the bandwidth requirements. Typically, the bandwidth requirement for IoT applications is less than 1 MHz. Due to their reliable propagation characteristics, sub-GHz bands (Europe utilizes 868 MHz band and the USA uses the 915 MHz band) are often preferred. Although these frequency bands see less traffic with respect to other bands in the spectrum (like

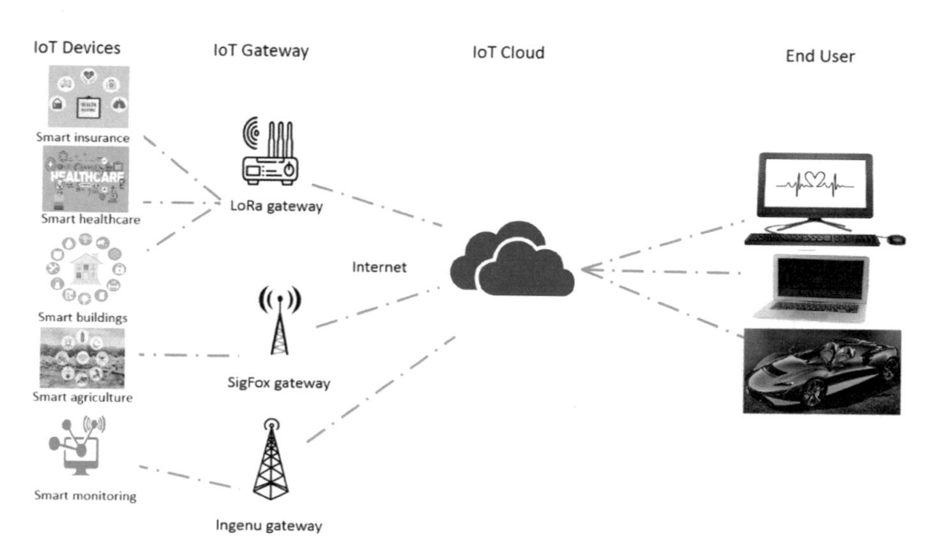

Fig. 23.1 A typical network infrastructure for 5G IoT applications [8]

2.4 GHz band), there is always a possibility of interference issues arising from other standards. The 5G communication networks provide faster, better and reliable communication networks for communicating systems. 5G network constitutes two parts, the Long-Term Evolution technology (FR1 – lower-frequency range, up to 6 GHz) and the millimetre wave (FR2 – higher-frequency range, from 24 to 52 GHz) [9]. The 5G antenna module for the IoT-based system is designed based on one or more of data rate, frequency of operation, range of communication and power consumption. The designer has several limitations to work with, namely, small footprint, minimum fading (interference) or negligible noise, higher gain and maximum radiation efficiency, and an enhanced version of multiplexing order is also required. The substrate integrated waveguide (SIW)-based antenna design provides a viable solution offering advantages like lightweight, high gain, maximum radiation efficiency, low fabrication and maintenance cost, high quality factor (Q) as well as improved power handling capacity than its counterparts μ-strip and CPW technology [10].

There are a large number of applications where we have a multiuser system environment requiring a reliable communication link (military applications) or a system having numerous antenna elements connected to a base station (medical or IoT applications) and the like(s). For a multiuser system environment connected with the base station consisting of large number of antenna elements, the SIW-based MIMO (multiple-input multiple-output) antenna for the 5G band application functional over the frequency range of 3.4–3.6 GHz is used. The MIMO system offers zero or negligible amount of mutual coupling over a single element. This is achieved by deploying an additional diplexer and triplexer decoupling network. A diplexer, triplexer, quadplexer or multiplexer is defined as a device that utilizes a single antenna for multiple transmitters operating at different frequencies. This decoupling network will consist of a diplexer, triplexer, quadplexer or multiplexer element, depending upon the number of input ports (feed points, di for two, tri for three, quad for four and multi for more than four transmitters), generally incorporated with antenna modules. These increase not only the antenna space, size and cost but the complexity of the proposed system as well, thus making these systems not suitable for portable IoT application or 5G communication systems.

This chapter provides a design to overcome the above problems. The SIW-based self-multiplexed antenna integrated with decoupling network working over 5G is proposed. This technique offers small footprint and compact size and is easily realized with highly demanding IoT systems. The isolation between two antennas having gain stability as well as unidirectional radiation pattern is more than −20 dB.

2 Evolution of Technology from 1G to 5G

In the era of advancement in the wireless communication system, 5G communication technology takes the outstanding part because of its prominent features. Previously, a decade ago from 1980 to 1990, the 1G-based communications was used, in which the basic voice service is used by the consumer with analog-based protocols. At this time only voice service was under consideration with PSTN (public switched telephone network) and PDMA (packet division multiple access) techniques. 1G technology is operated over the frequency of 150 MHz/900 MHz with the narrower bandwidth of 30 KHz and lower speed of 2.4 kbps. It has the several of disadvantages like bad voice quality, large size and poor battery life of cell phones. At this time, it is better than nothing; at least it's wireless and mobile. After that from 1991 to 2000, 2G communication had discovered using the new digital technique of GSM and CDMA. It is the first digital standard era of wireless communication with improved coverage area and capacity of no. of users than 1G technology. It is operated at the frequency of 1.8 GHz/900 MHz with the moderate bandwidth of 25 MHz and data rate of 64 kbps. It allows the text messaging service also, and the signals are also stronger than 1G. In this range, 2.5G comes with the GPRS cellular technique. It has the additional features of web browsing and e-mail services. At this time, the cell phones are combined with camera also. After the evolution takes place from 2.5G to 3G wireless system from 2000 to 2010, it has the frequency range of 1.6–2.0 GHz with bandwidth of 100 MHz. It is designed by CDMA, UMTS and EDGE techniques to provide the digital broadband and increased speed. It is the first mobile with broadband data services [11, 12]. At this time, the cell phones/mobile become the smart mobile with the extra features of fast communication, video call and broadcasting, mobile TV, etc. It also has the excellent value of data rates of 144 kbps to 2 Mbps with increased no. user capacity. These 3G mobiles are rather expensive than 2G. After that a new era has introduced with 4G communication from 2010 to 2020 with advent features of 3G. It is designed primarily for increased data rate from 100 Mbps to 1 Gbps. It is working at the frequency range of 2–8 GHz with the bandwidth of 100 MHz. It depends upon Wi-Fi and LTE technology and has the IP-based protocols and high speed and connectivity. It is the true mobile broadband service with MAGIC. The MAGIC term is defined as follows: M stands for mobile multimedia, A stands for anytime anywhere, G stands for global mobile support, I stands for integrated wireless solution, and C stands for customized personal service. 4G is a kind of multifunctional and flexible technology which depends upon OFDM (orthogonal frequency-division multiplexing) and OFDMA (orthogonal frequency-division multiple access) techniques. Nowadays, due to the 4G communication technique, we can easily and rapidly upload and download the high-definition (HD) video, movies, songs and other information. 4G system produces the high quality of services, super security and bigger battery life also. Then after 4G, the advanced and versatile features of 5G communication are presented with extreme high data rate of 1–100 Gbps, which is 100 times higher than 4G communication. The 5G communication has introduced from 2020 to 2030 with high capacity and speedy data rates. It also supports voice streaming, buckle up, high Internet, and interactive multimedia applications. It supports WWWW (World Wide Wireless Web) technology and defines the next version of mobile communication. After 5G, a novel version of communication technology is also taking place as 6G and 7G. 6G will integrate 5G with satellite global coverage with ultra-fast Internet access along with smart home and cities applications, while

7G relates world completely wireless with space roaming and artificial intelligence techniques. It's a fully wireless network which depends over artificial intelligence methodology [13].

2.1 5G Communication

Wireless networks have made an extraordinary development in the past few years. The demand of more bandwidth and lower latency has been a motivation to develop efficient systems. 5G is the fifth-generation mobile network, and this is the wireless communication technology which enhances user experience with the help of its features to personalize mobile communication experience. 5G is intended to offer higher data speeds (multi-Gbps), better reliability, increased availability, enormous network capacity, ultra-low latency and a better than before uniform user experience. Improved performance and enhanced efficiency allow new as well as better user experiences. 5G provides the infrastructure which will increase the performance and capabilities of the communication network. 5G includes high carrier frequencies, unprecedented number of antennas, massive bandwidths and device density. 5G emerged from orthogonal frequency-division multiplexing (OFDM). OFDM is the method of modulating a digital signal across numerous different channels to reduce interference. Sub-6 GHz and mm-wave which have wider bandwidth are used in 5G technology. The same mobile networking principles like 4G LTE are used in 5G OFDM. In addition to it, 5G air interface further increases OFDM to give a better scalability and flexibility. As a result, more people can have access to 5G technology [14].

5G provides a seamless compatibility with dense heterogeneous network. This satisfies the high demand of traffic and efficient connectivity to the users. 5G works smoothly even when the number of users connected to the Internet goes over billions in number. Basically, 5G uses the unused part of 3–300 GHz high-frequency mm-wave. This sub-spectrum of mm-wave spectrum can support improved data rate over present wireless system to over hundreds of times to satisfy end user needs. 5G is designed to efficiently use every bit of spectrum across a wide array of available bands and spectrum. 5G is invented to emerge into service areas say, for example, connecting the Internet of Things, artificial intelligence, virtual reality and critical communication systems. This is achieved by 5G NR air interface design techniques, like self-contained TDD sub-frame design. 5G is designed to be energy efficient; it consumes less energy as compared to prior communication devices. This helps in reduction in environmental issues as well as network maintenance issues. Current mobile network consumes about 15–20% of total power consumption on actual data traffic, and the rest of the energy is wasted [15].

5G will have a massive impact on businesses; it will provide high data speeds with higher network reliability. Businesses could use 5G to connect their devices to the same network. By doing this, we can make machines work with the help of mobile device. This can be applied where higher degree of precision is needed.

This will help in increasing the efficiency of businesses and giving users faster access to information. 5G technology can also help in intelligent movement and communication among vehicles. Traffic management can be achieved by designing a network of interconnected vehicles. Real-time data of traffic can be provided to the drivers so that vehicles on the road can choose less congested road to arrive at destination.

2.2 5G-Enabled IoT Applications

The upcoming years of 5G communication will totally improve the insight of lifestyle on society, industry and business field also. It is a radical technology with unique and massive research areas like in healthcare, smart city, robotics and virtual reality-based systems. The IoT technology is incorporated with 5G communication network to attain the tremendous outcomes in every perception of human life. IoT defines as the Internet of Things, in which the things are integrated with various new technologies, software and sensors for the result to obtain the exchange of data and information over the network. 5G with IoT applications defines as a new opening generation for integration of intelligence with comfort and security [16–18].

1. Transition from normal to smart communication
2. Excellence of services
3. Internet of Things to everything
4. Artificial intelligence and edge intelligence
5. Vehicular technology in 5G and beyond
    (a) Intelligent aerial vehicles
    (b) Intelligent car without man
    (c) Intelligent transportation
    (d) Intelligent robotics system

2.3 Blockchain

A system or policy for recording information in a secure and robust manner that it is very difficult, if not impossible, to hack, maliciously change, or cheat a system. Could this be true? Yes, through a system called blockchain. It is a digital ledger or an indefinitely growing list of cryptographically linked blocks (records) [19–23]. These blocks individually are constituted by a collection essentially of a cryptographic hash of the previous block, a timestamp and transaction data (generally represented as a Merkle tree). These are distributed as well as duplicated across the entire network of computer systems on the blockchain. It has supported the transition to a much required cryptographically secured and decentralized network from the existing centralized client-server systems (Internet). It facilitates the users to have a distributed P2P network, where information can be exchanged without the need of any trusted intermediary [24] among non-trusting members.

2.4 Usage of Blockchain in 5G-Enabled IoT

The fast-paced evolution of smart applications focused at improving the quality of life for us; Internet of Things is the forerunner in the digitization of services. As IoT infrastructure grows, there is an increase in the access points that need to access and share information. Cloud computing has played an important role in aiding these fast-paced developments related to the IoT domain. However, centralized structure like these as described in detail in [25] may lead to failure in maintaining data transparency. Blockchain provides loads of improved features in terms of security, decentralization and scalability, identity, autonomy, security, reliability [25–27] as well as privacy [28]; Fig. 23.2 depicts the advantages of blockchain deployment in industrial automation based on 5G-enabled IoT systems.

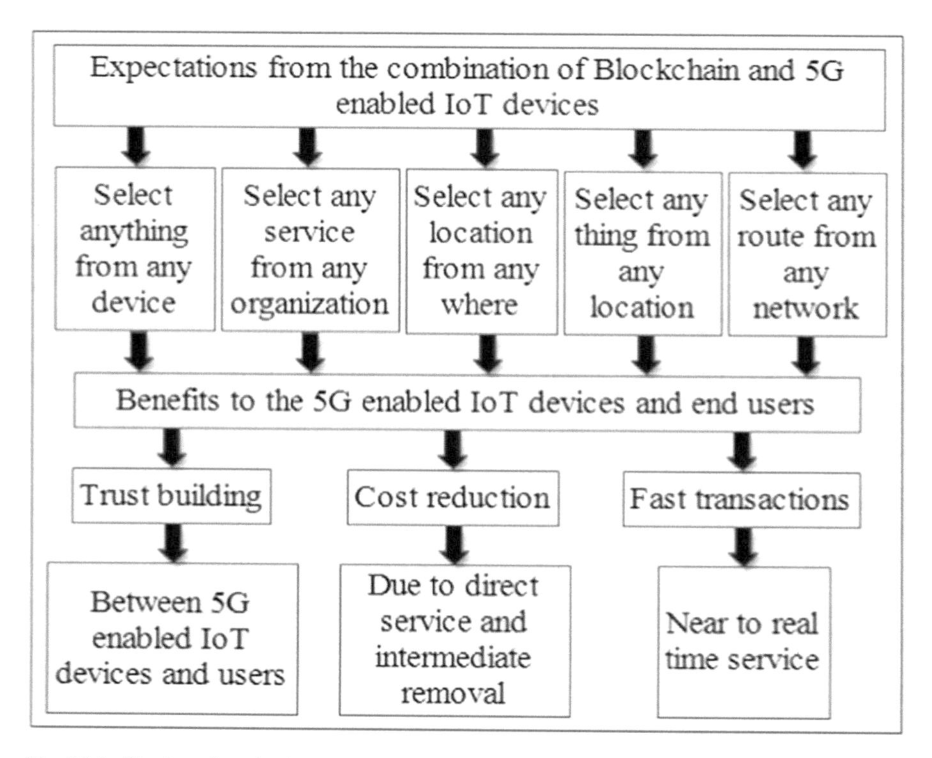

Fig. 23.2 Key benefits of using blockchain with 5G-enabled IoT for industrial automation [23]

2.5 *Role of an Antenna Design for 5G-Enabled IoT*

Antenna is an interface between the IoT device and the rest of the network or system. It is apparent that without a reliable antenna, none of the advantages as have been portrayed above are possible. A good antenna contributes to the success of the system by:

- Improving gains without increasing the cost of additional battery consumption
- Providing consistent signal propagation to facilitate on-the-go communication during IoT device movement
- Improving the signal level while simultaneously reducing noise reception (elimination of noise if possible) through optimum isolation (particularly from other antennas or components in a product having multiple elements, that is, antennas or components)

In a typical communication system, there is a requirement of two antennas – one that transmits and the other that receives. Thus, slight improvement in the antenna involved and the performance receives a twofold improvement in received signal-to-noise ratio, leading to improvement in the data rates, enhanced range and increased security. It can be concluded that there is an overall improvement in the user experience, thus resulting in the improved customer satisfaction.

3 Antenna

The antenna is the interface between the transmission line and space. The IEEE standard defines an antenna as "the antenna or aerial as a means for radiating or receiving the radio waves". Antennas are passive devices; the power radiated cannot be greater than the power entering from the transmitter side. The antennas are reciprocal in nature, i.e. one design works as both a receiver and a transmitter [29].

- Some techniques to implement the antenna:

  (a) Microstrip-based antenna
  (b) Waveguide-based antenna
  (c) Meta-material-based antenna

Figure 23.3 shows the category wise available in different antennas. Some of the antennas are suitable for IoT system applications.

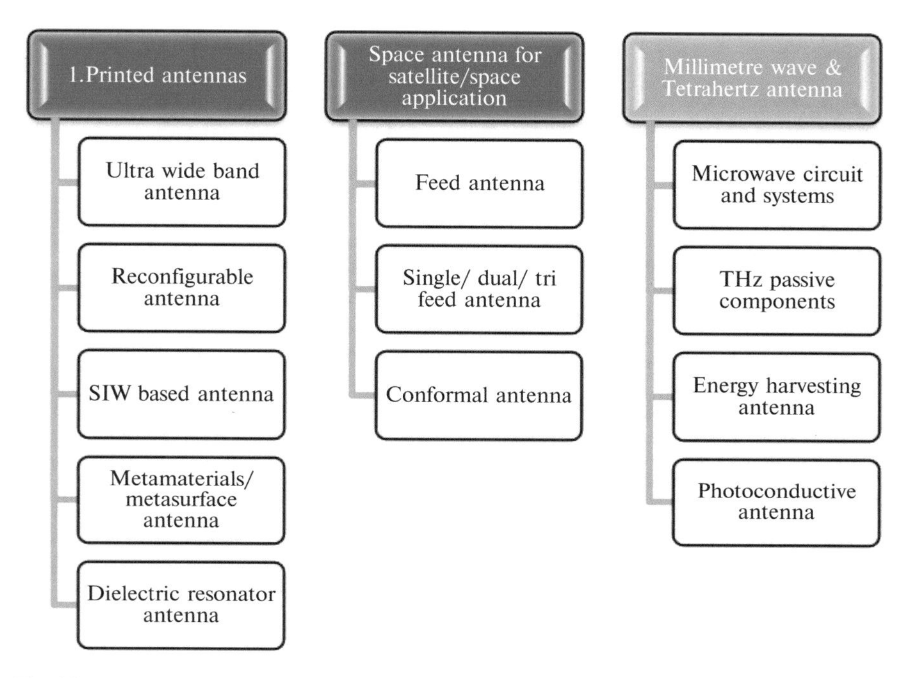

Fig. 23.3 Typical available antennas

3.1 *Yearly Configuration of Radio Engineering*

Radio engineering utilizing the waveguide, transmission line and antenna as a key element and concept of the electromagnetic field to produce signals within the radio band (frequency range – 20 kHz up to 300 GHz). Figure 23.4 shows the yearly evolution of the radio engineering field.

3.2 *Applications of Radio Engineering and Antennas*

There are a wide range of applications of antennas in radio engineering; some are listed below for ready reference.

1. The Commercial Application
   (a) 802.11x markets: (i). WLAN and (ii). WPAN
   (b) Automotive RADAR at 77/79 GHz
   (c) Telecommunications backhaul
   (d) Wireless last-mile connectivity

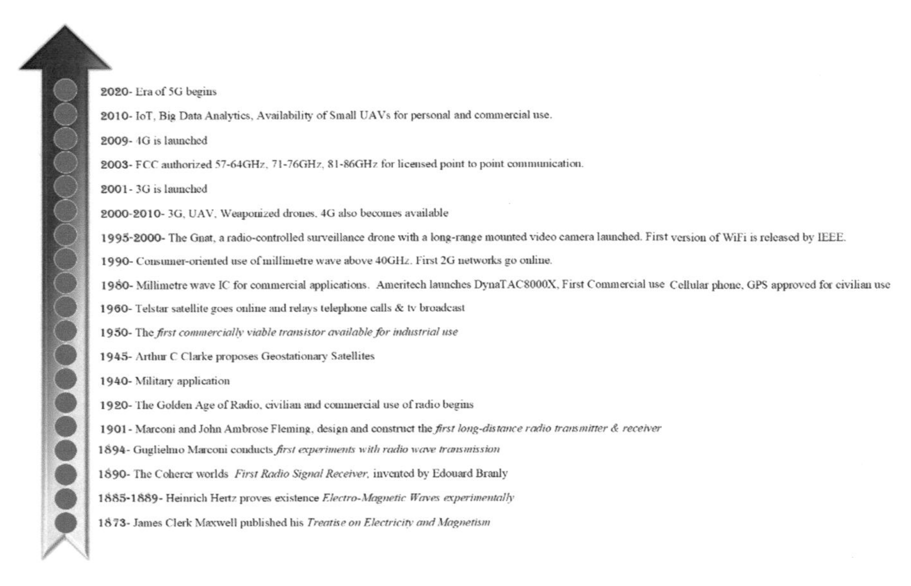

Fig. 23.4 Evolution of the radio engineering field

2. Military Markets (38, 60, 94 GHz)
    (a) Future combat systems
    (b) Secure communications
    (c) Satellite communications
    (d) Military phased array
    (e) Reconfigurable, software definable systems

To implement any kind of antenna, the transmission line takes the most critical role in designing it. So, several transmission lines with their properties and disadvantages are displayed in Table 23.1 [30–36].

4 Substrate Integrated Circuits (SICs)

The basic principle of SICs is to convert non-planar structures into their corresponding planar form and enable the planar fabrication processing of the non-planar and 3D design [37, 38].

- SIC Structures:
    1. SIW (SIW)
    2. Substrate integrated non-radiated dielectric guide (SINRD)
    3. Substrate integrated image guide (SIIG)

    In all of the above, SIW is the most popular structure.

Table 23.1 Planar transmission line properties

Transmission line	Microstrip line	Strip line	Suspended	Fin line	Slot line	Inverted microstrip line	Coplanar waveguide
Operating frequency (GHz)	≤110	≤60	≤220	≤220	≤110	≤220	≤110
Characteristics impedance range	10–100	20–150	20–150	20–400	60–200	25–130	40–150
Dimension	Small	Moderate	Moderate	Moderate	Small	Small	Small
Loss	High	Low	Low	Moderate	High	Moderate	High
Power handling	Low	Low	Low	Low	Low	Low	Low
Solid-state device mounting	Fair	Moderate	Moderate	Easy	Easy	Moderate	Very easy
Low-cost production	Good	Good	Fair	Fair	Good	Fair	Good

4.1 Substrate Integrated Waveguide (SIW) Technology

Recently, the microwave-based wireless components perform a significant role in the field of communication modules. It includes the microwave- and RF-based components for the development of different systems. However, previously, several techniques are combined with microstrip, metallic and coplanar waveguides to achieve the high Q-factor for active as well as passive components. As the microwave system is the emerging topic for the RF researchers to implement the device with high data rates for 5G/6G application and excellent sensing and power handling capabilities. These all are the basic requirement of wireless communication system, so to cover all above advantages, SIW plays the crucial role in the microwave system with easy fabrication topology. The upcoming wireless communication module with SIW technology provides the advancement towards the high reliability, increased performance, good stability and enhanced integration with systems [39].

SIW is the planar form of the rectangular waveguide. A dielectric is filled between two parallel metal plates. These two rows of the conducting vias/holes connect the two parallel plates through the substrate. To implement the millimetre wave integrate circuit, the SIW technology is mostly used. It offers advantages like cost-effectiveness and high-density integration and is most suitable for the mass production of the wireless system. It has widespread solution for millimetre wave applications. SIW was first proposed by **K. Wu and Deslandes**.

Definition *"The waveguide like structure fabricated by using two periodic rows of metallic holes or vias or slots connecting the top and bottom ground planes of dielectric substrate".*

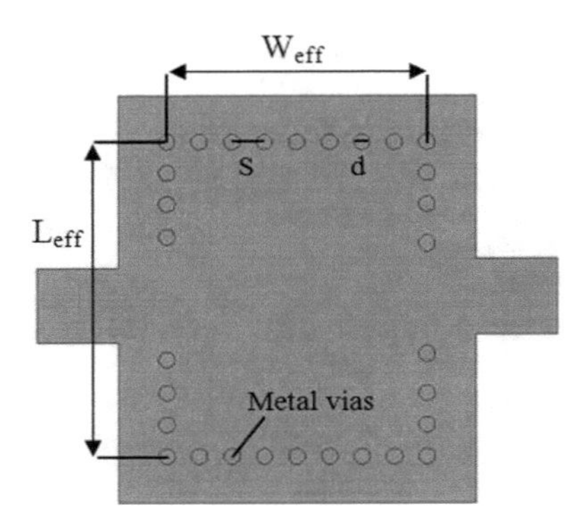

Fig. 23.5 Substrate integrated waveguide structure

It also combines the advantages of microstrip (low cost, easy fabrication, compact size, low weight) and metallic waveguide (low loss, complete shielding and high power handling capability) [40]. Figure 23.5 depicts a typical SIW structure.

- The effective width of the SIW is given as:

$$w_{\text{eff}} = w - \frac{d^2}{0.95s} \text{ or } w_{\text{eff}} = w - 1.08\frac{d^2}{s} + 0.1\frac{d^2}{w} \quad (23.1)$$

- To maintain loss-free radiation, choose and guided wavelength:

$$s \geq 2d \quad (23.2)$$

$$d \leq \lambda_g/5 \quad (23.3)$$

$$\lambda_g = \frac{2\pi}{\sqrt{\frac{\varepsilon_r(2\pi f)^2}{c^2} - \left(\frac{\pi}{a}\right)^2}} \quad (23.4)$$

where w is the width of the rectangular waveguide, s is the pitch, d is the diameter of vias and λ_g is the guided wavelength of SIW.

4.2 Substrate Dimensions

- Height $(h) \leq W_{\text{eff}}/2$.
- Thickness of copper $(t) = 5*$ skin depth.
- Skin depth $= \frac{1}{\sqrt{\pi * f * \mu * \sigma}} = 0.0105$.
- Substrate height is 0.508 mm and thickness is 0.035 mm.

Table 23.2 Previous reported work for the SIW-based antenna

Techniques	Advantages	Disadvantages
Array type	Less bandwidth and large gain	Complex design and large sizes
SIW feed	Less bandwidth and large gain	Large size
SIW cavity	More bandwidth and size miniaturization	Average gain
HMSIW cavity	More bandwidth and size miniaturization	Average gain
QMSIW cavity	More bandwidth and size miniaturization	Average gain
Super state	More bandwidth and moderate gain	Complex design and large size

Table 23.3 Comparative analysis of the SIW technique with other available techniques

Features	Loss	Power handling capability	Compactness	Cost	Self-shielded	Self-packaged
Waveguide	Less	High	Low	High	Yes	Yes
Planar transmission line	High	Low	Good	Less	No	No
SIW	Moderate	Moderate	Good	Less	Yes	Yes

4.3 *Advantages of SIW Technology*

1. Complete circuit in a planar form (including passive component, active component and antennas)
2. Low-cost and well-developed manufacturing process
3. High-density integration of millimetre wave components and systems
4. Complete shielding (no interference)
5. Low losses (energy saving)
6. High Q-factor
7. High power handling
8. High performance

A brief comparison among the different techniques and antenna technology available in literature is shown in Tables 23.2 and 23.3, respectively [41, 42].

In the last decades, SIW technology has reached incredible popularity

4.4 *SIW-Based Components*

SIW is fabricated by using planar circuitry as given below, and the comparison among different types of SIW-based antenna is described in Table 23.4.

1. Active component

   (a) Feedback oscillator

Table 23.4 Comparison of different types of SIW antenna

Features	Horn	Patch	Slot	Leaky-wave antenna
Frequency	<1 THz	<0.1 THz	<0.5 THz	<0.5 THz
Cost	Average+	Low	Low	High+
Gain	Average	Low	High	Moderate
Size	Large+	Average	Small	Large++
Bandwidth	Wide	Narrow	Wide	Narrow
Fabrication	Difficult	Easy	Average	Difficult

   (b) Mixer
   (c) Frequency selector
   (d) Power amplifier

2. Passive component
   (a) Filter
   (b) Directional coupler
   (c) Magic tee
3. Antennas
   (a) SIW horn antenna
   (b) SIW slot antenna
   (c) SIW leaky-wave antenna

4.5 *Future Scope of SIW Technology*

1. The SIW array-based slot antenna as the best solution for improving the gain as well as radiation efficiency
2. The meta-material-based slot loaded antenna also as improvement of gain and bandwidth
3. Extended to implement some or other unlicensed frequency applications:
   (a) 70 GHz band (E-band)
   (b) 79 GHz (automotive RADAR systems)
   (c) 80 GHz (E-band)
   (d) 94 GHz (millimetre wave imaging)
4. Extended to dual- and tri-band unlicensed frequency applications
5. Extended to implement high-frequency applications, i.e. THz applications [38–41]

5 Diplexer/Triplexer as Frequency-Selective Element

In the last few years, wireless communication systems have been mostly developed in diplexing and triplexing, maintaining additional and extraordinary functionality and compactness. The major part of conventional diplexing- or triplexing-based communication system is diplexer/triplexer as shown in Fig. 23.6, in which the uplink and downlink channels of transmitter and receiver signal are in neighbouring frequency bands. To enhance the isolation between the uplink and downlink bands, transmission noise rejection and low power transmission for the transceiver, we will propose planar SIW-based self-diplexed/triplexed antennas.

The function of this device is to combine and/or split RF transmitters to facilitate the use of single device by multiple transmitters or receivers on different frequencies reducing cost, space and requirement of addition circuit components. This device may either facilitate transmission from more than one transmitter over a single RF antenna, or it may be used to work as a transceiver for transmitting over one frequency band while receiving is done on another band of frequency [24].

The antenna diplexer has multiple applications. The most common application is to be used as a transceiver at a cellular base station for simultaneous inflow and outflow of signals. This device (self-duplexing antenna) provides high isolation between the used receiving and transmitting feeds, thus allowing the same device to operate for transmission/reception simultaneously blocking the signal. An alternate application for the device could be at a broadcast station, where signal is simultaneously transmitted with many a different frequencies with one antenna element. The device's operation prevents the output of any of the transmitters being fed back into the other's input, thus enabling the use of a single antenna.

In domestic environment, these devices may be used to couple the TV feed coming from terrestrial feeds as well as satellite transmissions into a common feed multiplexed over frequency passed down through the same lead. Figure 23.7 shows the mode of operation of the diplexer.

A diplexer/triplexer designer must keep the following in mind while designing the device and the overall system:

- To function in the adjacent frequency band used for reception and transmission
- To efficiently handle the output power (at the transmitter end)
- To provide adequate rejection of transmitter noise occurring at the reception frequency

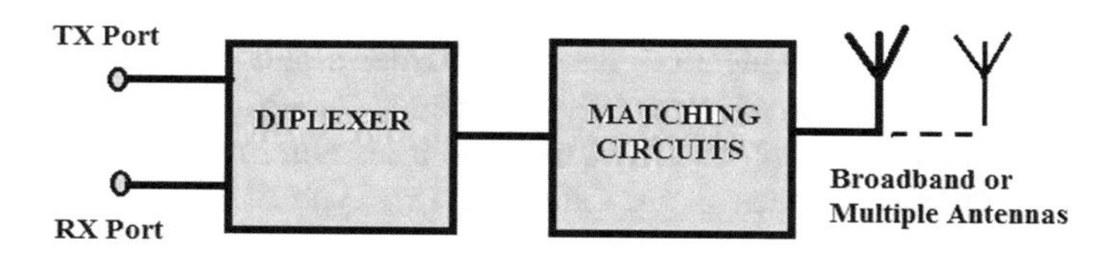

Fig. 23.6 Block diagrams of a traditional diplexer antenna system [27]

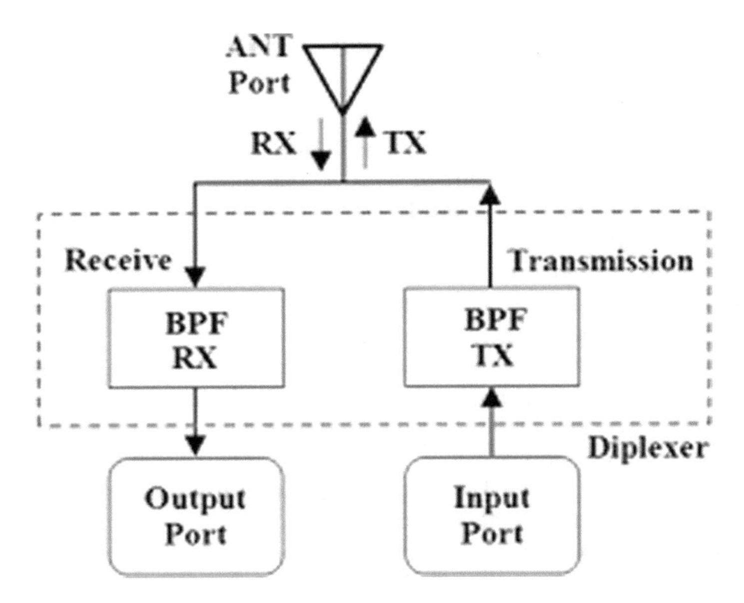

Fig. 23.7 Diplexer working for transmission and receiving antenna [24]

- To operate at, or less than, the frequency separation between the transmitter and receiver
- Sufficient isolation to be ensured to eliminate receiver desensitization [43]

5.1 Motivation

1. *Self-Diplexing Technique*

   In this technique, the antenna has two ports of different frequencies as uplink and downlink. With the help of self-diplexed antenna, we can reduce the complexity of high-order diplexer network, which leads to minimum cost and small size of overall RF (radio-frequency) front-end system. From the last few years, the state of the art of SIW technology becomes a very promising candidate to the implementation and realization of conventional waveguide and other non-planar circuits into their planar counterparts.
2. *Self-Triplexing Technique*

   In the field of communication, the higher-order diplexer and triplexer come under the types of frequency-selective elements, which provide ease of connectivity to multi-band antenna from multiple transmitter/receiver with the help of better isolation between them. Generally this type of circuit increases the complexity of system and limits its application also.
3. *Self-Quadplexing Technique*

In the RF front-end system, quad band antenna is connected to the external element as quadplexer to provide the isolation or low mutual coupling among the input ports, but it requires additional circuitry and in turn extra space for this circuitry in order to keep things compact by eliminating the external quadplexer from RF system, a new technique is proposed called a self-quadplexing technique used to implement the four-port single antenna with high isolation values.

5.2 Different Techniques to Achieve the Objectives

1. *Defected Ground Structure (DGS)*
   Recently, DGS plays a significant role in wireless as well as electromagnetic field. DGS can be found by etching off any shape over the ground plane, as its name describes itself. DGS depends upon the dimension and shape of the defect. Due to DGS method, the shielded current distribution is also disturbed, which results in the propagation of electromagnetic wave through the substrate layer. For getting the better performance of the system, the shape of defect may be altered from simple to complex. With the help of DGS, the isolation between the ports can be increased, and mutual coupling can be reduced also, e.g. dumbbell and bowtie-shaped, etc. [26, 44, 45].
2. *Electromagnetic Band Gap Structure (EBG)*
   EBG structures are used for improving the performances of many RF and microwave devices utilizing the surface wave suppression. They are inserted between the arrays of antenna to reduce the mutual coupling and increase the isolation between the input ports. It prevents some undesired operating mode and control harmonics, e.g. 3-D, 2-D and 1-D EBG, mushroom and uni-polar EBG, etc. [46, 47].
3. *Meander Lines*
   Meander lines are added between two input ports, and this feature supplies an extra input to the port to cancel the signal due to mutual coupling, by which mutual coupling can be reduced and isolation increased [48].
4. *Grounding Vias*
   It is the kind of efficient approach for implementing self-diplexing and self-triplexing antenna system. This approach is recently proposed in microstrip patch antenna for better polarization and controlling the resonant frequency of antenna, so this methodology can be proposed for SIW-based antennas also, where the loaded grounding vias are considered as shunt inductors and produce the inductive effect. This method can also enhance the isolation between ports and gives the better tuning of frequency response [37, 38].
5. *Shunt Component Between Transmission Lines*
   It provides the good isolation level between transmitter and receiver antenna but has the complex structure to fulfil the desire.

5.3 Flowchart for Simulating the Electromagnetic Structure

To simulate any kind of active or passive electromagnetic components, the proposed flowchart is used (shown in Fig. 23.8). To analyse the simulation of any microwave or terahertz components, different electromagnetic software are used based on their applications as defined in Table 23.5.

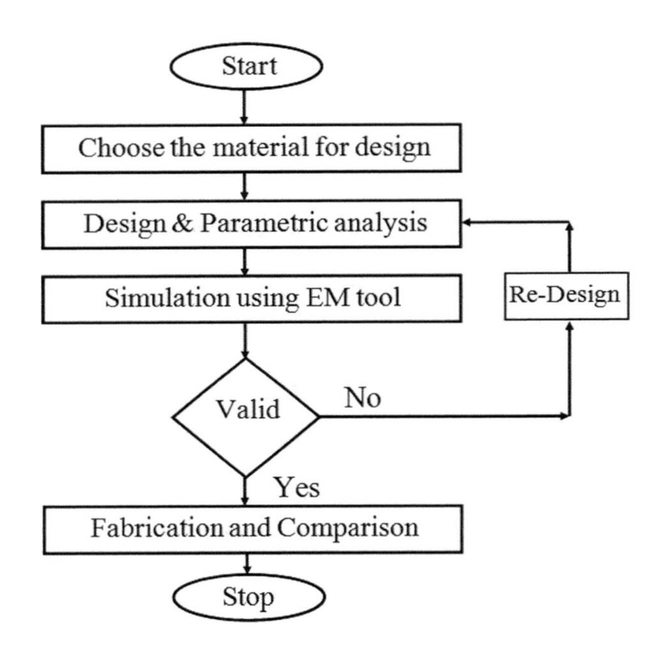

Fig. 23.8 Flowchart of simulating the electromagnetic structure

Table 23.5 Software for antenna design and simulation

Software	Numerical techniques used	Applications (microwave structures)
HFSS	FEM	EM simulation of passive 3-D μwave structures
CST	FIT/FEM/TLM	EM simulation of passive 3-D μwave structures
MW STUDIO	FDTD	EM simulation of passive 3-D μwave structures
ENSEMBLE	MOM	EM simulation of passive 2-D μwave structures
IE3D	MOM	EM simulation of passive 2-D μwave structures
TICRA	GO/GTD/PO/PTD	Reflector antenna analysis, shaping, optimization
MW WIZARD	MOM	Synthesis, analysis of 3-D passive μwave filter
WASPNET	MOM/FEM/FDTD	EM analysis tool for 3-D passive μwave filter
WIPL-D	MOM	EM simulation/analysis of EM structures
EMPIRE	FDTD	EM simulation of passive 3-D μwave structures
ADS	Equations	Microwave passive circuit simulation (feed networks)

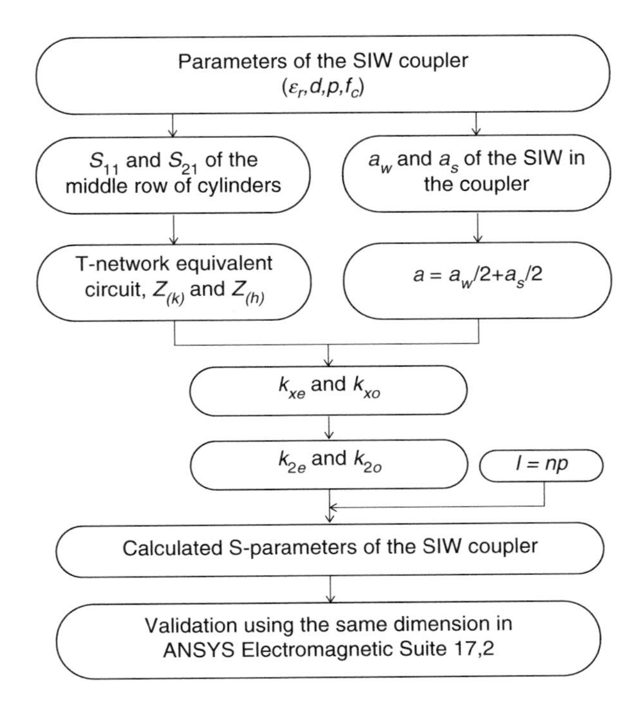

Fig. 23.9 SIW coupling analysis flowchart

5.4 SIW Coupling Analysis Method

The SIW coupling is based on the different features of antenna as its relative permittivity (ε_r), vias gapping (*d*), resonating/centre frequency (f_c) and pitch (*p*) values of SIW design. The row of the cylinder or conducting vias proposed the electromagnetic boundary conditions, and with the help of ADS software, "T"-network can be found out. By this even and odd impedance can be retrieved and at last |S|-parameters are calculated by HFSS software. The overall working is defined in the flowchart as shown in Fig. 23.9. The coupling analysis is done by S_{21} parameter for the two-port device.

6 IoT-Based Antennas

IoT-based communication mostly depends upon the sensors and artificial intelligence techniques. Data Communication Systems use a device called IoT antenna for data transmission and reception. These antennas are compact and flexible in nature to operate within 5G bands as in lower and upper both bands. These bands are defined in Table 23.6. An antenna is a passive electromagnetic device used to transfer the information from one place to another. For this transmitting

Table 23.6 5G communication bands for IoT applications

Frequency band	Frequency range of operation	Objective
Sub-1 GHz band	600 MHz band 700 MHz band	Wide area coverage
Sub-6 GHz	3.3–3.8 GHz (mostly allocated) 3.8–4.2 GHz, 4.5–5 GHz 2.3–2.5 GHz, 5.9–7.1 GHz	Good mixture of coverage and speed requirements
Millimetre wave band	28 GHz, 38 GHz bands May be 60 GHz band	Higher speed requirements

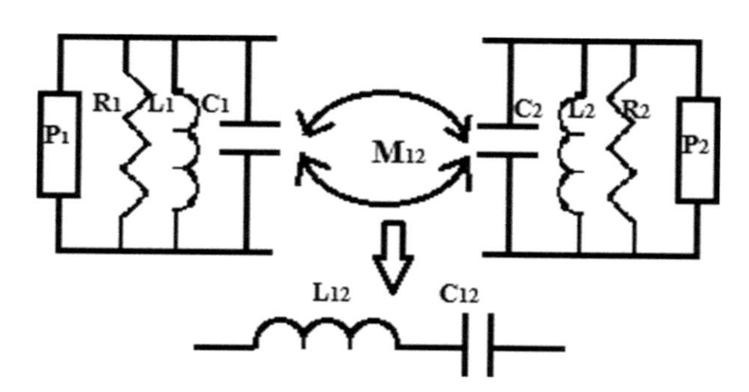

Fig. 23.10 The equivalent circuit model of self-diplexing antenna

as well as receiving, both antennas are used generally in communication and RF front-end systems. Previously, the antenna systems were combined with the diplexer and triplexer elements. These elements are kind of frequency selective or filter element and used to recognize the particular frequency. The self-diplexing, self-triplexing and self-quadplexing antenna, are mostly implemented by SIW technology, these antennas find applications for satellite/RADAR communication, and can be integrated with IoT applications best use. These are defined in sub-sections as below [49].

6.1 Equivalent Circuit Model of Self-Diplexing Circuit Modules

The equivalent circuit model for the self-diplexing antenna is shown in Fig. 23.10. The proposed equivalent circuit model is realized with the help of ADS (Advanced Design System) software. Here each parallel combination of RLC components produces the resonant frequencies. Likewise, parallel combination of R_1, L_1 and C_1 produces the first resonant frequency, while another parallel combination of R_2, L_2 and C_2 produces the second resonant frequency. The mutual coupling (M_{12}) between two frequencies is defined by series combination of LC network. Here L_{12} and C_{12} present the minimum isolation value for the proposed self-diplexing antenna.

The resonating frequency can be defined by:

$$f_0 = {}^{1}\big/_{2\pi\sqrt{LC}} \tag{23.5}$$

6.2 *Self-Triplexing Antenna for IoT Applications*

The self-triplexing antenna and triplexers are working at three distinct resonant frequencies. Previously the triplexers are used in RF front-end systems for selecting and filtering the required three frequencies used for further communication. This triplexer is used to provide the required isolation value among three ports; however, this geometry contains more space in overall system. So, the self-triplexing antennas are introduced in RF field to remove the extra circuitry as triplexer from communication system [50–56]. These proposed antennas are more in demand, and RF researchers are mostly working in this field to integrate it with other technology like artificial intelligence, sensors, etc. The comparison among self-triplexing antennas and triplexer is displayed in Table 23.7.

6.2.1 Equivalent Circuit Model of Self-Triplexing Circuit Modules

The equivalent circuit model for self-triplexing antenna is shown in Fig. 23.11. The triplexing antenna works at three resonating frequencies, and each resonating frequency is defined by the parallel section of RLC network connected with the port

Table 23.7 Comparison among triplexer and self-triplexing circuit modules

Features/ref.	[54]	[52]	[50]	[51]	[56]
Resonating frequencies (GHz)	7.8, 9.4, 9.8	11, 12.2, 13.1	4.14, 6.1, 8.32	8.5, 9.7, 12.0	0.85, 1.65, 2.45
Min. isolation	22.5	26.45	30.8	23	22.7
Permittivity (ε_r)	2.2	2.2	2.2	2.2	NA
Gain (dBi)	7.2, 7.2, 7.0	5.1, 5.54, 6.12	4.26, 4.41, 6.27	3.5, 4.7, 5.2	0.85, 4.0, 4.23
Frequency tunability	Yes	Yes	Yes	NA	Yes
Size (mm^2)	32 × 23	26 × 27.5	22 × 22	20.5 × 20.5	115 × 77
FTBR	>17.3	>15	>15	>19.5	NA
Multiplexing circuit	Not required	Not required	Not required	Not required	Required

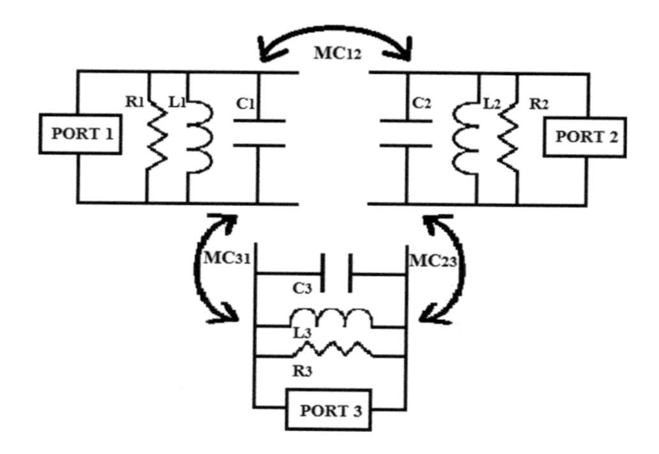

Fig. 23.11 The equivalent circuit model of the self-triplexing antenna

of 50 ohm line. At that time, only single port is ON and produces the corresponding resonating frequency. The first RLC network as R_1, L_1 and C_1 produces the first resonant frequency; likewise, the second and third parallel RLC sections give the second and third operating frequencies. The isolation is the main parameter of the proposed antenna, so the mutual coupling is defined by the M_{12}, M_{23} and M_{31} components, where M_{12} shows the mutual coupling between port 1 and port 2. Similarly, the M_{23} and M_{31} define the minimum mutual coupling between port 2 and port 3 and port 3 and port 1. The proposed equivalent model can be easily realized by ADS software and justified by HFSS also. The value of each resonating frequency can be obtained by Eq. (23.5).

6.3 *Self-Quadplexing Antenna for IoT Applications*

The frequency-selective element or filters are used to refine the frequency of interest and blocked the others, so quadplexer is also the same element used to find out the four required frequencies with the help of each port. This is mostly used in satellite-, RADAR- and RF-based systems to eliminate the noise and unwanted frequencies; only the desired range of frequency can be passed by the four ports of the quadplexer element. It requires the extra space in any of RF communication systems to provide the isolation or low coupling among the input ports, so the latest technology named as self-quadplexing antenna is introduced by RF researches. It is the most prominent technique used in RF front-end system with light weight, compact space and low losses [57–60]. Here several features of self-quadplexing antenna are explained in Table 23.8. Nowadays, this proposed quadplexing technique is implemented with the help SIW methodology, to get extreme better results than others.

Table 23.8 Comparison among multiplexer and self-quadplexing circuit modules

Features/ref.	[61]	[60]	[59]	[58]	[57]
Resonating frequencies (GHz)	3.5, 5.2, 5.5, 5.8	8.1, 8.78, 9.7, 11.0	5.14, 5.78, 6.74, 7.74	2.4, 3.5, 5.2, 5.8	1.2, 2.4, 3.5, 5.2
Min. isolation	23.6	22.6	28	NA	NA
Permittivity (ε_r)	2.2	2.2	2.33	NA	NA ($\mu_r = 1$)
Gain (dBi)	5.43, 4.1, 3.56, 3.6	5.5, 6.9, 7.47, 7.45	4.1, 4.96, 6.2, 6.1	2.8, 2.1, 3.5, 3.2	5.47, 5.88, 1.97, 3.56
Frequency tunability	Yes	Yes	Yes	No	No
Size (mm^2)	38.8 × 25.6	29 × 29	22 × 22	25 × 20	90 × 60
FTBR	NA	>17	>17.5	NA	NA
Multiplexing circuit	Not required	Not required	Not required	Required	Required

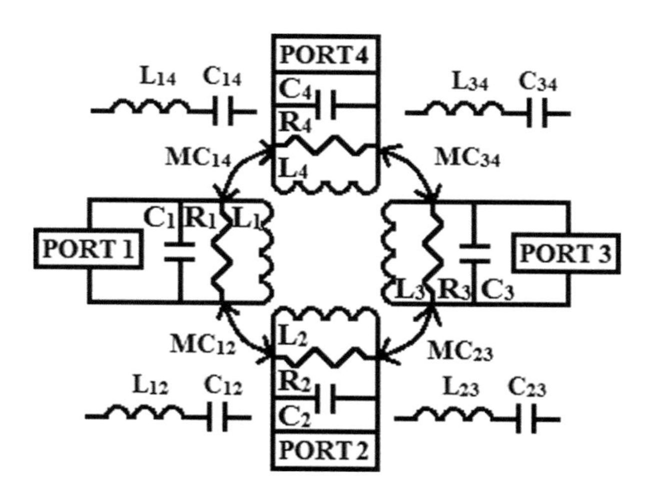

Fig. 23.12 The equivalent circuit model of the self-quadplexing antenna

6.3.1 Equivalent Circuit Model of Self-Quadplexing Circuit Modules

The equivalent circuit model for the self-quadplexing antenna is displayed in Fig. 23.12. The quadplexing antenna module contains the four input ports, and the isolation is defined among all ports in terms of mutual coupling. Here all four resonating frequencies are determined by the parallel combination of RLC network, and mutual coupling is attained by series combination of LC networks. One port is ON simultaneously and produces the particular resonant frequency. The mutual coupling is known in the form of interference, so the minimum mutual coupling is reciprocal to high isolation value and it is proportional to low interference among input ports. The mutual coupling among four ports is defined by M_{12}, M_{23}, M_{34}

and M_{41}. This model can be implemented by ADS software with the help of no. of equations, and electromagnetic structure can be executed by HFSS or CST software.

7 Conclusion

With the substantial development of the latest mobile and satellite communication, the multiple frequency antennas with high isolation and low mutual coupling are one of the particular interests. The SIW (SIW)-based single-layered self-diplexing, self-triplexing and self-quadplexing antenna has been proposed for mobile and satellite communication, the design techniques help to achieve high isolation and low mutual coupling between transmitting and receiving antenna. Additionally, low cross-polarization, high gain and maximum front-to-back ratio are obtained. These proposed antennas are used in RF front-end system without the need of extra circuitry like diplexer and triplexer. These antennas have low manufacturing cost, light weight and low losses; by virtue of this, it can be easily incorporated with 5G-enabled applications and IoT-based applications also. As the upcoming era will become fully 5G and IoT-enabled communication, then proposed antennas as self-diplexing, self-triplexing and self-quadplexing antennas are of great and fit choice for RF researchers and engineers.

References

1. I. Budhiraja, S. Tyagi, S. Tanwar, N. Kumar, J.J.P.C. Rodrigues, Tactile internet for smart communities in 5G: An insight for NOMA-based solutions. IEEE Trans. Ind. Inform. **15**(5), 3104–3112 (2019)
2. R. Gupta, A. Shukla, S. Tanwar, AaYusH: A smart contract-based telesurgery system for healthcare 4.0, in *2020 IEEE Int. Conf. Commun. Work. ICC Work. 2020 – Proc.*, (2020), pp. 1–6
3. S. Li, L. Da Xu, S. Zhao, 5G Internet of Things: A survey. J. Ind. Inf. Integr. **10**(February), 1–9 (2018)
4. D. Astely, E. Dahlman, G. Fodor, S. Parkvall, J. Sachs, LTE release 12 and beyond. IEEE Commun. Mag. **51**(7), 154–160 (2013)
5. D. Vukobratovic et al., CONDENSE: A reconfigurable knowledge acquisition architecture for future 5G IoT. IEEE Access **4**, 3360–3378 (2016)
6. Q. Wang, D. Chen, N. Zhang, Z. Qin, Z. Qin, LACS: A lightweight label-based access control scheme in IoT-based 5G caching context. IEEE Access **5**(c), 4018–4027 (2017)
7. I.F. Akyildiz, S. Nie, S.C. Lin, M. Chandrasekaran, 5G roadmap: 10 key enabling technologies. Comput. Netw. **106**, 17–48 (2016)
8. M. Marchese, A. Moheddine, F. Patrone, IoT and UAV integration in 5G hybrid terrestrial-satellite networks. Sensors (Switzerland) **19**(17), 3704 (2019)
9. R. Gupta, S. Tanwar, S. Tyagi, N. Kumar, Tactile internet and its applications in 5G era: A comprehensive review. Int. J. Commun. Syst. **32**(14), 1–49 (2019)
10. J. Vora, S. Kaneriya, S. Tanwar, S. Tyagi, N. Kumar, M.S. Obaidat, TILAA: Tactile internet-based ambient assistant living in fog environment. Futur. Gener. Comput. Syst. **98**, 635–649 (2019)

11. I. Syrytsin, S. Zhang, G.F. Pedersen, A.S. Morris, Compact quad-mode planar phased array with wideband for 5G mobile terminals. IEEE Trans. Antennas Propag. **66**(9), 4648–4657 (2018)
12. Y. Zhang, J.Y. Deng, M.J. Li, D. Sun, L.X. Guo, A MIMO dielectric resonator antenna with improved isolation for 5G mm-wave applications. IEEE Antennas Wirel. Propag. Lett. **18**(4), 747–751 (2019)
13. M. Ciydem, E.A. Miran, Dual-polarization wideband Sub-6 GHz suspended patch antenna for 5G base station. IEEE Antennas Wirel. Propag. Lett. **19**(7), 1142–1146 (2020)
14. M.M. Samadi Taheri, A. Abdipour, S. Zhang, G.F. Pedersen, Integrated millimeter-wave wideband end-fire 5G beam steerable array and low-frequency 4G LTE antenna in mobile terminals. IEEE Trans. Veh. Technol. **68**(4), 4042–4046 (2019)
15. Z. Ren, A. Zhao, S. Wu, MIMO antenna with compact decoupled antenna pairs for 5G mobile terminals. IEEE Antennas Wirel. Propag. Lett. **18**(7), 1367–1371 (2019)
16. N. Wang, P. Wang, A. Alipour-Fanid, L. Jiao, K. Zeng, Physical-layer security of 5G wireless networks for IoT: Challenges and opportunities. IEEE Internet Things J. **6**(5), 8169–8181 (2019)
17. A. Ijaz et al., Enabling massive IoT in 5G and beyond Systems: PHY radio frame design considerations. IEEE Access **4**(ii), 3322–3339 (2016)
18. L. Chettri, R. Bera, A comprehensive survey on Internet of Things (IoT) toward 5G wireless systems. IEEE Internet Things J. **7**(1), 16–32 (2020)
19. W.Z. Khan, M.H. Rehman, H.M. Zangoti, M.K. Afzal, N. Armi, K. Salah, Industrial internet of things: Recent advances, enabling technologies and open challenges. Comput. Electr. Eng. **81**, 106522 (2020)
20. R. Gupta, S. Tanwar, S. Tyagi, N. Kumar, Tactile-internet-based telesurgery system for healthcare 4.0: An architecture, research challenges, and future directions. IEEE Netw. **33**(6), 22–29 (2019)
21. R. Gupta, A. Kumari, S. Tanwar, N. Kumar, Blockchain-envisioned softwarized multi-swarming UAVs to tackle COVID-I9 situations. IEEE Netw., 1–8 (2020)
22. S. Tanwar, J. Vora, S. Tyagi, N. Kumar, M.S. Obaidat, A systematic review on security issues in vehicular ad hoc network. Secur. Priv. **1**(5), e39 (2018)
23. I. Mistry, S. Tanwar, S. Tyagi, N. Kumar, Blockchain for 5G-enabled IoT for industrial automation: A systematic review, solutions, and challenges. Mech. Syst. Signal Process. **135**, 106382 (2020)
24. H. Chu, L. Peng, C. Jin, J.X. Chen, An unbalanced-to-balanced diplexer based on substrate integrated waveguide cavity. Int. J. RF Microw. Comput. Eng. **25**(2), 173–177 (2015)
25. A. Kumar, D. Chaturvedi, S. Raghavan, Design of a self-diplexing antenna using SIW technique with high isolation. AEU – Int. J. Electron. Commun. **94**, 386–391 (2018)
26. N.C. Karmakar, S.M. Roy, I. Balbin, Quasi-static modeling of defected ground structure. IEEE Trans. Microw. Theory Tech. **54**(5), 2160–2168 (2006)
27. D. Silveira et al., Improvements and analysis of nonlinear parallel behavioral models. Int. J. RF Microw. Comput. Eng. **19**(5), 615–626 (2009)
28. S. Nandi, A. Mohan, SIW-based cavity-backed self-diplexing antenna with plus-shaped slot. Microw. Opt. Technol. Lett. **60**(4), 827–834 (2018)
29. C.A. Balanis, *Antenna Theory Analysis and Design* (Wiley, Hoboken, 2005), p. 811
30. C.Y.D. Sim, C.C. Chang, J.S. Row, Dual-feed dual-polarized patch antenna with low cross polarization and high isolation. IEEE Trans. Antennas Propag. **57**(10 PART 2), 3321–3324 (2009)
31. D.H. Schaubert, F.G. Farrar, A. Sindoris, S.T. Hayes, Microstrip antennas with frequency agility and polarization diversity. IEEE Trans. Antennas Propag. **29**(1), 118–123 (1981)
32. Y.M. Pan, P.F. Hu, X.Y. Zhang, S.Y. Zheng, A low-profile high-gain and wideband filtering antenna with metasurface. IEEE Trans. Antennas Propag. **64**(5), 2010–2016 (2016)
33. J. Ouyang, F. Yang, Z.M. Wang, Reducing mutual coupling of closely spaced microstrip MIMO antennas for WLAN application. IEEE Antennas Wirel. Propag. Lett. **10**, 310–313 (2011)

34. D. Deslandes, K. Wu, Integrated microstrip and rectangular waveguide in planar form. IEEE Microw. Wirel. Compon. Lett. **11**(2), 68–70 (2001)
35. D. Deslandes, Design equations for tapered microstrip-to-substrate integrated waveguide transitions. IEEE MTT-S Int. Microw. Symp. Dig., 704–707 (2010)
36. M. Bozzi, A. Georgiadis, K. Wu, Review of substrate-integrated waveguide circuits and antennas. IET Microw. Antennas Propag. **5**(8), 909–920 (2011)
37. H. Uchimura, T. Takenoshita, M. Fujii, Development of a 'laminated waveguide. IEEE Trans. Microw. Theory Tech. **46**(12 PART 2), 2438–2443 (1998)
38. G.L. Huang et al., Lightweight perforated waveguide structure realized by 3-D printing for RF applications. IEEE Trans. Antennas Propag. **46**(8), 3897–3904 (2017)
39. N. Ranjkesh, M. Shahabadi, Loss mechanisms in SIW and MSIW. Prog. Electromagn. Res. B **4**, 299–309 (2008)
40. L. Yan, W. Hong, G. Hua, J. Chen, K. Wu, T.J. Cui, Simulation and experiment on SIW slot array antennas. IEEE Microw. Wirel. Compon. Lett. **14**(9), 446–448 (2004)
41. F. Kuroki, R.J. Tamaru, Low-loss and low-cost solution for printed transmission lines at millimeter-wavelengths by using bilaterally metal-loaded tri-plate transmission line. IEEE MTT-S Int. Microw. Symp. Dig., 301–304 (2009)
42. J. Hirokawa, M. Ando, Single-layer feed waveguide consisting of posts for plane tem wave excitation in parallel plates. IEEE Trans. Antennas Propag. **46**(5), 625–630 (1998)
43. C.H. Liang, C.Y. Chang, Novel microstrip stepped-impedance resonator for compact wideband bandpass filters, in *APMC 2009 – Asia Pacific Microw. Conf. 2009*, (2009), pp. 941–944
44. R. Sharma, T. Chakravarty, S. Bhooshan, A.B. Bhattacharyya, Design of a novel 3 DB microstrip backward wave coupler using defected ground structure. Prog. Electromagn. Res. **65**, 261–273 (2006)
45. N.C. Karmakar, S.M. Roy, I. Balbin, Quasi-static modeling of defected ground structure in IEEE Trans. Microw. Theory Tech. **54**(5), 2160–2168 (2006)
46. H.H. Xie, Y.C. Jiao, L.N. Chen, F.S. Zhang, An effective analysis method for EBG reducing patch antenna coupling. Prog. Electromagn. Res. Lett. **21**(March), 187–193 (2011)
47. B. Mohajer-Iravani, S. Shahparnia, O.M. Ramahi, Coupling reduction in enclosures and cavities using electromagnetic band gap structures. IEEE Trans. Electromagn. Compat. **48**(2), 292–303 (2006)
48. J. Ghosh, S. Ghosal, D. Mitra, S.R.B. Chaudhuri, Mutual coupling reduction between closely placed microstrip patch antenna using meander line resonator. Prog. Electromagn. Res. Lett. **59**(April), 115–122 (2016)
49. G.A. Akpakwu, B.J. Silva, G.P. Hancke, A.M. Abu-Mahfouz, A survey on 5G networks for the internet of things: Communication technologies and challenges. IEEE Access **6**(c), 3619–3647 (2017)
50. S.K.K. Dash, Q.S. Cheng, R.K. Barik, N.C. Pradhan, K.S. Subramanian, A compact triple-fed high-isolation SIW-based self-triplexing antenna. IEEE Antennas Wirel. Propag. Lett. **19**(5), 766–770 (2020)
51. P. Nigam, The substrate integrated waveguide based self (2019), pp. 241–246
52. P. Nigam, R. Agarwal, A. Muduli, S. Sharma, A. Pal, Substrate integrated waveguide based cavity-backed self-triplexing slot antenna for X-Ku band applications. Int. J. RF Microw. Comput. Eng. **30**(4), 1–11 (2020)
53. A. Kumar, D. Chaturvedi, S. Raghavan, Low-profile substrate integrated waveguide (SIW) cavity-backed self-triplexed slot antenna. Int. J. RF Microw. Comput. Eng. **29**(3), 1–7 (2019)
54. K. Kumar, S. Dwar, Substrate integrated waveguide cavity-backed self-Triplexing slot antenna. IEEE Antennas Wirel. Propag. Lett. **16**(c), 3249–3252 (2017)
55. A. Kumar, S. Raghavan, A self-triplexing SIW cavity-backed slot antenna. IEEE Antennas Wirel. Propag. Lett. **17**(5), 772–775 (2018)
56. P. Cheong, K.F. Chang, W.W. Choi, K.W. Tam, A highly integrated antenna-triplexer with simultaneous three-port isolations based on multi-mode excitation. IEEE Trans. Antennas Propag. **63**(1), 363–368 (2015)

57. H. Liu, P. Wen, S. Zhu, B. Ren, X. Guan, H. Yu, Quad-band CPW-fed monopole antenna based on flexible pentangle-loop radiator. IEEE Antennas Wirel. Propag. Lett. **14**(c), 1373–1377 (2015)
58. X. Sun, G. Zeng, H.C. Yang, Y. Li, A compact quadband CPW-fed slot antenna for M-WiMAX/WLAN applications. IEEE Antennas Wirel. Propag. Lett. **11**, 395–398 (2012)
59. S.K.K. Dash, Q.S. Cheng, R.K. Barik, A compact substrate integrated waveguide backed self-quadruplexing antenna for C-band communication. Int. J. RF Microw. Comput. Eng. **30**(10), 1–9 (2020)
60. S. Priya, S. Dwari, K. Kumar, M.K. Mandal, Compact self-quadruplexing SIW cavity-backed slot antenna. IEEE Trans. Antennas Propag. **67**(10), 6656–6660 (2019)
61. A. Kumar, Design of self-quadruplexing antenna using substrate-integrated waveguide technique. Microw. Opt. Technol. Lett. **61**(12), 2687–2689 (2019)

Index

S. Tanwar (ed.), *Blockchain for 5G-Enabled IoT*,
https://doi.org/10.1007/978-3-030-67490-8

H

Printed in the United States
by Baker & Taylor Publisher Services